Lecture Notes in Computer Science 4545

Commenced Publication in 1973
Founding and Former Series Editors:
Gerhard Goos, Juris Hartmanis, and Jan van Leeuwen

Editorial Board

Hirokazu Anai   Katsuhisa Horimoto
Temur Kutsia (Eds.)

Algebraic Biology

Second International Conference, AB 2007
Castle of Hagenberg, Austria, July 2-4, 2007
Proceedings

 Springer

Volume Editors

Hirokazu Anai
CREST, Japan Science and Technology Agency
Honcho, Kawaguchi 332-0012, Japan
E-mail: anai@jp.fujitsu.com

Katsuhisa Horimoto
National Institute of Advanced Industrial Science and Technology (AIST)
Computational Biology Research Center (CBRC)
Tokyo, 135-0064, Japan
E-mail: k.horimoto@aist.go.jp

Temur Kutsia
Johannes Kepler University
Research Institute for Symbolic Computation
4040 Linz, Austria
E-mail: Temur.Kutsia@risc.uni-linz.ac.at

Library of Congress Control Number: 2007929551

CR Subject Classification (1998): F.3.1, F.4, D.2.4, I.1, J.3

LNCS Sublibrary: SL 1 – Theoretical Computer Science and General Issues

ISSN   0302-9743
ISBN   3-540-73432-5 Springer Berlin Heidelberg New York
ISBN   978-3-540-73432-1 Springer Berlin Heidelberg New York

Springer is a part of Springer Science+Business Media

springer.com

© Springer-Verlag Berlin Heidelberg 2007

Typesetting: Camera-ready by author, data conversion by Scientific Publishing Services, Chennai, India
Printed on acid-free paper   SPIN: 12086368   06/3180   5 4 3 2 1 0

Preface

This volume contains the proceedings of the 2nd International Conference on Algebraic Biology (AB2007). It was held during July 2–4, 2007 in the Castle of Hagenberg, Austria, and was organized by the Research Institute for Symbolic Computation (RISC) of the Johannes Kepler University, Linz.

Algebraic biology is the interdisciplinary forum for the presentation of research on all aspects of applications of symbolic computation (computer algebra, computational logic, and related methods) in biology. The first conference on algebraic biology (AB 2005) was held during November 28–30, 2005 in Tokyo, Japan.

The initiation of the series of algebraic biology conferences was motivated by the recent trends in symbolic computation and biology: In symbolic computation, the recent advances in computer performance and algorithmic methods have accelerated the extension of the scientific fields to which symbolic computation can be applied. In biology, the determination of complete genomic sequences and the subsequent improvements of experimental techniques have yielded large amounts of information about the biological molecules underlying various biological phenomena. Under these circumstances, the marriage of symbolic computation and biology is expected to generate new algebraic models for biological phenomena and new symbolic techniques for biological data analysis.

This remains the intended profile of the series of algebraic biology conferences, and it figured in the manuscripts published in AB 2007. The papers in the present volume are evidence of the healthy growth in the field of algebraic biology.

We received 40 submissions from 22 countries (Armenia, Australia, Bulgaria, Canada, Chile, France, Germany, Greece, Hungary, India, Italy, Japan, Portugal, Romania, Russia, Spain, South Africa, Switzerland, Taiwan, The Netherlands, UK, and USA), and 19 papers were accepted for publication. Each submission was assigned to at least three Program Committee members, who carefully reviewed the papers, in many cases with the help of external referees. The merits of the submissions were discussed by the Program Committee over one week through the Internet, by means of the EasyChair conference management system.

Besides the contributed papers, this volume also includes three invited papers, by Reinhard Laubenbacher (Discrete Models of Biochemical Networks: The Toric Variety of Nested Canalyzing Functions), Bud Mishra (Algebraic Systems Biology: Theses and Hypotheses), and Gheorghe Paun (Membrane Computing as a Framework for Bio-Modeling).

The tutorial session of the conference provided an opportunity for scientists in symbolic computation and biology to come together and learn about each others' research problems and problem-solving techniques. The session consisted of five symbolic computation and five biology tutorials. Four tutorial speakers submitted their papers to the proceedings. These papers are included in this volume.

We are pleased to start our collaboration with Springer, who agreed to publish the conference proceedings in the *Lecture Notes in Computer Science* series.

We, the AB Steering Committee, and the organizers of the conference, are grateful to the following sponsors for their financial contributions towards its operation and success: Austrian Grid, Linzer Hochschulfonds, MapleSoft, the National Institute of Advanced Industrial Science and Technology, Raiffeisen Landesbank Oberösterreich, RISC Software GmbH, Special Research Program SFB F013 of the Austrian Science Fund (FWF), and the Upper Austrian Government.

Our thanks are also due to the members of the Program Committee and the additional referees, and to those who ensured the effective running of the conference.

July 2007

Hirokazu Anai
Bruno Buchberger
Hoon Hong
Katsuhisa Horimoto
Temur Kutsia

Conference Organization

Conference Chairs

Hirokazu Anai   Fujitsu Laboratories Ltd., Japan
Bruno Buchberger   Johannes Kepler University of Linz, Austria
Hoon Hong   North Carolina State University, USA
Katsuhisa Horimoto   National Institute of Advanced Industrial Science
and Technology, Japan

Program Chairs

Hirokazu Anai   Fujitsu Laboratories Ltd., Japan
Katsuhisa Horimoto   National Institute of Advanced Industrial Science
and Technology, Japan
Temur Kutsia   Johannes Kepler University of Linz, Austria

Program Committee

Tatsuya Akutsu   Kyoto University, Japan
Armin Biere   Johannes Kepler University of Linz, Austria
Bruno Buchberger   Johannes Kepler University of Linz, Austria
Vincenzo Capasso   Università degli studi di Milano, Italy
Luca Cardelli   Microsoft Research, Cambridge, UK
Gautam Dasgupta   Columbia University, USA
François Fages   INRIA Rocquencourt, France
Shinji Hara   University of Tokyo, Japan
Sepp Hochreiter   Johannes Kepler University of Linz, Austria
Hoon Hong   North Carolina State University, USA
Hans Irschik   Johannes Kepler University of Linz, Austria
Erich Kaltofen   North Carolina State University, USA
Veikko Keränen   Rovaniemi University of Applied Sciences, Finland
James F. Lynch   Clarkson University, USA
Manfred Minimair   Seton Hall University, USA
Enno Ohlebusch   University of Ulm, Germany
Stanly Steinberg   University of New Mexico, USA
Bernd Sturmfels   University of California at Berkeley, USA
Carolyn L. Talcott   SRI International, USA
Ashish Tiwari   SRI International, USA

Jens Volkert   Johannes Kepler University of Linz, Austria
Dongming Wang   Beihang University, China and UPMC-CNRS, France
Kazuhiro Yokoyama   Rikkyo University, Japan
Ruriko Yoshida   Duke University, USA

Invited Speakers

Reinhard Laubenbacher   Virginia Bioinformatics Institute, USA
Bud Mishra   New York University, USA
Gheorghe Păun   Institute of Mathematics, Romanian Academy,
                        Romania

Tutorial Speakers

Sachiyo Aburatani   National Institute of Advanced Industrial Science
                        and Technology, Japan
Nobuhiro Go   Japan Atomic Energy Agency
John Harrison   Intel Corporation
Hoon Hong   North Carolina State University, USA
Hans Irschik   Johannes Kepler University of Linz, Austria
Veikko Keränen   Rovaniemi University of Applied Sciences, Finland
Francis Thackeray   Transvaal Museum, Northern Flagship Institution,
                        South Africa
Hiroyuki Toh   Kyushu University, Japan
Bridget S. Wilson   University of New Mexico, USA
Limsoon Wong   National University of Singapore

Local Organization

Betina Curtis   Johannes Kepler University of Linz, Austria
Temur Kutsia   Johannes Kepler University of Linz, Austria

External Reviewers

Andreas Deutsch   Irina Kogan
Kord Eickmeyer   Sung Koh
Martin Giese   Richard Mayr
Tomohisa Hayakawa   Andrew Millar
Steffen Heber   Stefan Müller
Monika Heiner   Jose Carlos Nacher
Meng Jin   Wei Niu
Manuel Kauers   Dirk Nowotka

Andy Poggio
Luis-Garcia Puente
Sven Rahmann
Georg Regensburger
Adrien Richard
Paul Ruet

Peter Ruoff
Eberhard Voit
Andreas Weber
Bican Xia
Hitoshi Yanami

Sponsors

Austrian Grid
Linzer Hochschulfonds
MapleSoft
National Institute of Advanced Industrial Science and Technology, Japan
Raiffeisen Landesbank Oberösterreich
RISC Software GmbH
Special Research Program SFB F013 of the Austrian Science Fund (FWF)
Upper Austrian Government

Table of Contents

Algebraic Systems Biology: Theses and Hypotheses (Invited Talk)...... 1
  Bud Mishra

Discrete Models of Biochemical Networks: The Toric Variety of Nested
Canalyzing Functions (Invited Talk).............................. 15
  Abdul S. Jarrah and Reinhard Laubenbacher

Membrane Computing as a Framework for Bio-modeling (An Informal
Glimpse) (Invited Talk)... 23
  Gheorghe Păun

Relating Attractors and Singular Steady States in the Logical Analysis
of Bioregulatory Networks 36
  Heike Siebert and Alexander Bockmayr

Translating Time-Course Gene Expression Profiles into Semi-algebraic
Hybrid Automata Via Dimensionality Reduction 51
  Alberto Casagrande, Kevin Casey, Rachele Falchi, Carla Piazza,
  Benedetto Ruperti, Giannina Vizzotto, and Bud Mishra

On Proving the Absence of Oscillations in Models of Genetic Circuits... 66
  François Boulier, Marc Lefranc, François Lemaire,
  Pierre-Emmanuel Morant, and Aslı Ürgüplü

Attenuation Regulation as a Term Rewriting System 81
  Eugene Asarin, Thierry Cachat, Alexander Seliverstov,
  Tayssir Touili, and Vassily Lyubetsky

Glucose-Insulin Control of Type1 Diabetic Patients in H_2/H_∞ Space
Via Computer Algebra 95
  Levente Kovács and Béla Paláncz

Exact Parameter Determination for Parkinson's Disease Diagnosis with
PET Using an Algebraic Approach 110
  Hiroshi Yoshida, Koji Nakagawa, Hirokazu Anai, and
  Katsuhisa Horimoto

Efficient Haplotype Inference with Pseudo-boolean Optimization 125
  Ana Graça, João Marques-Silva, Inês Lynce, and Arlindo L. Oliveira

An Algebraic Algorithm for the Identification of Glass Networks with
Periodic Orbits Along Cyclic Attractors........................... 140
  Igor Zinovik, Daniel Kroening, and Yury Chebiryak

Analyzing Pathways Using SAT-Based Approaches 155
  Ashish Tiwari, Carolyn Talcott, Merrill Knapp,
  Patrick Lincoln, and Keith Laderoute

Algorithmic Algebraic Model Checking IV: Characterization of
Metabolic Networks 170
  Venkatesh Mysore and Bud Mishra

Cascaded Games... 185
  Jittisak Senachak, Mun'delanji Vestergaard, and René Vestergaard

On Differential Algebraic Decision Methods for the Estimation of
Anaerobic Digestion Models 202
  Elena Chorukova, Sette Diop, and Ivan Simeonov

Protein Structure Prediction Using Residual Dipolar Couplings 217
  Ioannis Z. Emiris and Sotirios I. Pantos

A Stochastic Pi Calculus for Concurrent Objects 232
  Céline Kuttler, Cédric Lhoussaine, and Joachim Niehren

Modeling Static Biological Compartments with Beta-binders 247
  Maria Luisa Guerriero, Corrado Priami, and Alessandro Romanel

Deducing Interactions in Partially Unspecified Biological Systems 262
  Paolo Baldan, Andrea Bracciali, Linda Brodo, and Roberto Bruni

Reduction of Algebraic Parametric Systems by Rectification of Their
Affine Expanded Lie Symmetries 277
  Alexandre Sedoglavic

Prefix Reversals on Binary and Ternary Strings 292
  Cor Hurkens, Leo van Iersel, Judith Keijsper, Steven Kelk,
  Leen Stougie, and John Tromp

Toric Ideals of Phylogenetic Invariants for the General Group-Based
Model on Claw Trees $K_{1,n}$ 307
  Julia Chifman and Sonja Petrović

Inference of Protein-Protein Interactions by Using Co-evolutionary
Information (Tutorial Talk) 322
  Tetsuya Sato, Yoshihiro Yamanishi, Katsuhisa Horimoto,
  Minoru Kanehisa, and Hiroyuki Toh

A Short Survey of Automated Reasoning (Tutorial Talk) 334
  John Harrison

Inference of Complex Regulatory Network for the Cell Cycle System in
Saccharomyces Cerevisiae (Tutorial Talk) 350
  Sachiyo Aburatani

Manifestation and Exploitation of Invariants in Bioinformatics (Tutorial Talk) 365

  Limsoon Wong

Author Index 379

Algebraic Systems Biology: Theses and Hypotheses[*]

Bud Mishra[1,2]

[1] Courant Institute, New York University, New York, NY, U.S.A.
[2] NYU School of Medicine, New York University, New York, NY, U.S.A.
mishra@nyu.edu

Abstract. What is systems biology? What can biologists gain from an attempt to algebraize the questions in systems biology? Starting with plausible biological theses, can one algebraically model them and then manipulate them to suggest meaningful hypotheses? Using these hypotheses, can one measure and mine suitable experimental data to validate or refute these hypotheses? Through these intertwined processes of measuring, mining, modeling and manipulating biological systems, can one generate the set of theses and hypotheses upon which systems biology will be founded? This review provides one algorithmic-algebraist's somewhat idiosyncratic response to these and other related questions, but also aims to persuade young algebraists to examine the possible role they and algebra can play to enrich this subject.

1 Hypotheses Non Fingo: Hooke Meets Newton

Over the last few years, Sir Robert Hooke, a somewhat maligned, but still a very fascinating English experimental scientist, had begun to feature unexpectedly prominently in practically all my public presentations on Systems Biology. Initially, what had attracted me to the story of Hooke, was the uncanny resemblance he bore to many contemporary scientists in terms of their insistence on data, observations and hypotheses, their apparent non-rigorous and intuitive approaches to scientific questions, but most inexplicably, their protracted and debilitating open rivalries over the questions of recognition. But, as I learned more about Hooke's life and views, it also became clearer that his indirect influence on the way we think about science today is only surpassed by the opinions of only a handful of other contemporary thinkers, with some of whom Hooke fought bitter and hopeless semi-philosophical battles. They have, thus, unwittingly lent us a useful perspective that is worth examining with some care. How the emerging field of systems biology could establish itself, how it should face its trials and tribulations along the way, and how it could be a significant component of the "new new" biology, etc., could all be examined from the points of view of these 17th century scientists—a viewpoint that remains anachronically and peculiarly relevant even today.

Robert Hooke (1635-1703) was an experimental scientist, mathematician, architect, and astronomer. He was also the first Secretary of the Royal Society from 1677 to 1682, and because of his wide ranging interests, Hooke has been variously described as the "England's Da Vinci." His work Micrographia of 1665 contained his microscopical

---

[*] The work reported in this paper was supported by two grants from NSF ITR program.

H. Anai, K. Horimoto, and T. Kutsia (Eds.): AB 2007, LNCS 4545, pp. 1–14, 2007.

investigations, which included the first identification of biological cells, an enduring discovery that has maintained its central place in subsequent developments in biology for more than three centuries. In his drafts of Book II, Newton had referred to him as the most illustrious Hooke—"Cl[arissimus] Hookius." However, not long after, Hooke became involved in a bitter dispute with Sir Isaac Newton over the priority of the discovery of the inverse square law of gravitation. In a letter Hooke wrote to Halley, he complained about omission of credit given to his discovery of the properties of gravity, "which of late Mr. Newton has done me the favour to print and publish as his own inventions." In response Newton wrote back to Halley, "Now is this not very fine? Mathematicians that find out, settle & do all the business must content themselves with being nothing but dry calculators & drudges I beleive[sic] you would think him a man of a strange unsociable temper"—perhaps still a common protest of many unhappy mathematicians whose contributions have been ignored or forgotten. In a more well-known letter that Newton wrote directly to Hooke, he famously said, "If I have seen further[sic] than other men, it is because I have stood on the shoulders of giants"—where, of course, the giants Newton was alluding to were Kepler and Galileo, and not the dwarfish, small-minded and short-tempered likes of Hooke! When Christopher Wren was brought in to resolve this rather strangely English war-of-words, Wren diplomatically described the disagreement using Clairaut's characterization of "the great distance between a glimpsed truth and a demonstrated truth"—raising perhaps, the question of relative roles that should be ascribed to the inductive hypothesis-driven science with respect to the deductive principle-driven science—theses vs. hypotheses.

What is the nature of "TRUTH" in biology, and how is it to be sought? Hooke saw biology as an observational science; he wrote in Micrographia, "The truth is, the science of Nature has already been too long made only a work of the brain and the fancy. It is now high time that it should return to the plainness and soundness of observations on material and obvious things," —a view supporting hypothesis-driven experimentation that advances science through steps of falsification or validation. Newton, on the other hand, championed a search for deep and unifying principles. Newton shunned hypotheses; his motto stated in Principia was "Hypotheses non fingo." ("I feign no hypotheses.") Newton's viewpoints are probably best stated by his most ardent disciple, Halley; in his rather ornately titled essay 'The true Theory of the Tides, extracted from that admired Treatise of Mr. Issac Newton, Intituled, Philosophiae Naturalis Principia Mathematica,' he wrote the following: "Truth being uniform and always the same, it is admirable to observe how easily we are enabled to make out very abstruse and difficult matters, when once true and genuine Principles are obtained."

Biology still remains an observational science; it continues to move through the toils of a vast army of scientists each examining a small subsystem of a favored organism, as the scientists sharpen their intuitions, build upon guesses, conjectures, and hypotheses, and refine their ideas in many small steps—occasionally interrupted by a great leap, a grand vision or a comprehensive shift in paradigm. If subtle principles are to be brought to light, they must wait for serendipity. It has been argued that life is complex, it does not yield to few small neat explanations or pigeon-holing, and if there is a unifying principle in biology, it is that there is no unifying principle in biology.

Can ideas from algorithms and algebra be brought to bear to systematically hunt for principles and patterns that will reveal a grand unified theory of biology? Are their design rules at play in how these systems evolve, interact, and self-assemble? What algebraic tools must we build, if we wish to create a global view of biology? What can be automated to make computers work on tasks that are humanly impossible? Is algebraic systems biology the answer to the problems of biology?

2 Systems Biological Models

2.1 Processes

We start with the following taxonomy into which the cellular biochemical processes are typically organized, as described below.

GENETIC REGULATION: The oft-repeated "central dogma of biology" states that biochemical information in cells is encoded primarily in the Deoxyribo Nucleic Acid (DNA) molecules. DNA is *transcribed* into messenger Ribo Nucleic Acid (mRNA), and the mRNA then is *translated* into proteins at the ribosomes. Genetic regulation is the process of modulation of the expression of the relevant genes at the correct locations and times, and is keyed by specific proteins called transcriptional factors. Through transcriptional factors and other ancillary modulators, proteins, the products of genes, themselves partake in this genetic regulatory process, thus giving rise to complex interaction networks; such proteins interact with regions of the DNA to effect modulation of how genes are transcribed. The binding of the transcription machinery and the transcriptional factors to the DNA involves complex protein-DNA-protein interactions, where, more often than not, the structural modification of the DNA (such as euchromatin and heterochromatin regions) and the protein has to be accounted for.

The rate of gene transcription, the post-transcriptional mechanisms that affect mRNA half-life (i.e., stability) and the formation of the mRNA-ribosome complex are other aspects of genetic regulation. Similarly, there are post-translational mechanisms for protein modification such as phosphorylation of key residues, multimerization, chaperone-guided complex formation, protein-folding control, and genetic control by small interfering RNA (siRNA).

SIGNAL TRANSDUCTION: The cell responds to external signals through receptors, which may be on its surface or in its cytoplasm. The signal is transmitted to the interior through messengers, which induce the desired response to the external signal. Typically, a ligand binds to a trans-membrane receptor whose conformation subsequently changes. This change is detected by proteins bound to it (usually on the cytoplasmic side), or is manifested as a change in the receptor's chemical properties. Subsequently, second messenger molecules amplify the signal and communicate it to the target(s). Alternatively, the ligand can directly enter the cell through non-specific channels and then bind to the receptors inside the cell. Small molecules like calcium often participate in these pathways, where most of the reactants are enzymatic proteins. The net result of the signal transduction pathway is an appropriate response by the specific subcellular component. Very often, the signaling pathway results in the nuclear localization of transcription factors, leading to the transcription (or shutting down) of corresponding genes. The binding of the signaling molecule with the receptor, the modification of the structure of the receptor and associated proteins (with the receptor sometimes acting as an enzyme) and dispatching of second messengers are the activities near the cell membrane. Receptor desensitization, internalization and regeneration are other complex sub-processes, thus altering the physical properties of binding and diffusion.

METABOLISM: Metabolism represents almost all processes that are not genetic regulatory or signal transducing. The gigantic set of biochemicals needed by the cell are continuously produced and consumed by complex enzyme catalyzed pathways. These comprise the metabolic network. They essentially govern the matter and energy cycles of a cell— the way energy and matter are obtained, transformed and consumed by living organisms. Photosynthesis for example is the process by which light energy is converted into chemical energy during sugar (e.g., glucose) formation. During respiration, the oxidation of glucose transforms the energy into Adenosine Tri-Phosphate (ATP). While the ATP-cycle and photosynthesis comprise the well-known energy metabolism, carbohydrate metabolism deals with Glycolysis and Phosphates, lipid metabolism pertains to Triacyl Glycerol and Fatty Acids, and amino acid metabolism mostly refers to Glutamate and Urea.

OTHER PROCESSES: Biology is complex, and of course, there are still more aspects to cellular biology beyond this simple trichotomic characterization. These include the biophysics of DNA packaging, small interfering RNA (siRNA), protein folding and DNA-protein interaction, cell adhesion, non-transcriptional regulatory pathways, cellular compartments and related spatio-temporal phenomena, cell proliferation, and cell migration. While the modeling approaches suggested here, when further augmented with suitable stochastic and spatial formalisms, will generalize as well, I will not emphasize those applications directly in my discussion here.

2.2 Models

Algorithmic algebraic models of biological systems are created through a process of conceptual simplification. Models, created in this fashion, must strike a balance among fidelity, expressivity and ability to be manipulated algorithmically. For this purposes, the different component parts and processes in the biochemical domain may be represented at different levels of abstraction [22,37]. I summarize some of the major approaches below, but will guide the discussion towards hybrid automata representation, a very general and powerful model for these systems.

LOGICAL MODELING: The state of the reactant is captured through a finite number of abstract-states (where intermediate expression levels are assumed to have the same behavior), and functions are used to describe the new states (concentration range) of the chemical species, given their old states. The transitions between states can be assumed to occur synchronously or (more accurately) asynchronously. In the simplest case, only two states ("on" and "off") are used, and Boolean algebra is used to describe the dynamics. Literature on Concurrent Transition Systems [20,19] and Pathway (Rewrite)

Logic [25] provides good expositions of logical modeling. Kappler et al. [38] demonstrate how to extend simple Boolean networks by using ordinary differential equations to capture the concentration, while Boolean functions continue to determine the rates of the reactions. The probability of being in a state is sometimes a more reasonable measure to estimate, as in the case of Sachs et al. [57], who use Bayesian networks to model cell signaling pathways. Similarly, Shmulevich et al. [58] describe the use of probabilistic Boolean networks to model genetic regulatory networks and determine the long-term joint probabilistic behavior of a few selected genes. Platzer et al. [55] simulate the embryonic development of *C. elegans* by assuming Boolean states for the genes and synchronously updating at each time step based on an interaction matrix. Batt et al. [12] have applied model checking theory on biochemical systems modeled though qualitative simulation.

DIFFERENTIAL EQUATIONS: If instead the concentrations are represented exactly in the real continuous domain, the ordinary differential equations (ODEs) of the dynamics directly follow from the law of general mass action (GMA) [21,39,59]. For instance, in the reaction $aA + bB \longleftrightarrow cC + dD$, the rate of the forward reaction $v_f \equiv k_f[A]^a[B]^b$ and the rate of the backward reaction $v_b \equiv k_b[C]^c[D]^d$, where k_f and k_b are the forward and backward rate constants respectively and the rate of individual reactants is $\frac{1}{c}\dot{C} = \frac{1}{d}\dot{D} = -\frac{1}{a}\dot{A} = -\frac{1}{b}\dot{B} = (v_f - v_b)$. As a compromise between discrete and continuous representations, qualitative differential equations can be used with qualitative states corresponding to the different concentration ranges [12,23]. Partial differential equations are necessary for spatially distributed models, e.g., pde's, sde's, or reaction-diffusion equations.

HYBRID SYSTEMS: Many biological systems, such as the cell, follow a combination of discrete and continuous behaviors, which cannot be characterized in a proper way using either only discrete or only continuous models. On one hand, their evolution is ruled by a continuous dynamical law concerning substance concentrations and gradients, and, on the other hand, such a dynamical law may change discretely depending on the system status itself. Because of their hybrid nature, part discrete and part continuous, such systems are named hybrid systems. To model hybrid systems, Alur et al. introduced the notion of hybrid automata in [3]. Intuitively a hybrid automaton is a "finite-state" automaton with continuous variables, which evolve according to a set of continuous laws characterizing each discrete mode of the automaton itself. The use of hybrid automata for modeling biomolecular networks has been described by Alur et al. [1] and Mishra et al. [46]. Amonlirdviman et al. [7] demonstrated the utility of hybrid systems by modeling Drosophila planar cell polarity. Starting with the S-System formulation of Savageau and Voit [60], Antoniotti et al. [11] used an additional automaton to broaden the set of representable systems, subsequently using full-fledged hybrid automata [10]. Ghosh et al. presented both delta-notch [29,28] and protein signaling network [30] models based on the hybrid automaton formalism. Casagrande et al. [16] suggested a simple (and decidable) hybrid automaton model for the *E. coli* chemotaxis. Lincoln and Tiwari [43] detail hybrid automaton modeling of biochemical networks, while Hu et al. [36] describe stochastic hybrid system modeling of subtilin production in *Bacillus subtilis*. More recently, Drulhe et al. [24] have described piecewise-affine models of genetic regulatory networks.

ALGEBRAIC HYBRID AUTOMATA, TEMPORAL LOGIC AND ALGORITHMS: To create a comprehensive theoretical framework for systems biology, what is needed is an appropriate generalization of discrete-time systems, classical temporal logic, possible-world models of temporal logic given by Kripke (e.g., Kripke structures), model checking algorithms based on graph theoretic analysis, etc. to this richer and more powerful domain. However, the generalization must be suitably powerful to capture reasoning processes closely resembling what is used by the biologists, and yet it should also be appropriately constrained so that these systems can be reasoned by feasible computational means. At the least, the resulting problems should be decidable (computable). We seek such a framework below by a judicious amalgamation of symbolic algebra (using decision procedures of semi-algebraic geometry), sufficiently constrained dense-time logic and algebraic models based hybrid automata. We start with a discussion of such hybrid automata and their reachability problem.

3 Algebraic Systems and Biological Models

The subject *Algorithmic Algebraic Model Checking* was introduced to examine connections between systems biology, dynamical systems, modal logic and computability, and how they can be useful in the biological context. Towards this aim, one could begin by addressing the symbolic model checking problem for a new class of hybrid models arising in systems biology – *semi-algebraic hybrid systems*, introduced in the first paper of our "AAMC" (Algorithmic Algebraic Model Checking) series [53]. There, our goal was to characterize the widest range of automata that admit sound albeit expensive mathematical techniques, as opposed to focusing on a very narrow class of systems that often prematurely sacrifice genralizability for the sake of efficiency.

We built upon and integrated many existing ideas: e.g., semi-algebraic hybrid automata, the Blum-Shub-Smale model of "real" computation and TCTL (a powerful temporal logic formalism suitable for our setting)—more formally defined below.

Definition 1 Semi-Algebraic Set [45,47]. *Every quantifier-free boolean formula composed of polynomial equations and inequalities defines a semialgebraic set (i.e., unquantified first-order formulæ over the reals - $(\mathbb{R}, +, \times, =, <)$).* $\square$

Definition 2 Semi-Algebraic Hybrid Automata [53]. *A k-dimensional hybrid automaton is a 7-tuple, $H = (Z, V, E, Init, Inv, Flow, Jump)$, consisting of the following components:*

 - $Z = \{Z_1, ..., Z_k\}$ *a finite set of variables ranging over the reals $\mathbb{R}$;*
 - (V,E) *is a directed graph of discrete states and transitions;*
 - *Each vertex $v \in V$ is labeled by "Init"(initial), "Inv"(invariant) and "Flow" labels;*
 - *Each edge $e \in E$ is labeled by a "Jump" condition;*
 - *Init, Inv, Flow, and Jump are semi-algebraic.* $\square$

Definition 3 Semantics of Hybrid Automata. *Let $H = (Z, V, E, Init, Inv, Flow, Jump)$ be a hybrid automaton of dimension k.*

- *A location ℓ of H is a pair $\langle v, R \rangle$, where $v \in V$ is a state and $R \in \mathbb{R}^k$ is an assignment of values to the variables of Z. A location $\langle v, R \rangle$ is said to be* admissible, *if $Inv_v(R)$ is satisfied.*

- *The* continuous reachability transition relation $\xrightarrow[\mathscr{C}]{h}$ *forces the state invariant to hold at every point except the end-point along the evolution curve determined by the flow equations during the $h(> 0)$ time units from the current time t_0:*

$$\langle v, R \rangle \xrightarrow[\mathscr{C}]{h} \langle v, S \rangle \quad \textit{iff}$$

$$\left(Flow_v(R, S, t_0, h) \; \wedge \; \forall S', h' \in [0, h) \; Flow_v(R, S', t_0, h') \Rightarrow Inv_v(S') \right),$$

where $Flow_v(R, S, t, h)$ is a relation between the continuous state R at time t and the continuous state S after h time units in the discrete state v. It is "well-defined" in the sense that $\forall R, S, t, h \; Flow_v(R, S, t, h) \Rightarrow \{\forall h' \in [0, h) \; \exists S' \; Flow_v(R, S', t, h')\}$.

- *The* discrete reachability transition relation $\xrightarrow[\mathscr{D}]{0}$ *ensures that both parts of the zero-time jump[1] — the guard condition which needs to be satisfied just before the transition is taken, and the reset condition which determines the values after the transition, are satisfied.*

$$\langle v, R \rangle \xrightarrow[\mathscr{D}]{0} \langle u, S \rangle \quad \textit{iff} \quad \langle v, u \rangle \in E \; \wedge \; Jump_{v,u}(R, S).$$

- *The* transition relation $\mathscr{T}$ of H *connects the possible values of the system variables before and after one step — a discrete step for a time $h = 0$ or a continuous evolution for any time period $h > 0$:*

$$\mathscr{T}(\ell \xrightarrow{h} \ell') = \{h = 0 \wedge \ell \xrightarrow[\mathscr{D}]{0} \ell'\} \vee \{h > 0 \wedge \ell \xrightarrow[\mathscr{C}]{h} \ell'\}.$$

- *A* trace *of H is a sequence $\ell_0, \ell_1, \ldots, \ell_n, \ldots$ of admissible locations such that*

$$\forall i \geq 0, \; \exists h_i \geq 0, \; \mathscr{T}(\ell_i \xrightarrow{h_i} \ell_{i+1}). \qquad \square$$

Definition 4 Finite-Dimensional Machine Over $\mathscr{R}$: [13]. *A finite dimensional machine M over $\mathscr{R}$ consists of a finite directed connected graph with four types of nodes:* input, computation, branch *and* output.

In addition the machine has three spaces: input space $\mathscr{I}_M$, state space $\mathscr{S}_M$ *and* output space $\mathscr{O}_M$ *of the form $\mathscr{R}^n, \mathscr{R}^m, \mathscr{R}^l$, respectively, where n, m and l are positive integers.*

1. *Associated with the* input node *is a linear map $I : \mathscr{I}_M \to \mathscr{S}_M$ and a unique next node β_1.*

2. *Each* computation node η *has an associated* computation map, *a polynomial (or rational) map $g_\eta : \mathscr{S}_M \to \mathscr{S}_M$ given by m polynomials (or rational functions) $g_j : \mathscr{R}^m \to \mathscr{R}, j = 1, \cdots, m$, and a unique next node β_η.*

---

[1] $Jump_{v,u}(R, S) \equiv Guard_{v,u}(R) \wedge Reset_{v,u}(R, S)$.

3. *Each* branch node η *has an associated* branching function, *a nonzero polynomial function* $h_\eta : \mathscr{S}_M \to \mathscr{R}$.
4. *Each* output node η *has an associated linear map* $\mathscr{O}_\eta : \mathscr{S}_M \to \mathscr{O}_M$ *and no next node.*   □

Theorem 1 Path Decomposition Theorem: [13]. *For any machine $\mathscr{M}$ over $\mathscr{R}$ the following properties hold.*

1. *For any $T > 0$, the* time-T halting set *of $\mathscr{M}$:* $\Omega_T \ (= \bigcup_{\gamma \in \Gamma_T} v_\gamma)$ *is a finite disjoint union of basic semi-algebraic sets (respectively, basic quasi-algebraic sets, in the unordered case), where Γ_T is the set of time-T halting paths and v_γ is the initial path set.*
2. *The* halting set *of $\mathscr{M}$:* $\Omega_M \ (= \bigcup_{\gamma \in \Gamma_{M'}} v_\gamma)$ *is a countable disjoint union of basic semi-algebraic (respectively, basic quasi-algebraic) sets, where $\Gamma_{M'}$ is the set of minimal halting paths.*
3. *For $\gamma \in \Gamma_M$ (the set of halting paths of $\mathscr{M}$), the* input-output map *Φ_M restricted to $v_\gamma - \Phi_{M|v_\gamma}$ is a polynomial map, or a rational map if $\mathscr{R}$ is a field.*   □

Definition 5 The Mandelbrot Set [44]. $\mathscr{M}$ *is the subset of the set of complex numbers $\mathscr{C}$ that remains bounded when subject to the following iterative procedure: $f_0(c) = c$, $f_{n+1}(c) = f_n(z)^2 + c$. Formally, the complement $\mathscr{M}'$ of the Mandelbrot set is defined as*

$$\mathscr{M}' = \{ c \in \mathscr{C} | f_n(c) \to \infty \text{ as } n \to \infty \}.$$

It is to be noted that $f_i(c) \geq 2$ implies that eventually $f_n(c) \to \infty$.   □

Definition 6. The Mandelbrot Hybrid Automaton *consists of*

– *One discrete state with invariant* False *and two continuous variables x and y.*
– *Flow$_1$:* $\{ x' = x \wedge y' = y \}$ *(no continuous evolution).*
– *One Discrete State Transition:* $1 \to 1$ *with Jump$_1$:* $(x' = x^2 - y^2 + C_r) \wedge (y' = 2xy + C_i)$, *where C_r and C_i are two constants (real numbers).*
– *Only possible trace: zeno path of infinite self-loops.*   □

Theorem 2 Undecidability Of The Mandelbrot Set: [13]. *The Mandelbrot set[2] cannot be expressed as the countable union of semi-algebraic sets over $\mathscr{R}$, and hence not decidable over $\mathscr{R}$.*   □

Definition 7 TCTL[2]. *It has the following syntactic structure:*

$$\phi ::= p \mid \neg \phi \mid \phi_1 \vee \phi_2 \mid \phi_1 \exists \mathscr{U} \phi_2 \mid \phi_1 \forall \mathscr{U} \phi_2 \mid z.\phi.$$

Its associated semantics is described below:

– **z.:** *The freeze quantification "z." binds the associated variable z to the current time. Thus the formula z.$\phi(z)$ holds at time t iff $\phi(t)$ does.*

---

[2] The corresponding 2-dimensional set of real numbers.

- $\phi_1 \forall \mathcal{U} \, \phi_2$ and $\phi_1 \, \exists \mathcal{U} \, \phi_2$: *universal (on all paths) and existential (on at least one path) "until" operators. For $\phi_1 \, \mathcal{U} \, \phi_2$ to be true on a path, ϕ_2 is required to be true somewhere along the path, and ϕ_1 is required to be true all along the path up to (but not necessarily at) that point.* □

Remark 1. The basic notations are often extended by the following syntactic abbreviations [2].

1. $p \, \exists \mathcal{U}_{\leq max} \, q \equiv p \, \exists \mathcal{U} \, (q \wedge z.(z \leq max))$ and $p \, \forall \mathcal{U}_{\leq max} \, q \equiv p \, \forall \mathcal{U} \, (q \wedge z.(z \leq max))$: "subscripted" *Until* operators (*max* is the time-bound).
2. $(\forall \mathcal{F} \, p \equiv true \, \forall \mathcal{U} \, p)$ and $(\exists \mathcal{F} \, p \equiv true \, \exists \mathcal{U} \, p)$: "eventuality" operators.
3. $(\forall \mathcal{G} \, p \equiv \neg \exists \mathcal{F} \neg p)$ and $(\exists \mathcal{G} \, p \equiv \neg \forall \mathcal{F} \neg p)$: "invariance" operators.

Definition 8 Single-Step Until Operator, ▷, [35]. *The formula $p \triangleright q$ holds if $p \vee q$ is true all along "one step" of the hybrid system and q is true at the end of the transition.* □

Definition 9 $T\mu$-**Calculus Syntax: [35].** $\phi ::= X \mid p \mid \neg\phi \mid \phi_1 \vee \phi_2 \mid \phi_1 \triangleright \phi_2 \mid z.\phi \mid \mu X.\phi$, *where μ is the* least-fixpoint *operator*[3]. *Thus,*

- Existential Until: $p \, \exists \mathcal{U} \, q = \mu X.(q \vee (p \triangleright X))$
- Universal Until:[4] $p \forall \mathcal{U} q = \neg(\neg q \, \exists \mathcal{U} \, (\neg p \wedge \neg q))$ □

3.1 What Questions Can and Cannot Be Answered

One may now wish to devise algorithmic algebraic solutions to various kinds of queries (in TCTL) to examine interesting properties and invariants about the hybrid automata that model biochemical systems. The simplest and perhaps the most important question that one can ask about these systems is the symbolic state reachability problem: namely, can one reach a particular state from an initial state by following the dynamics of the hybrid automaton which may be described symbolically? A more relevant biological question could be to provide a symbolic description of the initial conditions (states) from which the biological system (modeled via a semi-algebraic hybrid automaton) can reach a desired state (say, apoptosis state for a cancer cell), or avoid certain unsafe states. In this sense, algebraic descriptions in systems biology can be a potent tool. However, the immediate answers to these questions are depressingly negative. Thus, our community needs to engage in many years of focused work to devise a mature algebraic systems biological toolset. We and others have made some progress by exploiting approximations, bounded reachability analysis, etc. or by suitably constraining the power of the family of hybrid automata studied [54,52,50,17,15,49,51,48,14]. But much more remains to be done!

Just to summarize few of the positive steps in this direction, we mention the following two different approaches: The first way is to identify hybrid automaton classes for which the problem is decidable and to use such classes to model hybrid systems. In the last ten years, many decidable classes have been discovered [3,6,56,40,41,18], but, because of the restrictions imposed on them to achieve decidability, often they cannot be

---

[3] The *greatest-fixpoint* ν can be expressed as $\neg\mu X.(\neg\phi[X := \neg X])$.

[4] This translation is valid only when q is "finitely variable" over all premodels [35].

easily applied in the analysis of real biological systems. The second way is approximate analysis, like bounded model checking [31,27], abstract interpretation [4,5], or quotient reduction [32,33,34], to obtain a partial (or approximate) result for the model checking problem (e.g., the property holds for at least ten seconds starting from the initial condition).

On other approaches that resemble the systems described here, we enumerate few recent results: Anai [8] and Fränzle [26] independently suggested the use of quantifier elimination for the verification of polynomial hybrid systems. Anai and Weispfenning subsequently expounded the use of quantifier elimination for the reachability analysis of continuous systems with parametric inhomogeneous linear differential equations [9]. Fränzle went on to prove that progress, safety, state recurrence and reachability are semi-decidable using quantifier elimination of semi-algebraic formulæ [26], and to develop proof engines for bounded model checking [27]. Lafferiere et al. [42] have described a quantifier-elimination-centric method for symbolic reachability computation of linear vector fields. Many of these powerful techniques remain to be fully integrated into the context that systems biology proposes.

We only present technical details of the following negative result, here. Rest can be found in the reference [52].

Theorem 3 General Undecidability Of Reachability. *For semi-algebraic hybrid systems, reachability is undecidable even in Blum et al.'s "real" Turing machine formalism.*

Proof. Consider the Mandelbrot hybrid automaton defined earlier, with the complex number $C = C_r + \iota.C_i$. Let $S(t) = x(t) + \iota.y(t)$. After 1 discrete state transition (self-loop), we get

$$S'(t) = \{x(t)^2 - y(t)^2 + C_r\} + \iota.\{2x(t)y(t) + C_i\} = \{x(t) + \iota.y(t)\}^2 + \{C_r + \iota.C_i\}$$

In other words, $S'(t) = S^2(t) + C$ which is the defining equation of the *Mandelbrot Set*. Clearly, if there exists an evolution where $|S(t)| \geq 2$ then we know that C does not belong to the Mandelbrot set i.e. if the reachability query[5] $(x^2 + y^2 \geq 4)$ is decidable, it would imply that the Mandelbrot set is decidable, thus resulting in a contradiction.   □

3.2 Final Thoughts

Lest some may mistakenly conclude that I have argued parochially in favor of theses over hypotheses (equivalently, Newton over Hooke), I conclude this review with the following beautiful quote from Hooke:

> "So many are the links, upon which the true Philosophy depends, of which, if any can be loose, or weak, the whole chain is in danger of being dissolved; it is to begin with the Hands and Eyes, and to proceed on through the Memory, to be continued by the Reason; nor is it to stop there, but to come about to the Hands and Eyes again, and so, by a continuall passage round from one Faculty to another, it is to be maintained in life and strength."

It is hoped that someday, algebra will serve its role as a strong link between biological theses and hypotheses— maintained in life and strength!

---

[5] *Reachable*$(p) \equiv \exists \mathscr{F}(p)$.

References

1. Alur, R., Belta, C., Kumar, V., Mintz, M., Pappas, G.J., Rubin, H., Schug, J.: Modeling and Analyzing Biomolecular Networks. Computing in Science and Engineering 4(1), 20–31 (2002)
2. Alur, R., Courcoubetis, C., Dill, D.: Model-Checking for Real-Time Systems. In: International Symposium on Logic in Computer Science, vol. 5, pp. 414–425. IEEE Computer Press, Los Alamitos (1990)
3. Alur, R., Courcoubetis, C., Halbwachs, N., Henzinger, T.A., Ho, P.-H., Nicollin, X., Olivero, A., Sifakis, J., Yovine, S.: The Algorithmic Analysis of Hybrid Systems. Theoretical Computer Science 138(1), 3–34 (1995)
4. Alur, R., Dang, T., Ivancic, F.: Reachability analysis of hybrid systems via predicate abstraction. In: Tomlin, C.J., Greenstreet, M.R. (eds.) HSCC 2002. LNCS, vol. 2289, pp. 25–27. Springer, Heidelberg (2002)
5. Alur, R., Dang, T., Ivancic, F.: Counter-example guided predicate abstraction of hybrid systems. In: Garavel, H., Hatcliff, J. (eds.) ETAPS 2003 and TACAS 2003. LNCS, vol. 2619, pp. 208–223. Springer, Heidelberg (2003)
6. Alur, R., Dill, D.L.: A theory of timed automata. Theoretical Computer Science 126(2), 183–235 (1994)
7. Amonlirdviman, K., Ghosh, R., Axelrod, J.D., Tomlin, C.J.: A Hybrid Systems Approach to Modeling and Analyzing Planar Cell Polarity. In: Proceedings of the 3rd International Conference on Systems Biology (2002)
8. Anai, H.: Algebraic Approach to Analysis of Discrete-Time Polynomial Systems. In: Proceedings of European Control Conference (ECC'99) (1999)
9. Anai, H., Weispfenning, V.: Reach set computations using real quantifier elimination. Technical Report MIP-0012, Fakultät für Mathematik und Informatik, Universität Passau (2000)
10. Antoniotti, M., Mishra, B., Piazza, C., Policriti, A., Simeoni, M.: Modelling Cellular Behavior with Hybrid Automata: Bisimulation and Collapsing. In: Priami, C. (ed.) CMSB 2003. LNCS, vol. 2602, pp. 57–74. Springer, Heidelberg (2003)
11. Antoniotti, M., Policriti, A., Ugel, N., Mishra, B.: XS-systems: eXtended S-Systems and Algebraic Differential Automata for Modeling Cellular Behavior. In: Proceedigs of the International Confernce on High Performance Computing, HiPC 2002, Bangalore, India (December 2002)
12. Batt, G., de Jong, H., Geiselmann, J., Page, M.: Qualitative Analysis of Genetic Regulatory Networks: A Model-Checking Approach. In: Bredeweg, B., Salles, P. (eds.) Working Notes of Seventeenth International Workshop on Qualitative Reasoning, QR-03, pp. 31–38 (2003)
13. Blum, L., Cucker, F., Shub, M., Smale, S.: Complexity and Real Computation. Springer, Heidelberg (1997)
14. Casagrande, A.: Hybrid Systems: A First-Order Approach to Verification and Approximation Techniques. PhD thesis, Department of Mathematics and Computer Science, University of Udine, Advisers - Policriti, A., Villa, T. (2006)
15. Casagrande, A., Mysore, V., Piazza, C., Mishra, B.: Independent Dynamics Hybrid Automata in System Biology. In: Proceedings of the First International Conference on Algebraic Biology, Tokyo (Japan), (November 28-30, 2005)
16. Casagrande, A., Mysore, V., Piazza, C., Mishra, B.: Independent Dynamics Hybrid Automata in Systems Biology. In: Proceedings of the First International Conference on Algebraic Biology (AB'05), November 2005, pp. 61–73, Tokyo, Japan, Universal Academy Press, Inc. (2005)

17. Casagrande, A., Piazza, C., Mishra, B.: Semi-Algebraic Constant Reset Hybrid Automata - SACoRe. In: Proceedings of the 44rd Conference on Decision and Control and European Control Conference (CDC-ECC'05), Seville, Spain, December 2005, pp. 678–683. IEEE Computer Society Press, Los Alamitos (2005)

18. Casagrande, A., Piazza, C., Mishra, B.: Semi-Algebraic Constant Reset Hybrid Automata - SACoRe. In: Proceedings of the 44rd Conference on Decision and Control and European Control Conference (CDC-ECC'05), Seville, Spain, December 2005, pp. 678–683. IEEE Computer Society Press, Los Alamitos (2005)

19. Chabrier, N., Chiaverini, M., Danos, V., Fages, F., Schächter, V.: Modeling and Querying Biochemical Interaction Networks. Theoretical Computer Science 325(1), 25–44 (2004)

20. Chabrier, N., Fages, F.: Symbolic Model Checking of Biochemical Networks. In: Proceedings of the First International Workshop on Computational Methods in Systems Biology, pp. 149–162 (2003)

21. Cornish-Bowden, A.: Fundamentals of Enzyme Kinetics, 3rd edn. Portland Press, London (2004)

22. de Jong, H.: Modeling and Simulation of Genetic Regulatory Systems: A Literature Review. Journal of Computational Biology 9(1), 69–105 (2002)

23. de Jong, H.: Modeling and Simulation of Genetic Regulatory Networks. Lectures Notes in Control and Information Sciences 294, 111–118 (2003)

24. Drulhe, S., Ferrari-Trecate, G., de Jong, H., Viari, A.: Reconstruction of Switching Thresholds in Piecewise-Affine Models of Genetic Regulatory Networks. In: Hespanha, J.P., Tiwari, A. (eds.) HSCC 2006. LNCS, vol. 3927, pp. 184–199. Springer, Heidelberg (2006)

25. Eker, S., Knapp, M., Laderoute, K., Lincoln, P., Meseguer, J., Sonmez, K.: Pathway Logic: Symbolic Analysis of Biological Signaling. In: Proceedings of the Pacific Symposium on Biocomputing, pp. 400–412 (January 2002)

26. Fränzle, M.: What will be eventually true of polynomial hybrid automata? In: Kobayashi, N., Pierce, B.C. (eds.) TACS 2001. LNCS, vol. 2215, pp. 340–359. Springer, Heidelberg (2001)

27. Fränzle, M., Herde, C.: Efficient proof engines for bounded model checking of hybrid systems. In: FMICS (2004)

28. Ghosh, R., Tiwari, A., Tomlin, C.: Automated Symbolic Reachability Analysis; with Application to Delta-Notch Signaling Automata. In: Maler, O., Pnueli, A. (eds.) HSCC 2003. LNCS, vol. 2623, pp. 233–248. Springer, Heidelberg (2003)

29. Ghosh, R., Tomlin, C.: Lateral Inhibition through Delta-Notch Signaling: A Piecewise Affine Hybrid Model. In: Di Benedetto, M.D., Sangiovanni-Vincentelli, A.L. (eds.) HSCC 2001. LNCS, vol. 2034, pp. 232–246. Springer, Heidelberg (2001)

30. Ghosh, R., Tomlin, C.: An Algorithm for Reachability Computations on Hybrid Automata Models of Protein Signaling Networks. In: Proceedings of the 44rd Conference on Decision and Control and European Control Conference (CDC-ECC'05), Seville, Spain, December 2005, pp. 2256–2261. IEEE Computer Society Press, Los Alamitos (2005)

31. Giorgetti, N., Pappas, G., Bemporad, A.: Bounded model checking of hybrid dynamical system. In: Proceedings of the 44rd Conference on Decision and Control and European Control Conference (CDC-ECC'05), Seville, Spain, December 2005, pp. 672–677. IEEE Computer Society Press, Los Alamitos (2005)

32. Girard, A., Pappas, G.J.: Approximate bisimulations for constrained linear systems. In: Proceedings of the 44rd Conference on Decision and Control and European Control Conference (CDC-ECC'05), Seville, Spain, December 2005, pp. 4700–4705. IEEE Computer Society Press, Los Alamitos (2005)

33. Girard, A., Pappas, G.J.: Approximate bisimulations for nonlinear dynamical systems. In: Proceedings of the 44rd Conference on Decision and Control and European Control Conference (CDC-ECC'05), Seville, Spain, December 2005, pp. 690–695. IEEE Computer Society Press, Los Alamitos (2005)

34. Girard, A., Pappas, G.J.: Approximate simulation relations for hybrid systems. In: Proceedings of Analysis and Design of Hybrid Systems (ADHA'06), June 2006, page Algero, Italy, June (to appear)
35. Henzinger, T.A., Nicollin, X., Sifakis, J., Yovine, S.: Symbolic Model Checking for Real-time Systems. In: 7th Annual IEEE Symposium on Logic in Computer Science IEEE, pp. 394–406. IEEE Computer Society Press, Los Alamitos (1992)
36. Hu, J., Wu, W.-C., Sastry, S.: Subtilin Production in Bacillus subtilis using Stochastic Hybrid Systems. In: Alur, R., Pappas, G.J. (eds.) HSCC 2004. LNCS, vol. 2993, pp. 417–431. Springer, Heidelberg (2004)
37. Ideker, T., Lauffenburger, D.: Building with a Scaffold: Emerging Strategies for High- to Low-Level Cellular Modeling. Trends in Biotechnology 21(6), 255–262 (2003)
38. Kappler, K., Edwards, R., Glass, L.: Dynamics in High Dimensional Model Gene Networks. Signal Processing 83, 789–798 (2002)
39. Keener, J.P., Sneyd, J.: Mathematical Physiology. Springer, New York (1998)
40. Kopke, P.W.: The Theory of Rectangular Hybrid Automata. PhD thesis, Faculty of the Graduate School, Cornell University, Adviser - Henzinger, T.A (1996)
41. Lafferriere, G., Pappas, G.J., Sastry, S.: O-minimal hybrid systems. Mathematics of Control, Signals, and Systems 13, 1–21 (2000)
42. Lafferriere, G., Pappas, G.J., Yovine, S.: Symbolic reachability computation for families of linear vector fields. Journal of Symbolic Computation 32(3), 231–253 (2001)
43. Lincoln, P., Tiwari, A.: Symbolic Systems Biology: Hybrid Modeling and Analysis of Biological Networks. In: Alur, R., Pappas, G.J. (eds.) HSCC 2004. LNCS, vol. 2993, pp. 660–672. Springer, Heidelberg (2004)
44. Mandelbrot, B.: The Fractal Geometry of Nature. Freeman Co, San Francisco (1982)
45. Mishra, B.: Algorithmic Algebra. Springer, New York (1993)
46. Mishra, B.: A Symbolic Approach to Modeling Cellular Behavior. In: Sahni, S.K., Prasanna, V.K., Shukla, U. (eds.) HiPC 2002. LNCS, vol. 2552, pp. 725–732. Springer, Heidelberg (2002)
47. Mishra, B.: Computational Real Algebraic Geometry, pp. 740–764. CRC Press, Boca Raton, FL (2004)
48. Mysore, V.: Algorithmic Algebraic Model Checking: Hybrid Automata & Systems Biology. PhD thesis, Department of Computer Science, New York University, Advisors - Mishra, B. (2006)
49. Mysore, V., Casagrande, A., Piazza, C., Mishra, B.: Tolque – A Tool for Algorithmic Algebraic Model Checking. In: The Ninth International Workshop on Hybrid Systems Computation & Control (HSCC06) Poster session (March 2006)
50. Mysore, V., Mishra, B.: Algorithmic Algebraic Model Checking III: Approximate Methods. In: Infinity 2005 – The 7th International Workshop on Verification of Infinite-State Systems, vol. 149(1) of Electronic Notes in Theoretical Computer Science, pp. 61–77 (February 2006)
51. Mysore, V., Mishra, B.: Algorithmic Algebraic Model Checking IV: Metabolic Networks. Journal of Mathematical Biology (to be submitted, 2006)
52. Mysore, V., Piazza, C., Mishra, B.: Algorithmic Algebraic Model Checking II: Decidability of Semi-Algebraic Model Checking and its Applications to Systems Biology. In: Peled, D.A., Tsay, Y.-K. (eds.) ATVA 2005. LNCS, vol. 3707, pp. 217–233. Springer, Heidelberg (2005)
53. Piazza, C., Antoniotti, M., Mysore, V., Policriti, A., Winkler, F., Mishra, B.: Algorithmic Algebraic Model Checking I: The Case of Biochemical Systems and their Reachability Analysis. CIMS-TR 2005-859, Courant Institute Of Mathematical Sciences (2005)
54. Piazza, C., Antoniotti, M., Mysore, V., Policriti, A., Winkler, F., Mishra, B · Algorithmic Algebraic Model Checking I: The Case of Biochemical Systems and their Reachability Analysis. In: Etessami, K., Rajamani, S.K. (eds.) CAV 2005. LNCS, vol. 3576, pp. 5–19. Springer, Heidelberg (2005)

55. Platzer, U., Meinzer, H.-P.: Simulation of Genetic Networks in Multicellular Context. In: Kim, D.J., Polani, M.T (ed.) Fifth German workshop on Artificial Life: Abstracting and Synthesizing the Principles of Living Systems, pp. 43–51 Akad. Verl.-Ges. (2002)
56. Puri, A., Varaiya, P.: Decidebility of hybrid systems with rectangular differential inclusions. Computer Aided Verification, pp. 95–104 (1994)
57. Sachs, K., Gifford, D., Jaakkola, T., Sorger, P., Lauffenburger, D.A.: Bayesian Network Approach to Cell Signaling Pathway Modeling. In: Sci. STKE (2002)
58. Shmulevich, I., Gluhovsky, I., Hashimoto, R.F., Dougherty, E.R., Zhang, W.: Steady-State Analysis of Genetic Regulatory Networks Modelled by Probabilistic Boolean Networks. Comparative and Functional Genomics 4, 601–608 (2003)
59. Voit, E.O.: Computational Analysis of Biochemical Systems. A Pratical Guide for Biochemists and Molecular Biologists. Cambridge University Press, Cambridge (2000)
60. Voit, E.O., Savageau, M.: Equivalence between S-systems and Volterra systems. Mathematical Biosciences 78, 47–55 (1986)

Discrete Models of Biochemical Networks: The Toric Variety of Nested Canalyzing Functions

Abdul S. Jarrah and Reinhard Laubenbacher*

Virginia Bioinformatics Institute, Virginia Tech, Blacksburg, VA 24061-0477, USA
{ajarrah,reinhard}@vbi.vt.edu

Abstract. This paper focuses on the class of nested canalyzing Boolean functions. This class has been introduced and studied recently as a possible source for models of biological networks with favorable dynamic properties. We provide a geometric model for this class in the form of a toric algebraic variety described by a set of binomial polynomial equations, each of whose rational points corresponds to a nested canalyzing function. Toric varieties have a rich geometric and combinatorial structure which provides a basis for a theoretical study of the properties of canalyzing functions. In particular, a good computational characterization of this class would facilitate their incorporation into network inference methods for discrete biochemical networks.

1 Introduction

Time-discrete dynamical systems with a finite state space have a long tradition as models for cellular biochemical networks, beginning with the use of Boolean networks as models of gene regulatory networks in [10]. Other discrete modeling frameworks that have been used in this context include the logical models introduced in [17] and multi-state polynomial models in [12], among others. The latter generalize Boolean networks to arbitrary finite fields, rather than just the field with two elements. The main result in [12] is a network inference algorithm using tools from computational algebra. It takes as input one or more time course measurements, such as gene expression measurements, and produces as output a most likely polynomial dynamical systems model over a suitable finite field that fits the given data set. We briefly describe a few details of the algorithm, since it provides the motivation for the main result of the paper.

Suppose that the biological system to be modeled contains n variables, e.g., genes, and we measure $r + 1$ time points $\mathbf{p}_0, \ldots, \mathbf{p}_r$, using, e.g., gene chip technology, each of which can be viewed as an n-dimensional real-valued vector. The first step is to discretize the entries in the $\mathbf{p}_i$ into a prime number of states, which

---

* The authors were supported partially by NSF Grant DMS-0511441. The second author was supported partially by NIH Grant RO1 GM068947-01, a joint computational biology initiative between NIH and NSF.

H. Anai, K. Horimoto, and T. Kutsia (Eds.): AB 2007, LNCS 4545, pp. 15–22, 2007.

are viewed as entries in a finite field k. If we choose to discretize into two states by choosing a threshold, then we will obtain Boolean networks as models. The discretization step is crucial in this process as it represents the interface between the continuous and discrete worlds. Other network inference methods, such as most dynamic Bayesian network methods, also have to carry out this preprocessing step. Unfortunately, there is very little work that has been done on this problem. We have developed a new discretization method which is described in [4]. It compares favorably to other commonly used discretization methods, using different network inference methods.

Given this data set, an *admissible model*

$$f = (f_1, f_2, \ldots, f_n) : k^n \longrightarrow k^n$$

consists of a dynamical system f which satisfies the property that

$$f(\mathbf{p}_j) = (f_1(\mathbf{p}_j), \ldots, f_n(\mathbf{p}_j)) = \mathbf{p}_{j+1}.$$

The algorithm then proceeds to select such a model f, which is the most likely one based on certain specified criteria. This is done by first reducing the problem to the case of one variable, that is, to the problem of selecting the f_i separately. For this purpose, we compute the set of all functions f_i such that $f_i(\mathbf{p}_j) = \mathbf{p}_{j+1}^i$, that is, all polynomial functions f_i whose value on $\mathbf{p}_j$ is the ith coordinate of $\mathbf{p}_{j+1}$. This set can be represented as the coset $f^0 + I$, where f^0 is a particular such function and $I \subset k[x_1, \ldots, x_n]$ is the ideal of all polynomials that vanish on the given data set, also known as the *ideal of points* of $\mathbf{p}_1, \ldots, \mathbf{p}_r$.

Modifications of the algorithm in [12] have been constructed. The algorithm in [6] starts with only data as input and computes all possible minimal wiring diagrams of polynomial models that fit the given data and outputs a most likely one, based on one of several possible model scoring methods. The algorithm in [3] uses the Gröbner fan of the ideal of points as a computational tool to find a most likely wiring diagram. Both of these algorithms circumvent the need for a particular choice of variable order needed for the algorithm in [12], which affects the structure of the resulting model.

Note that the model space $f^0 + I$ contains *all* possible polynomial functions that fit the given data. In order to improve the performance of model selection algorithms it would be very useful to be able to select certain subspaces of functions that have favorable properties as models of particular biological systems, thereby reducing the model space. For instance, one might consider imposing certain constraints on the structure of the polynomials. We might require that each f_i contains a single term. This amounts essentially to the assumption that the regulatory inputs for a given gene are multiplicative. Or one might add the restriction that the functions result in a dynamical system that has certain constraints on the possible dynamics, e.g., only fixed points as limit cycles. This amounts to the assumption that the system to be modeled does not show periodic behavior. For Boolean networks such constraints have been investigated previously (see, e.g., [2]), in particular the dynamic properties of so-called *canalyzing* Boolean functions.

Canalyzing functions were introduced by S. Kauffman [11] as appropriate rules in Boolean network models of gene regulatory networks. The definition is reminiscent of the concept of "canalisation" introduced by the geneticist C. H. Waddington [19] to represent the ability of a genotype to produce the same phenotype regardless of environmental variability. One important characteristic of canalyzing functions is that they exhibit a stabilizing effect on the dynamics of a system. For example, Moreira and Amaral [13], showed that the dynamics of a Boolean network which operates according to canalyzing rules is robust with regard to small perturbations.

A special type of canalyzing function, so-called *nested canalyzing functions* (NCFs), were introduced recently in [8], and it was shown in [9] that Boolean networks made from such functions show stable dynamic behavior. Nested canalyzing functions have received considerable attention recently. Other classes of functions have also been investigated, e.g., [14]. Certain post classes of Boolean functions have been studied in [16], chain functions in [5], and stabilizing functions in [15].

In order to restrict model selection algorithms to special classes such as nested canalyzing functions (or suitable multi-state generalizations), it is necessary to characterize them computationally in a way that can be integrated in the model selection process. This requirement is the motivation for the main result in this paper. We provide a parametrization of the class of all nested canalyzing functions as the rational points of an affine algebraic variety over the algebraic closure of k which is toric, that is, defined by a collection of binomial polynomial equations. (The set of rational points on the variety are those whose coordinates lie in k.) We also identify the irreducible components of the variety. Toric varieties have a particularly nice combinatorial structure and have been studied extensively. This result is to be interpreted as a first step in a program to parametrize interesting classes of polynomial functions by algebraic varieties, with the goal of studying them theoretically and characterizing them computationally.

2   Polynomial Form of Nested Canalyzing Functions

In this section we briefly recall some definitions and results in [7], where a polynomial form for nested canalyzing functions was derived. It was also shown there that nested canalyzing functions are the same as unate cascade functions, a class studied since the 1970's in electrical engineering and computer science.

We first recall the definitions of canalyzing and nested canalyzing functions from [9].

Definition 1. *A* canalyzing function *is a Boolean function with the property that one of its inputs alone can determine the output value, for either "true" or "false" input. This input value is referred to as the* canalyzing value, *while the output value is the* canalyzed value.

Example 1. The function $f(x, y) = xy$ is a canalyzing function in the variable x with canalyzing value 0 and canalyzed value 0. However, the function $f(x, y) = x + y$ is not canalyzing in either variable.

Nested canalyzing functions are a natural specialization of canalyzing functions. They arise from the question of what happens when the function does not get the canalyzing value as input but instead has to rely on its other inputs. Throughout this paper, when we refer to a function of n variables, we mean that f depends on all n variables. That is, for $1 \leq i \leq n$, there exists $(a_1, \ldots, a_n) \in \mathbb{F}_2^n$ such that $f(a_1, \ldots, a_{i-1}, a_i, a_{i+1}, \ldots, a_n) \neq f(a_1, \ldots, a_{i-1}, 1 + a_i, a_{i+1}, \ldots, a_n)$.

Definition 2. *A Boolean function f in n variables is a* nested canalyzing function *(NCF) in the variable order $x_1, x_2, \ldots, x_n$ with canalyzing input values $a_1, \ldots, a_n$ and canalyzed output values $b_1, \ldots, b_n$, respectively, if it can be expressed in the form*

$$
f(x_1, x_2, \ldots, x_n) =
\begin{cases}
b_1 & \text{if } x_1 = a_1, \\
b_2 & \text{if } x_1 \neq a_1 \text{ and } x_2 = a_2, \\
b_3 & \text{if } x_1 \neq a_1 \text{ and } x_2 \neq a_2 \text{ and } x_3 = a_3, \\
\vdots & \qquad \vdots \\
b_n & \text{if } x_1 \neq a_1 \text{ and } \cdots \text{ and } x_{n-1} \neq a_{n-1} \\
 & \text{and } x_n = a_n, \\
b_n + 1 & \text{if } x_1 \neq a_1 \text{ and } \cdots \text{ and } x_n \neq a_n.
\end{cases}
$$

Example 2. The function $f(x, y, z) = x(y - 1)z$ is nested canalyzing in the variable order x, y, z with canalyzing values $0, 1, 0$ and canalyzed values $0, 0, 0$, respectively. However, the function $f(x, y, z, w) = xy(z + w)$ is not a nested canalyzing function because if $x \neq 0$ and $y \neq 0$, then the value of the function is not constant for any input values for either z or w.

It is shown in [7] that the ring of Boolean functions is isomorphic to the quotient ring $R = \mathbb{F}_2[x_1, \ldots, x_n]/I$, where $I = \langle x_i^2 - x_i : 1 \leq i \leq n \rangle$. Indexing monomials by the subsets of $[n] := \{1, \ldots, n\}$ corresponding to the variables appearing in the monomial, we can write the elements of R as

$$
R = \left\{ \sum_{S \subseteq [n]} c_S \prod_{i \in S} x_i : c_S \in \mathbb{F}_2 \right\}.
$$

As a vector space over $\mathbb{F}_2$, R is isomorphic to $\mathbb{F}_2^{2^n}$ via the correspondence

$$
R \ni \sum_{S \subseteq [n]} c_S \prod_{i \in S} x_i \longleftrightarrow (c_\emptyset, \ldots, c_{[n]}) \in \mathbb{F}_2^{2^n}, \tag{1}
$$

for a given fixed total ordering of all square-free monomials. That is, a polynomial function corresponds to the vector of coefficients of the monomial summands. The main result in [7] is the identification of the set of nested canalyzing functions in R with a subset V^{ncf} of $\mathbb{F}_2^{2^n}$ by imposing relations on the coordinates of its elements.

Definition 3. *Let σ be a permutation of the elements of the set $[n]$. We define a new order relation $<_\sigma$ on the elements of $[n]$ as follows: $\sigma(i) <_\sigma \sigma(j)$ if and* *only if* $i < j$. *Let* r_S^σ *be the maximum element of a nonempty subset* S *of* $[n]$ *with respect to the order relation* $<_\sigma$. *For any nonempty subset* S *of* $[n]$, *the completion of* S *with respect to the permutation* σ, *denoted by* $[r_S^\sigma]$, *is the set* $[r_S^\sigma] = \{\sigma(1), \sigma(2), \ldots, \sigma(r_S)\}$.

Note that, if σ *is the identity permutation, then the completion is* $[r_S] := \{1, 2, \ldots, r_S\}$, *where* r_S *is the largest element of* S.

Theorem 1. *Let* $f \in R$ *and let* σ *be a permutation of the set* $[n]$. *The polynomial* f *is a nested canalyzing function in the order* $x_{\sigma(1)}, x_{\sigma(2)}, \ldots, x_{\sigma(n)}$, *with input values* $a_{\sigma(i)}$ *and corresponding output values* $b_{\sigma(i)}, 1 \leq i \leq n$, *if and only if* $c_{[n]} = 1$ *and, for any proper subset* $S \subseteq [n]$,

$$c_S = c_{[r_S^\sigma]} \prod_{\sigma(i) \in [r_S^\sigma] \setminus S} c_{[n] \setminus \{\sigma(i)\}}. \qquad (2)$$

Corollary 1. *The set of points in* $\mathbb{F}_2^{2^n}$ *corresponding to nested canalyzing functions in the variable order* $x_{\sigma(1)}, x_{\sigma(2)}, \ldots, x_{\sigma(n)}$, *denoted by* V_σ^{ncf}, *is defined by*

$$V_\sigma^{\mathrm{ncf}} = \Big\{ (c_\emptyset, \ldots, c_{[n]}) \in \mathbb{F}_2^{2^n} : \qquad (3)$$

$$c_{[n]} = 1, c_S = c_{[r_S^\sigma]} \prod_{\sigma(i) \in [r_S^\sigma] \setminus S} c_{[n] \setminus \{\sigma(i)\}}, \text{ for } S \subseteq [n] \Big\}.$$

It was also shown in [7] that

$$V^{\mathrm{ncf}} = \bigcup_\sigma V_\sigma^{\mathrm{ncf}}.$$

In the next section we study the variety V^{ncf}, whose points parametrize all nested canalyzing functions in n variables.

3   The Variety of Nested Canalyzing Functions

Since $\mathbb{F}_2$ is a finite field, V^{ncf} and V_σ^{ncf} are algebraic varieties for all permutations σ on $[n]$. We call V^{ncf} the *variety of nested canalyzing functions* and V_σ^{ncf} the variety of nested canalyzing functions in the variable order $x_{\sigma(1)}, \ldots, x_{\sigma(n)}$. As mentioned earlier, the ideal of relations satisfied by the coefficients of a nested canalyzing function can be viewed as an algebraic model of the class of nested canalyzing functions. If this algebraic model has an interesting structure, then it can be studied with the tools of algebra and algebraic geometry, and its properties can help gain insight into the class of nested canalyzing functions. Ideals and their corresponding varieties are best studied over algebraically closed fields. In this section we show that the variety V_σ^{ncf} defined by the ideal of relations over the algebraic closure $\overline{\mathbb{F}}$ of $\mathbb{F}$ is the union of irreducible components, each of which is defined by an ideal I_σ for a particular variable order. That is, I_σ is a binomial prime ideal and, therefore, a toric ideal [18, p. 31]. Toric ideals and their varieties have much interesting structure, which can shed light on the class of nested canalyzing functions.

We show that for all permutations σ on $[n]$, the ideal of the variety V_σ^{ncf}, denoted by $I_\sigma := \mathbb{I}(V_\sigma^{\mathrm{ncf}})$, is a binomial prime ideal in the polynomial ring $\overline{\mathbb{F}}_2[\{c_S : S \subseteq [n]\}]$, where $\overline{\mathbb{F}}_2$ is the algebraic closure of $\mathbb{F}_2$, see [1, p. 62].

By the correspondence (1) and Corollary 1, the ideal $I_\sigma \subseteq \dfrac{\overline{\mathbb{F}}_2[\{c_S : S \subseteq [n]\}]}{\langle c_S^2 - c_S : S \subseteq [n]\rangle}$ is generated by the relations (2). That is,

$$I_\sigma = \left\langle c_{[n]} - 1,\, c_S - c_{[r_S^\sigma]} \prod_{\sigma(i) \in [r_S^\sigma] \setminus S} c_{[n] \setminus \{\sigma(i)\}} : S \subset [n]\right\rangle.$$

It is enough to show that the ideal I_{id} is prime, since I_σ becomes I_{id} after permuting the indexing set $[n]$ by σ.

Let $T = \{\emptyset\} \cup \{[r] : 1 \le r \le n\} \cup \{[n] \setminus \{i\} : 1 \le i \le n\}$.

Theorem 2. *Let* $\phi : \overline{\mathbb{F}}_2[\{c_S : S \subseteq [n]\}] \longrightarrow \dfrac{\overline{\mathbb{F}}_2[\{c_S : S \in T\}]}{\langle c_{[n]} - 1\rangle}$ *such that, for* $S \subseteq [n]$,

$$c_S \mapsto c_{[r_S]} \prod_{i \in [r_S] \setminus S} c_{[n] \setminus \{i\}}.$$

Then

$$\ker(\phi) = \left\langle c_{[n]} - 1,\, c_S - c_{[r_S]} \prod_{i \in [r_S] \setminus S} c_{[n] \setminus \{i\}} : S \subset [n] \setminus T\right\rangle.$$

Proof. By [1, Theorem 2.4.2] and the fact $\overline{\mathbb{F}}_2[\{c_S : S \in T\}] \subset \overline{\mathbb{F}}_2[\{c_S : S \subseteq [n]\}]$, it follows that $\ker(\phi)$ is the binomial ideal generated by $c_{[n]} - 1$ and the set of all binomials $c_S - \phi(c_S)$ where $S \subseteq [n]$. That is,

$$\begin{aligned}\ker(\phi) &= \langle c_S - \phi(c_S) : S \subseteq [n]\rangle \\ &= \left\langle c_{[n]} - 1,\, c_S - c_{[r_S]} \prod_{i \in [r_S] \setminus S} c_{[n] \setminus \{i\}} : S \subset [n]\right\rangle.\end{aligned}$$

But $\phi(c_S) = c_S$ for all $S \in T$. Thus

$$\ker(\phi) = \left\langle c_{[n]} - 1,\, c_S - c_{[r_S]} \prod_{i \in [r_S] \setminus S} c_{[n] \setminus \{i\}} : S \subset [n],\text{ and } S \notin T\right\rangle. \qquad \square$$

Now, since $\dfrac{\overline{\mathbb{F}}_2[\{c_S : S \in T\}]}{\langle c_{[n]} - 1\rangle}$ is an integral domain, the following is straightforward.

Corollary 2. *The ideal* I_{id} *is a prime ideal in the ring* $\overline{\mathbb{F}}_2[\{c_S : S \subseteq [n]\}]$. *Consequently, for any permutation* σ *on* $[n]$, *the ideal* I_σ *is prime in the ring* $\overline{\mathbb{F}}_2[\{c_S : S \subseteq [n]\}]$. *In particular, the ideal of the variety* V^{ncf} *is*

$$\mathbb{I}(V^{\mathrm{ncf}}) = \bigcap_\sigma I_\sigma.$$

4   Discussion

Discrete models for cellular biochemical networks have shown considerable promise in applications. By comparison to dynamical systems described by differential equations, however, the mathematical theory of finite dynamical systems is not well developed yet. The point of view of polynomial dynamical systems over finite fields provides a well-developed computational and theoretical basis with tools and concepts from algebra and algebraic geometry. The network inference algorithm in [12] capitalized on these tools. The description of the class of nested canalyzing functions as an algebraic variety with a rich structure should be viewed as a "proof-of-concept" result that interesting computational objects that have been studied mostly from a statistical point of view can be modeled by interesting mathematical objects in a rich mathematical context.

References

1. Adams, W.W., Loustaunau, Ph.: An introduction to Gröbner bases, vol 3 of Graduate Studies in Mathematics. American Mathematical Society, Providence, RI (1994)
2. Colón-Reyes, O., Laubenbacher, R., Pareigis, B.: Boolean monomial dynamical systems. Ann. Comb. 8, 425–439 (2004)
3. Dimitrova, E., Jarrah, A.S., Laubenbacher, R., Stigler, B.: A.Gröbner fan-based method for biochemical network modeling, Submitted (2007)
4. Dimitrova, E., McGee, J., Vera-Licona, P., Laubenbacher, R.: A comparison of discretization methods for biochemical network inference, Submitted (2007)
5. Gat-Viks, I., Shamir, R., Karp, R.M., Sharan, R.: Reconstructing chain functions in genetic networks. Pacific Symp. Biocomputing 9, 498–509 (2004)
6. Jarrah, A., Laubenbacher, R., Stigler, B., Stillman, M.: Reverse-engineering of polynomial dynamical systems. Adv. Appl. Math. in press (2006)
7. Jarrah, A., Raposa, B., Laubenbacher, R.: Nested canalyzing, unate cascade, and polynomial functions. Physica D, page under revision (2006)
8. Kauffman, S., Peterson, C., Samuelsson, B., Troein, C.: Random Boolean network models and the yeast transcriptional network. PNAS 100(25), 14796–14799 (2003)
9. Kauffman, S., Peterson, C., Samuelsson, B., Troein, C.: Genetic networks with canalyzing Boolean rules are always stable. PNAS 101(49), 17102–17107 (2004)
10. Kauffman, S.A.: Metabolic stability and epigenesis in randomly constructed genetic nets. J. Theor. Biol. 22(3), 437–467 (1969)
11. Kauffman, S.A.: The Origins of Order: Self-Organization and Selection in Evolution. Oxford University Press, New York, Oxford (1993)
12. Laubenbacher, R., Stigler, B.: A computational algebra approach to the reverse engineering of gene regulatory networks. J. Theor. Biol. 229, 523–537 (2004)
13. Moreira, A.A., Amaral, L.A.N.: Canalizing Kauffman networks: Nonergodicity and its effect on their critical behavior. Physical Review Letters 94(21), 218–702 (2005)
14. Raeymaekers, L.: Dynamics of boolean networks controlled by biologically meaningful functions. J. Theor. Biol. 218, 331–341 (2002)
15. Rämö, P., Kesseli, J., Yli-Harja, O.: Stability of functions in boolean models of gene regulatory networks. Chaos, 15 (2005)

16. Shmulevich, I., Lädesmäki, H., Dougherty, E.R., Astola, J., Zhang, W.: The role of certain post classes of in boolean network models of genetic networks. PNAS 100(19), 10724–10739 (2003)
17. Snoussi, E.H., Thomas, R.: Logical identification of all steady states: the concept of feedback loop characteristic states. Bull. Math. Biol. 55, 973–991 (1993)
18. Sturmfels, B.: Gröbner bases and convex polytopes, vol 8 of University Lecture Series. American Mathematical Society, Providence, RI (1996)
19. Waddington, C.H.: Canalisation of development and the inheritance of acquired characters. Nature 150, 563–564 (1942)

Membrane Computing as a Framework for Bio-modeling (An Informal Glimpse)

Gheorghe Păun[1,2]

[1] Institute of Mathematics of the Romanian Academy
PO Box 1-764, 014700 Bucureşti, Romania
[2] Research Group on Natural Computing
Department of Computer Science and Artificial Intelligence
University of Sevilla
Avda. Reina Mercedes s/n, 41012 Sevilla, Spain
george.paun@imar.ro, gpaun@us.es

Abstract. Membrane computing is a well developed branch of natural computing, having as its goal to abstract computing models from the structure and the functioning of the living cell. A recent vivid direction of investigation in this area is devising models for and carrying out applications in biology/medicine and in other disciplines. We briefly present here the basic ideas of membrane computing, its interest for modeling in biology, as well as some applications.

1 An Quick Presentation of Membrane Computing

Membrane computing was initiated [10] as an attempt to learn something useful to computer science from the structure and the functioning of the living cell and from complexes of cells, such as tissues and neural nets. Thus, the resulting research area developed initially as a branch of natural computing, investigating ideas, models, data structures, computing devices/architectures of interest for (theoretical) computer science, especially in what concerns the *computing power* (in comparison with standard computing models, such as Turing machines and restrictions of them) and the *computing efficiency* (the possibility to solve computationally hard problems in a polynomial time). A large variety of models – called *P systems* – were introduced, starting will cell-like systems, passing then to tissue-like and population P systems, and, in the last time, also considering (spiking) neural P systems.

Essentially, a P system is a distributed computing device, handling in a parallel way multisets of objects in the compartments defined by a cell-like or a tissue-like architecture of membranes.

There are several important points here, most of them related to the biological origin of this area of research.

First, the basic data structure used in this framework is the *multiset* (a set with multiplicities associated with its elements). The starting point is the fact that

H. Anai, K. Horimoto, and T. Kutsia (Eds.): AB 2007, LNCS 4545, pp. 23–35, 2007.

in biochemistry "the multiplicity matters", the bio-chemical reactions deal with counted objects, swimming in a solution, without any spatial ordering of them. Mathematically speaking, a multiset can be seen as a string modulo permutation, and this has an important consequence: we do not have positional information, only the number of copies of elements carries information. Thus, numbers are represented in base one by means of multisets. The previous observation also suggests a natural and compact representation of multisets: as strings, with the convention that a string w actually represents the equivalence class of all permutations of w (a string and any of its permutations represent the same multiset and the number of occurrences of a symbol in a string indicates the multiplicity of that symbol in the multiset; the empty multiset, the one with no elements, is represented by the empty string, denoted here by λ).

Then, extending the bio-chemical metaphor, the multisets are processed by means of "reactions", written as multiset rewriting rules. Like in formal language theory, we represent such a rule in the form $u \rightarrow v$, where u and v are strings representing multisets. Using such a rule for rewriting a multiset w means removing ("consuming") the elements indicated by u and adding the elements indicated by v, in each case, with the corresponding multiplicities.

Third, the evolution of the process (the "computation") is localized, by means of membranes. Like in a cell, we have a hierarchical arrangement of membranes, which determine compartments, "protected reactors", where specific (multisets of) objects evolve by means of specific evolution rules. In each compartment, all objects which can evolve by the local rules should do it. Moreover, all compartments of the system evolve simultaneously (a global clock is assumed, marking the time for the whole system, that is, the system is synchronized). Thus, we have two levels of parallelism, one for compartments and one for the objects in each compartment. In each compartment, the rules to apply to the available objects are chosen in a non-deterministic way.

The objects not only evolve, but they can also pass across membranes, thus relating the processes taking place in adjacent compartments. In turn, also the membranes can evolve, for instance, they can be dissolved (the contents of a dissolved membrane is left free in the surrounding membrane), merged, divided, created, etc.

Figure 1, which became a sort of logo of this area, illustrates the idea of a cell-like membrane structure and the associated terminology.

The membrane architecture and the contents of its compartments describe the *configuration* of the system. Using the rules as suggested above, we can pass from a configuration to another one. This passage is called *transition*. A sequence of transitions is called *computation*.

When interpreting a P system as a computing device, we have to specify an input and an output, and, in most cases, the input and the output should be numbers, encoded in multisets, hence in the multiplicity of certain objects. Several possibilities exist: using a P system in the *generative* mode, like a grammar (we start from an initial configuration and we collect all results of halting computations), in the *accepting* mode, like an automaton (a number is introduced in

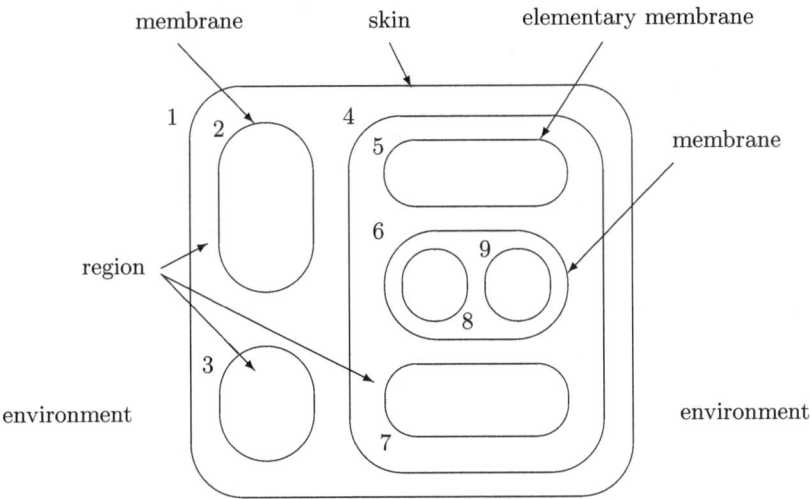

Fig. 1. A membrane structure

the system in the form of the multiplicity of a specified object and this number is accepted if the computation halts), or in the *functional* mode (an input is provided and an output is computed). A particular case of the last two possibilities is to use a P system in the *decidability* mode: an instance of a decision problem is introduced and the answer to this instance, **yes** or **no**, codified in a specified way, is computed.

Strings can also be processed, by considering the ordering of objects which enter and/or leave a system.

A lot of variants of P systems can be obtained by considering various forms of the membrane structure (hierarchical or described by graphs of arbitrary forms), types of objects (symbols as above, strings, arrays, and so on), forms of rules (besides multiset rewriting rules, we have mentioned rules for handling membranes, but we also have symport/antiport rules[1] like in biology, or other rules for moving objects across membranes; in the case of string objects we need specific rules for string manipulation), and ways to use the rules (maximally parallel, as suggested above, with a limited parallelism, sequentially, in the minimally parallel mode and so on).

---

[1] *Symport* is the process of moving simultaneously two or more chemicals across a membrane, through a protein channel, in the same direction, and *antiport* is the process of moving two chemicals across a membrane in opposite directions. Mathematically, a symport rule is written in the form (u, in) or (u, out), meaning that the objects from multiset u enter, respectively exit together the membrane, and an antiport rule is given in the form $(u, in; v, out)$, meaning that the objects of multiset u enter and those of multiset v exit simultaneously the membrane. Of course, each of these rules is applicable only if all the mentioned objects are present in the respective compartments.

We do not give here precise definitions, nor bibliographical details. The interested reader can consult the monograph [11], the collective volume [3] (it starts with a comprehensive introduction to membrane computing), as well as the up-to-dated bibliography from [15] (many downloadable papers are available at this address, in particular, the pre-proceedings of most workshops on membrane computing, as well as the proceedings of the brainstorming weeks on membrane computing held up to now). An example of a P system (with symbol objects, processed by multiset rewriting rules) is given in Appendix 1, both in the formal way and in a graphical representation.

It is important to note at this stage the generality of the approach. We start from the cell, but the abstract model deals with very general notions: membranes interpreted as separators of regions, objects and rules assigned to regions; the basic data structure is the multiset (a set with multiplicities associated with its elements); the rules are used in the (non-deterministic) parallel manner, and in this way we get sequences of transitions, hence computations. In such terms, membrane computing can be interpreted as a *bio-inspired framework for distributed parallel processing of multisets*.

2   Classes of Theoretical Results

As we have mentioned before, membrane computing was much developed as a branch of theoretical computer science, thus intensively investigating the power and the efficiency of P systems.

In what concerns the first direction of research, we may say that "the cell is a powerful computer" (and this is true also for cells organized in tissues or other higher order structures). Specifically, many classes of P systems, combining various ingredients (as described above or similar) are able of simulating Turing machines, hence they are *computationally complete*. Always, the proofs of results of this type are constructive, and this have an important consequence from the computability point of view: there are *universal* (hence *programmable*) P systems. In short, starting from a universal Turing machine (or an equivalent universal device, for instance, a universal Minsky register machine), we get an equivalent universal P system. Among others, this implies that in the case of Turing complete classes of P systems, the hierarchy on the number of membranes always collapses (at most at the level of the universal P systems). Actually, the number of membranes sufficient in order to characterize the power of Turing machines by means of P systems is always rather small. Universality results were obtained both in the case of P systems working in the generating and the accepting mode.

The computational power (the "competence") is only one of the important questions to be dealt with when defining a new (bio-inspired) computing model. The other fundamental question concerns the computing *efficiency*. Because P systems are parallel computing devices, it is expected that they can solve computationally hard problems in an efficient manner – and this expectation is confirmed for systems provided with ways for producing an exponential workspace in a linear time. Three main such biologically inspired possibilities have been considered so far in the literature, and *all of them were proven to lead to polynomial (often, linear) solutions to* **NP**-*complete problems (and even to still harder problems, such as* **PSPACE**-*complete problems).*

These three ideas are *membrane division, membrane creation,* and *string replication.* The standard problems addressed in this framework were decidability problems, starting with SAT, the Hamiltonian Path problem, the Node Covering problem, but also other types of problems were considered, such as the problem of inverting one-way functions, or the Subset-sum and the Knapsack problems (note that the last two are numerical problems, where the answer is not of the yes/no type, as in decidability problems).

There are a series of open problems in these areas, mainly related to the borderline between universality and non-universality and between efficiency (the possibility to solve computationally hard problems in polynomial time) and non-efficiency. How many membranes, objects, rules, and which ingredients we need in order to reach universality/efficiency? Many results were reported in the literature, but still many questions still wait to be solved. Because the focus of this note is not on the theoretical researches/results, but on applications, we do not enter into details and refer the reader to the titles mentioned in the bibliography and, in general, to those available at [15].

We have not mentioned above a recent direction of research in membrane computing inspired from neurology, namely, from the way neurons communicate among each other by means of *spikes; spiking neural P systems* are a class of P systems, introduced in [6] and much investigated in the last time (universality is again obtained, but the complexity matters were only briefly investigated in this case, while applications are still waited for).

3   Applications (In Biology)

As presented above, membrane computing does not deal with a system or a class of computing systems, but it is a general framework for devising such devices. Moreover, membrane computing proved to be a very useful framework for building models for biological applications. After a powerful abstract development, the domain returned to the area where it was originating, and although the initial goal was not to model the cell and the processes taking place in it, now this is a strong tendency and already a series of applications were reported proving the usefulness of this approach.

There are many features of membrane computing which make it attractive for applications in several disciplines, especially for biology.

First, there are several keywords which are genuinely proper to membrane computing and which are of interest for many applications: *distribution* (with the important system-part interaction, emergent behavior, non-linearly resulting from the composition of local behaviors), *algorithmicity* (hence easy programmability), *scalability/extensibility* (this is one of the main difficulties of using differential equations in biology), *transparency* (multiset rewriting rules are nothing else than reaction equations as customarily used in chemistry and bio-chemistry), *parallelism* (a dream of computer science, a common sense in biology), *non-determinism*, *communication* (with the marvelous and still not completely understood way the life is coordinating the many processes taking place in a cell, in contrast with the costly way of coordinating/synchronizing computations in parallel electronic computing architectures, where the communication time becomes prohibitive with the increase of the number of processors), and so on and so forth.

Then, for biology, besides the easy understanding of the formalism and the transparency of the (graphical and symbolic) representations, encouraging should also be the simple observation that membrane computing emerged as a bio-inspired research area, explicitly looking to the cell for finding computability models (though, not looking initially for models of relevance for the biological research), hence it is just natural to try to use these models in the study of the very originating ground. This should be put in contrast with the attempt to "force" models and tools developed in other scientific areas, e.g., in physics, to cover biological facts, presumably of a genuinely different nature as those of the area for which these models and tools were created and proven to be adequate/useful. (This does not mean that membrane computing should be seen as competing with differential equations as tools for bio-modeling, but only as a complement to them, especially adequate in cases when we deal with small populations of chemicals, hence the process is essentially discrete, or with slow reactions.)

Now, in what concerns the applications themselves reported up to now, they are developed at various levels. In many cases, what is actually used is the *language* of membrane computing, having in mind three dimensions of this aspect: (i) the long list of concepts either newly introduced, or related in a new manner in this area, (ii) the mathematical formalism of membrane computing, and (iii) the graphical language, the way to represent cell-like structures or tissue-like structures, together with the contents of the compartments and the associated evolution rules (the "evolution engine").

However, this level of application/usefulness is only a preliminary, superficial one. The next level is to use tools, techniques, results of membrane computing, and here there appears an important question: to which aim? Solving problems already stated, e.g., by biologists, in other terms and another framework, could be an impressive achievement, and this is the most natural way to proceed – but not necessarily the most efficient one, at least at a long term. New tools can suggest new problems, which either cannot be formulated in a previous framework (in plain language, as it is the case in biology, whatever specialized the specific jargon is, or using other tools, such as differential equations) or have no chance to be solved in the previous framework (see [8], [14] for more discussions about this idea).

Applications of all these types were reported in the literature of membrane computing. As expected and as natural, most applications were carried out in biology, but also applications in computer graphics (where the compartmentalization seems to add a significant efficiency to well-known techniques based on L systems), linguistics (both as a representation language for various concepts related to language evolution, dialogue, semantics, and making use of the parallelism, in solving parsing problems in an efficient way), economics (where many bio-chemical metaphors find a natural counterpart – see [12], with the mentioning that the "reactions" which take place in economics, for instance, in market-like frameworks, are not driven only by probabilities/stoichiometric calculations, but also by psychological influences, which makes the modeling still more difficult than in biology), computer science (in devising sorting and ranking algorithms), cryptography, approximate algorithms for optimization problems, etc.

These applications are usually based on experiments using programs for simulating/implementing P systems on usual computers, and there are already several such programs, more and more elaborated (e.g., with better and better interfaces, which allow for the friendly interaction with the program). We avoid to plainly say that we have "implementations" of P systems, because of the inherent non-determinism and the massive parallelism of the basic model, features which cannot be implemented, at least in principle, on the usual electronic computer – but which can be implemented on a dedicated, reconfigurable, hardware, on a local network, etc. This does not mean that simulations of P systems on usual computers are not useful; actually, such programs were used in all biological applications mentioned in this paper, and can also have important didactic and research applications. An overview of membrane computing software reported in literature (some programs are available in the web page [15]) can be found in [3]. Several applications are presented in detail – software included – at [16] and [17].

Of course, when using a P system for simulating a biological process we are no longer interested in its computing behavior (power, efficiency, etc.), but in its evolution in time; the P system is then interpreted as a dynamical system, and its trajectories are of interest, its "life". Moreover, the ingredients we use are different from those considered in theoretical investigations. For instance, in mathematical terms, we are interested in results obtained with a minimum of premises and with weak prerequisites, while the rules are used in ways inspired from automata and language theory (e.g., in a maximally or minimally parallel way), but when dealing with applications the systems are constructed in such a way to capture the features of reality (for instance, the rules are of a general form, they are applied according to probabilistic strategies, based on stoichiometric calculations, the systems are not necessarily synchronized, and so on).

Many applications in biology were presented at the seventh edition of Workshop on Membrane Computing, Leiden, The Netherlands, July 2006. The reader is refereed to [5] for details. Just to have a flavor of the diversity of these applications, we recall here some titles from this volume: Formalizing spherical membrane structures and membrane protein populations, A modeling approach based on P systems with bounded parallelism, Metabolic P approaches to biochemical dynamics: biological rhythms and oscillations, Modeling signal transduction using P systems, Towards a hybrid metabolic algorithm, Tau leaping stochastic simulation method in P systems, Mitotic oscillators as MP graphs, A protein substructure based P system for description and analysis of cell signalling networks.

The typical applications run as follows. One starts from a biological process described in general in graphical terms (chemicals are related by reactions represented in a graph-like manner, with special conventions for capturing the context-sensitivity of reactions, the existence of promoters or inhibitors, etc.) or already available in data bases in SBML (system biology mark-up language) form; these data are converted into a P system which is introduced in a simulator; the way the evolution rules (reactions) are applied is the key point in constructing this simulator (often, the classical Gilespie algorithm is used in compartments, or multi-compartmental variants of it are considered); as a result, the evolution in time of the multiplicity of certain chemicals is displayed, thus having a graphical representation of the interplay in time of certain chemicals, their growth and decay, and so on. Many illustrations of this scenario can be found in [5]. We also briefly recall in Appendix 2 a typical example of this kind, from [7]. Much more complex processes were approached; for instance, in [13] one works with 60 proteins, placed in three compartments (hence delimited by two membranes) and linked by 160 reactions, which is already a process of a real-life dimension. A similarly complex process is considered in [2]: 53 proteins, placed in four compartments (environment, cell surface, cytoplasm, mitochondria) and evolving by means of 99 reactions. A special case is that of investigations related to quorum sensing in bacteria (see, e.g., [1] and the references therein): simulations of populations of hundreds of bacteria were carried out (and the results are again consistent with the experimental observations), but in order to give relevant results to biologists it is necessary to scale-up to thousands of bacteria; in theory, this is straightforward, but from the computational point of view this needs much faster implementations than the existing ones.

This approach is especially interesting in the case of complex systems of biochemical equations, with an intricate behavior of objects, involving cycles. Usually, the cycles induce a non-linear behavior of the system, hard – if not impossibly – to predict when simply examining the system. However, because P systems of most types are universal, they are undecidable, hence no inquiry can be answered algorithmically, just examining mathematically the system, and this makes necessary the simulation on computer.

Most of the applications reported so far are of a "post-diction" type: one takes data from the literature, either based on laboratory experiments or on other types of models (e.g., differential equations), and one compares the results given by the P system simulation with those already available, thus checking whether the new approach is reliable (and this was always the case). The next step – already done in a few cases – is that of "pre-diction": working with research data, with hypotheses, and providing results which are not known by the biologist. A situation of this type is that from [9].

References

1. Bianco, L., Pescini, D., Siepmann, P., Krasnogor, N., Romero-Campero, F.J., Gheorghe, M.: Towards a P systems Pseudomonas quorum sensing model. In: [5], pp. 197–214
2. Cheruku, S., Păun, A., Romero-Campero, F.J. Pérez-Jiménez, M.J., Ibarra, O.H.: Simulating FAS-induced apoptosis by using P systems. In: Pre-proceedings of Conf. Bio-Inspired Computing – Theory and Applications, BIC-TA 2006, Wuhan, China. Volume of Membrane Computing Section, pp. 71–81 (2006)
3. Ciobanu, G., Păun, Gh., Pérez-Jiménez, M.J.: Applications of Membrane Computing. Springer, Berlin (2005)
4. Goldbeter, A.: A minimal cascade model for the mitotic oscillator involving cyclin and cdc2 kinase. PNAS 88(20), 9107–9111 (1991)
5. Hoogeboom, H.J., Păun, Gh., Rozenberg, G., Salomaa, A.: Membrane Computing. In: Hoogeboom, H.J., Păun, G., Rozenberg, G., Salomaa, A. (eds.) WMC 2006. LNCS, vol. 4361, Springer, Berlin (2006)
6. Ionescu, M., Păun, G., Yokomori, T.: Spiking neural P systems. Fundamenta Informaticae 71(2-3), 279–308 (2006)
7. Manca, V.: Metabolic P systems for biochemical dynamics. In: Pre-proceedings of Conf. Bio-Inspired Computing - Theory and Applications, BIC-TA 2006, Wuhan, China. Volume of Membrane Computing Section, pp. 15–26 (2006)
8. Mesarović, M.D.: System theory and biology – view of a theoretician. In: Mesarović, M.D. (ed.) System Theory and Biology, pp. 59–87. Springer, New York (1968)
9. Martegani, E., Tisi, R., Belotti, F., Colombo, S., Paiardi, C., Winderickx, J., Cazzaniga, P., Besozzi, D., Mauri, G.: Identification of an intracellular signalling complex for ras/camp pathway in yeast: experimental evidences and modelling. ISSY 25 Conf. Hanassari, Espo, Finland (2006)
10. Păun, Gh.: Computing with membranes. Journal of Computer and System Sciences, 61(1), pp. 108–143 (2000), (and Turku Center for Computer Science-TUCS Report 208, November 1998, www.tucs.fi)
11. Păun, Gh.: Computing with Membranes. An Introduction. Springer, Berlin (2002)
12. Păun, Gh., Păun, R.: Membrane computing as a framework for modeling economic processes. In: Proc. SYNASC 05, Timişoara, Romania, pp. 11–18. IEEE Press, NJ (2005)
13. Pérez-Jiménez, M.J., Romero-Campero, F.J.: A study of the robustness of the EGFR signalling cascade using continuous membrane systems. In: Mira, J.M., Álvarez, J.R. (eds.) IWINAC 2005. LNCS, vol. 3561, pp. 268–278. Springer, Berlin (2005)
14. Wolkenhauer, O.: Systems biology: The reincarnation of systems theory applied in biology? Briefings in Bioinformatics 2(3), 258–270 (2001)
15. The Web Page of Membrane Computing: http://psystems.disco.unimib.it
16. The Sheffield P Systems Applications Web Page: http://www.dcs.shef.ac.uk/~marian/PSimulatorWeb/P_Systems_applications.htm
17. The Web Page of Verona Research Group on Natural Computing: http://www.di.univr.it/dol/main?ent=arearic&id=21

Appendices

A1: One Simple Example

We consider here a simple P system, namely one which computes the function $n \longrightarrow n^2$, for any natural number $n \geq 1$. The initial configuration is given in

Figure 2. We use catalytic rules (of the form $ca \to cu$, where c is a catalyst, a the object which evolves with the help of the catalyst, and u the multiset of objects obtained from a) and non-cooperating rules (of the form $a \to u$, with a and u as above), as well as a rule with promoters, $b_2 \to b_2 e_{in}|_{b_1}$: the object b_2 evolves to $b_2 e$ only if at least one copy of object b_1 is present in the same region; b_1 can simultaneously evolve by means of other rules.

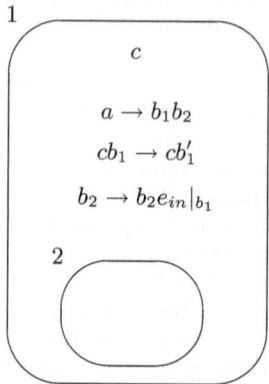

Fig. 2. A P system with catalysts and promoters

Formally, the system is given as follows (thus, we also introduce the usual way of presenting the components of a P system, in particular, the representation of a membrane structure by means of an expression of correctly matching labeled square brackets):

$$\Pi = (O, C, \mu, w_1, w_2, R_1, R_2, i_o), \text{ where:}$$
$$O = \{a, b_1, b'_1, b_2, c, e\} \text{ (the set of objects)}$$
$$C = \{c\} \text{ (the set of catalysts)}$$
$$\mu = [_1 \ [_2 \ ]_2 \ ]_1 \text{ (membrane structure)}$$
$$w_1 = c \text{ (initial objects in region 1)}$$
$$w_2 = \lambda \text{ (initial objects in region 2)}$$
$$R_1 = \{a \to b_1 b_2, \ cb_1 \to cb'_1, \ b_2 \to b_2 e_{in}|_{b_1}\} \text{ (rules in region 1)}$$
$$R_2 = \emptyset \text{ (rules in region 2)}$$
$$i_o = 2 \text{ (the output region).}$$

Note the target indication *in* present in the rule $b_2 \to b_2 e_{in}|_{b_1}$: the object e produced by using this rule has to immediately go to the inner membrane (another target indication can be *out*, meaning that the respective object has to immediately go outside the membrane where the rule was applied; having such an indication in the skin region will send the respective object into the environment; all objects without an explicit target indication are supposed to have the indication *here*, meaning that they remain in the same region where the rule is applied).

The rules are applied in the maximally parallel way: in each step, each object which can evolve must do it.

We start with only one object in the system, the catalyst c. If we want to compute the square of a number n, then we have to input n copies of the object a in the skin region of the system. In that moment, the system starts working, by using the rule $a \rightarrow b_1 b_2$, which has to be applied in parallel to all copies of a; hence, in one step, the n copies of object a are replaced by n copies of b_1 and n copies of b_2. From now on, the other two rules from region 1 can be used. The catalytic rule $cb_1 \rightarrow cb_1'$ can be used only once in each step, because the catalyst is present in only one copy. This means that in each step one copy of b_1 gets primed. Simultaneously (because of the maximal parallelism), the rule $b_2 \rightarrow b_2 e_{in}|_{b_1}$ should be applied as many times as possible and this means n times, because we have n copies of b_2. Note the important difference between the promoter b_1, which allows using the rule $b_2 \rightarrow b_2 e_{in}|_{b_1}$, and the catalyst c: the catalyst is involved in the rule, it is counted when applying the rule, while the promoter makes possible the use of the rule, but it is not counted; the same (copy of an) object can promote any number of rules. Moreover, the promoter can evolve at the same time by means of another rule (the catalyst is never changed).

In this way, in each step we change one b_1 to b_1' and we produce n copies of e (one for each copy of b_2); the copies of e are sent to membrane 2 (the indication *in* from the rule $b_2 \rightarrow b_2 e_{in}|_{b_1}$). The computation should continue as long as there are applicable rules. This means exactly n steps: in n steps, the rule $cb_1 \rightarrow cb_1'$ will exhaust the objects b_1 and after n steps neither this rule can be applied, nor $b_2 \rightarrow b_2 e_{in}|_{b_1}$, because its promoter does no longer exist. Consequently, in membrane 2, considered as the output membrane, we get n^2 copies of object e.

Note that the computation is deterministic, always the next configuration of the system is unique, and that, changing the rule $b_2 \rightarrow b_2 e_{in}|_{b_1}$ with $b_2 \rightarrow b_2 e_{out}|_{b_1}$, the n^2 copies of e will be sent to the environment, hence we can read the result of the computation outside the system, and in this case membrane 2 is useless.

A2: An Illustrative Application

We now briefly present an example of an application of membrane computing in modeling (and simulating) a biological phenomenon recalled from [7]. It deals with the mitotic oscillator in amphibian embryos. One starts from the representation of this oscillator given in [4], in the form presented in Figure 3.

One identifies here several objects linked through several reactions. These reactions can be formalized as multiset rewriting rules as shown in Table 1. We have both "non-cooperative" rules (having only one object in their left-hand side) and "cooperative" rules (with two or more objects reacting and getting transformed into other objects). With each rule one also specifies its reactivity, as given in [4]. (Actually, the passage from the description of the process as given in Figure 3 to the rules of the P system from Table 1 is done in [7] by means

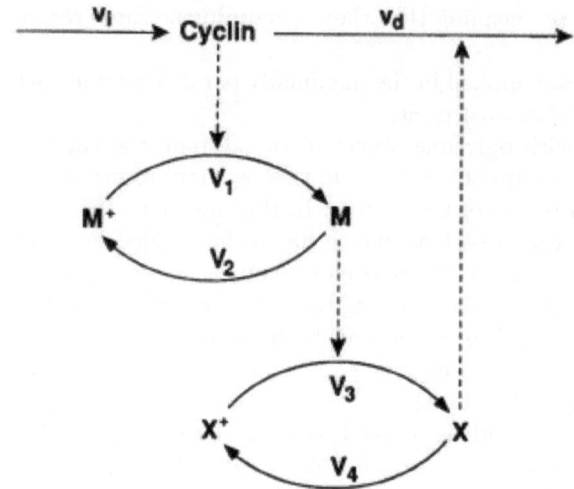

Fig. 3. The mitotic oscillator from [4]

Table 1. The evolution rules describing the reactions from Figure 3

$$
\begin{aligned}
&r_1 : \lambda \to C & &f_1 = v_i \\
&r_2 : XC \to X & &f_2 = v_d/kd + c \\
&r_3 : C \to \lambda & &f_3 = K_d \\
&r_4 : M^+C \to MC & &f_4 = V_{M1}/(K_c + c)(K_1 + m^+) \\
&r_5 : M \to M^+ & &f_5 = V_2/(K_2 + m) \\
&r_6 : X^+M \to XM & &f_6 = V_3/(K_3 + x^+) \\
&r_7 : X \to X^+ & &f_7 = V_4/(K_4 + x)
\end{aligned}
$$

Table 2. A non-cooperative system equivalent to the system from Table 1

$$
\begin{aligned}
&r_1 : \lambda \to C & &f_1 = v_i \\
&r_2' : C \to X & &f_2 = v_d \cdot x/k_d + c \\
&r_2'' : X \to \lambda & &f_2 = v_d \cdot c/k_d + c \\
&r_3 : C \to \lambda & &f_3 = K_d \\
&r_4' : C \to MC & &f_4 = V_{M1} \cdot m^+/(K_c + c)(K_1 + m^+) \\
&r_4'' : M^+ \to \lambda & &f_4 = V_{M1} \cdot c/(K_c + c)(K_1 + m^+) \\
&r_5 : M \to M^+ & &f_5 = V_2/(K_2 + m) \\
&r_6' : X^+ \to XM & &f_6 = V_3 \cdot m/(K_3 + x^+) \\
&r_6'' : M \to \lambda & &f_6 = V_3 \cdot x^+/(K_3 + x^+) \\
&r_7 : X \to X^+ & &f_7 = V_4/(K_4 + x)
\end{aligned}
$$

of a so-called MP-graph, a graph-based formalism for representing bio-chemical interactions introduced by Verona team and presented/investigated in a series of papers of this team; we skip the details, technical and bibliographical, and refer the reader to titles available at [15].)

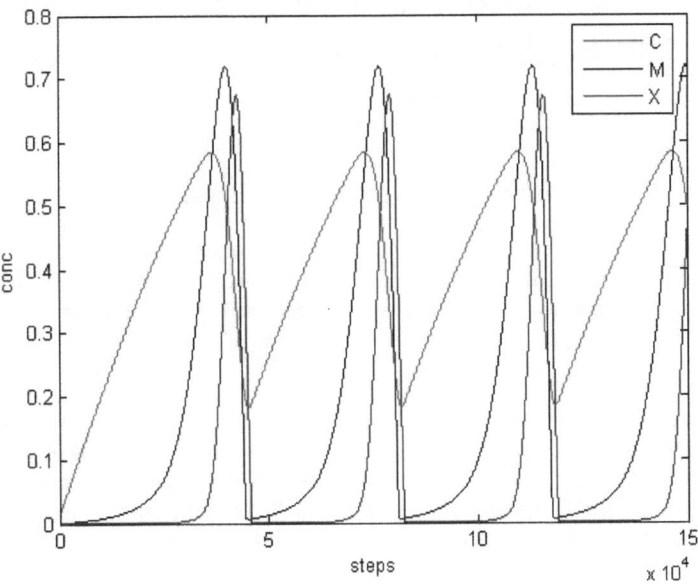

Fig. 4. The evolution of the P system from Tables 1, 2

We have here a P system with only one membrane, where objects $C, X, M,$ X^+, M^+ evolve according to rules $r_2, \ldots, r_7$; rule r_1 is meant to feed the system with copies of C, according to the reaction rate indicated by f_1.

This system is equivalent (in terms of its evolution in time) with the system having the non-cooperative rules given in Table 2.

The evolution of this system (simulated by means of a software realized in Verona by Luca Bianco – see his PhD thesis available at [15]) is given in Figure 4. The oscillations are rather similar to those found in [4] by means of handling a system of differential equations.

Relating Attractors and Singular Steady States in the Logical Analysis of Bioregulatory Networks

Heike Siebert and Alexander Bockmayr

DFG Research Center MATHEON,
Freie Universität Berlin, Arnimallee 3, D-14195 Berlin, Germany
siebert@mi.fu-berlin.de, bockmayr@mi.fu-berlin.de

Abstract. In 1973 R. Thomas introduced a logical approach to modeling and analysis of bioregulatory networks. Given a set of Boolean functions describing the regulatory interactions, a state transition graph is constructed that captures the dynamics of the system. In the late eighties, Snoussi and Thomas extended the original framework by including singular values corresponding to interaction thresholds. They showed that these are needed for a refined understanding of the network dynamics. In this paper, we study systematically singular steady states, which are characteristic of feedback circuits in the interaction graph, and relate them to the type, number and cardinality of attractors in the state transition graph. In particular, we derive sufficient conditions for regulatory networks to exhibit multistationarity or oscillatory behavior, thus giving a partial converse to the well-known Thomas conjectures.

1 Introduction

Suggested more than 30 years ago, the logical approach to modeling bioregulatory networks has become increasingly popular in the recent past. In the Boolean setting, components of the networks correspond to variables, which can take the values 0 and 1. Interactions between the components are described by logical equations capturing the evolution of the system. R. Thomas contributed a number of papers on the logical analysis of biological networks, starting with [10]. The distinctive feature of his method is the way he derives a representation of the dynamics from the given Boolean functions. Rather than executing all indicated changes in the components at the same time, an asynchronous updating rule is employed to obtain a non-deterministic state transition graph. It has been shown that this approach captures essential qualitative features of the dynamical behavior of complex biological networks, see [11] and [12] for an overview.

In the following years the framework was extended to allow not only for Boolean but multi-valued variables that describe different activity levels of the regulatory components in the network. Each interaction in the network was associated with a unique threshold value, which determines when the interaction becomes effective. Snoussi and Thomas realized that a closer inspection of the

H. Anai, K. Horimoto, and T. Kutsia (Eds.): AB 2007, LNCS 4545, pp. 36–50, 2007.
© Springer-Verlag Berlin Heidelberg 2007

impact of the threshold values, which they called singular values, would further improve the understanding of the system's dynamics. In [8] they introduced the notion of singular steady states and linked them to feedback circuits in the interaction graph describing the structure of the network. The importance of feedback circuits for the analysis of the dynamical behavior has long been recognized. Thomas conjectured in 1981 that the existence of a positive (resp. negative) circuit, in the interaction graph is a necessary condition for the existence of two distinct attractors (resp. a cyclic attractor) in the state transition graph. The conjectures have been proven in different settings (see e.g. [9], [4] and [5]). In [2] it is shown, that isolated elementary regulatory circuits result in fundamentally different dynamics depending on their sign. A positive circuit can be linked to the occurrence of two stable states, while a negative circuit causes an attractor comprising dynamical cycles. However, the situation becomes more difficult to grasp as soon as the circuits are embedded in larger and more complex networks.

When trying to incorporate Snoussi's and Thomas' idea of singular states in a Boolean framework, we are faced with several difficulties. On this level of abstraction, every interaction is associated with the same threshold value, a symbolic value between 0 and 1. Thus when crossing the threshold we do not have the advantage of knowing that one and only one interaction becomes effective. As a result we cannot link singular states to circuits in the interaction graph in a non-ambiguous way, while still preserving some essential features known from the multi-valued setting. Despite those complications and the high level of abstraction, this paper shows that the introduction of singular states in the Boolean case is a useful tool for refining our understanding of the relation between structure and dynamics of bioregulatory networks.

The organization of the paper is as follows. In Section 2 we give a short overview of the Boolean description of biological networks and introduce the notion of an attractor of a state transition graph. In Section 3 we extend the framework by establishing the concept of singular states. We give different characterizations of singular steady states using the notion of circuit characteristic states and regular adjacent states. In the main section of this paper, we prove several statements that allow us to derive information on the attractors of the state transition graph from the existence of singular steady states. Conversely, we can deduce the existence of a singular steady state if we have specific knowledge about the attractors of the state transition graph. We conclude by outlining ideas for future work.

2   Structure and Dynamics of Regulatory Networks

In the following we introduce the Boolean formalism of R. Thomas for modeling regulatory networks (see for example [11]). We mainly use the notation introduced in [1] and [6]. Throughout the text $\mathcal{B}$ will denote the set $\{0, 1\}$.

Definition 1. *An* interaction graph *(or bioregulatory graph)* $\mathcal{I}$ *is a labeled directed graph with vertex set* $V := \{\alpha_1, \ldots, \alpha_n\}$, $n \in \mathbb{N}$, *and edge set* E. *Each edge* $\alpha_j \to \alpha_i$ *is labeled with a sign* $\varepsilon_{ij} \in \{+, -\}$.

The only information on a regulatory component we incorporate in the model for now is whether or not it is active. A vertex α_i can be seen as a variable that adopts values in $\mathcal{B}$, where the value 1 indicates that α_i is active. To simplify notation, we identify each vertex α_i with its index i.

An edge $\alpha_j \rightarrow \alpha_i$ signifies that α_j influences α_i in a positive or negative way depending on the sign ε_{ij}. For each α_i we denote by $Pred(\alpha_i)$ the set of *predecessors* of α_i, i.e., the set of vertices α_j such that $\alpha_j \rightarrow \alpha_i$ is an edge in E.

We will be mainly interested in the following structures of the interaction graph. A tuple $(\alpha_{i_1}, \ldots, \alpha_{i_k})$ of distinct vertices of $\mathcal{I}$ is called a *circuit* if $\mathcal{I}$ contains an edge from α_{i_j} to $\alpha_{i_{j+1}}$ for all $j \in \{1, \ldots, k-1\}$ as well as an edge from α_{i_k} to α_{i_1}. The *sign* of a circuit is the product of the sign of its edges.

Definition 1 captures structural aspects of the network. Now we consider the corresponding dynamical behavior.

Definition 2. *Let $\mathcal{I}$ be an interaction graph comprising n vertices. A state of the system described by $\mathcal{I}$ is a tuple $s \in \mathcal{B}^n$. The set of (regular) resources $R_i(s) = R_i^{\mathcal{I}}(s)$ of α_i in state s is the set*

$$\{\alpha_j \in Pred(\alpha_i) \mid (\varepsilon_{ij} = + \wedge s_j = 1) \vee (\varepsilon_{ij} = - \wedge s_j = 0)\}.$$

Given a set

$$K(\mathcal{I}) := \{K_{i,\omega} \mid i \in \{1, \ldots, n\}, \ \omega \subseteq Pred(\alpha_i)\}$$

of (logical) parameters, which adopt values in $\mathcal{B}$, we define the Boolean function $f = f^{K(\mathcal{I})} : \mathcal{B}^n \rightarrow \mathcal{B}^n$, $s \mapsto (K_{1,R_1(s)}, \ldots, K_{n,R_n(s)})$. The pair $N := (\mathcal{I}, f)$ is called bioregulatory network.

The set of resources $R_i(s)$ provides information about the presence of activators and the absence of inhibitors for some regulatory component α_i in state s. It contains all genes that contribute to an activation of α_i in state s. Note that the absence of an inhibitor is interpreted as an activating influence on the target gene. The value of the parameter $K_{i,R_i(s)}$ indicates how the level of activity α_i will evolve. It will increase (resp. decrease) if the parameter value is greater (resp. smaller) than s_i. The activity level stays the same if both values are equal. Thus, the function f maps a state s to the state the system tends to evolve to. Snoussi and Thomas posed the following condition on the parameter values of the systems they considered:

$$\omega \subseteq \omega' \Rightarrow K_{i,\omega} \leq K_{i,\omega'} \tag{1}$$

for all $i \in \{1, \ldots, n\}$. The condition signifies that an effective activator or a non-effective inhibitor cannot induce the decrease of the activity level of α_i. In the following we always assume that this condition is valid.

The choice of parameters completes the definition of the model given by the graph $\mathcal{I}$. Depending on their values, edges in the graph may or may not be *functional* in the following sense. Clearly, if there is an edge $\alpha_j \rightarrow \alpha_i$ and $K_{i,M} = K_{i,M \setminus \{\alpha_j\}}$ for all $M \subseteq Pred(\alpha_i)$, then the edge $\alpha_j \rightarrow \alpha_i$ has no influence on the dynamics of the system. Eliminating this edge from the interaction graph

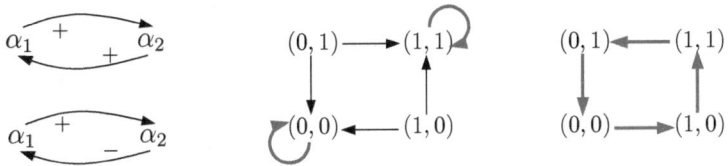

Fig. 1. Two interaction graphs consisting of a positive resp. a negative circuit. In both cases we choose $K_{1,\{2\}} = K_{2,\{1\}} = 1$ and $K_{1,\emptyset} = K_{2,\emptyset} = 0$. The state transition graph corresponding to the positive circuit is in the middle, the one corresponding to the negative circuit is on the right. Attractors are indicated by colored, fat lines.

does not change the function f. Thus we may assume for every $N = (\mathcal{I}, f)$ that whenever there is an edge $\alpha_j \to \alpha_i$ in $\mathcal{I}$, there exists a set $M \subseteq Pred(\alpha_i)$ such that $K_{i,M} \neq K_{i,M \setminus \{\alpha_j\}}$.

To derive the dynamics of the system from the function f we take the following consideration into account. In a biological system, the time delays corresponding to changes in the activity level of distinct components will most likely differ. Thus we may assume that in each state transition at most one component is modified. This procedure is called *asynchronous update* in Thomas' framework. We obtain the following definition.

Definition 3. *The state transition graph $\mathcal{S}_N$ describing the dynamics of the network N is a directed graph with vertex set $\mathcal{B}^n$. There is an edge $s \to s'$ if and only if $s' = f(s) = s$ or $s'_i = f_i(s)$ for some $i \in \{1, \ldots, n\}$ satisfying $s_i \neq f_i(s)$ and $s'_j = s_j$ for all $j \neq i$.*

In the following we introduce some basic structures in this graph that are of biological interest. In addition we use standard terminology from graph theory, such as paths and cycles.

Definition 4. *Let $\mathcal{S}_N$ be a state transition graph. An infinite path $(s_0, s_1, \ldots)$ in $\mathcal{S}_N$ is called* trajectory. *A nonempty set of states D is called* trap set *if every trajectory starting in D never leaves D. A trap set A is called* attractor *if for any $s^1, s^2 \in A$ there is a path from s^1 to s^2 in $\mathcal{S}_N$. A state s^0 is called* steady state, *if s^0 is a fixed point of f, that is, if there is an edge from s^0 to itself. A cycle $C := (s^1, \ldots, s^r, s^1)$, $r \geq 2$, is called a* trap cycle *if every s^j, $j \in \{1, \ldots, r\}$, has only one outgoing edge in $\mathcal{S}_N$, i.e., the trajectory starting in s^1 is unique.*

Thus, the attractors of $\mathcal{S}_N$ correspond to the terminal strongly connected components of the graph. It is easy to see that steady states and trap cycles are attractors. In Figure 1 we show two simple interaction graphs. The positive circuit generates a state transition graph with two steady states. The graph derived from the negative circuit consists of a trap cycle, that is, we find an attractor of cardinality greater than one. This corresponds to the typical behavior assigned to positive (resp. negative) circuits mentioned in the introduction.

Attractors represent regions of predictability and stability in the behavior of the system. It is not surprising that an attractor often has a biological interpretation. A fixed point in a gene regulatory network associated with cell differentiation, for example, may represent the stable state reached at the end of a developmental process. Attractors of cardinality greater than one imply cyclic behavior, and thus can often be identified with homeostasis of sustained oscillatory activity, as can be found in the cell cycle or circadian rhythm.

The following proposition is an easy observation concerning attractors.

Proposition 1. *Every state transition graph $\mathcal{S}_N$ contains at least one attractor.*

Proof. For $s \in \mathcal{B}^n$ we denote by $D(s)$ the set of states reachable from s by a path in $\mathcal{S}_N$. Then $D(s)$ is a trap set for every $s \in \mathcal{B}^n$. Fix $s \in \mathcal{B}^n$ and choose $A \subseteq D(s)$ a minimal trap set, i.e., every proper subset of A is not a trap set. Let $x, y \in A$. Then $D(x) \subseteq A$, since A is a trap set. Since A is minimal, we have $A = D(x)$. Consequently, there is a path from x to y. Thus, A is an attractor. □

Note that the above proof shows that for every state in the state transition graph there is a trajectory leading to an attractor.

The number of states in the state transition graph grows exponentially with the number of regulatory components in N. Thus our aim is to infer from restrictions of f to sets of vertices obtained by considering certain subgraphs of $\mathcal{I}$ as much information on the structure of $\mathcal{S}_N$ as possible.

3   Singular States

In the following, we incorporate threshold values of interactions into the formalism to get a more complete understanding of the dynamics of the system. We mainly use the framework introduced in [6].

Definition 5. *Set $\mathcal{B}_\theta := \{0, \theta, 1\}$, where θ is a symbolic representation of the threshold value and satisfies the order $0 < \theta < 1$. We allow each regulatory component α_i to take values in $\mathcal{B}_\theta$. The values 0 and 1 are called* regular values *and θ is called* singular value. *The elements of $\mathcal{B}_\theta^n$ are called* states. *If a state comprises only regular components it is called* regular *state. Otherwise it is called* singular *state. For every state s we define $J(s) := \{i \in \{1, \ldots, n\} \mid s_i = \theta\}$.*

To describe the dynamics of the system we have to extend the definition of resources.

Definition 6. *Let $s \in \mathcal{B}_\theta^n$. In addition to the set $R_i(s)$ of regular resources introduced in Definition 2, we define the set $R_i^\theta(s)$ of singular resources of α_i in s as the set*

$$R_i^\theta(s) := \{\alpha_j \in Pred(\alpha_i) \mid s_j = \theta\}.$$

The definition of a set of logical parameters $K(\mathcal{I})$ remains the same as in Definition 2. In particular, the logical parameters can only adopt regular values.

We call $\lfloor a, b \rfloor$ a *qualitative value* if $a, b \in \mathcal{B}$ and $a \leq b$. The qualitative value $\lfloor 0, 0 \rfloor$ is identified with the regular value 0, $\lfloor 1, 1 \rfloor$ with the regular value 1, and $\lfloor 0, 1 \rfloor$ with the singular value θ. The relations $<, >$, and $=$ are used with respect to this identification.

Definition 7. *Let $K(\mathcal{I})$ be a set of parameters. We define*

$$f^\theta = f^{K(\mathcal{I}),\theta} : \mathcal{B}_\theta^n \to \mathcal{B}_\theta^n \quad by \quad f_i^\theta(s) = |K_{i,R_i(s)}, K_{i,R_i(s)\cup R_i^\theta(s)}|$$

for all $i \in \{1,\ldots,n\}$.

The map f^θ is well defined since condition (1) ensures that $K_{i,R_i(s)} \leq K_{i,R_i(s)\cup R_i^\theta(s)}$ for all $i \in \{1,\ldots,n\}$. Note that whenever s is a regular state, then $f^\theta(s)$ is regular, too, since any set of singular resources in a regular state is empty. We have $f^\theta(s) = f(s)$ for all $s \in \mathcal{B}^n$. Thus the state transition graph corresponding to $N = (\mathcal{I}, f)$ is consistent with f^θ. Extending the definition in the previous section, we call $s \in \mathcal{B}_\theta^n$ a *steady state* if $f^\theta(s) = s$. The notion of functionality of an edge remains the same as in Section 2. We consider only those edges that effectively influence the dynamical evolution of the system.

We may relate a singular state s to structures in the interaction graph $\mathcal{I}$ by considering the subgraphs of $\mathcal{I}$ induced by the vertices α_j with singular values, that is $j \in J(s)$. The following definition proves useful and was first introduced by E. H. Snoussi in [8], albeit in a different framework. The remainder of this section adapts ideas presented in [8].

Definition 8. *Let $C = (\alpha_{i_1},\ldots,\alpha_{i_r})$ be a circuit in $\mathcal{I}$. A state $s \in \mathcal{B}_\theta^n$ is called characteristic state of C if $s_{i_l} = \theta$ for all $l \in \{1,\ldots,r\}$.*

A characteristic state of a circuit is not unique unless all the regulatory components of the network are contained in the circuit. In this case the state $(\theta,\ldots,\theta)$ is the unique characteristic state. Obviously, the state $(\theta,\ldots,\theta)$ is characteristic of each circuit in $\mathcal{I}$.

Another simple observation is the following. Whenever $R_j^\theta(s) \neq \emptyset$ holds for all singular components $j \in J(s)$, the state s is characteristic of some circuit in $\mathcal{I}$. This is due to the fact that every resource of some regulatory component α_i is a predecessor of α_i and that there are only finitely many components in the network. With that in mind we can easily prove the next statement.

Theorem 1. *Every singular steady state is characteristic of some circuit in $\mathcal{I}$.*

Proof. Let s be a singular state that is not characteristic of any circuit in $\mathcal{I}$. Then there is $i \in \{1,\ldots,n\}$ such that $s_i = \theta$ and $R_i^\theta(s) = \emptyset$. It follows that $f_i^\theta(s) = |K_{i,R_i(s)}, K_{i,R_i(s)}| = K_{i,R_i(s)} \neq \theta = s_i$, since the parameters take only regular values. Thus s is not a steady state. □

If the network consists of a single circuit, then the corresponding characteristic state is always steady under our standard assumption that every edge in the graph is functional. As mentioned before, such a circuit displays a characteristic behavior depending on its sign. In general, the existence of a steady characteristic state of a circuit does not always result in the corresponding dynamical behavior, as will be illustrated in the next section.

It is possible to give a characterization of the singular steady states using only regular states and the function f.

Definition 9. *Let $s \in \mathcal{B}_\theta^n$ and $k \in \{1, \ldots, n\}$. Let $s^{k,+}$ and $s^{k,-}$ be regular states that satisfy $s_i^{k,+} := s_i^{k,-} := s_i$ for all $i \notin J(s)$ and*

$$s_i^{k,+} := \begin{cases} 1 & , & \varepsilon_{ki} = + \\ 0 & , & \varepsilon_{ki} = - \end{cases} \quad and \quad s_i^{k,-} := \begin{cases} 1 & , & \varepsilon_{ki} = - \\ 0 & , & \varepsilon_{ki} = + \end{cases} \tag{2}$$

for all $i \in J(s)$ satisfying $\alpha_i \in R_k^\theta(s)$. Then $s^{k,+}$ and $s^{k,-}$ are called a maximal resp. minimal adjacent state of s with respect to k.

There are generally many states $s^{k,+}$, $s^{k,-}$ that satisfy the above conditions. If the sets $R_k^\theta(s)$, $k \in \{1, \ldots, n\}$, are disjoint, then we can define states s^+ and s^- which are maximal resp. minimal adjacent states of s with respect to every $k \in \{1, \ldots, n\}$. If, in addition, the union of all sets $R_k^\theta(s)$ is equal to the set $\{\alpha_j \, ; \, j \in J\}$, then s^+ and s^- are unique and are called the maximal resp. minimal adjacent state of s.

Theorem 2. *A state $s \in \mathcal{B}_\theta^n$ is steady iff for all $k \in \{1, \ldots, n\}$ there is some choice of $s^{k,+}$, $s^{k,-}$ such that $f_k(s^{k,+}) = s_k^{k,+} = s_k^{k,-} = f_k(s^{k,-})$, if $k \notin J(s)$, and $f_k(s^{k,-}) < \theta < f_k(s^{k,+})$, if $k \in J(s)$.*

Proof. We show that $R_k(s^{k,+}) = R_k(s) \cup R_k^\theta(s)$ and $R_k(s^{k,-}) = R_k(s)$ for all $k \in \{1, \ldots, n\}$. First, let $\alpha_i \in R_k(s^{k,+})$. Then α_i is a predecessor of α_k. If $i \notin J := J(s)$, then $s_i = s_i^{k,+}$, and thus $\alpha_i \in R_k(s)$. If $i \in J$, we have $s_i = \theta$, and thus $\alpha_i \in R_k^\theta(s)$. Now, let $\alpha_i \in R_k(s) \cup R_k^\theta(s)$. Again $\alpha_i \in Pred(\alpha_k)$. If $\alpha_i \in R_k(s)$, then $i \notin J$. It follows that $s_i = s_i^{k,+}$, and thus $\alpha_i \in R_k(s^{k,+})$. If $\alpha_i \in R_k^\theta(s)$, then $\alpha_i \in R_k(s^{k,+})$ according to (2). Analogous reasoning provides the second statement.

Now, suppose that the last condition of the theorem is true. Then $f_k^\theta(s) = |K_{k,R_k(s)}, K_{k,R_k(s) \cup R_k^\theta(s)}| = |K_{k,R_k(s^{k,-})}, K_{k,R_k(s^{k,+})}| = |f_k(s^{k,-}), f_k(s^{k,+})|$ for all $k \in \{1, \ldots, n\}$. According to the assumption we have $|f_k(s^{k,-}), f_k(s^{k,+})| = s_k^{k,+} = s_k$ for $k \notin J$, and $|f_k(s^{k,-}), f_k(s^{k,+})| = |0, 1| = s_k$ for all $k \in J$. Thus s is a steady state. Similar reasoning can be used to show the inverse statement. $\square$

The theorem and the definition of $s^{k,+}$ and $s^{k,-}$ imply that whenever every regulatory component in the network can be influenced in its behavior by some other regulatory components, the state containing only singular entries is a steady state. In other words, if for every α_k we have $K_{\alpha_k, \emptyset} = 0$ and $K_{\alpha_k, Pred(\alpha_k)} = 1$, then the state $(\theta, \ldots, \theta)$ is a steady state.

4   Relating Singular Steady States and Attractors

We have seen that singular steady states can be characterized by regular states and that they are closely related to circuits in the interaction graph. In the following we show what kind of information on the state transition graph can be inferred from the existence of a singular steady state. First, we need some additional notations.

Let $s \in \mathcal{B}_\theta^n$ be a singular state. Recall that $J(s)$ is the set of indices corresponding to the singular values of s and that we identify each vertex α_i with its index i. With $\mathcal{I}^\theta(s)$ we denote the graph with vertex set $V^\theta(s) := J(s)$ and edge set $E^\theta(s)$ consisting of those $\{\alpha_i, \alpha_j\}$ with $i, j \in J(s)$ such that $\alpha_i \to \alpha_j$ or $\alpha_j \to \alpha_i$ is an edge in $\mathcal{I}$. The graph $\mathcal{I}^\theta(s)$ is undirected. It represents the existence of a dependency between singular components, without specifying the type of interaction. A (connected) *component* of $\mathcal{I}^\theta(s)$ is a maximal connected subgraph of $\mathcal{I}^\theta(s)$. By abuse of notation we denote the vertex set of a component Z of $\mathcal{I}^\theta(s)$ also with Z. Vertices of different components of $\mathcal{I}^\theta(s)$ represent regulatory components in $\mathcal{I}$ that do not influence each other directly. Figure 2 illustrates the concept on a small example. Let C be a circuit composed of vertices in $J(s)$. Then there is a component of $\mathcal{I}^\theta(s)$ which contains the vertices of C. We denote this component by $J_C(s)$.

The next lemma shows that for a singular steady state s value changes in a component of $\mathcal{I}^\theta(s)$ do not influence the image $f^\theta(s)$ outside that component. It will play an important role in all the following considerations.

Lemma 1. *Let s be a singular steady state, and let $Z_1, \ldots, Z_m$ be the components of $\mathcal{I}^\theta(s)$. Consider a union Z of arbitrary components Z_j. Let $\tilde{s} \in \mathcal{B}_\theta^n$ such that $\tilde{s}_i = s_i$ for all $i \notin Z$. Then $f_i^\theta(\tilde{s}) = f_i^\theta(s) = s_i = \tilde{s}_i$ for all $i \notin Z$.*

Proof. For $i \in J(s) \setminus Z$ we know that $R_i(s) = R_i(\tilde{s})$ and $R_i^\theta(s) = R_i^\theta(\tilde{s})$, since no element of Z is a predecessor of α_i. Thus $f_i^\theta(\tilde{s}) = f_i^\theta(s) = s_i$ for all $i \in J(s) \setminus Z$. For $i \notin J(s)$ we have $R_i(s) \subseteq R_i(\tilde{s})$, since a singular resource of α_i may have turned into a regular resource. In addition, $R_i(\tilde{s}) \cup R_i^\theta(\tilde{s}) \subseteq R_i(s) \cup R_i^\theta(s)$, since a singular resource of α_i might have been eliminated by turning its value to a regular value not contributing to activation. In summary we obtain $R_i(s) \subseteq R_i(\tilde{s}) \subseteq R_i(\tilde{s}) \cup R_i^\theta(\tilde{s}) \subseteq R_i(s) \cup R_i^\theta(s)$ and with condition (1) we derive

$$K_{i,R_i(s)} \leq K_{i,R_i(\tilde{s})} \leq K_{i,R_i(\tilde{s}) \cup R_i^\theta(\tilde{s})} \leq K_{R_i(s) \cup R_i^\theta(s)}.$$

Moreover, $K_{i,R_i(s)} = K_{i,R_i(s) \cup R_i^\theta(s)}$, since $f_i^\theta(s) = s_i$. Thus the above inequality becomes an equality and $f_i^\theta(\tilde{s}) = K_{i,R_i(s)} = s_i = \tilde{s}_i$ for all $i \notin J(s)$. □

The above lemma allows us to focus on the possible dynamical behavior in the isolated parts of the biological network corresponding to the components $Z_1, \ldots, Z_m$ and leads us to the following theorem.

Theorem 3. *For every singular steady state s there is an attractor A in $\mathcal{S}_N$ such that $u_i = s_i$ holds for all $u \in A$ and $i \notin J(s)$.*

Proof. Set $P := \{x \in \mathcal{B}^n \mid \forall i \notin J(s) : x_i = s_i\}$. Then $f_i(x) = x_i = s_i$ for all $i \notin J(s)$ according to Lemma 1, i.e., $f(x) \in P$. Thus all successors of x in $\mathcal{S}_N$ are also in P. It follows that P is a trap set. Like in the proof of Prop. 1 we deduce that P contains an attractor A, and $u_i = s_i$ for all $u \in A$ and $i \notin J(s)$. □

It is not difficult to see that we can derive such an attractor A from attractors $A_1, \ldots, A_k$ arising in the system's dynamical behavior restricted to the components $Z_1, \ldots, Z_k$ of $\mathcal{I}^\theta(s)$. To illustrate this we examine the example given in

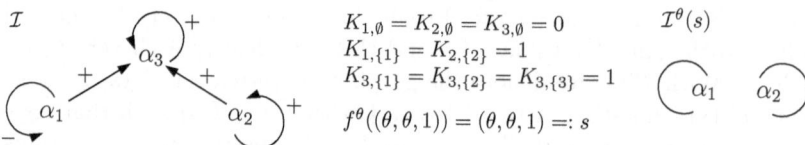

Fig. 2. An interaction graph $\mathcal{I}$ and a specification of the parameters. Missing parameter values follow from condition (1). The graph $\mathcal{I}^\theta(s)$ for $s := (\theta, \theta, 1)$ has two components.

Figure 2. The state $(\theta, \theta, 1)$ is steady, the components of $\mathcal{I}^\theta(s)$ are $Z_1 = \{\alpha_1\}$ and $Z_2 = \{\alpha_2\}$. We consider the dynamics restricted to Z_1 given by the projection $f^{(Z_1)} : \mathcal{B} \to \mathcal{B}, x \mapsto f_1^\theta(x, \theta, 1)$. It generates a state transition graph that consists of a cycle comprising the states 0 and 1. Thus it has a single attractor $A_1 = \{0, 1\}$. The state transition graph corresponding to the analogously defined function $f^{(Z_2)}$ consists of the two attractors $A_2^1 = \{0\}$ and $A_2^2 = \{1\}$. According to Lemma 1, the value of the third component of s will remain fixed, regardless of the values of the first two components. Thus we can derive two attractors in $\mathcal{S}_N$, namely $A^1 = A_1 \times A_2^1 \times \{s_3\} = \{(0,0,1), (1,0,1)\}$ and $A^2 = A_1 \times A_2^2 \times \{s_3\} = \{(0,1,1), (1,1,1)\}$.

We have seen above that we can link a singular steady state to a regular attractor. However, different singular steady states s^1 and s^2 may give rise to the same regular attractor. The above proof shows that this possibility is precluded if s^1 and s^2 differ in a component $i \notin J(s^1) \cup J(s^2)$.

A more precise analysis of the correspondence of attractors and singular steady states is possible if we take into account structural information on the underlying interaction graph $\mathcal{I}$. In the preceding section we have seen that every singular steady state s is characteristic of some circuit C of the interaction graph $\mathcal{I}$. If we know in addition that s is not characteristic of any other circuit in $\mathcal{I}$ with vertices in the connected component $J_C(s)$ of $\mathcal{I}^\theta(s)$, we can derive information on the singular valued predecessors of vertices belonging to C. This is shown in the next lemma.

Lemma 2. Let $C = (\alpha_{i_1}, \ldots, \alpha_{i_m})$ be a circuit in $\mathcal{I}$ and let s be a steady characteristic state of C. Assume that C is the only circuit in $\mathcal{I}$ with all its vertices contained in $J_C(s)$. Then $R_{i_j}^\theta(s) = \{\alpha_{i_{j-1}}\}$ for all $j \in \{1, \ldots, m\}$ with indices taken modulo m.

Proof. Set $J := J(s)$ and $J_C := J_C(s)$. Clearly, $\alpha_{i_{j-1}} \in R_{i_j}^\theta(s)$ for all $j \in \{1, \ldots, m\}$. Assume that there is $k \in \{1, \ldots, m\}$ such that there exists $l \in J$ satisfying $\alpha_l \neq \alpha_{i_{k-1}}$ and $\alpha_l \in R_{i_k}^\theta(s)$. Then $\alpha_l \in Pred(\alpha_{i_k})$ and thus $l \in J_C$. If $l = i_j$ for some $j \neq k - 1$, then $(\alpha_{i_j}, \alpha_{i_k}, \ldots, \alpha_{i_{j-1}})$ is a circuit other than C in J_C. This contradicts the hypothesis. Thus α_l is not a vertex of C.

Since s is a steady state, we know that $R_j^\theta(s) \neq \emptyset$ for all $j \in J$. Furthermore, $R_j^\theta(s) \subseteq J_C$ for all $j \in J_C$. Thus for every $j \in J_C$ we find $i \in J_C$, such that $\alpha_i \to \alpha_j$ is an edge in $\mathcal{I}$. Since there are only finitely many vertices in J_C, there is a circuit in $\{\alpha_j \in J_C ; \exists$ path from α_j to α_l in $\mathcal{I}\}$ that differs from C. Again, this leads to a contradiction. $\qquad\square$

Note that there may be vertices in $J_C(s)$ that have more than one singular resource. Lemma 2 allows us to represent $J_C(s)$ by a chain of nested sets.

Lemma 3. *Under the hypotheses of Lemma 2, there exist sets $M_1, \ldots, M_l \subseteq J_C(s)$ such that $M_1 = \{i_1, \ldots, i_m\}$, $M_l = J_C(s)$, $M_i \subsetneq M_{i+1}$ and $R_j^\theta(s) \subseteq M_i$ for all $j \in M_{i+1}$ and $i \in \{1, \ldots, l-1\}$.*

Proof. Set $M_1 := \{i_1, \ldots, i_m\}$. If $J_C(s) \setminus M_1 \neq \emptyset$, then there exists at least one element $j \in J_C(s) \setminus M_1$ such that $R_j^\theta(s) \subseteq M_1$. Otherwise for every $j \in J_C(s) \setminus M_1$ there is $k_j \in J_C(s) \setminus M_1$ such that α_{k_j} is a predecessor of α_j in $\mathcal{I}$. That would imply the existence of a circuit other than C in $J_C(s)$, since $J_C(s) \setminus M_1$ is finite. Thus by defining $M_2 := \{j \in J_C(s); R_j^\theta(s) \subseteq M_1\}$ we obtain a set strictly containing M_1. Since $J_C(s)$ is finite, we can repeat the procedure until we get $M_l := \{j \in J_C(s); R_j^\theta(s) \subseteq M_{l-1}\} = J_C(s)$. $\square$

In the following we make use of the information on the sign of the circuit C.

Theorem 4. *Let C be a positive circuit in $\mathcal{I}$ and let s be a steady characteristic state of C. Assume that C is the only circuit in $\mathcal{I}$ with all its vertices contained in $J_C(s)$. Then f^θ has at least three fixed points.*

Proof. Set $J := J(s)$ and $J_C := J_C(s)$. Without loss of generality we may assume that $C = (\alpha_1, \ldots, \alpha_r)$ for some $r \in \{1, \ldots, n\}$. We determine states $s^0, s^1 \in \mathcal{B}_\theta^n$ by an iterative process such that s, s^0 and s^1 are fixed points of f^θ. Initially, we set $s_i^0 := s_i^1 := s_i$ for all $i \notin J_C$ and choose the other components of s^0 and s^1 arbitrary.

From Lemma 1 it follows that $f_i^\theta(s^0) = s_i^0$ and $f_i^\theta(s^1) = s_i^1$ for all $i \notin J_C$. Next, we define the values s_i^0 and s_i^1 for $i \in \{1, \ldots, r\}$. We set $s_1^0 := 0$, $s_1^1 := 1$, and for $l \in \{0, 1\}$

$$s_{i+1}^l := \begin{cases} 0 & , \quad (s_i^l = 0 \wedge \varepsilon_{i+1,i} = +) \vee (s_i^l = 1 \wedge \varepsilon_{i+1,i} = -) \\ 1 & , \quad (s_i^l = 1 \wedge \varepsilon_{i+1,i} = +) \vee (s_i^l = 0 \wedge \varepsilon_{i+1,i} = -) \end{cases}$$

for all $i \in \{1, \ldots, r-1\}$. This definition amounts to setting $s_{i+1}^l = 1$ iff the value of s_i^l characterizes α_i as regular resource of α_{i+1}. As is easy to see we also have

$$s_{i+1}^l = \begin{cases} s_i^l & , \quad \varepsilon_{i+1,i} = + \\ 1 - s_i^l & , \quad \varepsilon_{i+1,i} = - \end{cases}.$$

It follows for all $i \in \{1, \ldots, r-1\}$ that $s_{i+1}^l = s_1^l$ if $\prod_{j=1}^i \varepsilon_{i+1,j}$ is positive, and $s_{i+1}^l = s_1^l$ if $\prod_{j=1}^i \varepsilon_{i+1,j}$ is negative. Since C is a positive circuit, the value of s_1^l is consistent with the value we obtain by using the above definition for $i = r$, that is we do not contradict the definition of s^l if we use the above iterative formula modulo r. Note that s_1^0, s_1^1 and s_1 are distinct.

According to Lemma 2 we have $R_i^\theta(s) = \{\alpha_{i-1}\}$ for all $i \in \{1, \ldots, r\}$, indices again taken modulo r. Thus $R_i^\theta(s^l) = \emptyset$ for all $i \in \{1, \ldots, r\}$. Moreover, we have

$$R_i(s^l) = \begin{cases} R_i(s) & , \quad (s_{i-1}^l = 0 \wedge \varepsilon_{i,i-1} = +) \vee (s_{i-1}^l = 1 \wedge \varepsilon_{i,i-1} = -) \\ R_i(s) \cup R_i^\theta(s) & , \quad (s_{i-1}^l = 1 \wedge \varepsilon_{i,i-1} = +) \vee (s_{i-1}^l = 0 \wedge \varepsilon_{i,i-1} = -) \end{cases}$$

for all $i \in \{1, \ldots, r\}$. Since $f_i^\theta(s) = |K_{i,R_i(s)}, K_{i,R_i(s) \cup R_i^\theta(s)}| = |0, 1|$, it follows from the definition of s_i^l and condition (1) that

$$f_i^\theta(s^l) = K_{i,R_i(s^l)} = \begin{cases} K_{i,R_i(s)} = 0 & , \quad s_i^l = 0 \\ K_{i,R_i(s) \cup R_i^\theta(s)} = 1 & , \quad s_i^l = 1 \end{cases}.$$

Thus, we have $f_i^\theta(s^0) = s_i^0$ and $f_i^\theta(s^1) = s_i^1$ for all $i \in \{1, \ldots, r\}$, not depending on the values of the components in $J_C \setminus \{1, \ldots, r\}$.

Finally, we have to specify s_i^l for all $i \in J_C \setminus \{1, \ldots, r\}$ and $l \in \{0, 1\}$. According to Lemma 3 we find sets $M_1, \ldots, M_k \subseteq J_C$ satisfying $M_1 = \{1, \ldots, r\}$, $M_k = J_C$, $M_j \subsetneq M_{j+1}$ and $R_i^\theta(s) \subseteq M_j$ for all $i \in M_{j+1}$ and $j \in \{1, \ldots, k-1\}$. Thus we can deduce that α_i, $i \in M_2$, has no predecessors in $J_C \setminus M_1$, since otherwise they would be in $R_i^\theta(s)$. Furthermore, for every $i \in M_2$ we have $R_i^\theta(s^l) = \emptyset$ since all components corresponding to vertices in C have regular values. Now we set $s_i^l := K_{i,R_i(s^l)}$ for all $i \in M_2$. Note that this parameter depends only on components previously specified, i.e., on the values s_i^l for $i \notin J_C \setminus \{1, \ldots, r\}$. Since α_i does not have singular resources in state s^l for all $i \in M_2$, we have $f_i^\theta(s^l) = K_{i,R_i(s^l)} = s_i^l$ for all $i \in M_2$. Because the sets M_j are nested, we can repeat the above procedure for consecutive sets without encountering contradictions. Thus we are able to specify all components s_i^l for $i \in J_C \setminus \{1, \ldots, r\}$, such that $f_i^\theta(s^l) = s_i^l$.

We have shown that the resulting states s^0 and s^1 are fixed points of f^θ. Since s, s^0, and s^1 are distinct, f^θ has at least three fixed points. $\qquad \square$

The proof shows that at least two fixed points of f^θ differ in a regular component. Applying Theorem 3 and the subsequent observations we immediately obtain the following statement.

Corollary 1. *Under the hypotheses of Theorem 4 there are at least two distinct attractors in the corresponding state transition graph.*

The corollary is illustrated in Figure 3 (a) and (c). The singular steady state $(1, \theta, 0)$ is characteristic of the positive circuit comprising α_2 and of no other circuit. The resulting state transition graph shows two distinct attractors. The importance of the condition concerning the circuit C and the component $J_C(s)$ is demonstrated in Figure 3 (b). The state (θ, θ, θ) is steady and characteristic of the positive circuit comprising α_2. Moreover, the state $(\theta, 0, \theta)$ is steady and characteristic of the positive circuit comprising α_1 and α_3. In both cases the states are characteristic of further circuits in the same component, and the state transition graph has only one attractor. Figure 4 shows the importance of C being the only circuit with vertices in $J_C(s)$ for the validity of Theorem 4. The interaction graph given in (a) contains a positive circuit with characteristic state $s := (\theta, \theta, \theta, \theta)$. Together with the parameters given in (b) it gives rise to a system that has no regular fixed point. Moreover, from the logical implications in (d) we can easily deduce that s is the only singular steady state.

The network in Figure 4 (b), together with the parameters given in (c), illustrates that the sufficient condition of Theorem 4 is not necessary. The given

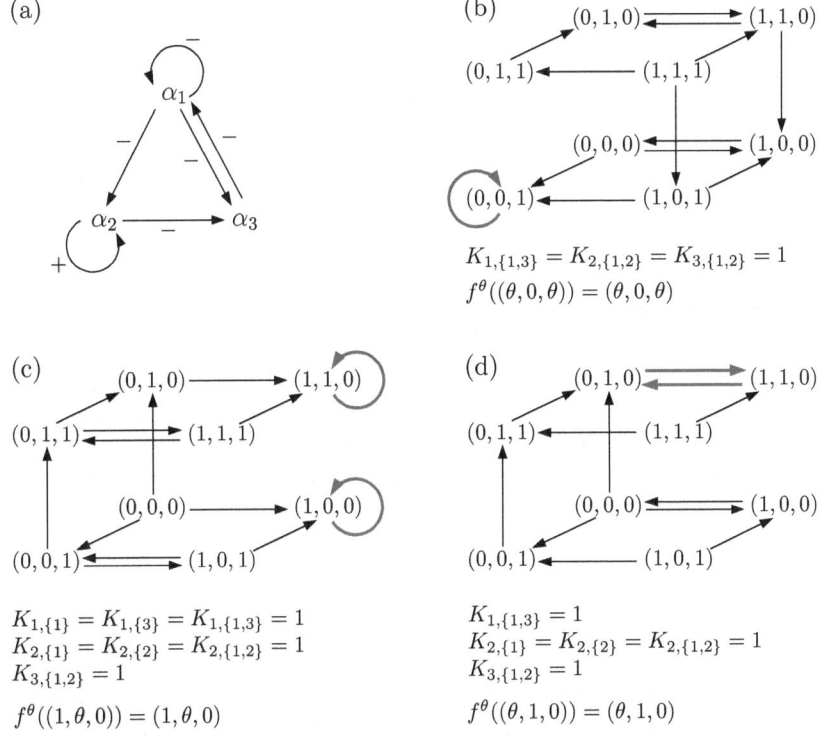

(a)

(b)
$$K_{1,\{1,3\}} = K_{2,\{1,2\}} = K_{3,\{1,2\}} = 1$$
$$f^{\theta}((\theta,0,\theta)) = (\theta,0,\theta)$$

(c)
$$K_{1,\{1\}} = K_{1,\{3\}} = K_{1,\{1,3\}} = 1$$
$$K_{2,\{1\}} = K_{2,\{2\}} = K_{2,\{1,2\}} = 1$$
$$K_{3,\{1,2\}} = 1$$
$$f^{\theta}((1,\theta,0)) = (1,\theta,0)$$

(d)
$$K_{1,\{1,3\}} = 1$$
$$K_{2,\{1\}} = K_{2,\{2\}} = K_{2,\{1,2\}} = 1$$
$$K_{3,\{1,2\}} = 1$$
$$f^{\theta}((\theta,1,0)) = (\theta,1,0)$$

Fig. 3. An interaction graph comprising three components is given in (a). Figures (b)-(d) show the state transition graphs corresponding to the chosen parameter values. We only listed the non-zero parameters. Attractors are indicated by colored, fat lines. For each choice of parameters one singular steady state other than (θ,θ,θ) is given.

system has two regular fixed points, (0,0,0,0) and (1,1,1,1). However, the only steady characteristic state is $s := (\theta,\theta,\theta,\theta)$, as easy to see from the implications in (d). Its components comprise the vertices of all three cycles of the network.

The next theorem clarifies the impact of a negative circuit.

Theorem 5. *Let C be a negative circuit in $\mathcal{I}$ and let s be a steady characteristic state of C. Assume that C is the only circuit in $\mathcal{I}$ with all its vertices contained in $J_C(s)$. Then there exists an attractor with cardinality greater than one.*

Proof. Again set $J := J(s)$ and $J_C := J_C(s)$ and assume that $C = (\alpha_1, \ldots, \alpha_r)$ for some $r \in \{1, \ldots, n\}$. By P_j, $j \in \{1, \ldots, r\}$, we denote the set of all regular states x satisfying $x_k = s_k$ for all $k \notin J$ and

$$x_{i+1} = \begin{cases} x_i &, \quad \varepsilon_{i+1,i} = + \\ 1 - x_i &, \quad \varepsilon_{i+1,i} = - \end{cases} \quad \text{for all } i \in \{1, \ldots, r\} \setminus \{j\},$$

with indices i taken modulo r. Choose $j \in \{1, \ldots, r\}$ and $x \in P_j$. Lemma 1 implies that $f_i(x) = s_i$ for all $i \notin J$. Now set $\tilde{x} = f(x)$ and let $i \in \{1, \ldots, r\}$. Again,

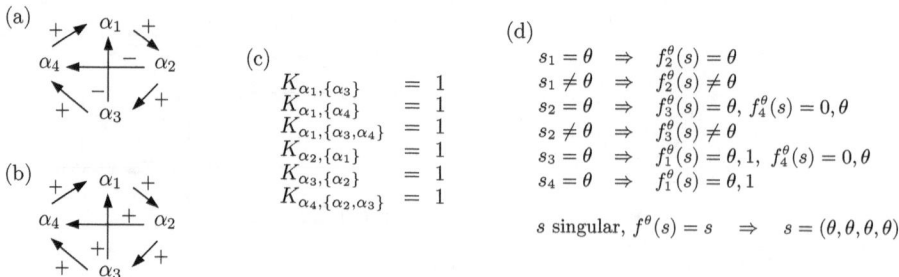

Fig. 4. Interaction graphs and parameter values of networks with only one singular steady state. Given are the non-zero logical parameters. For details see the text.

consider indices modulo r. According to Lemma 2 the only singular resource of α_{i+1} in s is α_i. Furthermore, we know $f^\theta_{i+1}(s) = s_{i+1} = \theta = |0,1|$. Thus, with reasoning similar to that in the proof of Theorem 4, we can deduce that

$$\tilde{x}_{i+1} = K_{i+1,R_{i+1}(x)} = \begin{cases} 0 & , \quad (x_i = 0 \wedge \varepsilon_{i+1,i} = +) \vee (x_i = 1 \wedge \varepsilon_{i+1,i} = -) \\ 1 & , \quad (x_i = 1 \wedge \varepsilon_{i+1,i} = +) \vee (x_i = 0 \wedge \varepsilon_{i+1,i} = -) \end{cases},$$

that is

$$\tilde{x}_{i+1} = \begin{cases} x_i & , \quad \varepsilon_{i+1,i} = + \\ 1 - x_i & , \quad \varepsilon_{i+1,i} = - \end{cases}.$$

Now, if $i \neq j+1$, we can express x_i in terms of x_{i-1}, since x is in P_j. Furthermore, we can then express x_{i-1} in terms of $\tilde{x}_i$ according to the observation above, which is valid for all $i \in \{1, \ldots, r\}$. Some easy substitutions yield firstly

$$\tilde{x}_{i+1} = \begin{cases} x_{i-1} & , \quad (\varepsilon_{i+1,i} = + \wedge \varepsilon_{i,i-1} = +) \vee (\varepsilon_{i+1,i} = - \wedge \varepsilon_{i,i-1} = -) \\ 1 - x_{i-1} & , \quad (\varepsilon_{i+1,i} = + \wedge \varepsilon_{i,i-1} = -) \vee (\varepsilon_{i+1,i} = - \wedge \varepsilon_{i,i-1} = +) \end{cases},$$

and secondly that $\tilde{x}_{i+1} = \tilde{x}_i$, if $\varepsilon_{i+1,i} = +$, and $\tilde{x}_{i+1} = 1 - \tilde{x}_i$, if $\varepsilon_{i+1,i} = -$. It follows that $\tilde{x} = f(x)$ is an element of P_{j+1}. Furthermore, in case $\varepsilon_{i,i-1} = +$ and $i \neq j+1$, we have $\tilde{x}_i = x_{i-1}$ as seen above and $x_{i-1} = x_i$, since $x \in P_j$. This shows $f_i(x) = \tilde{x}_i = x_i$. The same reasoning leads to $f_i(x) = \tilde{x}_i = x_i$ for $\varepsilon_{i,i-1} = -$ and $i \neq j+1$. It follows that every successor x' of x in the state transition graph is either in P_j, in case $x'_{j+1} = x_{j+1}$, or in P_{j+1}, in case $x'_{j+1} \neq x_{j+1}$. Since our reasoning is true for indices modulo r, we can deduce that the union P of the sets P_j, $j \in \{1, \ldots, r\}$, is a trap set and thus contains an attractor A (see the proof of Proposition 1).

Finally, we show that each state in P, and thus in A, has a successor other than itself. For $x \in P_j$ we have

$$x_j = \begin{cases} x_{j+1} & , \quad \varepsilon_{j+2,j+1} \cdots \cdot \varepsilon_{j,j-1} = + \\ 1 - x_{j+1} & , \quad \varepsilon_{j+2,j+1} \cdots \cdot \varepsilon_{j,j-1} = - \end{cases}.$$

Furthermore, we know that $\tilde{x}_j = x_j$ with $\tilde{x} := f(x)$ and $\tilde{x} \in P_{j+1}$. It follows that $\tilde{x}_{j+1} = x_j$, if $\varepsilon_{j+1,j} = +$, and $\tilde{x}_{j+1} = 1 - x_j$, if $\varepsilon_{j+1,j} = -$. Thus we obtain

$$\tilde{x}_{j+1} = \begin{cases} x_{j+1} & , \quad \prod^r_{k=1} \varepsilon_{k+1,k} = + \\ 1 - x_{j+1} & , \quad \prod^r_{k=1} \varepsilon_{k+1,k} = - \end{cases},$$

with indices k taken modulo r. Since C is negative, we know $\prod_{k=1}^{r} \varepsilon_{k+1,k} = -$, and thus $f_{j+1}(x) \neq x_{j+1}$. Thus x has a successor other than itself in the state transition graph. It follows that the cardinality of A is greater than one.   □

Figure 3 illustrates the theorem. In (d) we give a parameter specification that allows the state $(\theta, 1, 0)$ to be steady. This state is characteristic of the negative circuit comprising α_1. The resulting state transition graph contains the attractor $\{(0, 1, 0), (1, 1, 0)\}$. As for Theorem 4, Figure 3 (b) illustrates the importance of C being the only circuit in $J_C(s)$. Although $(\theta, 0, \theta)$ is characteristic of the negative circuit comprising α_1, and (θ, θ, θ) is characteristic of the negative circuit comprising α_1, α_2 and α_3, the only attractor in the state transition graph consists of a single state. Figure 4 (a) and (c) specify a system that illustrates that the sufficient condition in Theorem 5 is not necessary. By calculating the corresponding state table we can see that there is no regular steady state of the system. Thus there has to be an attractor with cardinality greater than one. However, from the logical implications given in (d), it follows easily that the only singular steady state is $(\theta, \theta, \theta, \theta)$, which is characteristic for all circuits in the interaction graph given in (a).

The proofs of Theorems 4 and 5 show that the situation is easy to grasp in case that the only components with singular values are those of the circuit C. In the context of Theorem 4, we then obtain two regular fixed points, that is two steady states in the state transition graph. Those can be explicitly constructed as shown in the proof of Theorem 4. If C is a negative circuit, we find a trap cycle in the state transition graph. It is composed of the states in the set P introduced in the proof of Theorem 5.

If we detect the above mentioned structures in the state transition graph, we can conversely derive singular steady states. The proofs of the next two propositions are omitted for lack of space. They can be found in [7].

Proposition 2. *Let $x, y \in \mathcal{B}^n$ be steady states in the state transition graph $\mathcal{S}_N$. Let I be the set of components i satisfying $x_i \neq y_i$. Then there exists a singular steady state s such that $s_i = \theta$ for all $i \in I$.*

Proposition 3. *Let $C := (x^1, \ldots, x^r, x^1)$ be a trap cycle in the state transition graph $\mathcal{S}_N$. Let I be the set of components i such that there exists j_1, j_2 satisfying $x_i^{j_1} \neq x_i^{j_2}$. Then there is a singular steady state such that $s_i = \theta$ for all $i \in I$.*

The proofs in [7] show how to derive singular steady states satisfying the statements of Prop. 2 and 3. However, those singular steady states may coincide with $(\theta, \ldots, \theta)$, even when $I \neq \{1, \ldots, n\}$.

5 Perspectives

We have seen in this paper that it is possible to relate systematically singular steady states to attractors in the state transition graph. To do so, we often exploit knowledge about the structure of the associated interaction graph. The results obtained illustrate the possibilities of studying the dynamical behavior of the system without the explicit use of the state transition graph. However, we have focussed on a coarse description, characterizing state transition graphs by the number of their attractors, and distinguishing attractors by their cardinality. In order to tap the full potential of this approach to analyzing the system's dynamics, it should be refined further. A promising starting point for future work is the concept of local interaction graphs introduced in [3]. The authors associate every state of the system with an interaction graph, the union of which is the global interaction graph. This approach allows for a better understanding of what structures in the interaction graph influence the system's behavior in a given state. Combining this local view with our understanding of singular steady states may yield a more detailed description of the resulting dynamical behavior.

Acknowledgment

We would like to thank an anonymous referee for suggestions to simplify several proofs in an earlier version of the paper.

References

1. Bernot, G., Comet, J.-P., Richard, A., Guespin, J.: Application of formal methods to biological regulatory networks: extending Thomas' asynchronous logical approach with temporal logic. J. Theor. Biol. 229, 339–347 (2004)
2. Remy, É., Mossé, B., Chaouiya, C., Thieffry, D.: A description of dynamical graphs associated to elementary regulatory circuits. Bioinform. 19, 172–178 (2003)
3. Remy, É., Ruet, P., Thieffry, D.: Graphic requirements for multistability and attractive cycles in a boolean dynamical framework. (prépublication 2005)
4. Remy, É., Ruet, P., Thieffry, D.: Positive or negative regulatory circuit inference from multilevel dynamics. In: Positive Systems: Theory and Applications. LNCIS, vol. 341, pp. 263–270. Springer, Heidelberg (2006)
5. Richard, A., Comet, J.-P.: Necessary conditions for multistationarity in discrete dynamical systems. Rapport de Recherche (2005)
6. Richard, A., Comet, J.-P., Bernot, G., Thomas, R.: Modeling of biological regulatory networks: introduction of singular states in the qualitative dynamics. Fundamenta Informaticae 65, 373–392 (2005)
7. Siebert, H., Bockmayr, A.: Relating attractors and singular steady states in the logical analysis of bioregulatory networks. preprint 373, DFG Research Center MATHEON (2007)
8. Snoussi, E.H., Thomas, R.: Logical identification of all steady states: the concept of feedback loop characteristic states. Bull. Math. Biol. 55, 973–991 (1993)
9. Soulé, C.: Graphical requirements for multistationarity. ComPlexUs 1, 123–133 (2003)
10. Thomas, R.: Boolean formalization of genetic control circuits. J. Theor. Biol. 42, 563–585 (1973)
11. Thomas, R., d'Ari, R.: Biological Feedback. CRC Press (1990)
12. Thomas, R., Kaufman, M.: Multistationarity, the basis of cell differentiation and memory. II. Logical analysis of regulatory networks in terms of feedback circuits. Chaos 11, 180–195 (2001)

Translating Time-Course Gene Expression Profiles into Semi-algebraic Hybrid Automata Via Dimensionality Reduction*

Alberto Casagrande[1,2], Kevin Casey[4], Rachele Falchi[3], Carla Piazza[1],
Benedetto Ruperti[3], Giannina Vizzotto[3], and Bud Mishra[4,5]

[1] Dept. of Math. and Computer Science, University of Udine, Udine, Italy
[2] Institute of Applied Genomics, Udine, Italy
[3] Dept. of Crop Science and Agricultural Engineering
[4] Courant Institute of Mathematical Science, NYU, New York, U.S.A.
[5] NYU School of Medicine, 550 First Avenue, New York, 10016 U.S.A.
{casa,piazza}@dimi.uniud.it, {rachele.falchi,ruperti,vizzotto}@uniud.it,
{mishra,kjc261}@nyu.edu

Abstract. Biotechnological innovations which sample gene expressions allow to measure the gene expression levels of a biological system with varying degree of accuracy, cost and speed. By repeating the measurement steps at different sampling rates, one can both infer relations among the genes and define a dynamic model of the underlying biological system. When a very large number of genes and measurements are involved, they raise several difficult algorithmic questions, as accurate model-building, checking and inference tasks. Semi-algebraic hybrid automata were proposed as a modeling formalism for biological systems (see, e.g., [17,6]), and demonstrated their abilities to handle complex biochemical pathways. This paper proposes an automatic procedure to build semi-algebraic hybrid automata from gene-expression profiles. In order to reduce the size of the resulting automata and to minimize their analysis computational complexity, our approach exploits various dimensionality reduction techniques. The paper concludes with several experimental results about peach fruit.

1 Introduction

It is often said that progress in science is characterized by successive steps of measurement, arithmetization, algorithmization, and algebraization—each step representing in a succinct manner the intuitions collected in the earlier step. In biology, various breakthrough in biotechnology, e.g., sequencing, DNA synthesis, DNA amplification with PCR, high-throughput measurement of DNA/RNA

* This work is developed within the framework of the HYCON Network of Excellence, contract number FP6-IST-511368 and partially supported by the projects PRIN 2005 2005015491 and PRIN 2004 2004079422_004 (Role of sugar signalling in peach fruit quality development) and by the regional project BioCheck. B.M. has been supported by funding from two NSF ITR grants and one NSF EMT grant.

H. Anai, K. Horimoto, and T. Kutsia (Eds.): AB 2007, LNCS 4545, pp. 51–65, 2007.
© Springer-Verlag Berlin Heidelberg 2007

abundance through real time PCR [5,10,14], SAGE or microarrays [18,12], etc., have made it possible to obtain a numerical picture of the transcriptomic state of a cell at a certain instant and under certain conditions. Equipped with such a collection of numerical pictures of these states, one may organize them into a state-diagram for further statistical and algorithmic study of the dynamics implied by the state-transitions (see, e.g., [3,9,16,4]). Computational systems biology has come to represent the many varied efforts within this framework, and yet, it shies away from the final step of the algebraization of biology. It may even not be clear what such a final step would entail.

Here, we propose a framework for the algebraization of biology, by examining the question of translating time-course data of numerical biological measurements into the well-studied structures of semi-algebraic hybrid automata [17,7,6]. We concede that this is a first step in this direction, and would require much additional collaboration with biologists, algebraists and computer scientists to establish its final theoretical foundation. In particular, we believe that this new field will need to borrow many ideas originally developed in the context of rate-distortion theory in communication engineering, where the notion of lossy-compression was rigorously studied by Shannon and Kolmogorov [19].

This paper highlights many such connections and provides several heuristic algorithms that can be used for practical data analysis. It concludes with a discussion of the possible future paths of the emerging area of "Algebraic Biology."

1.1   Semi-algebraic Hybrid Automata

The notion of *Hybrid Automata* was first introduced [1] as a model and specification language for systems with both continuous and discrete dynamics, i.e., for systems consisting of a discrete program within a continuously changing environment. The simplest class of such models studied in computer science was the class of timed-automata to model asynchronous systems with many local clocks evolving at different but constant rates, while the system made discrete state transitions according to the local time. Subsequently, the field has seen many interesting and nontrivial generalizations (see, e.g., [2,15,7]). Here, we focus on one that is motivated by our interest in modeling biochemical processes.

First we introduce some notations and conventions. Capital letters Z_m, Z'_m, where $m \in \mathbb{N}$, denote variables ranging over $\mathbb{R}$. Analogously, Z denotes the vector of variables $\langle Z_1, \ldots, Z_k \rangle$ and Z' denotes the vector $\langle Z'_1, \ldots, Z'_k \rangle$; and Z^n denotes the vector $\langle Z^n_1, \ldots, Z^n_k \rangle$. The temporal variables T and T' model time and range over $\mathbb{R}^+$. We use the small letters $p, q, r, s, \ldots$ to denote k-dimensional vectors of real numbers. Occasionally, we will use the notation $\varphi[X_1, \ldots, X_m]$ to stress the fact that the set of free variables of the first-order formula φ, denoted by *Free*(φ), is included in the set of variables $\{X_1, \ldots, X_m\}$. By extension, if $\{X^1, \ldots, X^n\}$ is a set of variable vectors, $\varphi[X^1, \ldots, X^n]$ indicates that the free variables of φ are included in the set of components of $X^1, \ldots, X^n$. Moreover, given a formula $\varphi[X^1, \ldots, X^i, \ldots, X^n]$ and a vector p of the same dimension as the variable vector X^i, the formula obtained by component-wise substitution of X^i with p is denoted by $\varphi[X^1, \ldots, X^{i-1}, p, X^{i+1}, \ldots, X^n]$. If in φ the free variables are just the components of X^i, we can compute the truth value of $\varphi[p]$.

We are now ready to formally introduce semi-algebraic hybrid automata already presented in [17] and further studied in [7,6]. For each node of a graph, we have an invariant condition and a dynamic law. This dynamic law may depend on the initial conditions, i.e., on the values of the continuous variables at the beginning of the evolution in the state. The jumps from one discrete state to another are regulated by the activation and reset conditions. All these conditions are defined through first-order formulæ over the reals, i.e., over the first-order language of $(\mathbb{R}, 0, 1, +, \times, =, >)$.

Definition 1 ((Semi-Algebraic) Hybrid Automata - Syntax). *A hybrid automaton* $H = (Z, Z', \mathcal{V}, \mathcal{E}, Inv, \mathcal{F}, Act, Reset)$ *of dimension k consists of the following components:*

1. $Z = \langle Z_1, \ldots, Z_k \rangle$ *and* $Z' = \langle Z'_1, \ldots, Z'_k \rangle$ *are two vectors of variables ranging over the reals* $\mathbb{R}$;
2. $\langle \mathcal{V}, \mathcal{E} \rangle$ *is a directed graph; the objects,* $v \in \mathcal{V}$, *are called* locations;
3. *Each vertex* $v \in \mathcal{V}$ *is labeled by the formulæ* $Inv(v)[Z]$ *and* $Dyn(v)[Z, Z', T]$;
4. *Each edge* $e \in \mathcal{E}$ *is labeled by the formulæ* $Act(e)[Z]$ *and* $Reset(e)[Z, Z']$.

We say that H is semi-algebraic *if the constraints Inv, Dyn, Act, and $Reset$ are first-order formulæ over the reals (i.e., over $(\mathbb{R}, 0, 1, +, \times, =, >)$).*

The semantics of hybrid automata is given in terms of continuous and discrete transitions.

Definition 2 (Hybrid Automata - Semantics). *A state ℓ of H is a pair* $\langle v, r \rangle$, *where* $v \in \mathcal{V}$ *is a location and* $r = \langle r_1, \ldots, r_k \rangle \in \mathbb{R}^k$ *is an assignment of values for the variables of Z. A state $\langle v, r \rangle$ is said to be* admissible *if $Inv(v)[r]$ is true.*

The continuous reachability transition relations $\xrightarrow{t}_C$, *where $t > 0$ is the transition elapsed time, between admissible states is defined as follows:*

$$\langle v, r \rangle \xrightarrow{t}_C \langle v, s \rangle \iff \begin{array}{l} \textit{The equation } s = f_v(r, t) \textit{ holds, and for each} \\ t' \in [0, t] \textit{ the formula } Inv(v)[f_v(r, t')] \textit{ is true.} \end{array}$$

The discrete reachability transition relation $\to_D$ *between admissible states is defined as follows:*

$$\langle v, r \rangle \to_D \langle u, s \rangle \iff \begin{array}{l} \textit{The relation } \langle v, u \rangle \in \mathcal{E} \textit{ holds, and the formulæ} \\ Act(\langle v, u \rangle)[r] \textit{ and } Reset(\langle v, u \rangle)[r, s] \textit{ are true.} \end{array}$$

Building upon continuous and discrete transitions, we can introduce the notions of *trace* and *reachability*. A trace is a sequence of continuous and discrete transitions. A point s is reachable from a point r, if there is a trace from r and to s. We use $\ell \to \ell'$ to denote that either $\ell \xrightarrow{t}_C \ell'$, for some t, or $\ell \to_D \ell'$.

Definition 3 (Hybrid Automata - Reachability). *Let I be either $\mathbb{N}$ or an initial interval of $\mathbb{N}$. A trace of H is a sequence $\ell_0, \ell_1, \ldots, \ell_i$, with $i \in I$, of admissible states such that $\ell_{i-1} \to \ell_i$ holds for each $i \in I$ with $i > 0$ and* *continuous and discrete transitions are alternating. Such a trace is also denoted by $(\ell_i)_{i \in I}$. A point $r \in \mathbb{R}^k$ reaches a point $s \in \mathbb{R}^k$ (in time t), if there exists a trace $\ell_0, \ldots, \ell_n$ of H such that $\ell_0 = \langle v, r \rangle$ and $\ell_n = \langle u, s \rangle$, for some $v, u \in \mathcal{V}$ (and t is the sum of the elapsed continuous transition times).*

In [17] we defined first-order formulæ over the reals which allow one to study the reachability problem over semi-algebraic hybrid automata. The problem is undecidable in the general case, since it is necessary to consider an infinite number of formulæ. However, in [7,6] we introduced two classes of semi-algebraic automata over which we demonstrated the decidability of the reachability problem showing that it is sufficient to consider a finite number of formulæ. Moreover, we showed that this decidability result for the reachability problem is also the basis for the decidability of model checking with other more complex temporal logic formulæ, which can be used to analyze biochemical pathways.

However, this earlier work was based on the assumption that the hybrid automaton model was available and accurately captured the dynamics of the underlying biochemical system. Its means of construction, however, were left unspecified. Construction of such models from experimental time-course data is the subject of this paper. Specifically, this paper deals with a suitable approach for identifying a semi-algebraic hybrid automaton representation of a biochemical dynamic system, where the biochemical system is initially represented as a matrix of gene expression data sampled at many discrete time instants.

2   From Time-Courses to Semi-algebraic Automata

We would like to capture the activity of a biological system using the formalism of hybrid automata. Specifically, we aim to represent the concerted activity of an organism's gene expression and regulation using the discrete and continuous dynamics of semi-algebraic hybrid automata as defined above.

One of the main problems that arises when hybrid automata are used to represent biological systems is that each component (e.g. gene) is modeled with a continuous variable. As a consequence the resulting automaton has a high computational complexity. In particular, when first-order formulæ are used to study reachability, the number of variables occurring in the formulæ are multiples of the number of continuous variables of the automaton (see, e.g., [7]). Thus, we would like to reduce the complexity of the system under study by grouping genes that have similar dynamics together and then considering this compressed representation when building our automata. Of course, there will be some loss of information when one clusters the data in such a way, as such, it is important to handle any data compression in a responsible manner.

When we define an automaton to represent a biological system, another difficulty consists in the identification of the locations. If the system is represented as a system of differential equations, then we can immediately define a trivial automaton having just one location whose dynamical law is an algebraic approximation of the solutions to the differential equations. However, this dynamical law could be very complex, and it may be convenient to split the single location into multiple locations in order to get simpler dynamical laws. In our case, the system is represented as a set of time series data that captures the temporal evolution of the genes' expression. One would like to find points in time at which elements in the data substantially change their behavior and then consider locations that correspond to the intervals of time between these critical time points. That is, we would like a temporal partition of the time series such that the data is broken into a number of disjoint temporal windows, each of which represents some coordinated biological activity, and the boundaries of which correspond to significant reorganization of gene expression. One could then identify locations with individual temporal windows, thus building an automata whose discrete transitions correspond to significant organizational events in the system's gene regulation, and whose continuous dynamics correspond to periods of concerted co-expression. The construction of such an automata requires a long preprocessing or clustering phase which results in an automaton with a number of locations proportional to the number of distinct temporal windows.

The two problems stated above are both related to the creation of a compact or compressed representation of the biological system. On the one hand, grouping like genes together and considering the collective continuous dynamics of clusters of co-expressed genes allows one to reduce the complexity of the resulting automata by simplifying the dynamical laws. On the other hand, generating temporal windows allows one to reduce the number of locations from the order of the number of time points down to the order of the number of time windows. These considerations are directly related to compressing the original time series data both in the number of genes and in the number of experiments, and a variety of bi-clustering techniques have been explored for this purpose [16]. We will discuss two methods: one directly exploiting correlation among gene expressions and consecutive time points (through Principal Component Analysis, or PCA), and the other method emphasizing "lossy compression" of hybrid automata by building on rate distortion theory and graph search; these approaches show how one can ultimately go from clustered data of reduced dimension to a "reasonably faithful" hybrid automata model. First however, we step through a number of intuitively simple, but successively more complex, examples of representations of our time series data using hybrid automata. Next, we introduce several key ideas from information theory as well as our clustering algorithm. Finally, we will present a hybrid automata constructed from our time series that represents a significantly compressed version of the original data.

Let M be an $m \times n$ matrix of biological time series data, where $G_1,\ldots,$ G_n are the genes under consideration and $D_1,\ldots,D_m$ are the dates that the samples were captured on. We can define a semi-algebraic hybrid automaton H representing M in various ways; we illustrate them in ascending order of complexity, beginning with a couple of trivial examples.

The most simple way to construct a hybrid automata from our time series data is to have a single location and a continuous representation of each gene's expression profile. Thus, for our our $m \times n$ matrix of expression values, we have n polynomials, where each of the n genes is represented by a polynomial of degree m (i.e. the number of time points). In this way we completely capture all of the information in our expression matrix without loss, in fact we can reconstruct our matrix of expression measurements exactly from this representation. Note that in this case we get an automaton without edges, since there is only a single location. Thus, there are no guard or reset conditions. This construction is close to the classical approach of using one system differential equations.

Rather than having a single location with polynomial representations of degree m for each gene, we could instead have m locations, one for each time point, and each of these could have linear dynamics for each gene. Clearly this is also completely equivalent to our original data and represents it without loss, in fact the representation can be seen as a simple distribution of the rows of our expression data across the m locations. In this case the discrete graph underlying the automaton is simply a chain and the automaton is linear.

The two examples above either use completely continuous or completely discrete representations of the dynamics, and are incapable of taking advantage of the hybrid nature of the dynamics, where it exists. Thus, one may be able to avoid automata of prohibitive complexity, by using a representation that reduces the dimensionality of the underlying data and yields automata with fewer locations and simpler dynamical laws. We can progress toward this goal by considering coordinated genes within suitably sized (initially, uniform length) windows of time and by letting the number of locations in our automata equal the number of windows under consideration. Next, we describe such an example with uniformly sized time windows; a version with nonuniform adaptively sized windows will be discussed below.

- the continuous variables are $G = \langle G_1, \ldots, G_n \rangle$ and $G' = \langle G'_1, \ldots, G'_n \rangle$;
- the directed graph $\langle \mathcal{V}, \mathcal{E} \rangle$ has $L = \lceil \frac{m}{h} \rceil$ locations $v_1, \ldots, v_L$ and its edges are defined as $\mathcal{E} = \{\langle v_i, v_{i+1} \rangle \mid 1 \leq i < L\}$;
- for each $v_i \in \mathcal{V}$ and each $e_i = \langle v_i, v_{i+1} \rangle \in \mathcal{E}$ we have: $Inv(v_i)[G] \overset{\text{def}}{=} \text{true}$; $Dyn(v_i)[G, G', T] = \wedge_{j=1}^n G'_j = p_{(i,j)}(T)$, where $p_{(i,j)}$ is the polynomial of degree at most h connecting the values of G_j at $D_{h*(i-1)+1}, \ldots, D_{h*i+1}$; $Act(e_i)[G, G'] = \wedge_{j=1}^n G_j = g_{(i,j)}$, where $g_{(i,j)}$ is the expression level of G_j at D_{h*i+1}; $Reset(e_i)[G, G'] = \wedge_{j=1}^n G'_j = G_j$ is the identity.

In the activation conditions we have implicitly assumed that the biological system has no memory. In fact, the activation considers only the final values of a state and not the trajectory which leads to these values. More sophisticated constraints are necessary to model systems with memory. The proposed automaton has a number of locations which depends on the number of dates and on the degree of the dynamical laws, and a number of variables which is proportional to the number of genes under consideration. Again, the discrete graph underlying the automaton is simply a chain, but the automaton is not linear.

Example 1. Let us consider three genes G_1, G_2, and G_3 for which we have measured the following expression levels:

D	0	1	2
G_1	0.25	0.20	0.42
G_2	0.49	0.41	0.80
G_3	0.10	0.20	0.30

In Figure 1, we depict the hybrid automaton, built by applying the naïve methods, described above, with $h = 1$ and without time windows. The dynamics are written inside the locations, while the resets and activations are represented on the edges. The incoming edge on the left provides the initialization conditions which can be used to obtain the trace which corresponds to the expression levels measured in the matrix.

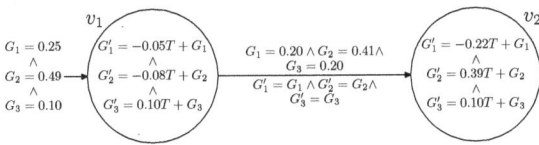

Fig. 1. The automaton of Example 1

Notice, that the fact that we are using only polynomial constraint is not too restrictive since: non-polynomial functions can be approximated with polynomials; polynomials can always interpolate finite sets of data.

In the following sections, we will explore methods to improve the automaton construction discussed above, by exploiting correlations, performing dimensionality reduction via correlation coefficients, Principal Component Analysis (PCA) [13], information theory [19], and exploiting clustering techniques [11].

2.1   Reductions Via Correlations on Genes

Given a gene expression level matrix M having m rows corresponding to the dates and n columns corresponding to the genes we can interpret each column of M as a random variable and compute the correlation coefficients between pairs of genes. As a result we get an $n \times n$ symmetric matrix $Corr$, such that $Corr[i, j]$ is the correlation between G_i and G_j ranging in the interval $[-1, 1]$. We can now use the absolute values of the elements of $Corr$ as similarity measures (or equivalently set the distance between G_i and G_j to $d(G_i, G_j) = 1 - |Corr[i, j]|$). These similarity measures can be used to to cluster the genes. There are different clustering techniques which can be used (see, e.g., [11,9]) leading to different results. However, we do not use clustering techniques to infer properties of the biological system under investigation, but only to build a compact hybrid automaton representing it. The analysis of the automaton will then help us to understand the system behaviors. Another possibility is that of clustering the genes using

PCA, i.e., using their correlation coefficients with respect to the new coordinate system. In both cases we obtain classes of (highly correlated) genes.

In each class of correlated genes we can choose a representative gene which is closer to all of the elements of the class and construct a hybrid automaton whose continuous variables are only the representative genes. The values of the non-representative genes can be approximated at any time from the representative ones exploiting their linear relationships. Alternatively, we could compute a cluster average as a continuous variable of a new fictitious representative gene, and use it to approximate the behavior of the non-representative genes, encoded through linear relationships. For the sake of simplicity, we focus on the former representation, here, and relegate the more complex treatment to the full paper.

Example 2. Let us consider the genes G_1, G_2, and G_3 of Example 1. The correlation coefficient of G_1 and G_2 is 0.99, while G_3 is less correlated both with G_1 and G_2 (0.74 and 0.75, respectively). If we apply hierarchical clustering on correlations we obtain a class with G_1 and G_2 together and another class containing G_3 only. Applying a clustering based on PCA in this case we obtain the same result. Hence, we can construct a hybrid automaton which has only G_1 and G_3 as continuous variables, i.e., the automaton of Figure 1 with G_2 deleted, and at any time we can infer the value of G_2 from G_1 ($G_2 \approx 2 * G_1$).

2.2   Reductions Via Correlations on Dates

We would next like to exploit dates-correlations to cluster dates, yielding better adaptive time windows for the construction of our automaton . However, if we analyze what happens if we transpose our gene expression matrix M, i.e., we consider the dates as random variables, and compute the correlation coefficients, we notice that this not only provides a better time segmentation, but also, a compact symbolic representation of the transcriptomic dynamics of genes.

Considering the dates as random variables means that each observation represents the values of a gene at each date. We have to imagine a coordinate system in which each axis corresponds to a date. In this system we can plot a point for each gene: the coordinates of this point are the expression levels of the gene at the different dates. When two dates are highly linearly correlated it is sufficient to know the expression levels of the genes at the first data to approximate the levels at the second one. If more than two dates are highly correlated, then the levels at one of them are sufficient to reconstruct the levels at all the other dates. In particular, if the random variables (dates) $D_i, \ldots, D_{i+r}$ are highly correlated, then we can relate them through a linear system of the form

$$\begin{cases} \widehat{G}(D_i) & = f_0(q) \\ \widehat{G}(D_{i+1}) = f_1(q) \\ \ldots \\ \widehat{G}(D_{i+r}) = f_r(q) \end{cases} \tag{1}$$

where $\widehat{G}$ is a symbolic variable gene expression, q is a parameter and the f_j's are linear function in q. If we know that the expression level of the gene G_j at

D_{i+s} is $g_{s,j}$, then we can use it to determine the corresponding value of q, i.e., $q_{s,j} = f_s^{-1}(g_{s,j})$. Now by substituting the $q_{s,j}$ to q in the equation corresponding to the date D_u, we can approximate the expression level $g_{u,j}$ of G_j at D_{i+u}, i.e., $g_{u,j} \approx f_u(q_{s,j})$.

Since we are not only interested in the expression levels at the measured data, but we would like to reconstruct all the genes time evolution, we can apply interpolation techniques to obtain a dynamical law. To keep the presentation simple we discuss here the case of linear interpolation (see [3] for more sophisticated interpolation methods). We have that the expression level of the gene G_j at time t, where $D_{i+a} \leq t \leq D_{i+a+1}$, for some $a \in [0, r-1]$ can be approximated with:

$$\frac{f_{i+a+1}(q_{s,j}) - f_{i+a}(q_{s,j})}{D_{i+a+1} - D_{i+a}}(t - D_{i+a}) + f_{i+a}(q_{s,j}) \tag{2}$$

Hence, we can construct our hybrid automaton by using a single location for dates which are highly correlated and in these locations the dynamical laws are the same for all the genes and are given by system (1) together with expression (2). This means that our automata will now have a single $\widehat{G}$ variable able to represent all the genes. In the case in which there are blocks of non consecutive dates which are correlated we can still use one location for all of them and introduce a loop in the discrete topology of the automaton. In order to simplify the notation we present only the definition for the case of adjacent correlated dates (the general case is presented in Example 3). Let M be a gene expression matrix of dimension $m \times n$. Let us assume that we cluster the dates exploiting their correlation coefficients as follows: $Cl_1 = \{D_1, \ldots, D_{d_1}\}$, $Cl_2 = \{D_{d_1+1}, \ldots, D_{d_2}\}$, $\ldots$, and $Cl_{cl} = \{D_{d_{(cl-1)}+1}, \ldots, D_m\}$. The *dates reduced* automaton H representing M is $HD(M) = (G, G', \mathcal{V}, \mathcal{E}, Inv, \mathcal{F}, Act, Reset)$, where:

- $G = \langle \widehat{G} \rangle$ and $G' = \langle \widehat{G'} \rangle$;
- $\langle \mathcal{V}, \mathcal{E} \rangle$ has cl locations $v_1, \ldots, v_{cl}$ and $\mathcal{E} = \{\langle v_i, v_{i+1} \rangle \mid 1 \leq i < cl\}$;
- for each v_i corresponding to $Cl_i = \{D_a, \ldots, D_b\}$, where for each $a \leq c \leq b$ it holds $\widehat{G}(D_c) = f_c(q)$ and for each $e_i = \langle v_i, v_{i+1} \rangle$ we have: $Inv(v_i)[\widehat{G}] = \text{true}$, $Act(e_i)[\widehat{G}] = \vee_{j=1}^n \widehat{G} = M[b, j]$, $Reset(e_i)[\widehat{G}, \widehat{G'}] = \vee_{j=1}^n (\widehat{G} = M[b, j] \wedge \widehat{G'} = M[b+1, j])$, and

$$Dyn(v_i)[\widehat{G}, \widehat{G'}, T] = \bigvee_{a \leq c < b}(D_c - D_a \leq T \leq D_{c+1} - D_a \wedge \\ \widehat{G'} = \frac{f_{c+1}(f_a^{-1}(\widehat{G})) - f_c(f_a^{-1}(\widehat{G}))}{D_{c+1} - D_c}T + f_c(f_a^{-1}(\widehat{G})))$$

In the above automaton we have reduced the states from m to cl without increasing the complexity of the involved formulæ. Notice that since the f's are linear, their inverses, f^{-1}'s, are still linear and the automaton is semi-algebraic. Moreover, it is important to notice that inside each state/location the continuous dynamics of the genes are all regulated by a single law. In fact, what changes from one gene to another is only the value of $\widehat{G}$. This drastically reduces the complexity of the analysis in many cases. Imagine for instance that we wish to check the following property: *Each time a gene reaches an expression level lower than low it never increases again enough to reach the expression level low'*. In each state v_i we can check this property at the same time for all the genes. We only have to write a first-order formula representing the values of $\widehat{G}$ which violate the property, and then check that all the initialization values of the genes that are outside of this set. In this sense we can say that the reduction based on dates correlation reduce both the number of states and of variables, which were our main objectives.

Due to the non-determinism introduced in the discrete jumps the date reduced automaton H correctly approximates the behaviors observed in M only, provided that in M there is not a date in which two genes have the same value. This assumption is not restrictive in the real cases.

Example 3. Let us consider the following transposed gene matrix:

D	0	1	2	3	4	5	6
G_1	1	2	3	4	8	4	5
G_2	2	3	4	1	2	3	4
G_3	3	4	5	3	6	1	2

This is a toy example in which we have a perfect correlation on the dates 0, 1, 2, 5, and 6 and a perfect correlation on the dates 3 and 4. The automaton we can build generalizing the technique to use also clusters of non adjacent dates is depicted in Figure 2. We label each edge only with the reset constraint, since the activation one can be reconstructed from it.

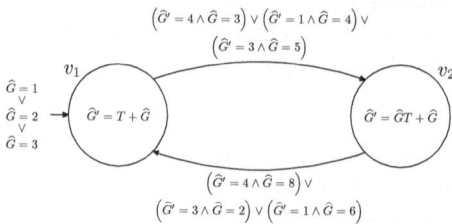

Fig. 2. The automaton of Example 3

3   Rate Distortion Theory and Extensions

In the above discussion we considered various ways of reducing the dimensionality of the data and deriving an automaton that captured the dynamics of this new compressed data set. As stated above, this can take the form of clustering the genes and subsequently using one gene from each cluster to approximate the others, or of considering windows of time to reduce both the number of locations and the number of variables in our hybrid automata. Finally, one could also reduce the complexity of the model used to represent the continuous dynamics, for example, one could use lower order polynomials or splines rather than polynomials of high degree. Each of these methods of simplifying our hybrid automata results in a distortion or disagreement between our model and the raw data. For instance, clustering forces us to live with discrepancies between the approximated profiles and the actual data vectors. What we really desire is a formalism to represent such distortions precisely, allowing us to specify an objective function that we can minimize, thus obtaining an optimal partition of our data and a low complexity automaton. We look to information theory for such a formalism and find it in the rate distortion theory of Shannon and Kolmogorov [8].

In rate distortion theory, one desires a compressed representation Z of a random variable X that minimizes some measure of distortion between the data elements $x \in X$ and their prototypes $z \in Z$. Taking $I(Z; X)$, the mutual information between Z and X, to be a measure of the compactness or degree of compression of the new representation, and defining a distortion measure $d(x, z)$ that measures distance between cluster prototypes and data elements, one can frame the problem as a trade-off between compression and average distortion. The main idea is that one balances the desire to achieve a compressed description of the data with the precision of the clustering, as measured by the average distortion, and finds the appropriate balance that maintains enough information while eliminating noise and inessential details.

In rate distortion theory, this trade-off is characterized mathematically as an optimization problem: $\mathcal{F}_{min} = I(Z; X) + \beta \langle d(x, z) \rangle$, where average distortion is defined as $\langle d(x, z) \rangle = \sum_{x,z} p(x)p(z|x)d(x, z)$ and is simply the weighted sum of the distortions between the data elements and their prototypes. More recently, Slonim et al. [20] have discussed a modification to rate distortion clustering for which only relations between data elements are used in the distortion function, rather than explicit mention of cluster prototypes. We have used a similar approach as a component in our graph search based approach to the time course segmentation problem.

Moving beyond classical rate distortion theory, we will need to generalize the problem further. In this generalized picture, we are presented with a family of time-course data all sampled from the same dynamical system; for example, k matrices of dimension $m \times n$. These matrices may be thought of as describing essentially the same dynamics, but corrupted by measurement noise, or affected by unmodeled/unmodelable environmental conditions. We may wish to introduce a notion of "distorted bisimulation", generalizing the idea of classical bisimulation, by allowing for certain constraints on allowable bisimulation. In this setting, it makes perfect sense to ask for a minimal complexity hybrid-automata representation of the datasets, subject to a constrained "distorted bisimulation".

This general notion will be explored in more detail in the full paper, but here, we focus on the most immediate problem of compressing (with loss) a given time-course data set by means of a semi-algebraic hybrid automaton. Returning to our earlier discussion, in the specialized setting, we note that the functional above captures the compression-precision trade-off inherent in the clustering problem and when combined with a shortest path graph search algorithm (as described below), it allows one to use an iterative method to find a numerical solution to the time course segmentation problem. The trade-off is controlled by the Lagrange parameter β that mitigates the trade-off between compression and preservation of relevant information, as β becomes large we focus on precision, as β tends to zero we focus more on compression. Setting the clustering problem up in this way allows us to find both an optimal windowing of our data, and optimal clusters of genes within the windows. From this compressed representation, we can create a hybrid automaton having minimal disagreement with the original data.

3.1   Reductions Via Rate Distortion

We would like to cluster our data in both the genes and in time, that is, we would like a procedure that yields windows in time and that captures intervals of concerted gene activity in which the genes are clustered into a small number of groups of co-expressed elements. From such a compressed representation, we can produce an automaton whose number of locations is the number of time windows, and for which the dynamical laws are less complex because we derive our continuous dynamics from the clustered data rather than from individual genes. We briefly discuss a method that performs this type of compression.

Let $D = \{D_1, D_2, \ldots, D_m\}$ be the time points at which a given system is sampled, and l_{min} and l_{max} be the minimum and maximum window lengths respectively. For each time point $D_a \in D$, we define a candidate set of windows starting from D_a as $S_{D_a} = \{W_a^b | l_{min} \leq D_b - D_a \leq l_{max}\}$, where W_a^b is the time window containing the dates $D_a, D_{a+1}, \ldots, D_b$. Each of these windows may then be clustered and labeled with a score based on its length and the cost associated with the clustering functional defined in (3). Following scoring, we formulate the problem of finding the lowest cost windowing of our time series in terms of a graph search problem and use a shortest path algorithm to generate the final set of (non-overlapping) time windows that fully cover the original series.

To score the windows, we use a variant of rate distortion clustering, using a distortion function defined between pairs of data elements. We aim to maximize compression (by minimizing the mutual information between the clusters and data elements), while at the same time forcing our clusters to have minimal distortion (as described in [20]). We perform rudimentary model selection by iterating over the number of clusters while optimizing (line search) over beta. This procedure, while somewhat expensive, results in a fairly complete sampling of the rate-distortion curves. Essentially, we trace the various solutions for different model sizes while tuning β, and choose the simplest model that achieves minimal cost in the target functional. In this way we obtain for each window a score that is the minimum cost in terms of model size and model fit, based on the trade-off between compression and precision. This method is computationally expensive and run times can be substantial, for this reason we have developed an implementation that can take advantage of parallel hardware.

Once the scores are generated, we pose the problem of finding the lowest cost tiling of the time series as a graph search problem. We consider a graph $G = (V, E)$ for which the vertices are time points $V = \{D_1, D_2, \ldots, D_m\}$, and the edges represent windows with associated scores. Each edge $e_{ab} \in E$ represents the corresponding window W_a^b from time point D_a to time point D_b, and has an initially infinite positive cost. The edges are labeled with the costs for the windows they represent, each edge e_{ab} gets assigned a cost $(F_{ab} * length)$ where F_{ab} is the minimum cost found by the clustering procedure and length is the length of the window $(b - a)$. Our original problem of segmenting the time series into an optimal sequence of windows can now be formulated as finding the minimal cost path from the vertex D_1 to the vertex D_m. The vertices on the path with minimal cost represent the points at which our optimal windows begin and end. We use a shortest path algorithm and generate a windowing that segments our original time series data into a sequence of optimal windows which perform maximal compression in terms of the clustering cost functional. We are now in a position to sketch one possible way to construct hybrid automata that have a compact representation in time and that reflect clusters with respect to gene expression. Further, we construct our models to have minimal distortion with respect to the original data. We accomplish this by clustering using the method just discussed and then building an automata with the same number of locations as windows and simplified dynamical laws constructed from the clustered genes.

Hence, for each cluster we can choose a representative gene which minimizes the distance to all of the other genes in the cluster and construct a hybrid automaton whose continuous variables correspond to those of the representative genes. Further, our time windows naturally provide a means to simplify the dynamics of our model by exploiting correlations in time. Our graph based approach allows for a convenient method of locating repeated segments in the data that are correlated, i.e., loops in our automata can be readily located. We will provide a complete characterization of this construction in the forthcoming paper, but note that our clustering procedure provides a method to optimally partition the data such that minimum distortion hybrid automata may be constructed.

4   Experimental Results and Conclusions

We now apply the techniques presented in previous sections to build a simple model of the metabolism of peach fruit. We measured the expression profiles of two classes of genes, ARF and RAB, along a period of 42 days, starting 72 days after flowering and sampling the genes every week. Gene expressions profiles were collected using real time PCR [5,10,14]. In particular, we considered 13 and 20 genes for the ARF and RAB families, respectively. Each sample consists in the average of 3 measurements normalized with respects to Ubiquitin Conjugating Enzyme level. We analyze the data applying the techniques described in Sections 2.1 and 2.2. A hierarchical clustering based on the function $d(X, Y) = (1 - |Corr(X, Y)|)$ for ARF genes is reported in Figure 3. We choose as distance between two clusters, C_1 and C_2, the minimum distance between $X \in C_1$ and $Y \in C_2$. The label of the circle shaped nodes represent the distance between subgraphs. Requiring a correlation of at least 70% we obtained 3 and 5 gene clusters for ARF and RAB, respectively. Applying the clustering on the date correlations we noticed a higher correlation: requiring a correlation of at

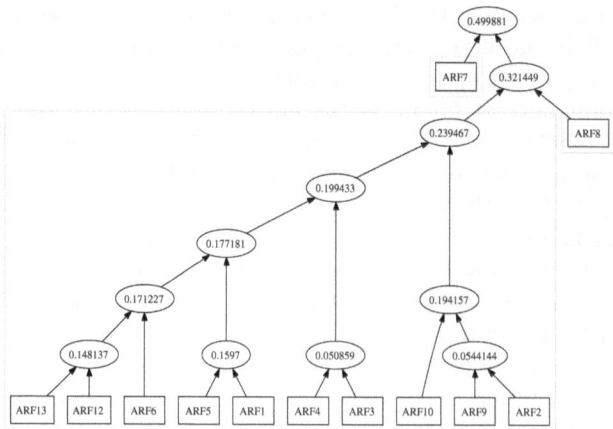

Fig. 3. The cluster hierarchy of the ARF gene correlations

least 98% we obtained 1 date cluster for ARF genes, while requiring a correlation of at least 93% for RAB genes, we got 2 date clusters. Hence, in the case of RAB genes we can built an automaton having just 2 discrete locations and the variables $\widehat{G}$, $\widehat{G}'$ and T which represent the evolution along 7 dates of 20 genes.

In conclusion, we emphasize that we have only established the preliminary foundations of a theory, aiming at the questions of how experimental data collected in biology may be treated rigorously within a semi-algebraic hybrid automata framework. We have hinted at its deep connection to dimensionality reduction and classical rate-distortion theory, but have relegated its complete treatment to the full paper. However, once such a framework has been created, it opens the field to many new questions. Namely, the following: How does one compare the dynamics of several closely related systems, e.g., a wild-type, mutant and a double-mutant? Can one factor the dynamics so that the final automaton may be viewed as product of several component automata, where most of the component modules remain unchanged over evolutionary time? How can the interaction between two or more biological systems (e.g., host-pathogen, host-vector-parasite, or an ecology) be modeled as products of hybrid automata constructed from different datasets?

References

1. Alur, R., Courcoubetis, C., Henzinger, T.A., Ho, P.H.: Hybrid Automata: An Algorithmic Approach to the Specification and Verification of Hybrid Systems. In: Grossman, R.L., Nerode, A., Ravn, A.P., Richel, H. (eds.) Hybrid Systems. LNCS, pp. 209–229. Springer, Heidelberg (1992)
2. Anai, H.: Algebraic Approach to Analysis of Discrete-Time Polynomial Systems. In: European Control Conference (ECC'99) (1999)

3. Bar-Joseph, Z., Gerber, G., Gifford, D.K., Jaakkola, T.S., Simon, I.: A New Approach to Analyzing Gene Expression Time Series Data. In: Proc. of Int. Conference on Computational biology (RECOMB'02), pp. 39–48. ACM Press, New York (2002)

4. Bar-Joseph, Z.: Analyzing time series gene expression data. Bioinformatics 20(16), 2493–2503 (2004)

5. Bustin, S.A.: Absolute quantification of mRNA using real-time reverse transcription polymerase chain reaction assays. Journal of Mol. Endoc. 25, 169–193 (2000)

6. Casagrande, A., Mysore, V., Piazza, C., Mishra, B.: Independent dynamics hybrid automata in systems biology. In: Proc. of the First International Conference on Algebraic Biology (AB'05), pp. 61–73. Universal Academy Press, Inc. (2005)

7. Casagrande, A., Piazza, C., Mishra, B.: Semi-Algebraic Constant Reset Hybrid Automata - SACoRe. In: Proc. of Conference on Decision and Control (CDC'05), pp. 678–683. IEEE Computer Society Press, Los Alamitos (2005)

8. Cover, T.M., Thomas, J.A.: Elements of information theory. Wiley-Interscience, New York, NY, USA (1991)

9. Datta, S., Datta, S.: Comparisons and validation of statistical clustering techniques for microarray gene expression data. Bioinformatics 19(4), 459–466 (2003)

10. Gachon, C., Mingam, A., Charrier, B.: Real-time PCR: what relevance to plant studies?. Journal of Experimental Botany 55(402), 1445–1454 (2004)

11. Jain, A.K., Murty, M.N., Flynn, P.J.: Data clustering: a review. ACM Comput. Surv. 31(3), 264–323 (1999)

12. Jekins, R., Pennington, S.: Arrays for protein expression profiling: towards a viable alternative to two-dimensional gel electrophoresis?. Prot. 1(1), 13–29 (2001)

13. Jolliffe, I.T.: Principal component analysis. Series in statistics. Springer, Heidelberg (1986)

14. Kubista, M., Andrade, J.M., Bengtsson, M., Forootan, A., Jonák, J., Lind, K., Sindelka, R., Sjöback, R., Sjögreen, B., Strömbom, L., Ståhlberg, A., Zoric, N.: The real-time polymerase chain reaction. Mol. Aspects of Medicine 27, 95–125 (2006)

15. Lafferriere, G., Pappas, G.J., Sastry, S.: O-minimal Hybrid Systems. Mathematics of Control, Signals, and Systems 13, 1–21 (2000)

16. Madeira, S., Oliveira, A.: Biclustering algorithms for biological data analysis: a survey. IEEE/ACM Trans. on Comp. Biology and Bioinformatics 1, 24–45 (2004)

17. Piazza, C., Antoniotti, M., Mysore, V., Policriti, A., Winkler, F., Mishra, B.: Algorithmic algebraic model checking i: The case of biochemical systems and their reachability analysis. In: Etessami, K., Rajamani, S.K. (eds.) CAV 2005. LNCS, vol. 3576, Springer, Heidelberg (2005)

18. Schena, M., Shalon, D., Davis, R.W.: Quantitative monitoring of gene expression patterns with a complementary dna microarray. Science 270(5235), 467–470 (1995)

19. Shannon, C.E.: A Mathematical Theory of Communication. The Bell System Technical Journal 27, 379–423 (1948)

20. Slonim, N., Atwal, G.S., Tkacik, G., Bialek, W.: Information-based clustering. In: Proc Natl Acad Sci USA (2005)

On Proving the Absence of Oscillations in Models of Genetic Circuits

François Boulier[1], Marc Lefranc[2], François Lemaire[1],
Pierre-Emmanuel Morant[2], and Aslı Ürgüplü[1]

[1] University Lille I, LIFL, 59655 Villeneuve d'Ascq, France
{boulier,lemaire,urguplu}@lifl.fr
http://www.lifl.fr/~{boulier,lemaire,urguplu}
[2] University Lille I, PHLAM, 59655 Villeneuve d'Ascq, France
Marc.Lefranc@univ-lille1.fr, morant@phlam.univ-lille1.fr
http://www-phlam.univ-lille1.fr/perso/lefranc

Abstract. Using computer algebra methods to prove that a gene regulatory network cannot oscillate appears to be easier than expected. We illustrate this claim with a family of models related to historical examples.

1 Introduction

The authors belong to a pluridisciplinary working group whose goal is to model the gene regulatory network controlling the circadian clock of a unicellular green alga [1]. See [2] for a survey on circadian rhythms and [3, Chapter 9] or [4,5] for more general texts about oscillations in biology. In doing so, they have gained some experience in designing models of oscillating gene regulatory networks.

One of the main problems faced by our working group can be formulated as follows: given a system of parametric ordinary differential equations built using mass action law kinetics, does there exist ranges of values for the model parameters and variables which are both meaningful from a biological point of view and where oscillating trajectories, i.e. limit cycles, can be found ?

This issue is theoretically very difficult. It is related to the unsolved Hilbert sixteenth problem. Indeed, systems of parametric ordinary differential equations which oscillate may do so only for very restricted ranges of parameters values. The difficulty is strengthened by the number of parameters arising in biochemical models, which can quickly become very large.

A related but easier problem consists of searching for the existence of parameter and variable values which are both meaningful from a biological point of view and give rise to a Hopf (more precisely Poincaré–Andronov–Hopf) bifurcation. See [6, Chapter 11], [5, Section 3.5] or [7, Section I.16]. In the neighborhood of a Hopf bifurcation indeed, a stable steady point of the model under study gives birth to a small stable limit cycle under some general hypotheses. Note that searching for Hopf bifurcations is not as general as searching for limit cycles: first, some Hopf bifurcations (e.g. the subcritical ones) do not strictly imply the existence of stable limit cycles; second, there may exist limit cycles not related to

H. Anai, K. Horimoto, and T. Kutsia (Eds.): AB 2007, LNCS 4545, pp. 66–80, 2007.
© Springer-Verlag Berlin Heidelberg 2007

Hopf bifurcations; third, a model may involve a Hopf bifurcation for parameters and variables values which are *close* to but *outside* of the biologically meaningful parameter domain values and generate limit cycles inside this domain.

There exist software packages such as AUTO or XPPAUT [8,9] which locate Hopf bifurcations by means of numerical calculations. They allow one to evidence the existence of Hopf bifurcations but not to prove their absence, and thus cannot be used to discard a model. Theoretically, the existence or the absence of Hopf bifurcations can be decided algebraically. See e.g. [10,11,12,13]. In particular, it can be decided by means of computer algebra methods which rely on Sturm sequences computations and algebraic elimination. Practitioners usually seem to avoid these methods because of their huge complexity in the worst case. In particular, we tried[1] the QEPCAD [14] package, which is based on quantifier elimination methods. However, we could not solve the problems addressed in this paper with it. We did not try the REDLOG package [15] and the software described in [10] for they rely on QEPCAD for the quantifier elimination process. An attempt to solve the addressed problem using the RAGLib library [16] is in progress, with the help of its author.

By comparison, the computer algebra methods described in this paper are very light. They take advantage of the special structure of the equations and of the biological constraints. This indicates that if used carefully, computer algebra methods may apply on more complex examples than one might expect.

In order to illustrate the core ideas of this paper in a simple setting, we do not study realistic models of circadian clocks but focus on a simple family of models depending on an integer parameter n and featuring a negative feedback loop, one of the core ingredients for generating oscillations [3]. These abstract models are closely related to models studied by Goodwin and Griffith in the 60's [17,18,19]. In particular, Griffith considered a model of a gene regulated by a polymer formed of n copies of its own protein. We study the same problem, but in a slightly more general case, where gene activation is not assumed to be fast. We conclude with the absence of Hopf bifurcation in our family of models for $n \leq 8$ and their existence for $n \geq 9$. Although we do focus here on biology, it should be stressed that a cooperativity of order 9 is not as unrealistic as it may seem. In particular, gene regulation by an octamer has been reported [20]. Moreover, an effective cooperativity of order 9 may also be obtained as a consequence of reducing a higher-dimensional, more realistic, model to a three-variable one. Finally, our conclusions are consistent with those of Griffith [3, Pages 244–246] and of other works devoted to more sophisticated variants of the Goodwin model [21,22,23], and thus we believe that the interest of the present paper goes beyond illustrating computer algebra methods. The application of these methods to more realistic biological models is in progress and is left for a future paper.

---

[1] QEPCAD was downloaded from www.cs.usna.edu/~qepcad and installed on a computer endowed with a Pentium 4 and 512 MB of RAM. Tests were performed by increasing the default number of cells up to its maximal limit: 200 of millions of cells.

Our paper is organized as follows. Section 2 describes the family of models we study. Some basic facts about Hopf bifurcations are recalled in section 3. In section 4, we use computer algebra methods to prove the absence of Hopf bifurcation in our model for $n \leq 8$ and the occurrence of Hopf bifurcations for $n \geq 9$. The methods involved are Gröbner basis theory [24,25] and Descartes' rule of signs [26]. Computations are performed using the MAPLE 9 computer algebra software. Our proofs were constructed after carrying out intensive numerical simulations which strongly suggested the results.

2   Our Family of Models

Figure 1 displays a gene regulated by a polymer obtained by combining n times a protein. The model variables are the state G of the gene, the mRNA concentration M and the concentration P of the protein translated from the mRNA. Greek letters represent parameters. The initial model involves $n + 2$ differential equations depending on $2\,n + 5$ parameters. By means of a suitable quasi-steady state approximation, described in section 2.1, one obtains the following reduced model, involving only three equations:

$$
\begin{aligned}
\dot{G} &= \theta\,(\gamma_0 - G - G\,P^n), \\
\dot{P} &= n\,\alpha\,(\gamma_0 - G - G\,P^n) + \delta\,(M - P), \\
\dot{M} &= \lambda\,G + \gamma_0\,\mu - M.
\end{aligned}
\tag{1}
$$

All variables and parameters are positive apart λ, which is allowed to be negative. The protein P reacts with itself, forming a polymer. Gene activity is regulated by the polymer as it binds to the gene promoter. Depending on the sign of λ, the polymer is an activator or a repressor: if $\lambda < 0$ then mRNA transcription

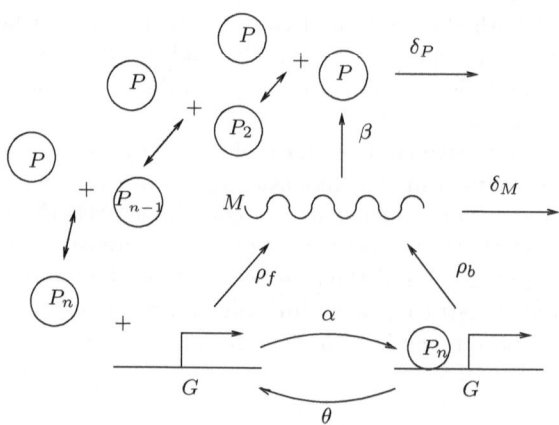

Fig. 1. A gene regulated by a polymer of its protein

is enhanced when polymer is bound to promoter ; if $\lambda > 0$ then mRNA transcription is reduced. The G variable takes values in the range $[0, \gamma_0]$ and can be viewed as an averaged gene activity. The values $G = 0$ and $G = \gamma_0$ correspond repectively to a polymer being bound to the gene promoter or not.

2.1 Model Reduction

The chemical system involves $n + 5$ reactions, described below. Denote P_i the polymer obtained by combining i proteins P with the convention $P_1 = P$.

$$G + P_n \underset{\theta}{\overset{\alpha}{\rightleftharpoons}} G : P_n, \quad G \xrightarrow{\rho_f} G + M, \quad G : P_n \xrightarrow{\rho_b} G : P_n + M,$$

$$M \xrightarrow{\beta} M + P, \quad M \xrightarrow{\delta_M} \emptyset, \quad P \xrightarrow{\delta_P} \emptyset, \quad P_i + P \underset{k_i^-}{\overset{k_i^+}{\rightleftharpoons}} P_{i+1} \quad (1 \le i \le n - 1).$$

The dynamics of these reactions is governed by the following equations, where $A_i = (1/\varepsilon)(k_{i+1}^- P_{i+1} - k_{i+1}^+ P_i P)$:

$$\dot{G} = \theta\,(\gamma_0 - G) - \alpha\,G\,P_n,$$
$$\dot{M} = \rho_f\,G + \rho_b\,(\gamma_0 - G) - \delta_M\,M,$$
$$\dot{P} = \beta\,M - \delta_P\,P + 2\,A_1 + A_2 + \cdots + A_{n-1},$$
$$\dot{P}_i = -A_{i-1} + A_i \quad (2 \le i \le n - 1),$$
$$\dot{P}_n = -A_{n-1} + \theta\,(\gamma_0 - G) - \alpha\,G\,P_n$$

The $1/\varepsilon$ factor is introduced to express the fact that the various steps of protein polymerization are assumed to be fast compared to other reactions (transcription, translation, degradation, binding of polymer to the gene). Eliminating the A_i by means of quasi steady state assumptions ($\dot{P}_2, \ldots, \dot{P}_n$ are assumed to be small), one reformulates the third differential equation as:

$$\dot{P} = \beta\,M - \delta_P\,P + n\,(\theta\,(\gamma_0 - G) - \alpha\,G\,P_n).$$

The P_n variable can be reexpressed as

$$P_n = \frac{k_1^+ \cdots k_{n-1}^+}{k_1^- \cdots k_{n-1}^-}\,P^n + \text{a term multiplied by } \varepsilon.$$

Neglecting the term multiplied by ε and introducing a new parameter $\bar{\alpha}$, one is thereby led to the following system of three differential equations:

$$\dot{G} = \theta\,(\gamma_0 - G) - \alpha\,\bar{\alpha}\,G\,P^n,$$
$$\dot{P} = n\,\theta\,(\gamma_0 - G) - n\,\alpha\,\bar{\alpha}\,G\,P^n + \beta\,M - \delta_P\,P,$$
$$\dot{M} = \rho_f\,G + \rho_h\,(\gamma_0 - G) - \delta_M\,M.$$

The model can now be simplified by rescaling all parameters and indeterminates G, M and P. Since the reduction involves many (easy) intermediate computations, we only sketch it here. First replace $\alpha\,\bar{\alpha}$ by α. Then replace P by

$(\theta/\alpha)^{1/n} P$. Expand the equation in $\dot{M}$, replace $\rho_f - \rho_b$ by λ and $\rho_b \gamma_0$ by μ. This implies that λ may be *positive or negative*. Then apply rescale time by replacing t by $\delta_M t$. This last transformation simplifies the term $-\delta_M M$ into $-M$. At this stage, one gets the following system:

$$\dot{G} = \frac{\theta}{\delta_M} (\gamma_0 - G - G P^n), \quad \dot{M} = \frac{\lambda}{\delta_M} G + \frac{\mu}{\delta_M} - M,$$

$$\dot{P} = n \left(\frac{\alpha}{\theta}\right)^{\frac{1}{n}} \frac{\theta}{\delta_M} (\gamma_0 - G - G P^n) + \left(\frac{\alpha}{\theta}\right)^{\frac{1}{n}} \frac{\beta}{\delta_M} M - \frac{\delta_P}{\delta_M} P.$$

Then discard all the δ_M by replacing β/δ_M, δ_P/δ_M, λ/δ_M and μ/δ_M by β, δ, λ and μ. Then replace α^n/θ^{n-1} by α. Then replace θ/δ_M and α/δ_M by θ and α. Using the fact that M occurs only in linear terms, renormalize last M so that $\beta = \delta_P$ and update λ and μ. One finally gets our reduced model (1).

Comments. Gene activity is regulated by P^n. The reduced model is designed so that the steady state depends only on parameters λ, μ and γ_0 while θ, α and δ control time scales. Note that Griffith model is recovered by letting θ and α tend towards $+\infty$, keeping the ratio θ/α constant. When translation is equal to degradation i.e. $\delta = 0$, $n\, G - (\theta/\alpha)\, P$ is constant which expresses the fact that DNA binding and unbinding do not modify the total quantity of proteins.

3   Hopf Bifurcations

3.1   Hurwitz Determinants

Let $\dot{x} = F(x)$ be a differential system in m dependent variables. The steady points of the differential system are the zeros of the system (that we assume to be polynomial or rational) $F(x) = 0$. To each steady point, one may associate a linear system $\dot{x} = J x$ where J is the $m \times m$ jacobian matrix of the differential system, evaluated over the steady point. The stability of the steady state is determined by the eigenvalues of J. It is stable if and only if all eigenvalues have negative real parts. Thus to each steady point, one may associate the characteristic polynomial $C(\sigma) = \sigma^m + a_1 \sigma^{m-1} + \cdots + a_m$ $(a_0 = 1)$ of J. Thanks to the Routh-Hurwitz criterion, the stability of the steady points can be studied by analyzing the sign of the Hurwitz determinants $c_{k,0}$. These ones can be directly computed from the coefficients of the characteristic polynomial, as shown below. Following [7, Section I.13], compute the Sturm sequence:

$$p_0(\omega) = \Re\left(\frac{C(i\,\omega)}{i^m}\right), \quad p_1(\omega) = -\Im\left(\frac{C(i\,\omega)}{i^m}\right) \tag{2}$$

$$p_{k+2}(\omega) = -\operatorname{rem}(p_k, p_{k+1}, \omega) \quad (k \geq 0).$$

Denote $p_k(\omega) = c_{k,0}\,\omega^{m-k} + c_{k,1}\,\omega^{m-k-2} + c_{k,2}\,\omega^{m-k-4} + \cdots$ Observe that the computation of p_k must be performed carefully (e.g. using subresultant sequences) to ensure that $c_{k,0}$ actually is a Hurwitz determinant. See [10]. Indeed,

$$c_{0,0} = 1, \quad c_{1,0} = a_1, \quad c_{2,0} = a_1 a_2 - a_3, \quad \ldots, \quad c_{m,0} = a_m\, c_{m-1,0}.$$

The two following propositions are well known. The first one is nearly a corollary to the Routh Theorem [7, Theorem 13.4].

Proposition 1. *With the same notations, if all the Hurwitz determinants $c_{k,0}$ are positive, apart perhaps $c_{m,0}$, then J has no pure imaginary eigenvalue.*

Proof. If all the Hurwitz determinants $c_{k,0}$ are positive ($0 \leq k < m$) then they are *a fortiori* nonzero. Assume J has pure imaginary eigenvalues $\pm i\,\bar{\omega}$ (they are necessarily conjugate). These values $\pm\bar{\omega}$ are then common zeros of p_0 and p_1. The gcd of p_0 and p_1 has thus degree greater than or equal to 2. This gcd is the last nonzero polynomial in the sequence $p_0, \ldots, p_{m-1}$. Thus one polynomial p_k with $0 \leq k < m$ must vanish identically. Therefore the corresponding Hurwitz determinant $c_{k,0}$ must vanish also.

Proposition 2. *With the same notations, if all the Hurwitz determinants $c_{k,0}$ are positive ($0 \leq k \leq m-2$) and $c_{m-1,0} = 0$ and $c_{m-2,1} < 0$ then all the eigenvalues of J have negative real parts except a purely imaginary conjugate pair.*

Proof. The polynomial p_{m-1} has the special form $p_{m-1} = c_{m-1,0}\,\omega$. We have $c_{m-1,0} = 0$. Then p_0 and p_1 have a degree two gcd, p_{m-2}, which has the special form $p_{m-2} = c_{m-2,0}\,\omega^2 + c_{m-2,1}$. We have $c_{m-2,1} < 0$ and $c_{m-2,0} > 0$ thus, the common roots $\pm\bar{\omega}$ of p_0 and p_1 are real. Therefore J has one pair of purely imaginary conjugate eigenvalues $\pm i\,\bar{\omega}$. Now, compute the Sturm sequence (2) over the polynomial $\bar{C}(\sigma) = C(\sigma)/(\sigma^2 + \bar{\omega}^2)$. This Sturm sequence $\bar{p}_0, \bar{p}_1, \ldots, \bar{p}_{\bar{m}}$ can actually be derived from that of C:

$$\bar{p}_0(\omega) = \frac{p_0}{\sigma^2 + \bar{\omega}^2}, \quad \bar{p}_1(\omega) = \frac{p_1}{\sigma^2 + \bar{\omega}^2}, \quad \ldots, \quad \bar{p}_{\bar{m}}(\omega) = c_{m-2,0}.$$

All the corresponding Hurwitz determinants are positive. According to the Routh Theorem [7, Theorem 13.4], all the roots of $\bar{C}$ have negative real parts. This concludes the proof of the proposition.

For $m = 3$ we have $c_{m-2,1} = -a_3$. For $m = 4$ we have $c_{m-2,1} = -a_1\,a_4$.

3.2 Hopf Bifurcations

The differential systems encountered in biological modelling involve parameters. Let $\dot{x} = F(x, \theta)$ be a differential system in m variables and p parameters θ. If some real values are assigned to the parameters then one gets a system such as the one described in section 3.1. If these real values continuously vary then the steady points and their associated eigenvalues continuously vary also.

Definition 1. *With notations as above, a Hopf bifurcation arises for a steady point when all the eigenvalues associated to the steady point have negative real parts except one complex conjugate pair, which crosses the imaginary axis because of a variation in the system parameters.*

3.3 In Computer Algebra

In computer algebra, an important point is to avoid to compute the steady points, i.e. not to solve the system $F(x, \theta) = 0$. The Hurwitz determinants can be computed generically. They depend on the system parameters. Their sign is studied modulo the ideal I generated by the polynomial system $F(x, \theta) = 0$. The absence of Hopf bifurcation is established, thanks to proposition 1 and definition 1, by proving that the Hurwitz determinants $c_{0,0}, \ldots, c_{m-1,0}$ are positive for all x and θ, considering that x and θ satisfy $F(x, \theta) = 0$ plus, usually, some extra (positivity) conditions such as $x, \theta > 0$.

The Hurwitz determinants $c_{k,0}$ get reformulated by computing their normal forms $\bar{c}_{k,0}$ w.r.t. any Gröbner basis of the ideal I. Reference books for the Gröbner basis theory are [24,25]. Indeed, the difference $c_{k,0} - \bar{c}_{k,0}$ belongs to I. Over any steady point of the differential system, it is thus zero, thus the two polynomials $c_{k,0}$ and $\bar{c}_{k,0}$ have the same value hence the same sign.

In practice moreover, Gröbner bases can be computed in dimension zero. Computing in dimension zero corresponds to some generic computation, which may be false for particular values of the system variables and parameters. However, in biological models, parameters (and thus variables) have no accurate values and zero dimensional computing makes sense.

4   Application to Our Models

To permit the reader to reproduce our computations, we provide the sequence of MAPLE 9 commands which prove that no Hopf bifurcation may arise in our models for positive values of the system variables and parameters (apart λ).

The *LinearAlgebra* package, the *Groebner* package and the *Jacobian* function of the *VectorCalculus* package are loaded. The list of the model variables is assigned to *vars*.

```
with (LinearAlgebra):
with (VectorCalculus,Jacobian):
with (Groebner):
vars := [G, P, M]:
```

The list of the right–hand sides of the model equations is assigned to the *equilibria* variable. The zeroes of this polynomial system provide the steady points of the model.

```
equilibria := [
    theta*(gamma0 - G - G*P^n),
    n*alpha*(gamma0 - G - G*P^n) + delta*(M-P),
    lambda*G + gamma0*mu - M]:
```

In general, one cannot compute a Gröbner basis of the ideal I generated by such a system if a symbolic n is left as an exponent. But in our case, generic Gröbner bases exist, at least w.r.t. some admissible orderings. Let us fix the pure lexicographical ordering given by $\lambda > M > \gamma_0$. The other model variables and parameters are considered as algebraically independent elements of the base field of the equations. The Gröbner basis is thus computed in dimension zero i.e. in the polynomial ring $\mathbb{K}[\lambda, M, \gamma_0]$ where $\mathbb{K}$ denotes the field obtained by adjoining all the remaining variables and parameters to the field of the rational numbers. Here are the Gröbner basis elements. The leading terms appear on the left–hand side of the equations:

$$\gamma_0 = G + G\,P^n, \quad M = P, \quad \lambda = \frac{P - \mu\,G - \mu\,G\,P^n}{G}.$$

Observe that for any particular value of n, the Gröbner basis can be computed by the following sequence of MAPLE commands (these commands do not permit to obtain the generic form directly):

```
ordre := plex (lambda, M, gamma0):
basis := gbasis (equilibria, ordre):
seq (leadterm (basis [i], ordre) =
    solve (basis [i], leadterm (basis [i], ordre)),
    i = 1 .. nops (basis));
```

The computed Gröbner basis has two striking properties: its leading terms are plain variables ; apart for λ which is allowed to be negative, the right–hand sides of the three other equations of the Gröbner basis are necessarily positive. The first property implies that the quotient ring is a free algebra: a polynomial ring. In particular, the product of two normal forms is itself a normal form. The second property implies that there are no constraints on the values that can be assigned to the model variables and parameters occuring in the right–hand sides of the equations since positivity is the only requirement for the values of γ_0 and M. Therefore, to evaluate the Routh–Hurwitz criterion over the model steady point, it is sufficient to replace each element by its normal form in the Jacobian matrix J of the model.

```
J := Jacobian (equilibria, vars):
```

$$J := \begin{pmatrix} -\theta\,(1 + P^n) & -n\,\theta\,G\,P^{n-1} & 0 \\ -n\,\alpha\,(1 + P^n) & -n^2\,\alpha\,G\,P^{n-1} - \delta & \delta \\ \lambda & 0 & -1 \end{pmatrix}$$

The generic normal form of J is:

$$J = \begin{pmatrix} -\theta\,(1 + P^n) & -n\,\theta\,G\,P^{n-1} & 0 \\ -n\,\alpha\,(1 + P^n) & -n^2\,\alpha\,G\,P^{n-1} - \delta & \delta \\ \dfrac{P - \mu\,G - \mu\,G\,P^n}{G} & 0 & -1 \end{pmatrix}$$

From now on, J is assumed to be under normal form. This implies in particular that all the expressions computed from J are free of the parameter λ. These expressions thus only involve positive variables. For any particular value of n, the normal form of the jacobian can be computed by the following command:

```
J := map (normalf, J, basis, ordre):
```

The characteristic polynomial of J writes: $\sigma^3 + a_1 \sigma^2 + a_2 \sigma + a_3$. Its coefficients are stored in indexed variables:

```
pol := CharacteristicPolynomial (J, sigma):
for i from 1 to nops (vars) do
    a[i] := coeff (pol, sigma, nops (vars) - i)
od:
```

The Hurwitz determinants $c_{k,0}$ can now be computed from the coefficients a_k of the characteristic polynomial.

```
c[0,0] := 1:
c[1,0] := a[1]:
c[2,0] := a[1]*a[2]-a[3]:
c[3,0] := a[3]*(a[1]*a[2]-a[3]):
```

In order to apply propositions 1 and 2, one needs to study the positivity of the Hurwitz determinants $c_{k,0}$ for $0 \le k \le m-1 = 2$. The coefficient $c_{0,0}$ is obviously positive. So is the coefficient $c_{1,0}$ (below), for it is generically equal to a sum of monomials involving positive variables and coefficients:

$$1 + n^2 \alpha G P^{n-1} + \delta + \theta + \theta P^n$$

The coefficient $c_{2,0}$ is more complicated. However it has a very special form. It is equal to a sum of monomials with positive coefficients minus the single following monomial: $n \theta \delta P^{n+2}$. Readers who would like to reproduce the computations may want to use our *negterms* function, described in section 4.4:

```
negterms(c[2,0]);
```

$$[[-n \theta \delta P^{n+2}], []]$$

4.1   Cases $n = 1$ and $n = 2$

Those two cases are easy. Indeed, for these values of n, the negative term in $c_{2,0}$ is cancelled by another coefficient (namely $2 \theta \delta P^{n+2}$). This implies that $c_{2,0}$ is always positive. Thanks to proposition 1 no Hopf bifurcation may occur.

```
negterms(subs(n=1, c[2,0]));
```

$$[[], []]$$

```
negterms(subs(n=2, c[2,0]));
```

$$[[], []]$$

4.2   Cases $3 \le n \le 8$

In these cases, $c_{2,0}$ is also always positive. The proof here is less straightforward than above. It relies on Descartes' rule of signs. See [26].

Proposition 3. *(Descartes' rule of signs)*
Let $p = a_0 \, x^d + \cdots + a_{d-1} \, x + a_d$ be a polynomial in one indeterminate x and real coefficients. Denote $r(p)$ the number of positive real roots of p, counted with multiplicities, and $v(p)$ the number of sign changes in the sequence $a_0, \ldots, a_d$ (the zero coefficients must be removed). Then $v(p) \geq r(p)$. In particular, if $v(p) = 0$ then $r(p) = 0$; if $v(p) = 1$ then $r(p) = 1$.

To simplify the positivity proof of $c_{2,0}$, one first cancels all the monomials not depending on θ, δ and P (i.e. the variables and parameters occuring in the negative term). One then clears the denominator P^2 out. One is thus led to prove the positivity of the *cond* polynomial computed below.

```
X := indets (c[2,0]) minus indets (negterms (c[2,0]));
```
$$X := \{\alpha, \, G, \, \mu\}$$

```
cond := numer (subs (seq (X[i]=0, i=1..nops(X)), c[2,0]))/P^2:
```

The polynomial *cond* is actually a polynomial in P^n. By a change of variables, one is led to study the positivity of the following degree 2 polynomial in P:

```
cond := collect (subs (P^n=P, cond), P):
```

$$(\theta^2 \, \delta + \theta^2) \, P^2 + (2 \, \theta^2 \, \delta + \theta + 2 \, \theta \, \delta + \delta^2 \, \theta + 2 \, \theta^2 - \theta \, \delta \, n) \, P + (2 \, \theta \, \delta + \delta^2 \, \theta + \delta^2 + \theta^2 \, \delta + \theta + \delta + \theta^2)$$

Let us write *cond* as $A \, P^2 + B \, P + C$. The coefficients A and C are positive for they only involve monomials with positive coefficients. Therefore, the conditions $B^2 - 4 \, A \, C \geq 0$ (to ensure the existence of real roots) and $B < 0$ (to ensure the existence of positive roots by Descartes' rule of signs) are necessary and sufficient to have *cond* $= 0$ for $P > 0$. Condition $B^2 - 4 \, A \, C \geq 0$ leads to $\theta \leq \theta_0$ where

```
theta0 := solve (discrim (cond, P) / theta^2, theta):
```

$$\theta_0 := \frac{\delta^4 - 2 \, \delta^3 \, n + \delta^2 \, n^2 - 4 \, \delta^2 \, n + 1 - 2 \, \delta^2 - 2 \, \delta \, n}{4 \, \delta \, n \, (1 + \delta)}.$$

The condition $B < 0$ leads to $\theta < \theta_1$, where

```
theta1 := solve (coeff (cond, P, 1) / theta, theta);
```

$$\theta_1 := \frac{-1 - 2 \, \delta - \delta^2 + \delta \, n}{2 \, (1 + \delta)}.$$

One is thus led to prove that the two conditions

$$0 < \theta \leq \theta_0, \quad 0 < \theta < \theta_1 \tag{3}$$

cannot be satisfied at the same time. One may convince oneself by plotting curves (see Figure 2). Let us continue the analysis more algebraically. For $\delta = 0$ we have $\theta_1 < 0$. Therefore, $\theta_1(\delta) > 0$ only if δ lies in the interval bounded by the two real roots of θ_1 i.e.

$$\frac{n - \sqrt{n^2 - 4 \, n}}{2} - 1 < \delta < \frac{n + \sqrt{n^2 - 4 \, n}}{2} - 1.$$

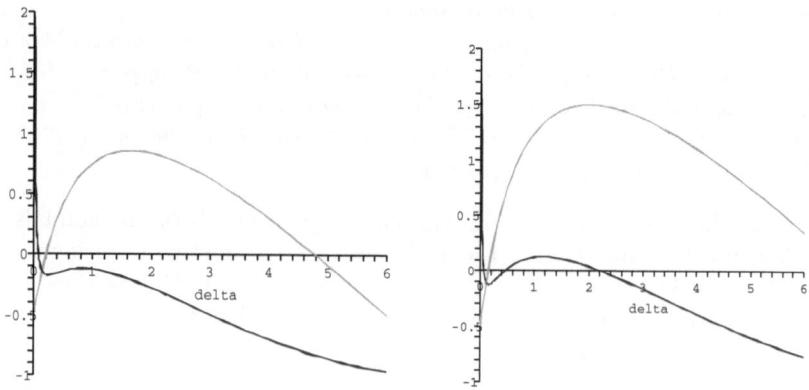

Fig. 2. The curves $\theta_0(\delta)$ and $\theta_1(\delta)$ are not simultaneously positive for $n = 7$ (left) while they are simultaneously positive for $n = 9$ (right). The curve for $\theta_1(\delta)$ starts with a negative value.

which only holds for $n > 4$. Hence no Hopf bifurcation may occur for $n \leq 4$, thanks to proposition 1.

To solve the remaining cases $5 \leq n \leq 8$, it is sufficient to compute a table of variations. For each value of n, one may isolate the positive real roots of $\theta_0(\delta)$ and $\theta_1(\delta)$ in arbitrary small intervals, ensuring that all the intervals are disjoint. Evaluating these two expressions for values of δ outside these intervals, one easily proves that conditions (3) cannot hold simultaneously. Hence no Hopf bifurcation may occur for $5 \leq n \leq 8$. Here is such a table for $n = 7$:

$$\begin{array}{c|ccccccccc} \delta & & 0.07 & & 0.21 & & 4.79 & & 12.58 & \\ \hline \theta_0 & + & 0 & - & - & - & - & - & 0 & + \\ \theta_1 & - & - & - & 0 & + & 0 & - & - & - \end{array}$$

The *realroot* function of MAPLE may be used to isolate the roots, the following commands show. This function implements the algorithm described in [27]. Its is based on Descartes' rule of signs.

```
readlib (realroot):
realroot (subs (n=7, numer (theta0)), 1e-3);
```

$$\left[\left[\frac{81}{1024}, \frac{41}{512}\right], \left[\frac{12879}{1024}, \frac{805}{64}\right]\right]$$

```
realroot (subs (n=7, numer (theta1)), 1e-3);
```

$$\left[\left[\frac{213}{1024}, \frac{107}{512}\right], \left[\frac{2453}{512}, \frac{4907}{1024}\right]\right]$$

4.3   Case $n = 9$

Hopf bifurcations may arise for $n = 9$. One can be found using proposition 2. Using a table of variations or looking at Figure 2, we see that $\delta = 1$ and $\theta = 1/10$ satisfy conditions (3). Replace α, μ and G, which were set to zero in *cond*, by small positive values (say 1/1000). The Hurwitz determinant $c_{2,0}$ is then a polynomial in P with real roots. The following commands compute one of them.

```
vals := n=9,delta=1,theta=1/10,alpha=1/1000,mu=1/1000,G=1/1000:
c20 := normal (subs (vals, c[2,0])):
valP := realroot (c20, 1e-6) [1];
```

$$valP = \left[\frac{41381}{32768}, \frac{1324193}{1048576}\right]$$

The above values cancel the Hurwitz determinant $c_{2,0}$. The following command checks that $c_{1,1} = -a_3 < 0$. Observe that the computation could be performed more carefully by means of interval arithmetics.

```
evalf(subs(vals,P=valP[1],-a[3])), evalf(subs(vals,P=valP[2],-a[3]));
```

$$-8.268768356, \quad -8.268823875$$

According to proposition 2, a Hopf bifurcation should occur. Numerical simulations show that this Hopf bifurcation gives birth to oscillations.

4.4   Comments

A general analysis. It is possible to prove the absence of Hopf bifurcation for $n \leq 8$ without discussing all the cases. The idea, which is only sketched here, starts by noticing that the real roots of $\theta_0(\delta)$ belong to the range[2] $]1/n, n-2[$.

One first computes the resultant w.r.t. δ between $\theta_1(\delta)$ and its first derivative w.r.t. δ. It vanishes only for $n = 8$. Thus, apart for this case which may be handled separately, all the roots of $\theta_1(\delta)$ are simple.

One then proves that $\theta_1(\delta)$ admits one real root in the range $]0, 1/n[$. This can be done by performing a change of variables over $\theta_1(\delta)$, mapping the range $]0, 1/n[$ to the range $]0, +\infty[$ in order to apply Descartes' rule of signs. Here are the corresponding MAPLE commands:

```
z  := collect (numer (theta0), delta):
z1 := collect (numer (subs (delta=1/(n*(sigma+1)), z)), sigma);
```

$$z_1 := n^4\sigma^4 + 2n^4\sigma^3 + (n^4 - 4n^3 - 2n^2)\sigma^2 - (8n^3 + 6n^2)\sigma - 4n^3 - 4n^2 + 1$$

For any $n \geq 1$, the number of changes of signs is equal to 1. By Descartes' rule of signs, $z_1(\sigma)$ admits exactly one positive real root whence[3] $\theta_1(\delta)$ admits exactly one positive root smaller than $1/n$. Therefore, $\theta_1(\delta) < 0$ after crossing this root.

---

[2] One denotes $]a, b[$ the open interval $[a, b]$.

[3] This is the very argument applied by the *realroot* function!

Performing similar changes of variables and using Descartes' rule of signs, one easily studies the existence of real roots of $\theta_1(\delta)$ in the ranges such $]1/n, 1[$ and $]1, n-2[$. The respective changes of variables are:

$$\delta = \frac{n+\sigma}{n\,(\sigma+1)} \quad \text{and} \quad \delta = 1 + \frac{n-3}{\sigma+1}.$$

Applying Descartes' rule of signs over the resulting expressions, one concludes that $\theta_1(\delta)$ has no root located in the range defined by the two real roots of $\theta_0(\delta)$ for $n < 8$ and always two roots located in that range for $n \geq 9$. This is sufficient to prove that no Hopf bifurcation occurs for $n < 8$.

To prove that Hopf bifurcations occur for $n \geq 9$ by using proposition 2, one still needs to prove that $c_{1,1} = -a_3$ (below) may be negative for all these values. The variables μ and G may be assigned arbitrary small positive values. Therefore $c_{1,1}$ may be negative for each $n \geq 9$. This concludes the (sketched) proof that Hopf bifurcation arise for any $n \geq 9$.

$$\theta\,\delta\,(\mu\,G\,(n\,P^{n-1} + n\,P^{2\,n-1}) - (n+1)\,P^n - 1)$$

The γ_0 parameter. It is tempting to avoid the γ_0 parameter in the model and use the value 1 instead. However, over some of the models we tried, the shape of the Gröbner basis was nicer, naming this 1 and eliminating it.

The Gröbner basis. The choice of the ordering is important. It is at least necessary to eliminate the problematic λ parameter and, more generally, every variable or parameter which is allowed to be negative. Other Gröbner bases can be used to prove that $c_{2,0}$ is positive over the model steady points (at least for $n \leq 2$). A such example is obtained by replacing γ_0 by G in the ordering.

The *negterms* function. The MAPLE code of the *negterms* function is provided here. It gathers as input a rational fraction *expr*. It returns a pair of lists L_1 and L_2. The list L_1 (resp. L_2) is the list of the monomials of the numerator (resp. denominator) of *expr* which have negative coefficients.

```
negterms := proc (expr)
    local f, p, koeffs, terms, result;
    f := proc (x,y) if x < 0 then x*y else NULL fi end:
    result := NULL;
    for p in [expand (numer (expr)), expand (denom (expr))] do
        if indets (p) <> {} then
            koeffs := coeffs (p, indets (p), 'terms');
            result := result, zip (f, [koeffs], [terms])
        else
            result := result, []
        fi;
    od;
    [result]
end:
```

5   Conclusion

We have studied a simple system depending on an integer n, which describes the regulation of a gene by a polymer of order n of its protein. We have shown that no Hopf bifurcation may occur for $n \leq 8$ and that Hopf bifurcations arise for $n \geq 9$, taking into account that biologically relevant values of most of the model variables and parameters must be positive. Strictly speaking, this is not sufficient to prove the absence of limit cycles for $n \leq 8$. However, our analysis is confirmed by extensive numerical simulations.

Our study led us to study the positivity of complicated rational fractions modulo the ideal I generated by the steady points equations. This problem is in general a difficult problem in computer algebra though it is theoretically solved [28]. Our study was however much simplified by the fact that we could compute a Gröbner basis of the ideal I having a very nice shape and by the fact that most of model variables and parameters are positive.

We believe that these simplifying properties occur more often that expected and that they imply that, at least in the domain of biological modeling, computer algebra methods are not necessarily restricted to academic problems.

Thanks. We would like to thank Thomas Erneux for stimulating discussions about the Routh–Hurwitz criterion. We would like also to thank all the members of the circadian rythms working group in Lille.

References

1. Morant, P.E., Vandermoere, C., Parent, B., Lemaire, F., Corellou, F., Schwartz, C., Bouget, F.Y., Lefranc, M.: Oscillateurs génétiques simples. Applications à l'horloge circadienne d'une algue unicellulaire. In: proceedings of the Rencontre du non linéaire, Paris (2007), http://nonlineaire.univ-lille1.fr
2. McClung, C.R.: Plant Circadian Rhythms. The Plant Cell 18, 792–803 (2006)
3. Fall, C.P., Marland, E.S., Wagner, J.M., Tyson, J.J. (eds.): Computational Cell Biology. Interdisciplinary Applied Mathematics, vol. 20. Springer, Heidelberg (2002)
4. Goldbeter, A.: Biochemical Oscillations and Cellular Rhythms: The Molecular Bases of Periodic and Chaotic Behaviour. Cambridge University Press, Cambridge (2004)
5. Françoise, J.P.: Oscillations en biologie. Mathématiques et Applications, vol. 46. Springer, Heidelberg (2005)
6. Hale, J.K., Koçak, H.: Dynamics and Bifurcations. Texts in Applied Mathematics, vol. 3. Springer, New York (1991)
7. Hairer, E., Norsett, S.P., Wanner, G.: Solving ordinary differential equations I. Nonstiff problems, 2nd edn. Springer Series in Computational Mathematics, vol. 8. Springer, New York (1993)
8. Doedel, E.: AUTO software for continuation and bifurcation problems in ODEs (1996), http://indy.cs.concordia.ca/auto
9. Ermentrout, B.: Simulating, Analyzing, and Animating Dynamical Systems: A Guide to XPPAUT for Researchers and Students. vol 14 of Software, Environments, and Tools, vol. 14, SIAM (2002)

10. El Kahoui, M., Weber, A.: Deciding Hopf bifurcations by quantifier elimination in a software–component architecture. Journal of Symbolic Computation 30(2), 161–179 (2000)
11. Wang, D., Xia, B.: Stability Analysis of Biological Systems with Real Solution Classification. In: proceedings of ISSAC 2005, Beijing, China 354–361 (2005)
12. Gatermann, K., Hosten, S.: Computational algebra for bifurcation theory. Journal of Symbolic Computation 40, 1180–1207 (2005)
13. Gatermann, K., Eiswirth, M., Sensse, A.: Toric Ideals and graph theory to analyze Hopf bifurcations in mass action systems. Journal of Symbolic Computation 40, 1361–1382 (2005)
14. Brown, C.W.: QEPCAD B: a program for computing with semi-algebraic sets using CADs. SIGSAM Bulletin 37(4), 97–108 (2003)
15. Dolzmann, A., Sturm, T.: Redlog: computer algebra meets computer logic. SIGSAM Bulletin 31(2), 2–9 (1997)
16. El Din, M.S.: RAGLib (Real Algebraic Library Maple package) (2003), http://www-calfor.lip6.fr/~safey/RAGLib
17. Goodwin, B.C.: Temporal Organization in Cells. Academic Press, London (1963)
18. Goodwin, B.C.: In: Advances in Enzyme Regulation, vol. 3, p. 425. Pergamon Press, Oxford (1965)
19. Griffith, J.S.: Mathematics of Cellular Control Processes. I. Negative Feedback to One Gene. Journal of Theoretical Biology 20, 202–208 (1968)
20. Selleck, W., Howley, R., Fang, Q., Podolny, V., Fried, M.G., Buratowski, S., Tan, S.: A histone fold TAF octamer within the yeast TFIID transcriptional coactivator. Nature Structural Biology 8(8), 695–700 (2001)
21. Ruoff, P., Rensing, L.: The Temperature–Compensated Goodwin Model Simulates Many Circadian Clock Properties. Journal of Theoretical Biology 179, 275–285 (1996)
22. Ruoff, P., Vinsjevik, M., Mohsenzadeh, S., Rensing, L.: The Goodwin Model: Simulating the Effet of Cycloheximide and Heat Shock on the Sporulation Rhythm of Neurospora crassa. Journal of Theoretical Biology 196, 483–494 (1999)
23. Kurosawa, G., Mochizuki, A., Iwasa, Y.: Comparative Study of Circadian Clock Models, in Search of Processes Promoting Oscillation. Journal of Theoretical Biology 216, 193–208 (2002)
24. Cox, D., Little, J., O'Shea, D.: Ideals, Varieties and Algorithms. An introduction to computational algebraic geometry and commutative algebra. Undergraduate Texts in Mathematics. Springer, New York (1992)
25. Becker, T., Weispfenning, V.: Gröbner Bases: a computational approach to commutative algebra. Graduate Texts in Mathematics, vol. 141. Springer, Heidelberg (1991)
26. Basu, S., Pollack, R., Roy, M.F.: Algorithms in Real Algebraic Geometry. Algorithms and Computation in Mathematics, vol. 10. Springer, Heidelberg (2003)
27. Collins, G.E., Akritas, A.G.: Polynomial real root isolation using Descartes'rule of signs. In: proceedings of ISSAC 1976, Yorktown Heights NY 272–275 (1976)
28. Collins, G.E.: Quantifier Elimination for the Elementary Theory of Real Closed Fields by Cylindrical Algebraic Decomposition. In: Brakhage, H. (ed.) Automata Theory and Formal Languages. LNCS, vol. 33, pp. 134–183. Springer, Heidelberg (1975)

Attenuation Regulation
as a Term Rewriting System[*]

Eugene Asarin[1], Thierry Cachat[1], Alexander Seliverstov[2],
Tayssir Touili[1], and Vassily Lyubetsky[2]

[1] LIAFA, CNRS and University Paris Diderot
{asarin,txc,touili}@liafa.jussieu.fr
[2] IITP, Russian Academy of Science
{slvstv,lyubetsk}@iitp.ru

Abstract. The classical attenuation regulation of gene expression in
bacteria is considered. We propose to represent the secondary RNA structure in the leader region of a gene or an operon by a term, and we give a
probabilistic term rewriting system modeling the whole process of such
a regulation.

1 Introduction

Modeling the mechanisms of regulation of gene expression, allowing prediction
of quantitative characteristics of this expression (such as estimation of the level
of expression and concentration of the substrate) is an important research challenge. In a previous work [LRSP06, LPRS07], a model of one particular kind
of regulation, the classical attenuation regulation, has been suggested. In that
model, the evolution of the secondary RNA structure in the leader region of a
gene, and the progress of the ribosome and the polymerase along the RNA/DNA
strands, are represented by a very special, elaborated in detail, Markov chain.
In this chain the transition probability corresponding to the progress of the ribosome depends on a "control variable" — the concentration of charged tRNA
molecules in the cell. All the other probabilities do not depend on the control
variable, they can be determined from energy-based considerations. Termination
and antitermination (of gene expression) correspond to particular random events
in the Markov chain. In [LRSP06], a Monte-Carlo simulation of this Markov chain
led to biologically realistic dependence of termination probability from the control variable. Due to a large size and a complex structure of the Markov chain,
its simulation is a heavy computational task, but it was successfully solved, and
a software tool called RNAMODEL simulates one trajectory in fractions of a second [LRSP06, RNA]. However, the approach based on the direct description of
the Markov chain and its simulation has some limitations, especially for a theoretical analysis. Biologically, it would be nice to have a more structured and
compact representation of the Markov chain and its instantaneous probability

[*] The support of CNRS-RAS cooperation agreement 19122 EVOLVER is gratefully
acknowledged.

H. Anai, K. Horimoto, and T. Kutsia (Eds.): AB 2007, LNCS 4545, pp. 81–94, 2007.

distributions over all states at every instant, or only for sufficiently large time, or only probabilities of the two biologically important events — termination and antitermination.

Note that the problem of modeling the classical attenuation regulation, as stated in [LRSP06] and in the current article, is related to the representation of the transient behavior of the secondary structure on a sliding window on the RNA strand between the ribosome and the polymerase (see below for details). This differs from the kinetics of the secondary RNA structure on a fixed nucleotide sequence for unlimited time, i.e. unlimited number of steps, investigated in many papers. The structure that appears after a large amount of time is called equilibrium secondary RNA structure, it corresponds to a minimum of energy, see e.g. [Zuk03, FFHS00]. The tool RNAMODEL has also the function of determining this equilibrium structure and its energy as a special part of the full model in [LRSP06]. However, real structures that appear on the RNA strand during the regulation process are far from the equilibrium and their energies are far from minimal.

In this article we discover a regular internal structure of the Markov chain describing the classical attenuation regulation. We show that it can be represented as a probabilistic term rewriting system for a particular type of terms. The set of rewriting rules can be large, but all of them are generated by a small set of (five) metarules. In fact we give the full description of the metarules and explain how to generate all the rules for the case of classical attenuation regulation.

Potential benefits of such a representation are multiple:

- easier and more precise modeling of regulation mechanisms depending on the dynamics of the secondary structure;
- compact description of such mechanisms, perhaps in dedicated languages, and hence a better biological understanding of regulation processes;
- convenient representation of secondary structures by terms;
- specific analysis and simulation methods for rewriting systems.

This article is structured as follows. In section 2 we describe shortly the biological phenomenon that we want to model: the mechanism of classical attenuation regulation (CAR). In section 3 we introduce a class of terms and probabilistic term rewriting systems. In section 4 we represent a qualitative metamodel of the biological mechanism of CAR by a term rewriting system. In section 5 we refine the previous system and decorate its transitions with rates, thus obtaining a representation of the Markov chain by a probabilistic term rewriting system. In section 6 we show some simulation results. In section 7 we discuss some related work on term rewriting and its applications. In section 8 we conclude with a discussion of perspectives of the rewriting approach to modeling the mechanisms involving RNA secondary structures, especially regulation.

2 Classical Attenuation Regulation

To begin with, we recall some well-known biological facts about the biological phenomenon playing the central role in this article.

The expression of a group of structural genes (that is synthesis of the corresponding proteins, which are ferments for a chemical reaction) can be regulated by a sequence of nucleotides placed on the DNA upstream inside the so called *leader region* of the genes [SB91]. This subsequence of the leader region is called the *regulatory region*. In this article we deal with one particular type of regulation, *classical attenuation regulation (CAR)* in bacteria. This regulation mechanism concerns structural genes (groups of genes — operons) that produce proteins which catalyze the synthesis of amino acids. The classical attenuation allows to activate such an operon when the cell contains a small concentration of the amino acid, to deactivate the operon whenever this concentration increases, and to do it fast. The mechanism of CAR involves several actors: the regulatory region on the DNA, its copy on the RNA, the ribosome, and a ferment called RNA polymerase (see Fig.1).

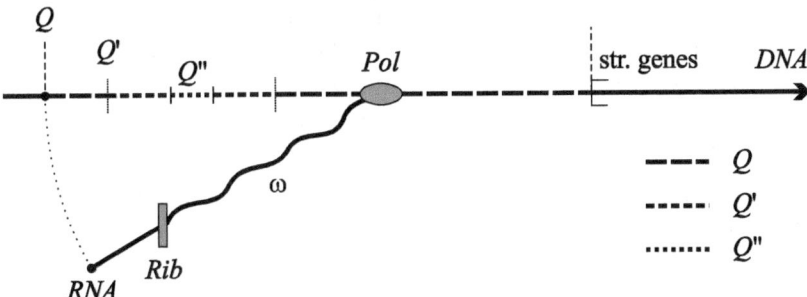

Fig. 1. Classical attenuation regulation. The RNA polymerase *Pol* transcribes the regulatory region Q, the ribosome *Rib* translates the leader peptide gene Q'. The movement of *Rib* on regulatory codons Q'' is controlled by the concentration of charged tRNA. The secondary RNA structure ω between *Rib* and *Pol* brakes *Pol* and pushes it off the chain. If *Pol* reaches the structural genes, then they are expressed, i.e. transcribed and then translated. Note that in both the DNA and the RNA, we use Q, Q' and Q'' to denote the regulatory region, the leader peptide gene, and the regulatory codons, respectively.

For structural genes to be expressed two concurrent processes should succeed: the regulatory region Q should be transcribed creating an RNA by RNA polymerase. At the same time the ribosome should be bound to the very beginning of the freshly created segment Q' (called the *leader peptide gene*) in the regulatory region Q on the RNA and starts translation of this leader peptide gene to an auxiliary protein. The essential part of the regulation process takes place when the ribosome moves on Q' on the RNA and the polymerase moves somewhere downstream of the ribosome on Q on the DNA.

The ribosome moves "rightwards" (formally speaking, in the direction from the 5' to the 3' end) on a segment Q' of the sequence Q. Its speed is constant except on a subsequence Q'' (*regulatory codons*) where it depends directly on the concentration of the amino acid (via charged tRNA concentration). To the right of the ribosome and independently of it, the polymerase moves rightwards on Q. Between the ribosome and the polymerase a secondary structure ω is formed on the RNA. This structure consists in pairing of some nucleotides, and it changes very fast. An important effect of the secondary structure ω consists in slowing down the movement of the polymerase. There are two possible scenarios:

- When ω is strong enough, its "braking" action on the polymerase increases, and moreover, the polymerase can slip off the DNA (this can only happen on so-called *T-rich sequence*, where the connection of the polymerase and the DNA weakens). Such an event is called *termination*, and in this case the structural genes are not expressed: the transcription of the regulatory region is aborted, the structural genes are not transcribed and therefore not translated.
- Another possibility is that the ribosome moves fast enough to weaken or partly destroy most of the structure ω. In this case the polymerase safely traverses the T-rich sequence, and arrives to the end of the leader region Q. Next, the polymerase enters the structural genes, and their transcription, followed by translation are unavoidable. This event is called *antitermination* and in this case the structural genes are expressed.

In the rest of this article we build a qualitative and a quantitative models of the regulation process described above.

3   Terms and Rewriting Systems

3.1   Unranked Unordered Terms

Let Σ be a finite set of function symbols and $\mathcal{X}$ an enumerable set of *variables* (standing for sets of terms). The set $T_\Sigma[\mathcal{X}]$ of terms over Σ and $\mathcal{X}$ is the smallest set that satisfies:

- $\Sigma \subseteq T_\Sigma[\mathcal{X}]$,
- $\{f(x) \mid f \in \Sigma \wedge x \in \mathcal{X}\} \subseteq T_\Sigma[\mathcal{X}]$,
- if $f \in \Sigma$ and $s \subseteq T_\Sigma[\mathcal{X}]$ is a *set* of terms, then $f(s)$ is in $T_\Sigma[\mathcal{X}]$.

By definition we also put $f(\emptyset) = f$ for $f \in \Sigma$. For convenience we write $f(g, h(e))$ instead of $f(\{g, h(\{e\})\})$. However one should remember that the coma-separated terms are unordered.

Example 1. Let $\Sigma = \{e, f, g, h\}$ and $\mathcal{X} = \{x, y, z, \dots\}$, then the followings are terms in $T_\Sigma[\mathcal{X}]$: $f(g, h(e))$, $f(f(x))$ and $e(g, f)$.

Note that we consider function symbols of variable arity. T_Σ stands for $T_\Sigma[\emptyset]$. Terms in T_Σ are called *ground terms*. Variables are used only to define substitution and rewriting rules. The "real" terms are ground terms. A *substitution* σ is a mapping from $\mathcal{X}$ to $2^{T_\Sigma[\mathcal{X}]}$, written as $\sigma = \{x_1 \to T_1, \dots, x_n \to T_n\}$, where T_i, $1 \leq i \leq n$, is a finite set of terms that substitutes the variable x_i. The term obtained by applying the substitution σ to a term t is written $t\sigma$. We call it an *instance* of t.

Let R be a rule of the form $l \to r$, where l and r are terms in $T_\Sigma[\mathcal{X}]$. For ground terms t, t' we write $t \to_R t'$ if there exists a substitution σ such that t' can be obtained from t by replacing an occurrence of the subterm $l\sigma$ by $r\sigma$. $\to_R$ defines a relation between ground terms. Let $\to_R^*$ be the reflexive transitive closure of $\to_R$.

Example 2. Let $R = l \to r$ with $l = f(x, e)$, $r = f(g(x), e)$ and $t = e(f(h, e))$, then $t \to_R t'$ where $t' = e(f(g(h), e))$.

A *term rewriting system (TRS)* is a finite set of rules of the form $l \to r$. Given a TRS $\mathcal{R}$ and a set of terms $I \subset T_\Sigma$, the language $\mathcal{R}^*(I)$ is defined as the set of all ground terms that can be obtained from the terms in I by applying a finite number of times the rules from $\mathcal{R}$, i.e., $\mathcal{R}^*(I) = \{t \in T_\Sigma \mid \exists t' \in I, t' \to_\mathcal{R}^* t\}$.

Example 3. Let $\mathcal{R} = \{f(x) \to g(f(x))\}$ and $I = \{f(e, h)\}$, then

$$\mathcal{R}^*(I) = \{g^n(f(e, h)) \mid n \in \mathbb{N}\}.$$

3.2 Probabilistic Term Rewriting Systems

A *Continuous Time Markov Chain* is a pair (S, ρ), where S is a finite or enumerable set of states and $\rho : S \times S \to [0, \infty)$ is the rate matrix. For $s, s' \in S$, $\rho(s, s') > 0$ means that there is a transition between states s and s', and that the probability for moving from s to s' within t time units is equal to $1 - e^{-\rho(s,s')\cdot t}$. If a state s has more than one outgoing transition (i.e., if there exist more than one state s' for which $\rho(s, s') > 0$) there exists a *race* between these transitions and the probability for moving from s to s' within t time units is equal to $\frac{\rho(s,s')}{E(s)}\left(1 - e^{-E(s)\cdot t}\right)$, where $E(s) = \sum_{s' \in S} \rho(s, s')$.

A (continuous time) *Probabilistic term rewriting system (PTRS)* over $\Sigma \cup \mathcal{X}$ is a (finite) set of rules of the form $l \xrightarrow{\Lambda} r$, where l and r are terms in $T_\Sigma[\mathcal{X}]$, and $\Lambda \in (0, \infty)$ is a rate.

A PTRS $\mathcal{R}$ over $\Sigma \cup \mathcal{X}$ defines a continuous time Markov chain on ground terms $M = (T_\Sigma, \rho)$, where $\rho(t, t') = \Lambda$ iff there exists a rule $l \xrightarrow{\Lambda} r \in \mathcal{R}$ such that $t \to_R t'$, where R is the "non probabilistic" rule $l \to r$.

Remark 1. If there are several rules (or several instances of the same rule) that lead from t to t', then $\rho(t, t') = \sum \Lambda$, where the sum is taken over all such rules or instances.

4   Metamodel

We want to model the phenomenon of the classical attenuation regulation described in section 2.

We suppose that a *regulatory region* Q (see Fig. 1) is given and fixed in the sequel, it is a sequence (word) $Q \in \{\mathbf{A}, \mathbf{C}, \mathbf{G}, \mathbf{T}\}^*$, the letters of this alphabet are called *nucleotides*. We denote by $|x|$ the length of any word x and x_i the ith letter of x, so $x = x_1 x_2 \ldots x_{|x|}$. The sequence Q can be folded[1] in a way that some nucleotides of Q are paired: $\mathbf{A}$ with $\mathbf{T}$ and $\mathbf{C}$ with $\mathbf{G}$. The complement of a nucleotide is written using a bar: $\mathbf{A} = \overline{\mathbf{T}}, \mathbf{T} = \overline{\mathbf{A}}, \mathbf{C} = \overline{\mathbf{G}}, \mathbf{G} = \overline{\mathbf{C}}$. We look in Q for subwords ("stems") of the form

$$Q_A Q_{A+1} \ldots Q_B \quad \text{and} \quad Q_C Q_{C+1} \ldots Q_D \quad \text{such that}$$
$$B - A = D - C, \quad A + 3 \leq B, \quad B + 3 \leq C \qquad (1)$$
$$Q_A = \overline{Q_D}, \quad Q_{A+1} = \overline{Q_{D-1}}, \ldots \quad Q_B = \overline{Q_C} .$$

Any pair of such stems forms a *hypohelix* (see Figure 2, where the labels A_i, B_i, C_i and D_i are positions in the word Q).

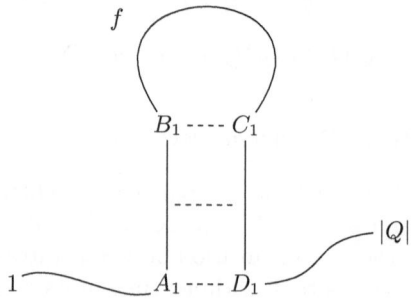

Fig. 2. One hypohelix f

We describe a hypohelix f by a tuple of its stems' extremities $f = (A, B, C, D)$, and we introduce the following notations:

$$stem(f) = [A, B] \cup [C, D], \quad loop(f) = [B + 1, C - 1], \quad supp(f) = [A, D].$$

There is a *ribosome* at some position on Q' and an *RNA polymerase* somewhere to the right of it. Both move to the right, in one step the ribosome moves by three successive nucleotides and the polymerase by one nucleotide. The *window* $w = (R, P)$ represents the segment of RNA from the first position R after the end of the ribosome to the last position P before the beginning of the polymerase. In fact the folding of the RNA sequence Q can only happen within the current window, i.e. between positions R and P. When the ribosome advances to the right, it can destroy the leftmost hypohelix of a current configuration, because it consumes the first three letters of the window. On the other hand any polymerase move adds one new letter to the window.

---

[1] Only on its "active" part called *window*, as we will see below.

Formally a window has the form $w = (R, P)$ with $R, P \in \mathbb{N}$. The following constraints should be satisfied:

$$13 \leq R \leq P \leq |Q| \tag{2}$$

Thus, the window is moving and changing its length.

Let $W = \{w = (R, P) \mid$ conditions (2) are satisfied$\}$ be the alphabet of all windows. We define

$$stem(w) = \emptyset, \; loop(w) = [R, P], \; supp(w) = [R, P] \; .$$

We will write terms over the alphabet Σ of all hypohelices and all windows:

$$\Sigma = H \cup W \text{ where } H = \{f = (A, B, C, D) \mid \text{conditions (1) are satisfied}\}.$$

We consider only terms of the form $w(\dots)$ for some $w \in W$ (rooted by some window w). According to the conditions that we will define next, a symbol $f = (A, B, C, D)$ can appear in a term $w(\dots)$ only if $R \leq A$ and $D \leq P$, where $w = (R, P)$.

We say that a hypohelix f is *embedded* in g (which can be a hypohelix or a window), written $f \prec g$, if $supp(f) \subseteq loop(g)$. Two hypohelices f and g are *disjoint*, written $f \bowtie g$, if $supp(f) \cap supp(g) = \emptyset$. We call f and g *unknotted* if either one of them is embedded in the other or they are disjoint. We say that $g = (A_2, B_2, C_2, D_2)$ is an *extension* of $f = (A_1, B_1, C_1, D_1)$, denoted $f \sqsubseteq g$, if $[A_1, B_1] \subseteq [A_2, B_2]$ and $B_2 - B_1 = C_1 - C_2$, hence $[C_1, D_1] \subseteq [C_2, D_2]$, and the pairing in g is an extension of that in f. See Figure 4.

We call a term t over Σ *well-formed* if it satisfies the following conditions:

(compatibility) any f and g appearing in t are unknotted, in particular any f can appear at most once,

(ordering) if f and g occur in t, then $f \prec g$ iff f is in the scope of g.

The combination of two hypohelices in Figure 4 is biologically feasible, but according to our rules these hypohelices are incompatible. We believe that this restriction (crucial for representation by terms) does not undermine significantly the accuracy of the model.

Notice, that a well-formed term of the form $w(\dots)$ (rooted by some window w) contains only hypohelices from

$$\Sigma_w = \{f \in H \mid f \prec w\}.$$

This simple observation greatly simplifies the simulation process.

In [LRSP06] an additional *maximality* condition is imposed. Using the terminology of this article, it requires that no hypohelix f in t can be replaced by its proper extension without creating an overlapping. Here we do not impose this restriction.

Each well-formed term represents a possible secondary RNA structure in a window in Q: the set of hypohelices that are present in this window. It could be

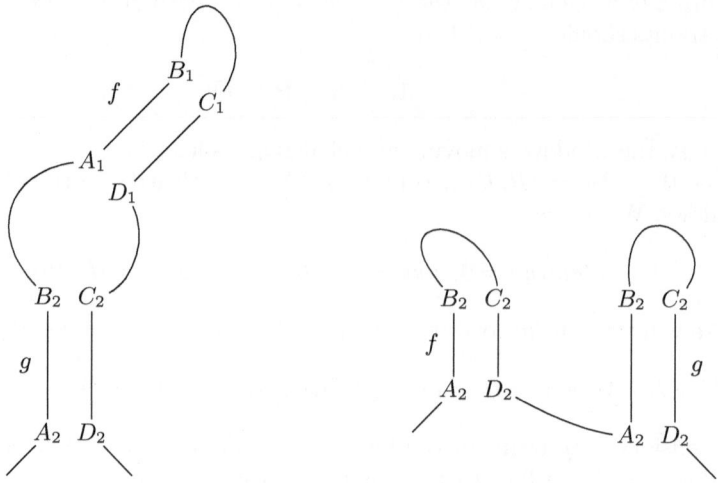

Fig. 3. Relative positions of two hypohelices f and g: $f \prec g$ and $f \bowtie g$. Here $f = (A_1, B_1, C_1, D_1)$ and $g = (A_2, B_2, C_2, D_2)$. On the left $B_2 < A_1$ and $D_1 < C_2$, on the right $D_2 < A_2$.

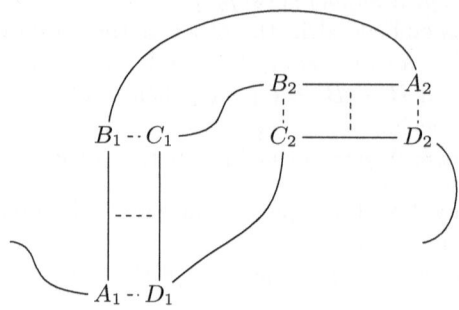

Fig. 4. Pseudo-knot: $A_1 < B_1 < A_2 < B_2 < C_1 < D_1 < C_2 < D_2$. Such configurations are not allowed in our model.

possible to allow knotted hypohelices, and hypohelices of length less than 3, but here we do not consider them.

We extend the definitions of $\bowtie$ and $\prec$: let f be a term and c a set of terms,

$$c \bowtie f \text{ iff } \forall g \in c \, (g \bowtie f),$$
$$c \prec f \text{ iff } \forall g \in c \, (g \prec f).$$

In the former case we say that f and c are *disjoint*, in the latter that c is *embedded* into f.

We start from a sequence Q without any pairing of nucleotides, this structure is described by a term $w()$ — "an empty window", where $w = (13, 13)$. Our aim is to represent the evolution of the secondary structure in the window, as well as the progress of the ribosome and the polymerase, through rewriting terms starting from $w()$. Our rewriting system will generate only well-formed terms.

On the whole, there are five rewriting *Meta*-rules:

- Binding and decomposition of a hypohelix f:

$$\big(\omega = g(c, d)\big) \longleftrightarrow \big(\omega' = g(c, f(d))\big) \quad \text{with } c \bowtie f, \ d \prec f, \ f \prec g, \quad (3)$$

where c and d are sequences of terms. The *concrete* rewriting rules — and their rates — depend on c and d, as explained below.
- Extension and reduction of a hypohelix

$$\big(\omega = f\big) \longleftrightarrow \big(\omega' = g\big) \quad \text{with } f \sqsubseteq g. \quad (4)$$

- The window movement can be described by the following rules, where $w = (R, P)$:

$$(R, P)(\omega) \longrightarrow (R + 3, P)(\omega') \,, \quad (5)$$
$$(R, P)(\omega) \longrightarrow (R, P + 1)(\omega) \,, \quad (6)$$
$$w(\omega) \longrightarrow \perp \,. \quad (7)$$

In the last rule, $\perp$ is a special symbol denoting termination. Rules (5) describe the movement of the ribosome. In these rules, ω' is obtained from ω by removing only the possible symbol that is incompatible with the new window $(R + 3, P)$, or replacing it by a "shorter" hypohelix. Indeed, if the leftmost hypohelix in ω starts at a position between R and $R + 3$, then the movement of the ribosome by three positions to the right will destroy this hypohelix. More formally, if $\omega \prec (R + 3, P)$, then $\omega' = \omega$. Otherwise the ribosome destroys the leftmost hypohelix. In this case, there is a single symbol f in ω such that $f \not\prec (R + 3, P)$. Suppose the subterm rooted by f is $f(c)$. Then, ω' is obtained by replacing in ω $f(c)$ by either $f'(c)$ or c, depending on the size of f, where $f' \sqsubseteq f$.

Rules 6 describe the movement of the polymerase. Note that if the polymerase reaches a position $P + 1$ where the structural genes are expressed, then we reach antitermination and the gene is expressed.

5   Quantitative Model

Now, we introduce the rates of the five rewriting rules.

Let $h(f_1(*), \ldots, f_n(*))$ be a term. Then the *free loop length* of the hypohelix h in this term is

$$l_h = |\,loop(h)| - \sum_{i=1}^{n} |\,supp(f_i)| \,.$$

This numeric characteristic corresponds to the number of nucleotides in the loop of the hypohelix h that do not participate in inner hypohelices.

In order to define the rate, we have to consider the concrete rule corresponding to the Metarule (3). For any $f, g, \mathbf{c} = c_1(x_1), \ldots, c_m(x_m)$ and $\mathbf{d} = d_1(y_1), \ldots, d_n(y_n)$ such that $\mathbf{c} \bowtie f$, $\mathbf{d} \prec f$, $f \prec g$ there is a concrete rule

$$\big(\omega = g(c_1(x_1), \ldots, c_m(x_m), \ d_1(y_1), \ldots, d_n(y_n))\big)$$
$$\longmapsto \big(\omega' = g(c_1(x_1), \ldots, c_m(x_m), \ f(d_1(y_1), \ldots, d_n(y_n)))\big) \qquad (8)$$

Recall that the subterms are unordered. Similarly the concrete rule corresponding to (4) is

$$\big(\omega = a(c_1(x_1), \ldots, c_m(x_m), \ f(d_1(y_1), \ldots, d_n(y_n)))\big)$$
$$\longmapsto \big(\omega' = a(c_1(x_1), \ldots, c_m(x_m), \ g(d_1(y_1), \ldots, d_n(y_n)))\big) \qquad (9)$$

Note that this transformation can change the free loop length of the hypohelix a. The rate of the rules (8-9) is denoted $K(\omega \to \omega')$, given by

$$K(\omega \to \omega') = \kappa \cdot \exp\left(\frac{1}{2}\left(E(\omega) - E(\omega')\right)\right), \qquad (10)$$

where the energy $E(\omega) = G_{hel}(\omega) + G_{loop}(\omega)$, κ is a parameter — usually $\kappa = 10^3$ — and

$$G_{hel}(\omega) = \frac{1}{RT} \cdot \sum_h E_h \ \text{ and } \ G_{loop}(\omega) = \sum_h 1.77 \cdot \ln(l_h + 1) + B, \qquad (11)$$

and h varies over all hypohelices from ω. E_h represents the total stacking energy along the hypohelix h. It is the sum of stacking bond energies of the adjacent base pairs of h. B can take three different values depending on the three possible types of the loop of the hypohelix g: terminal loop, single-strand bulge and double-strand bulge.

A *codon* is a triple of successive nucleotides. For a sequence Q', each codon is fixed to be either regulatory or non-regulatory. Analogously, each nucleotide in Q is fixed to be either non T-rich or T-rich [LRSP06]. Let s_0 be the "radius" of a ribosome — distance from P-site to the end of the ribosome — usually $s_0 = 12$, and let s_1 be the "radius" of a polymerase — distance from the 5' end of a polymerase to its transcription center — usually $s_1 = 9$. The rate of the rule (5) is denoted λ_{rib} and is constant when $R - s_0$ is a position of a non-regulatory codon, and otherwise λ_{rib} depends on an external parameter c — the concentration of charged tRNA [SB91]. The rate of the rule (6) is denoted ν and depends on secondary structure ω in the window. The rule (7) applies only when $P + s_1$ is a position of a T-rich nucleotide and its rate is denoted μ.

In [LRSP06] the rate of the rule (5) was denoted λ_{rib} and

$$\lambda_{rib}(c) = \frac{45\,c}{1+c}. \qquad (12)$$

The rate of the rule (6) was denoted ν and

$$\nu = 40 - F(\omega). \qquad (13)$$

The rate of the rule (7) was denoted μ and

$$\mu = \frac{1}{4}F(\omega) \ . \tag{14}$$

The function $F(\omega)$ in (13-14) for $\omega = f_1(*), \dots, f_n(*)$ depends only on functional symbols (hypohelices) $f_1, \dots, f_n$, and not on the structure of their arguments denoted by $*$. More precisely $F(\omega) = \max_i F(f_i)$, where

$$F(f) = \frac{\delta \cdot \exp\left(-\frac{r(f)}{r_0}\right)}{(L_2)^2 \cdot (p(f) - p_0)^2 + 1} \ , \tag{15}$$

with $p(f) \approx \frac{\pi}{\lceil supp(f) \rceil}$, and $r(f)$ the "free distance" from f to the end P of the window: for $f = (A, B, C, D)$ and $w = (R, P)$, we have

$$r(f) = R - D - \sum_i |\, supp(f_i)| \ . \tag{16}$$

Other symbols in equation (15) denote constants: $r_0 = 1, \delta = 30, L_2 = 27.1, p_0 = 0.18$, see [LRSP06].

Note that the rates of the rules depend only on the local configuration as explained above and not on the outside context. In particular it does not depend on instantiations of $x_1, \dots, x_m, y_1, \dots, y_n$.

6   Simulation Results

We have adapted the simulator described in [LRSP06] and available at [RNA] to obtain sequences of terms. As an example in Figure 6 we give one (slightly shortened and simplified) terminating trajectory of the regulation process for the *trpE* genes (responsible for the synthesis of tryptophan) in *E. coli*. The regulatory region itself is presented in Figure 5.

ATGAAAGCAATTTTCGTACTGAAAGGTTGGTGGCGCACTTCCTGAAACGGGCAGTGT
ATTCACCATGCGTAAAGCAATCAGATACCCAGCCCGCCTAATGAGCGGGCTTTTTTTTG

Fig. 5. A regulatory region for *trpE* genes in *E. coli*

7   Related Work

References to the literature on RNA regulation mechanisms can be found in [LRSP06, LPRS07].

Term rewriting systems have been used in the so called *Regular Model Checking* framework [KMM+01, BT02, AJMd02, ALdR05]. They have been successfully applied to the analysis of parameterized systems [BT02, AJMd02, ALdR05] and multithreaded programs [BT02, BT03, Tou05]. However, in the regular

$\langle 13, 13 \rangle () \to \langle 16, 27 \rangle (a) \to \langle 19, 27 \rangle () \to \langle 40, 50 \rangle (b) \to \langle 40, 51 \rangle (c) \to \cdot (b) \xrightarrow{*} \langle 40, 61 \rangle (c) \to \cdot (d) \xrightarrow{*} \cdot (g) \to \cdot (f(e)) \to \cdot (d) \to$
$\langle 40, 62 \rangle (c) \xrightarrow{*} \cdot (b) \to \cdot (b) \to \langle 40, 63 \rangle (b, h) \to \cdot (c) \xrightarrow{*} \cdot (h) \to \cdot (c, h) \to \cdot (c) \to \langle 40, 64 \rangle (c, h) \to \cdot (h) \to \cdot (b, h) \to \cdot (b) \xrightarrow{*} \cdot (g) \to$
$\cdot (f(e)) \to \cdot (b, h) \to \langle 40, 65 \rangle (h) \to \cdot (b) \to \cdot (b, h) \to \langle 40, 66 \rangle (h) \to \cdot (b) \to \cdot (b, i) \to \cdot (i) \to \cdot (c, i) \to \cdot (c, h) \to \cdot (c) \xrightarrow{*} \langle 40, 67 \rangle (c, h) \to$
$\cdot (c, i) \to \cdot (c, j) \to \cdot (c, k) \to \cdot (b) \xrightarrow{*} \cdot (k) \to (f(e)) \to \cdot (b, h) \to \cdot (b, i) \to \cdot (b, j) \to \cdot (b, k) \to \cdot (b, i) \to \langle 40, 68 \rangle (i) \to \cdot (b) \to \cdot (b, l) \to \cdot (b, h) \to$
$\cdot (b, j) \to \cdot (b, k) \to \cdot (h) \to \cdot (b, l(h)) \to \cdot (k) \to \cdot (l(h)) \to \cdot (l) \to \cdot (l(h)) \to \langle 40, 69 \rangle (h) \to \cdot (l) \to \cdot (b, l(h)) \to \cdot (b, h) \to \cdot (b, l) \to \cdot (b) \to$
$\cdot (l(h)) \to \langle 40, 70 \rangle (h) \to \cdot (l) \to \cdot (b, l(h)) \to \cdot (b, h) \to \cdot (b, l) \to \cdot (c, h) \to \cdot (m(h)) \to \cdot (b) \to \cdot (b, i) \to \cdot (b, j) \to \cdot (b, k) \to \cdot (m) \xrightarrow{*} \cdot (k) \to$
$\cdot (b, l(h)) \to \langle 40, 71 \rangle (l(h)) \to \cdot (b, h) \to \cdot (b, l) \to \cdot (h) \to \cdot (l) \to \cdot (b) \to \cdot (b, l(h)) \to \langle 40, 72 \rangle (l(h)) \to \cdot (b, h) \to \cdot (b, l) \to \cdot (h) \to \cdot (l) \to$
$\cdot (b) \to \cdot (b, i) \to \cdot (b, j) \to \cdot (b, n) \to \cdot (b, k) \to \cdot (b, o) \to \cdot (b) \to \langle 40, 73 \rangle (b, l) \xrightarrow{*} \cdot (b, o) \to \cdot (l) \to \cdot (b, l(h)) \to \cdot (h) \to \cdot (l(h)) \to \cdot (b, l(h)) \to$
$\langle 40, 74 \rangle (l(h)) \to \cdot (b, h) \to \cdot (b, l) \to \cdot (h) \to \cdot (l) \to \cdot (b) \to \cdot (b, h) \to \langle 40, 75 \rangle (h) \to \cdot (b) \to \cdot (b, l(h)) \to \cdot (l(h)) \to \cdot (b, l) \to \cdot (l) \to \cdot (b, i) \to$
$\cdot (b, j) \to \cdot (b, n) \to \cdot (b, k) \to \cdot (b, o) \to \cdot (k) \to \cdot (c, h) \to \cdot (m(h)) \to \cdot (c) \to \cdot (l(h)) \to \langle 40, 76 \rangle (h) \to \cdot (l) \to \cdot (b, l(h)) \to \cdot (b, h) \to \cdot (b, l) \to$
$\cdot (b) \to \cdot (b, h, p) \to \cdot (h, p) \to \cdot (b, p) \to \cdot (p) \to \cdot (c, h, p) \to \cdot (c, p) \to \cdot (d, p) \to \cdot (f, p) \to \cdot (e, p) \to \cdot (g, p) \to \cdot (b, i) \to \cdot (b, j) \to \cdot (b, n) \to$
$\cdot (b, q) \to \cdot (b, k) \to \cdot (b, o) \to \cdot (c, h) \to \cdot (m(h)) \to \cdot (b, h, p) \to \langle 40, 77 \rangle (h, p) \to \cdot (b, p) \to \cdot (b, p) \to \cdot (b) \to \cdot (h) \to \cdot (c, h, p) \to$
$\cdot (c, p) \to \cdot (d, p) \to \cdot (f, p) \to \cdot (e, p) \to \cdot (g, p) \to \cdot (b, l(h)) \to \cdot (l(h)) \to \cdot (b, l) \to \cdot (l) \to \cdot (b, i) \xrightarrow{*} \cdot (b, o) \to \cdot (q) \to \cdot (l(h)) \to \langle 40, 78 \rangle (h) \to$
$\cdot (l) \to \cdot (b, l(h)) \to \cdot (b, h) \to \cdot (b, l) \to \cdot (c, h) \to \cdot (m(h)) \to \cdot (h, p) \to \cdot (m) \to \cdot (m(i)) \to \cdot (m(j)) \to \cdot (m(k)) \to \cdot (m(h)) \to \langle 40, 79 \rangle (h) \to$
$\cdot (m) \to \cdot (c, h) \to \cdot (b, h) \to \cdot (l(h)) \to \cdot (h, p) \to \cdot (l) \to \cdot (b, l(h)) \to \cdot (b, l) \to \cdot (b) \to \cdot (b, h, p) \to \cdot (b, p) \to \cdot (p) \to \cdot (c, h, p) \to \cdot (b, h, p) \to$
$\langle 40, 80 \rangle (h, p) \to \cdot (b, p) \to \cdot (b, h) \to \cdot (p) \to \cdot (h) \to \cdot (c, h, p) \to \cdot (r(h, p)) \to \cdot (b) \to \cdot (c, p) \to \cdot (c, h) \to \cdot (c) \to \cdot (d, p) \to \cdot (r(p)) \to \cdot (f, p) \to$
$\cdot (e, p) \to \cdot (g, p) \to \cdot (b, l(h)) \to \cdot (b, h, p) \to \langle 40, 81 \rangle (h, p) \to \cdot (b, p) \to \cdot (b, h) \to \cdot (p) \to \cdot (b) \to \cdot (b, l) \xrightarrow{*} \cdot (b, s) \to \cdot (h) \to \cdot (c, h, p) \to$
$\cdot (r(h, p)) \to \cdot (b, l(h)) \to \cdot (l(h)) \to \cdot (b, h, p) \to \langle 40, 82 \rangle (h, p) \to \cdot (b, p) \to \cdot (b, h) \to \cdot (p) \to \cdot (b) \to \cdot (c, p) \to \cdot (d, p) \to \cdot (r(p)) \to$
$\cdot (f, p) \to \cdot (e, p) \to \cdot (g, p) \to \cdot (r) \to \cdot (r(e, p)) \to \cdot (r(g, p)) \to \cdot (r(h, p)) \to \cdot (r(e)) \to \cdot (r(h)) \to \cdot (h) \to \cdot (r(m(h))) \to \cdot (m(h)) \to$
$\cdot (r(m)) \to \cdot (m) \to \cdot (r(m(i))) \to \cdot (r(m(j))) \to \cdot (r(m(k))) \to \cdot (r(m(h))) \to \langle 40, 83 \rangle (m(h)) \to \cdot (r(h)) \to \cdot (r(m)) \to \cdot (r(m(h))) \to$
$\langle 40, 84 \rangle (m(h)) \to \cdot (r(h)) \to \cdot (r(m)) \to \cdot (r(m(h))) \to \langle 40, 85 \rangle (m(h)) \to \cdot (r(h)) \to \cdot (r(m)) \to \cdot (h) \to \cdot (r) \to \cdot (r(h, p)) \to \cdot (m) \to$
$\cdot (r(m(i))) \to \cdot (r(m(j))) \to \cdot (r(m(k))) \to \cdot (m(i)) \to \cdot (r(i)) \to \cdot (i) \to \cdot (r(m(h))) \to \langle 40, 86 \rangle (m(h)) \to \cdot (r(h)) \to \cdot (r(m)) \to \cdot (m) \to$
$\cdot (r) \to \cdot (r(m(i))) \to \cdot (r(m(j))) \to \cdot (r(m(k))) \to \cdot (r(m(h))) \to \langle 40, 87 \rangle (m(h)) \to \cdot (r(h)) \to \cdot (r(m)) \to \cdot (m) \to \cdot (r) \to \cdot (r(m(i))) \to$
$\cdot (r(m(j))) \to \cdot (r(m(k))) \to \cdot (m(i)) \to \cdot (r(i)) \to \cdot (h) \to \cdot (r(h, p)) \to \cdot (m(j)) \to \cdot (m(k)) \to \cdot (r(m(h))) \to \langle 40, 88 \rangle (m(h)) \to$
$\cdot (r(h)) \to \cdot (r(m)) \to \cdot (h) \to \cdot (m) \to \cdot (r) \to \cdot (r(h, p)) \to \cdot (h, p) \to \cdot (r(p)) \to \cdot (p) \to \cdot (r(e, p)) \to \cdot (r(g, p)) \to \cdot (e, p) \to \cdot (r(e)) \to$
$\cdot (r(m(i))) \to \cdot (r(m(j))) \to \cdot (r(m(k))) \to \cdot (r(m(h))) \to \langle 40, 89 \rangle (m(h)) \to \cdot (r(h)) \to \cdot (r(m)) \to \cdot (h) \to \cdot (m) \to \cdot (r) \to \cdot (r(h, p)) \to$
$\cdot (r(m(h))) \to \langle 40, 90 \rangle (m(h)) \to \cdot (r(h)) \to \cdot (r(m)) \to \cdot (m) \to \cdot (r) \to \cdot (r(m(i))) \to \cdot (r(m(j))) \to \cdot (r(m(k))) \to \cdot (h) \to \cdot (r(h, p)) \to$
$\cdot (c, h) \to \cdot (b, h) \to \cdot (t(h)) \to \cdot (l(h)) \to \cdot (h, p) \to \cdot (h, u) \to \cdot (t) \to \cdot (t(h, p)) \to \cdot (t(p)) \to \cdot (p) \to \cdot (t(v(p))) \to \cdot (c, h, p) \to \cdot (b, h, p) \to$
$\cdot (r(p)) \to \cdot (r(e, p)) \to \cdot (r(g, p)) \to \cdot (r(m(h))) \to \langle 40, 91 \rangle (m(h)) \to \cdot (r(h)) \to \cdot (r(m)) \to \cdot (m) \to \cdot (r) \to \cdot (r(m(i))) \to \cdot (r(m(j))) \to$
$\cdot (r(m(k))) \to \cdot (m(i)) \to \cdot (r(i)) \to \cdot (r(m(h))) \to \langle 40, 92 \rangle (m(h)) \to \cdot (r(h)) \to \cdot (r(m)) \to \cdot (h) \to \cdot (m) \to \cdot (r) \to \cdot (r(m(i))) \to$
$\cdot (r(m(j))) \to \cdot (r(m(k))) \to \cdot (r(h, p)) \to \cdot (h, p) \to \cdot (r(p)) \to \cdot (r(m(h))) \to \langle 40, 93 \rangle (m(h)) \to \cdot (r(h)) \to \cdot (r(m)) \to \cdot (r(m(h))) \to$
$\langle 40, 94 \rangle (m(h)) \to \cdot (r(h)) \to \cdot (r(m)) \to \cdot (h) \to \cdot (m) \to \cdot (r) \to \cdot (r(h, p)) \to \cdot (r(m(i))) \to \cdot (r(m(j))) \to \cdot (r(m(k))) \to \cdot (r(m(h))) \to$
$\langle 40, 95 \rangle (m(h)) \to \cdot (r(h)) \to \cdot (r(m)) \to \cdot (h) \to \cdot (r) \to \cdot (r(h, p)) \to \cdot (r(m(h))) \to \langle 40, 96 \rangle (m(h)) \to \cdot (r(h)) \to \cdot (r(m)) \to$
$\cdot (h) \to \cdot (m) \to \cdot (r) \to \cdot (r(h, p)) \to \cdot (r(m(i))) \to \cdot (r(m(j))) \to \cdot (r(m(k))) \to \cdot (h, p) \to \cdot (r(p)) \to \cdot (p) \to \cdot (c, h, p) \to \cdot (b, h, p) \to$
$\cdot (t(h, p)) \to \cdot (w(h, p)) \to \cdot (w(p)) \to \cdot (w(h)) \to \cdot (w) \to \cdot (w(l(h))) \to \cdot (w(h, u)) \to \cdot (w(v(p))) \to \cdot (v(p)) \to \cdot (w(v)) \to \cdot (w(v(p))) \to$
$\langle 40, 97 \rangle (v(p)) \to \cdot (w(p)) \to \cdot (w(v)) \to \cdot (p) \to \cdot (w) \to \cdot (w(h, p)) \to \cdot (w(v(p))) \to \langle 40, 98 \rangle (v(p)) \to \cdot (w(p)) \to \cdot (w(v)) \to \cdot (p) \to \cdot (w) \to$
$\cdot (w(h, p)) \to \cdot (v) \to \cdot (w(v(k))) \to \cdot (w(v(o))) \to \cdot (w(v(s))) \to \cdot (w(v(p))) \to \langle 40, 99 \rangle (v(p)) \to \cdot (w(p)) \to \cdot (w(v)) \to \cdot (w(v(p))) \to$
$\langle 40, 100 \rangle (v(p)) \to \cdot (w(p)) \to \cdot (w(v)) \to \cdot (p) \to \cdot (w) \to \cdot (w(h, p)) \to \cdot (v) \to \cdot (w(v(k))) \to \cdot (w(v(o))) \to \cdot (w(v(s))) \to \cdot (w(v(p))) \to$
$\langle 40, 101 \rangle (v(p)) \to \cdot (w(p)) \to \cdot (w(v)) \to \cdot (w(v(p))) \to \langle 40, 102 \rangle (v(p)) \to \cdot (w(p)) \to \cdot (w(v)) \to \cdot (w(v(p))) \to \langle 40, 103 \rangle (v(p)) \to$
$\cdot (w(p)) \to \cdot (w(v)) \to \cdot (w(v(p))) \to \langle 40, 104 \rangle (v(p)) \to \cdot (w(p)) \to \cdot (w(v)) \to \cdot (w(v(p)), x) \to \cdot (v(p), x) \to \cdot (w(p), x) \to \cdot (w(v, x) \to$
$\cdot (v) \to \cdot (w) \to \cdot (w(v(k))) \to \cdot (w(v(o))) \to \cdot (w(v(s))) \to \cdot (p) \to \cdot (w(h, p)) \to \cdot (w(v(p))) \to \langle 40, 105 \rangle (v(p)) \to \cdot (w(p)) \to \cdot (w(v)) \to$
$\cdot (w(v(p)), x) \to \cdot (p) \to \cdot (w) \to \cdot (w(p), x) \to \cdot (w(h, p)) \to \cdot (h, p) \to \cdot (w(h)) \to \cdot (w(h, p), x) \to \cdot (v) \to \cdot (w(v, x) \to \cdot (w(v(k))) \to$
$\cdot (w(v(o))) \to \cdot (w(v(s))) \to \cdot (v(p), x) \to \cdot (p, x) \to \cdot (v, x) \to \cdot (c, v(p), x) \to \cdot (b, v(p), x) \to \cdot (d, v(p), x) \to \cdot (t(v(p)), x) \to \cdot (b, p, x) \to$
$\cdot (b, v, x) \to \cdot (b, v(p)) \to \perp$

Fig. 6. A simulation result: one typical terminating trajectory for classical attenuation regulation of *trpE* genes in *E. coli*. **Notations:** $\to$ means one rewriting; $\xrightarrow{*}$ means several similar rewritings; repeated window positions (e.g. repetitions of $\langle 40, 51 \rangle$) are replaced by a $\cdot$ symbol; $\perp$ means termination. There are 24 helices, denoted by letters from *a* to *x*.

model checking framework, the rewriting rules are not probabilistic. This work constitutes the first step towards the extension of the regular model checking framework with probabilistic rewriting rules. This would allow for example the analysis of probabilistic parameterized systems and probabilistic multithreaded programs.

Rewriting systems have also been used in articles [BIK06, BCC+03] to model chemical reactions. Compared to our work, the rewriting systems considered in [BIK06, BCC$^+$03] are not probabilistic. Moreover, these works consider the modeling of chemical reactions whereas we consider modeling of RNA secondary structure.

Finally, probabilistic term rewriting systems have also been considered in [BH03, BK02, KSMA03]. But in these works, the symbols are of fixed arities and the terms are ordered, whereas in our framework, the symbols have arbitrary arities and the terms are not ordered. Moreover, as far as we know, this is the first time that probabilistic term rewriting systems are used to model attenuation regulation.

8 Conclusions and Perspectives

We have established that the framework of probabilistic term rewriting systems provides compact and structured description of detailed models of RNA regulation.

We intend to continue exploration of this framework. The most important task consists in the development of adequate data structures and algorithms, as well as approximation and abstraction methods for analysis of this kind of models. The next step would be a massive computational experimentation, the biological interpretation of results and validation of results by real biological data.

Acknowledgments

The authors are thankful to Sergey Pirogov, Konstantin Gorbunov and Lev Rubanov for a valuable discussion. Lev Rubanov has also provided assistance in use of the RNAMODEL tool. Oleg Zverkov has helped us in preparing computer graphics for this article.

References

[AJMd02]   Abdulla, P.A., Jonsson, B., Mahata, P.: Regular tree model checking. In: Brinksma, E., Larsen, K.G. (eds.) CAV 2002. LNCS, vol. 2404, pp. 555–568. Springer, Heidelberg (2002)

[ALdR05]   Abdulla, P.A., Legay, A., d'Orso, J., Rezine, A.: Simulation-based iteration of tree transducers. In: Halbwachs, N., Zuck, L.D. (eds.) TACAS 2005. LNCS, vol. 3440, pp. 30–44. Springer, Heidelberg (2005)

[BCC$^+$03]   Bournez, O., Côme, G.-M., Conraud, V., Kirchner, H., Ibanescu, L.: A rule-based approach for automated generation of kinetic chemical mechanisms. In: RTA'03. LNCS, vol. 2706, pp. 30–45. Springer, Heidelberg (2003)

[BH03]   Bournez, O., Hoyrup, M.: Rewriting logic and probabilities. In: Nieuwenhuis, R. (ed.) RTA 2003. LNCS, vol. 2706, pp. 61–75. Springer, Heidelberg (2003)

[BIK06]   Bournez, O., Ibanescu, L., Kirchner, H.: From chemical rules to term rewriting. In: 6th International Workshop on Rule-Based Programming, vol 147(1) of ENTCS, pp. 113–134 (2006)

[BK02]   Bournez, O., Kirchner, C.: Probabilistic rewrite strategies: Applications to ELAN. In: Tison, S. (ed.) RTA 2002. LNCS, vol. 2378, pp. 252–266. Springer, Heidelberg (2002)

[BT02]   Bouajjani, A., Touili, T.: Extrapolating tree transformations. In: Brinksma, E., Larsen, K.G. (eds.) CAV 2002. LNCS, vol. 2404, pp. 539–554. Springer, Heidelberg (2002)

[BT03]   Bouajjani, A., Touili, T.: Reachability analysis of process rewrite systems. In: Pandya, P.K., Radhakrishnan, J. (eds.) FST TCS 2003: Foundations of Software Technology and Theoretical Computer Science. LNCS, vol. 2914, pp. 73–87. Springer, Heidelberg (2003)

[FFHS00]   Flamm, C., Fontana, W., Hofacker, I.L., Schuster, P.: RNA folding at elementary step resolution. RNA 6(3), 325–338 (2000)

[KMM$^+$01]   Kesten, Y., Maler, O., Marcus, M., Pnueli, A., Shahar, E.: Symbolic model checking with rich assertional languages. Theoretical Computer Science 256, 93–112 (2001)

[KSMA03]   Kumar, N., Sen, K., Meseguer, J., Agha, G.: A rewriting based model for probabilistic distributed object systems. In: FMOODS'03. LNCS, vol. 2884, pp. 32–46. Springer, Heidelberg (2003)

[LPRS07]   Lyubetsky, V., Pirogov, S., Rubanov, L., Seliverstov, A.: Modeling classic attenuation regulation of gene expression in bacteria. Journal of Bioinformatics and Computational Biology, 5(1), in print (2007)

[LRSP06]   Lyubetsky, V., Rubanov, L., Seliverstov, A., Pirogov, S.: Model of gene expression regulation in bacteria via formation of RNA secondary structures. Molecular Biology 40(3), 440–453 (2006)

[RNA]   RNAmodel. Model of RNA-related regulation in bacteria, http://lab6. iitp.ru/rnamodel/rnamodee.html

[SB91]   Singer, M., Berg, P.: Genes & genomes. University Science Books Mill Valley, Calif (1991)

[Tou05]   Touili, T.: Dealing with communication for dynamic multithreaded recursive programs. In: 1st VISSAS workshop, IOS Press, Amsterdam (2005)

[Zuk03]   Zuker, M.: Mfold web server for nucleic acid folding and hybridization prediction. Nucleic Acids Research 31(13), 3406–3415 (2003)

Glucose-Insulin Control of Type1 Diabetic Patients in H_2/H_∞ Space Via Computer Algebra

Levente Kovács[1,*] and Béla Paláncz[2]

[1] Department of Control Engineering and Information Technology
Budapest University of Technology and Economics, Magyar tudósok krt. 2., H-1117,
Budapest, Hungary
lkovacs@iit.bme.hu
[2] Department of Photogrammetry and Geoinformatics
Budapest University of Technology and Economics, Muegyetem rkp. 3., H-1111,
Budapest, Hungary
palancz@epito.bme.hu

Abstract. This article presents the H_2/H_∞ control (disturbance rejection LQ method) of the Bergman minimal model [2] for Type1 diabetic patients under intensive care using computer algebra. To design the optimal controller, the disturbance rejection LQ method based on the minimax differential game is applied. The critical, minimax value of the scaling parameter γ_{crit} is determined by using the Modified Riccati Control Algebraic (MCARE) equation employing reduced Gröbner basis solution on rational field. The numerical results are in good agreement with those of the Control Toolbox of MATLAB. It turned out, that in order to get positive definite solution stabilizing the closed loop, γ should be greater than γ_{crit}. The obtained results are compared with the classical LQ technique on the original non-linear system, using a standard meal disturbance situation. It is also demonstrated that for $\gamma \gg \gamma_{crit}$, the gain matrix approaches the traditional LQ optimal control design solution. The symbolic and numerical computations were carried out with *Mathematica* 5.2, and with the CSPS Application 2, as well as with MATLAB 6.5.

1 Introduction

Diabetes mellitus is one of the most serious diseases which need to be artificially regulated. The newest statistics of the World Health Organization (WHO) predate an increase of adult diabetes population from 4% (in 2000, meaning 171 million people) to 5.4% (366 million worldwide) by the year 2030 [35]. This warns that diabetes could be the "disease of the future".

In many biomedical systems, external controller provides the necessary input, because the human body could not ensure it. The outer control might be partially

* Supported by the Hungarian National Scientific Research Foundation (OTKA T042990, F046726, T69055) and by the Hungarian National Office for Research and Technology (RET-04/2004).

H. Anai, K. Horimoto, and T. Kutsia (Eds.): AB 2007, LNCS 4545, pp. 95–109, 2007.

or fully automatized. The self-regulation has several strict requirements, but once it has been designed it permits not only to facilitate the patient's life suffering from the disease, but also to optimize (if necessary) the amount of the used dosage.

The blood-glucose control is one of the most difficult control problems to be solved in biomedical engineering. One of the main reasons is that patients are extremely diverse in their dynamics and in addition their characteristics are time-varying. Due to the inexistence of an outer control loop, replacing the partially or totally deficient blood-glucose-control system of the human body, patients are regulating their glucose level manually. Based on the measured glucose levels (obtained from extracted blood samples), they decide on their own what is the necessary insulin dosage to be injected. Although this process is supervised by doctors (diabetologists), mishandled situations often appear. Hyper- (deviation over the basal glucose level) and hypoglycemia (deviation under the basal glucose level) are both dangerous cases, but on short term the latter is more dangerous, leading for example to coma.

Starting from the late Sixties lot of researchers investigate the problem of the glucose-insulin interaction and control. The closed-loop glucose regulation as it was several times formulated [13,32,33], requires three components and the current paper focuses on the last component of them:

- Glucose sensor (already realized even for 10 min. frequent readings: MiniMed [27], Glucowatch [11]);
- Insulin pump, for insulin injection (MiniMed [28], Disetronic [8]);
- Control algorithm, which based on the glucose measurements, is able to determine the necessary insulin dosage.

To design an appropriate control, an adequate model is necessary. In the last 50 years several models appeared. The mostly used and also the simplest one proved to be the minimal model of Bergman, [1,2], but its shortcoming is its big sensitivity to variance in the parameters. Henceforward, the plasma insulin concentration must be known as a function of time. Therefore, extensions of this minimal model have been proposed [24,9,29,7] trying to capture the changes in patient dynamics of the glucose-insulin interaction, particularly with respect to insulin sensitivity. Other more general, but more complicated models appeared in the literature, [15,34].

Regarding the control strategies applied, the palette is very wide [31], starting form classical control strategies like PID control [5], optimal control [17], to the modern control techniques like adaptive control [24], neuro-fuzzy algorithms [6,16], model predictive control [13,14,25], but also post-modern control strategies, like H_∞ control [32,33,21], H_2/H_∞ control [22,20], μ-synthesis [18], Linear Parameter Varying (LPV) technique [19].

The investigations of [13] discourage the use of a low complexity control such as PID, if high level of performance is desired. However, probably the best way to approach the problem is to consider the system model and the applied control technique together [31,26].

This article presents the H_2/H_∞ control (disturbance rejection LQ method) of the Bergman minimal model [2] in symbolic way using Mathematica 5.2 together with its Control System Professional Suite (CSPS). The paper is structured firstly of a brief description of the model, which then is reduced to a two-state model eliminating the unmeasurable slow state variable. Secondly, the LQ and disturbance rejection LQ (minimax control, LQR) methods are presented and the symbolic computations are performed to determine a general solution of the considered model for these control methods. Finally, the obtained results are compared on a standard meal disturbance situation for the original non-linear system and checked under MATLAB 6.5 as well.

2   Preliminaries

2.1   Model Equations

Several different models of diabetic systems exist in the literature including, for example, the very detailed 21st-order metabolic model of Sorensen [34]. However, to have a system that on one hand, can be readily handled from the point of view of control design, but on the other hand is able to represent properly the biological process, we consider the three-state minimal patient model of Bergman [2]:

$$\dot{G}(t) = -p_1 G(t) - X(t)(G(t) + G_B) + h(t),$$
$$\dot{X}(t) = -p_2 X(t) + p_3 Y(t),$$
$$\dot{Y}(t) = -p_4(Y(t) + Y_B) + \frac{i(t)}{V_L},$$

where the three state variables are:

- $G(t)$ - Plasma glucose deviation, [mg/dL];
- $X(t)$ - Remote compartment insulin utilization, [1/min];
- $Y(t)$ - Plasma insulin deviation, [mU/dL].

The control variable ($i(t)$) is the exogenous insulin infusion rate ($[\frac{mU}{min}]$), while the disturbance ($h(t)$) represents the exogenous glucose infusion rate ($[\frac{mg}{dL\,min}]$). The physical parameters are:

- G_B - Basal glucose level, [mg/dL];
- Y_B - Basal insulin level, [mU/dL];
- V_L - Insulin distribution volume, [dL].

The model parameters are p_1 ($\frac{1}{min}$), p_2 ($\frac{1}{min}$), p_3 ($\frac{dL}{mU\,min^2}$), p_4 ($\frac{1}{min}$). The values of the model and physical parameters are from Furler et al. [10]: $p_1 = \frac{28}{1000}$, $p_2 = \frac{25}{1000}$, $p_3 = \frac{13}{100000}$, $p_4 = \frac{5}{54}$, $G_B = 110$, $Y_B = 1.5$, $V_L = 120$.

2.2   Model Reduction

Assuming that the unmeasurable variable $X(t)$, remote compartment insulin utilization, is a slow variable [23], namely $\dot{X}(t) \approx 0$, then $X(t) = \frac{p_3}{p_2}Y(t)$ can be eliminated by substituting it into the first equation. Now our reduced model becomes:

$$\dot{G}(t) = -p_1 G(t) - \frac{p_3}{p_2}Y(t)(G(t) + G_B) + h(t),$$

$$\dot{Y}(t) = -p_4(Y(t) + Y_B) + \frac{i(t)}{V_L}.$$

2.3   Linearized Model

To design optimal control, the first step is the linearization of the nonlinear model. In order to linearize our model, the *Control System Application* of *Mathematica* should be loaded, [30]:

    << ControlSystems'

The linearization as control object in state space is carried out at the equilibrium point, namely at (G_0, Y_0, h_0, i_0). The obtained linearized model in its symbolic form is:

$$\dot{x}(t) = \begin{pmatrix} -p_1 - \dfrac{p_3 Y_0}{p_2} & -\dfrac{(G_0 + G_B)p_3}{p_2} \\ 0 & -p_4 \end{pmatrix} x(t) + \begin{pmatrix} 1 & 0 \\ 0 & \dfrac{1}{V_L} \end{pmatrix} u(t),$$

$$y(t) = \begin{pmatrix} 1 & 0 \\ 0 & 1 \end{pmatrix} x(t) + \begin{pmatrix} 0 & 0 \\ 0 & 0 \end{pmatrix} u(t),$$

where $x(t)$, $u(t)$ and $y(t)$ represent the general state, input and output vectors of the state-space model.

While the steady state values are $h_0 = 0$, $i_0 = p_4 Y_B V_L = 16.667$, $G_0 = 0$, $Y_0 = 0$, the linearized system becomes:

$$\dot{x}(t) = \begin{pmatrix} -p_1 & -\dfrac{G_0 p_3}{p_2} \\ 0 & -p_4 \end{pmatrix} x(t) + \begin{pmatrix} 1 & 0 \\ 0 & \dfrac{1}{V_L} \end{pmatrix} u(t),$$

$$y(t) = \begin{pmatrix} 1 & 0 \\ 0 & 1 \end{pmatrix} x(t) + \begin{pmatrix} 0 & 0 \\ 0 & 0 \end{pmatrix} u(t).$$

The linearized system proved to be controllable:

    Controllable[ControlObjectSS]
    True

3   Classical LQ Method

It is well-known, that the dynamic of a linear time invariant system can be described in the following way:

$$\dot{x}(t) = Ax(t) + Bu(t),$$
$$y(t) = Cx(t) + Du(t).$$

where A, B, C and D are constant matrices.

Using the classical LQ control method, the requirement is to minimize the following quadratic cost functional:

$$J(u(t)) = \frac{1}{2} \int_0^\infty y^T(t)Qy(t) + u^T(t)Ru(t)dt.$$

In other words, LQ attempts to find an optimal control $u^*(t)$, $t \in [0, \infty]$ based on the CARE (Control Algebraic Riccati Equation) such that

$$J(u^*(t)) \leq J(u(t))$$

for all $u(t)$ on $t \in [0, \infty]$ under properly chosen R and Q matrices.

The first component of input vector $u(t)$, the exogenous glucose ($h(t)$) stands for disturbance, therefore its effect on the output should be minimized. Meanwhile, in terms of material, it is cheaper than insulin. As a result the disturbance (glucose) should be overweighted in the discussion of Q and R matrix (both are 2x2 matrices as the reduced model has two inputs and two outputs), while the "expensive" insulin should be underweighted. Consequently, R_{11} should be considerable greater than R_{22}. On the other hand, in the objective function the portion of the injected insulin is important and the values of the state variables play not an important role. Therefore, we choose the following matrices:

$$R = \begin{pmatrix} 1000 & 0 \\ 0 & \dfrac{1}{1000} \end{pmatrix}, \quad Q = \begin{pmatrix} \dfrac{1}{1000} & 0 \\ 0 & \dfrac{1}{1000} \end{pmatrix}.$$

As a result the optimal gain matrix is

`KLQ=LQRegulatorGains[ControlObjectSS/.NumericalValues,Q,R]`

$$\begin{pmatrix} 0.0000139581 & -0.0000560483 \\ -0.467069 & 2.62107 \end{pmatrix} \approx \begin{pmatrix} 0 & 0 \\ -0.467069 & 2.62107 \end{pmatrix}$$

In order to check this result, the KLQ was calculated with MATLAB (`lqr` command) too and the same result was obtained:

$$\begin{pmatrix} 0 & 0 \\ -0.4671 & 2.6211 \end{pmatrix}$$

4 Disturbance Rejection LQ Control (Minimax Control)

The disturbance rejection LQ method represents a generalization of the classical LQ method and is based on the minimax criteria, where the system dynamics is generally described as before. However, now the disturbance $d(t)$ is separated from active control input $\bar{u}(t)$ and could be considered as unmeasurable, namely:

$$\dot{x}(t) = Ax(t) + B\bar{u}(t) + Ld(t),$$
$$y(t) = Cx(t) + Du(t).$$

Therefore, in this case the quadratic cost functional will be modified with the disturbance explicitly [36]:

$$J(u(t)) = \frac{1}{2} \int_0^\infty y^T(t)y(t) + \bar{u}^T(t)\bar{u}(t) - \gamma^2 d^T(t)d(t)dt.$$

Now, the disturbance, while it appears with negative sign, attempts to maximize the cost, while we want to find a control $(\bar{u})(t)$ that minimizes the maximum cost achievable by the disturbance (by worst case disturbance). This is a case of so-called "worst-case" design and leads to the formulation of a min-max differential game [3]:

$$\max_{d(t)} J\left(\bar{u}(t), d(t)\right) \rightarrow \min_{\bar{u}(t)} J(u(t), d(t))$$

where $\bar{u}(t)$ and $d(t)$ are satisfying the state equation. It can be demonstrated that the unique solution of the differential game, $\{\bar{u}^*(t),\ d^*(t)\}$ exists and satisfies the saddle point condition,

$$J\left(\bar{u}^*(t), d(t)\right) \leq J\left(\bar{u}(t), d(t)\right) \leq J\left(\bar{u}(t), d^*(t)\right)$$

where $\bar{u}^*(t)$ is the optimal control and $d^*(t)$ is the worst-case disturbance. These functions can be computed as:

$$\bar{u}^*(t) = -B^T Px(t),$$
$$d^*(t) = \frac{1}{\gamma^2} L^T x(t),$$

where P is the positive definite symmetric solution of the Modified Control Algebraic Riccati Equation (MCARE):

$$PA + A^T P + C^T C - P\left(BB^T - \frac{1}{\gamma^2}LL^T\right)P = 0. \tag{1}$$

It was demonstrated [36], but also one can easily conclude from (1) that for big values of γ the disturbance rejection LQ problem becomes the classical LQ one.

5   Solution of MCARE Using Computer Algebra

Based on Fig. 1 the input matrix of the system B, should be divided in two parts: one for the control input, namely the insulin (this will be from now on the B matrix) and one for the glucose disturbance (L):

$$B = \begin{pmatrix} 0 & 0 \\ 0 & 1 \\ & \frac{1}{V_L} \end{pmatrix}, \quad L = \begin{pmatrix} 1 & 0 \\ 0 & 0 \end{pmatrix}.$$

We are looking for the symmetric solution matrix of the modified Riccati equations (P) in the form $\begin{pmatrix} p_{1,1} & p_{1,2} \\ p_{2,1} & p_{2,2} \end{pmatrix}$. The complete step-by-step solution in *Mathematica* notebook form can be found on the Wolfram Research site [30]. Due to the page limitations of this article, we focus only on the main steps.

Due to the fact that the solution matrix of the Riccati equation is a symmetric matrix, three equations should be solved, where γ represents the parameter:

$$1 - 2p_1 p_{1,1} + \frac{p_{1,1}^2}{\gamma^2} - \frac{p_{1,2}^2}{V_L^2} = 0,$$

$$p_{1,1}\left(-\frac{G_B p_3}{p_2} + \frac{p_{1,2}}{\gamma^2}\right) - \frac{p_{1,2}((p_1 + p_4)V_L^2) + p_{2,2}}{V_L^2} = 0,$$

$$1 - \frac{2G_B p_3 p_{1,2}}{p_2} + \frac{p_{1,2}^2}{\gamma^2} - p_{2,2}\left(2p_4 + \frac{p_{2,2}}{V_L^2}\right) = 0.$$

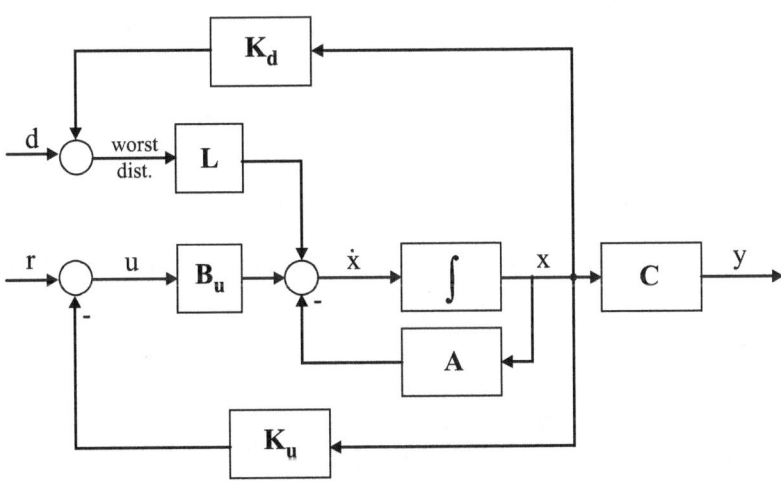

Fig. 1. General representation of the disturbance rejection LQ (minimax) control method

The critical solution of this system belongs to the critical value of the parameter γ. Crossing with γ this critical value, the solution which was real becomes imaginary and vice versa. This critical solution numerically is an ill-conditioned problem, therefore, the computation was carried out using reduced Gröbner basis [4] on rational field, which provides solution for the unknown variables with infinite precision avoiding round-off errors. Substituting the values of the system parameters in rational form and determining the solution for the $p_{1,1}$, the result was obtained as a fourth order monomial with γ as parameter [30]. Creating a function providing the first root of this monomial, $Im(p_{1,1}(\gamma))$, the critical value is the smallest nonzero positive root, namely

$$\gamma_{crit} = \min_{\gamma > 0} \left(Im(p_{1,1}(\gamma)) \right) = 0)$$

see the corner point on Fig. 2. A lower bound for this critical value can be computed using NMinimize function of *Mathematica*:

$$\gamma_{crit} = 17.0862,$$

see from Fig. 2. Then, starting from this lower bound using step size Δ, the location of critical value can be approximated with error $\epsilon \leq \Delta$. The critical value with ten significant digits is:

$$\gamma_{crit} = 17.11742594,$$

see Fig. 3. Employing MATLAB minimax control design, by using the interval halving technique together with the built-in care() function, we have got the same result, namely, the optimal, minimal γ, $\gamma_{min} = 17.11743$ is equal to the critical value γ_{crit} computed above. This value of γ means that this is the value where the effect of the disturbance is maximized, or by other words becomes the worst-case disturbance. The corresponding controller for this worst-case (obtained with *Mathematica* and checked by MATLAB) is:

$$KLQrej = \begin{pmatrix} -0.5785 & 4.2671 \\ 10.4192 & -74.5880 \end{pmatrix} \tag{2}$$

This result clearly shows that there is a compensation on the disturbance (glucose) part too, which in our case has no physical meaning, because in this way the result can be interpreted as if one improves the glucose control by exhausting glucose (some kind of negative injection, or glucose extraction form the body) from the system (see the structure of matrix B in MCARE). In real situation only insulin injection is possible. Therefore, for the considered application only the second row of the controller's matrix can be used.

However, if we are considering as feedback only the control input effect, at this value $\gamma = \gamma_{min} = \gamma_{crit}$ there is no positive definite solution, namely solving numerically the Ricatti equation and choosing from the two solutions the one also chosen by MATLAB, $P = \begin{pmatrix} -169.446 & 1249.84 \\ 1249.84 & -8947.39 \end{pmatrix}$, the eigenvalues of P are $\{-9125.15, 5.05453\}$.

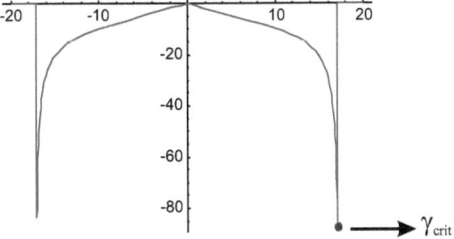

Fig. 2. The imaginary part of $p_{1,1}$ depending on γ

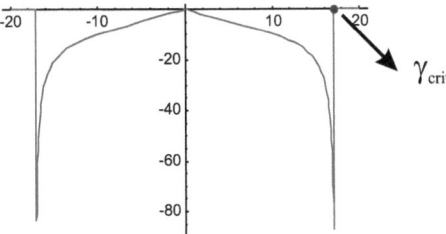

Fig. 3. Proper solution for γ_{crit}, realizing the worst-case disturbance effect

Although this solution satisfies the modified Riccati equation (the error is $7.404 * 10^{-16}$) [30], based on the earlier remarks the gain matrix should be:

$$\text{KLQrej} = \begin{pmatrix} 0 & 0 \\ 10.4192 & -74.5880 \end{pmatrix}.$$

With this controller the system remains an unstable one (the poles are $\{0.0669, -0.1062\}$), demonstrating that in the worst-case the effect of the disturbance should also be considered. It means that in our case, if we want to control the system only by the control input (because this is the physiologically interpretable case), this control is not stabilizing the system. However, let us increasing the value of γ up to the value providing positive definite solution. In this way we would like to investigate the situation if in the close neighborhood of the obtained (worst-case) critical γ is there a convenient solution, which also gave us better results than LQ did and in this case even the physiological case is interpretable. We have obtained that the positive definite solution could be obtained by increasing somewhat the value of γ by 0.25 % of γ_{crit}, so:

$$\gamma_{\text{crit}} = 17.1602. \tag{3}$$

In [30] it is presented how sensitive the solution is. Visually, the two values of γ are represented by the same point, but in this case the closed-loop system's eigenvalues, $\{-3.4739, -0.1105\}$, represent a stable system, where the control matrix:

$$KLQrej = \begin{pmatrix} 0 & 0 \\ -59.619 & 415.648 \end{pmatrix} \qquad (4)$$

and the modified Ricatti equation is also satisfied (the error is $5.9068 * 10^{-15}$).

6   Discussion and Simulation Results

6.1   Comparing Classical LQ and Disturbance Rejection LQ Control Design

The gain matrix KLQRej, provided by the disturbance rejection LQ design (4), depends on the actual value of the scaling parameter γ. In case of $\gamma \to \infty$, we get the gain matrix designed by LQ method [20,22], namely,

$$\lim_{\gamma \to \infty} KLQrej = KLQ \qquad (5)$$

To demonstrate this we have considered $\gamma = 100\gamma_{crit}$, and we repeated the computations for solving the Riccati equation

$$P = \begin{pmatrix} 13.9606 & -56.0634 \\ -59.0634 & 314.623 \end{pmatrix}.$$

The error in this case is $4.665 * 10^{-17}$ and the solution is once again in good agreement with MATLAB. Having the solution of the Riccati equation, the corresponding gain matrix (representing the controller) can be computed as:

$$KLQ = \begin{pmatrix} 0 & 0 \\ -0.4671 & 2.6218 \end{pmatrix}.$$

This result is very similar to the LQ control solution (see Section 3), demonstrating the theoretical concepts (5).

6.2   Meal Disturbance

The performance of the control is tested by using a standard meal disturbance with about six hour duration, modeled by Lehmann and Deutsch [23]. The following data pairs are considered:

$$time(t)[\min] = (0, 25, 50, 75, 100, 125, 150, 175, 200, 225, 250,$$
$$275, 300, 325, 350, 375, 400, 450, 1000)$$
$$glucose(h(t))[\mathrm{mg}/(\mathrm{dL\,min})] = (0, 0.185, 0.495, 0.765, 0.975, 1.15, 1.07, 0.82,$$
$$0.575, 0.335, 0.225, 0.145, 0.098, 0.06, 0.035,$$
$$0.02, 0.01, 0.005, 0)$$

The interpolated meal disturbance function can be seen in Fig. 4.

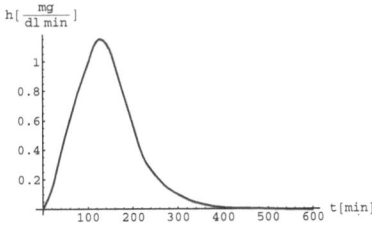

Fig. 4. Exogenous glucose infusion, $h(t)$

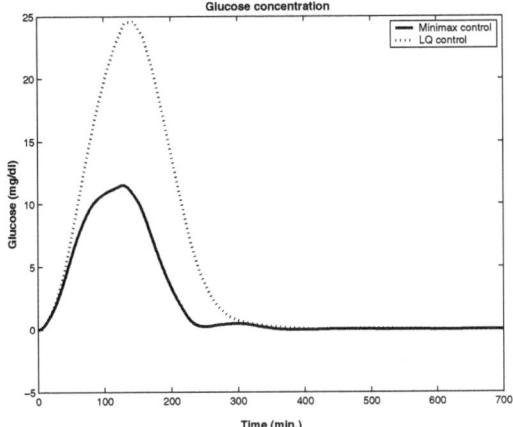

Fig. 5. The comparison of blood glucose concentration, $G(t)$ in case of LQ and disturbance rejection LQ control

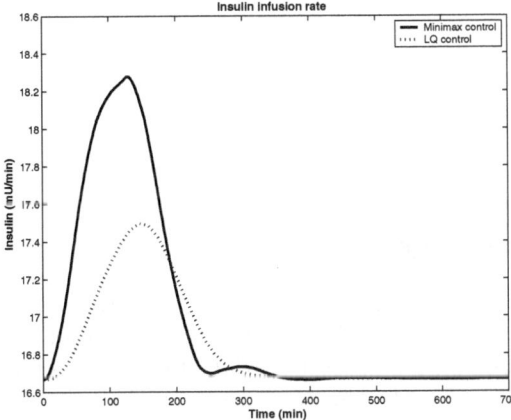

Fig. 6. The corresponding insulin infusion rate, $i(t)$ in case of LQ and disturbance rejection LQ control

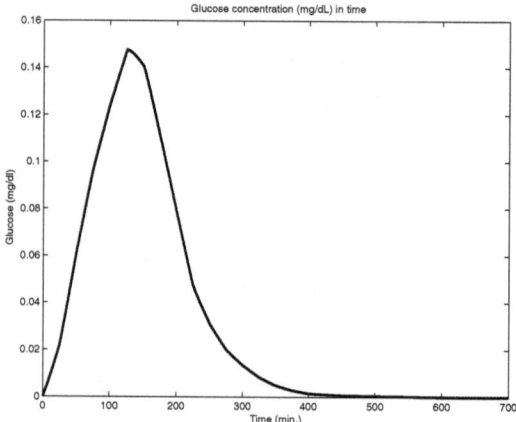

Fig. 7. The controlled dynamics of blood glucose concentration, $G(t)$ in case of disturbance rejection LQ control

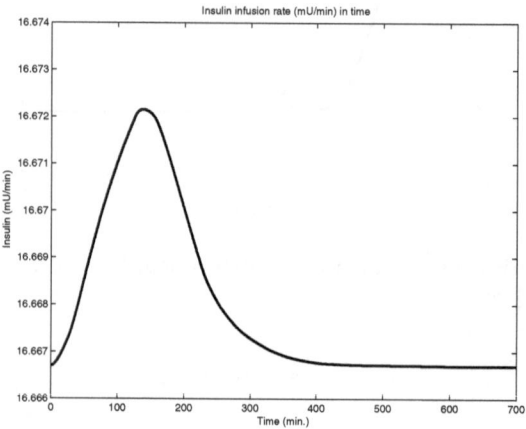

Fig. 8. The corresponding insulin infusion rate, $i(t)$ in case of disturbance rejection LQ control

6.3   Simulation Results

Although, the controller design was carried out for the reduced 2-states linear model described in section 2.2, the system is simulated for the original 3-states nonlinear model. Comparing the results of the two controls (classical LQ control (section 3) and minimax control for γ_{crit} corresponding to (3) and controller given in (4)), it can be seen that even in this considered case, the disturbance rejection LQ control is more efficient than the classical LQ, Fig. 5 and Fig. 6. However, just for comparison the best result is obtained in the case when the original, "worst-case" controller (given by (2)) is used, Fig. 7, Fig. 8. The considerably better result is due to that "negative injection" of glucose, what cannot be physiologically realized.

7   Conclusion

It turned out that the critical value of γ (where the value for imaginary parts of the solution of MCARE disappear) together with the physically realizable interpretation of a control system, will not ensure automatically a positive definite solution, if one needs a physically interpretable solution. This means that the minimax method determines the worst-case solution, but this depends on the concrete problem if it could be or could be not physically interpreted. However, even in this case it is possible to obtain a better solution than LQ does.

Furthermore, "blind" numerical solution of minimax problem can be misleading, and not always a positive definite solution is provided. This problem can be detected by using symbolic-numeric solution and increasing the value of γ up to the value, which already ensures positive eigenvalues of the Ricatti matrix. The other advantage of the symbolic-numeric solution (whenever it is possible) is its robustness concerning round-off errors.

However, this problem can be avoided by using another numerical technique for minimax control based on robust control on frequency domain proposed by Helton [12], and its application for glucose-insulin control [21]. It was also illustrated, that minimax control could provide better control quality than LQ does, see Fig. 7. Minimax control is interacting faster as well as employing higher infusion rate (in the considered case) than the classical control, see Fig. 8.

References

1. Bergman, B.N., Ider, Y.Z., Bowden, C.R., Cobelli, C.: Quantitive estimation of insulin sensitivity. American Journal of Physiology 236, 667–677 (1979)
2. Bergman, R.N., Philips, L.S., Cobelli, C.: Physiologic evaluation of factors controlling glucose tolerance in man. Journal of Clinical Investigation 68, 1456–1467 (1981)
3. Bokor, J.: Modern Control Theory II, PhD lecture notes (2003)
4. Buchberger, B.: Introduction to Gröbner Bases, In Gröbner Bases and Applications. London Mathematical Society Lecture Notes Series 25(1), 3–31 (1998)
5. Chee, F., Fernando, T.L., Savkin, A.V., van Heeden, V.: Expert PID control system for blood glucose control in critically ill patients. IEEE Engineering in Medicine and Biology 28(5), 189–195 (2003)
6. Dazzia, D., Taddei, F., Gavarini, A., Uggeri, E., Negro, R., Pezzarossa, A.: The control of blood glucose in the critical diabetic patient—a neuro-fuzzy method. Elsevier Journal of Diabetes and its Complications 15, 80–87 (2001)
7. de Gaetano, A., Arino, O.: Some considerations on the mathematical modeling of the intra-venous glucose tolerance test. Journal of Mathematical Biology 40, 136–168 (2000)
8. Disetronic. Insulin pump products. Available from the Web: http://www.disetronic-usa.com

9. Fernandez, M., Acosta, D., Villasana, M., Streja, D.: Enhancing parameter precision and the minimal modeling approach in Type1 diabetes. In: Proc. of 26th Ann. Int. Conf. of IEEE Eng. in Biomedicine Society (EMBC'04), pp. 797–800, San Francisco, USA (2004)
10. Furler, S.M., Kraegen, E.W., Smallwood, R.H., Chisolm, D.J.: Blood glucose control by intermittent loop closure in the basal mode: Computer simulation studies with a diabetic model. Diabetes Care 8, 553–561 (1985)
11. GlucoWatch. Available from the Web: http://www.glucowatch.com
12. Helton, J.W., Merino, O.: Classical Control Using H_∞ Methods—Theory, Optimization and Design. SIAM, 1998.
13. Hernjak, N., Doyle III, F.J.: Glucose control design using nonlinearity assessment techniques. AIChE Journal 51(2), 544–554 (2005)
14. Hovorka, R., Canonico, V., Chassin, L.J., Haueter, U., Massi-Benedetti, M., Orsini, M.: Nonlinear model predictive control of glucose concentration in subjects with Type1 diabetes. Physiological measurement 25, 905–920 (2004)
15. Hovorka, R., Shojaee-Moradie, F., Carroll, P.V., Chassin, L.J., Gowrie, I.J., Jackson, N.C., Tudor, R.S., Umpleby, A.M., Jones, R.H.: Partitioning glucose distribution/transport, disposal, and endogenous production during IVGTT. American Journal of Physiology – Endocrinology and Metabolism 282, 992–1007 (2002)
16. Ibbini, M.: A PI-fuzzy logic controller for the regulation of blood glucose level in diabetic patients. Journal of Medical Engineering and Technology 30(2), 83–92 (2006)
17. Ibbini, M.S., Masadeh, M.A., Bani Amer, M.M.: A semiclosed-loop optimal control system for blood glucose level in diabetics. Journal of Medical Engineering and Technology 28(5), 189–195 (2004)
18. Kovács, L., Kulcsár, B., Benyó, Z.: On the use of robust servo control in diabetes under intensive care. In: 3rd Romanian-Hungarian Joint Symposium on Applied Computational Intelligence (SACI'06), pp. 236–247, Timisoara, Romania (2006)
19. Kovács, L., Kulcsár, B., Bokor, J., Benyó, Z.: LPV fault detection of glucosinsulin system. In: 14th Mediterranean Conference on Control and Automation (MED'06), pp. TLA2–4, Ancona, Italy, electronic publication (2006)
20. Kovács, L., Paláncz, B., Almássy, Zs., Benyó, Z.: Optimal glucose-insulin control in $\mathcal{H}_2$ space. In: Proc. of 26th Ann. Int. Conf. of IEEE Eng. in Biomedicine Society (EMBC'04), pp. 762–765, San Francisco, USA (2004)
21. Kovács, L., Paláncz, B., Benyó, B., Török, L., Benyó, Z.: Robust blood-glucose control using Mathematica. In: Proc. of 28th Ann. Int. Conf. of IEEE Eng. in Biomedicine Society (EMBC'06), pp. 451–454, New York, USA, 2006.
22. Kovács, L., Paláncz, B., Benyó, Z.: Classical and modern control strategies in glucose-insulin stabilization. In: 16th IFAC World Congress, pp. 041–65, Prague, Czech Republic, electronic publication (2005)
23. Lehmann, E.D., Deutsch, T.A.: A physiological model of glucose-insulin interaction in Type1 diabetes mellitus. Journal of Biomedical Engineering 14, 235–242 (1992)
24. Lin, J., Chase, J.G., Shaw, G.M., Doran, C.V., Hann, C.E., Robertson, M.B., Browne, P.M., Lotz, T., Wake, G.C., Broughton, B.: Adaptive bolus-based setpoint regulation of hyperglycemia in critical care. In: Proc. of 26th Ann. Int. Conf. of IEEE Eng. in Biomedicine Society (EMBC'04), pp. 3463–3466, San Francisco, USA (2004)
25. Lynch, S.M., Bequette, B.W.: Model predictive control of blood glucose in Type1 diabetics using subcutaneous glucose measurements. In: American Control Conference (ACC'02) Anchorage, USA, vol. 5, pp. 4039–4043 (2002)

26. Makroglou, A., Li, J., Kuang, Y.: Mathematical models and software tools for the glucose—insulin regulatory system and diabetes: an overview. Elsevier Applied Numerical Mathematics 56(3–4), 559–573 (2006)
27. Mini Med. Continuous glucose monitoring system overview (CGMS). Available from the Web: http://www.minimed.com/professionals/products/cgms/
28. Mini Med. Real-time insulin pump and continuous glucose monitoring system. Available from the Web: http://www.minimed.com/products/insulinpumps/
29. Morris, H.C., O' Reilly, B., Streja, D.: A new biphasic minimal model. In Proc. of 26th Ann. Int. Conf. of IEEE Eng. in Biomedicine Society (EMBC'04), pp. 782–785, San Francisco, USA (2004)
30. Paláncz, B., Kovács, L.: Control in $\mathcal{H}_2/\mathcal{H}_\infty$ space via computer algebra. Available from the Web: http://library.wolfram.com/infocenter/MathSource/6628/
31. Parker, R., Doyle III, F.J., Peppas, N.A.: The intravenous route to blood glucose control. IEEE Engineering in Medicine and Biology 20(1), 65–73 (2001)
32. Parker, R.S., Doyle III, F.J., Ward, J.H., Peppas, N.A.: Robust $\mathcal{H}_\infty$ glucose control in diabetes using a physiological model. AIChE Journal 46(12), 2537–2549 (2000)
33. Ruiz-Velazquez, E., Femat, R., Campos-Delgado, D.U.: Blood glucose control for Type1 diabetes mellitus: A robust tracking $\mathcal{H}_\infty$ problem. Control Engineering Practice 12, 1170–1195 (2004)
34. Sorensen, J.T.: A Physiologic Model of Glucose Metabolism in Man and Its Use to Design and Assess Improved Insulin Therapies for Diabetes. PhD thesis, Dept. of Chemical Engineering, MIT, Cambridge, MA, USA (1985)
35. Wild, S., Roglic, G., Green, A., Sicree, R., King, H.: Global prevalence of diabetes—estimates for the year 2000 and projections for 2030. Diabetes Care 27(5), 1047–1053 (2004)
36. Zhou, K.: Robust and Optimal Control. Prentice Hall, New Jersey (1996)

Exact Parameter Determination for Parkinson's Disease Diagnosis with PET Using an Algebraic Approach

Hiroshi Yoshida[1], Koji Nakagawa[2], Hirokazu Anai[3], and Katsuhisa Horimoto[2]

[1] Faculty of Mathematics, Organization for the Promotion of Advanced Research, Kyushu University, Hakozaki 6-10-1, Higashi-ku, Fukuoka 812-8581 Japan
phiroshi@math.kyushu-u.ac.jp
[2] Computational Biology Research Centre (CBRC), National Institute of Advanced Industrial Science and Technology (AIST), Aomi 2-42, Koto-ku, Tokyo 135-0064, Japan
{nakagawa-koji,k.horimoto}@aist.go.jp
[3] IT Core Laboratories, FUJITSU LABORATORIES LTD./CREST, JST., Kamikodanaka 4-1-1, Nakahara-ku, Kawasaki 211-8588, Japan
anai@jp.fujitsu.com

Abstract. The mechanism of Parkinson's disease can be investigated at the molecular level by using radio-tracers. The concentration of dopamine in the brain can be observed by using a radio-tracer, 6-[^{18}F]fluorodopa (FDOPA), with positron emission tomography (PET), and the dopamine kinetics can be described as compartmental models for tissues of the brain. The models for FDOPA kinetics are solved explicitly, but the solution shows a complicated form including several convolutions over time domain. Owing to the complicated form of the solution, graphical analyses such as Logan or Patlak analysis have been utilized as conventional methods over past decades. Because some kinetic constants for Parkinson's disease are estimated in the graphical analyses with the slope or intercept of the line obtained under various assumptions, only a limited set of parameters have approximately been estimated. We have analysed the compartmental models by using the Laplace transformation of differential equations and by algebraic computation with the aid of Gröbner base constructions. We have obtained a rigorous solution with respect to the kinetic constants over the Laplace domain. Here, we first derive a rigorous solution for the parameters, together with a discussion about the merits of the derivation. Next, we describe a procedure to determine the kinetic constants with the observed time–radioactivity curves. Last, we discuss the feasibility of our method, especially as a criterion for diagnosing Parkinson's disease.

1 Introduction

Radio-tracers are often used to analyse metabolic systems in biomedical research. Usually the kinetics of metabolism are described as compartmental models, and kinetic constants are numerically estimated using the measurement of radio-tracers to diagnose the disease. In particular, the positron emission tomography (PET) has been developed to measure the details of metabolic events hitherto unavailable, and is especially useful to determine the kinetic constants to assist the diagnosis of various diseases.

H. Anai, K. Horimoto, and T. Kutsia (Eds.): AB 2007, LNCS 4545, pp. 110–124, 2007.

Parkinson's disease, which is due to abnormal levels of dopamine in the brain, is one of the diseases that can be diagnosed by radio-tracer measurement with PET, and by determination of kinetic constants in compartmental models for plasma and brain tissue [12]. There are two approaches to measuring the activity of radio-tracers. One is a combination of the measurement of a radio-tracer, L-3,4-dihydroxy-6-[18F]fluoro-phenylalanine (FDOPA) in the brain, and sampling the blood to measure the total activity of FDOPA (approach with blood sampling), and the other is the measurement of the FDOPA activity in two brain tissues (approach without blood sampling). In both approaches, the kinetics can be described as sets of compartmental models. Fortunately, a system of differential equations in the two sets of models can be solved explicitly , but unfortunately the solutions for estimating the kinetic constants are highly complicated. Indeed, the solutions are expressed by a few convolutions of complicated equations.

To overcome analytical difficulties in determining kinetic constants, there are two conventional methods of kinetic constant estimation in the compartmental models, Patlak Analysis [18, 19] and Logan Analysis [15]. In both methods, the combination of some parameters with various approximations is assumed to form a straight line as metabolism approaches an equilibrium. By plotting the observed data around the metabolic equilibrium (graphical analysis), the combined parameters can be estimated using the slope or intercept in the plotted line [14, 18].

By using graphical analysis, Parkinson's disease has been extensively studied in the two approaches with and without the blood sampling. In the approach using blood sampling, the kinetic constants with respect to plasma are calculated from the blood-sampling data, and then, using these constants, measurements for Parkinson's disease such as the constants describing FDOPA kinetics in brain tissue are calculated [10, 11, 12, 20]. In addition, Martin et al. [16] considered L-3,4-dihydroxy-6-[18F]fluoro-3-O-methylphenylalanine (3-OMFD) in compartmental models for FDOPA metabolism, because FDOPA is converted to 3-OMFD [1, 17], which has an influence on the total radioactivity observed in plasma and in the brain tissue by crossing the blood-brain barrier (BBB). In an approach without blood sampling, using the time–radioactivity curves of two distinct brain tissues, the constants for Parkinson's disease diagnosis are calculated [9, 14].

The diagnosis of Parkinson's disease with PET depends on graphical analysis, a simple presentation of the relationships between the kinetic constants of the FDOPA kinetics. However, the present analyses require further improvement for precise diagnoses. For example, graphical analysis using blood sampling is cumbersome, partly because the sampling requires a load to the patients, and partly because the separate measurement of radio-tracer from the blood provides an obstacle for the precise estimation of parameters such as the time delay and contamination of the samples in the tubing. Graphical analysis without blood sampling produces a highly complicated model and therefore requires various assumptions and approximations to estimate kinetic constants. Thus, the choice between the two approaches involves a trade-off between cumbersome blood sampling and difficult efficient parameter determination. Indeed, by considering the pitfalls described above, some methods have also designed by using different radio-tracers. For instance, Ichise et al. [8] proposed a method without blood sampling by using [123I]iodobenzofuran, and Lammertsma et al. [13] and

Logan et al. [14] proposed a method without blood sampling by using [^{11}C]raclopride, in which the cerebellum or cerebral cortex was used as a reference tissue and analysed as a single-tissue compartment. Although the pitfalls have partially been overcome, a rigorous solution has not yet been analysed.

In this paper, we propose radical deliverance from the aforementioned difficulty. We present an efficient method for determining kinetic constants for FDOPA kinetics with PET using an algebraic approach. The compartmental models are rigorously solved by the Laplace transformation of differential equations into algebraic equations, and by the following symbolic computation with the aid of Gröbner bases. Such usage of symbolic computation has overcome the analytical difficulties in the previous study [6], where general theory of compartmental models was derived over the Laplace domain for PET, but the analysis or determination of kinetic constants still required the system's equilibrium or steady state. In our method, by contrast, the derivation of a relationship between the observed concentrations without blood sampling by PET does not need any approximations and assumptions for the kinetic constants. Here, we first derive rigorous relationships between the parameters, and we discuss the merits of the derivation, in comparison with graphical analyses. Second, we describe an efficient procedure for determining the kinetic constants with observed time–radioactivity curves. Last, we discuss the feasibility of our method, especially as a criterion for diagnosing Parkinson's disease.

2   Model and Method

In this section, we introduce three compartmental models to describe the metabolism of the radio-tracer FDOPA and its metabolites with respect to two brain tissues and plasma. Differential equations corresponding to the kinetic model are derived, and the equations are transformed into a system of algebraic equations. Surprisingly, the rigorous solution is of a simple form over the Laplace domain. Finally, we describe a procedure to determine the kinetic constants of the models, which is performed over the Laplace domain.

2.1   Compartmental Model

Compartmental models (A) and (B) are introduced for the radio-tracer FDOPA and its metabolite 3-OMFD as shown in Figs. 1 (a) and (b). For simplicity, let A- and B-tissues denote tissues in which the radio-tracer kinetics can be described as shown in Figs. 1 (a) and (b), respectively. In the actual brain, A- and B-tissues correspond to the striatum (putamen/caudate) and the cerebellum/cerebral cortex in the brain [4, 7]. Furthermore, it is assumed that the relationships between plasma FDOPA, 3-OMFD, and extra-vascular 3-OMFD can be described as the compartmental model (C) as shown in Fig. 1 (c).

2.2   Kinetic Equations

According to the kinetic model in Fig. 1, the following system of differential equations has been obtained:

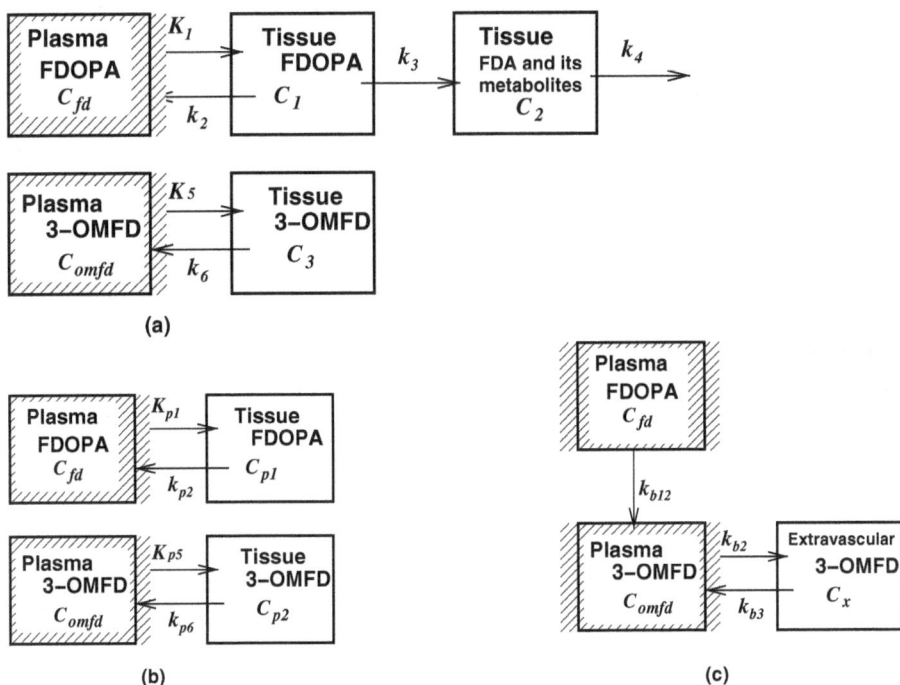

Fig. 1. Compartmental models for describing the radio-tracer kinetics in this paper, which were originally introduced by Huang et al. [7]. The shaded boxes represent the kinetics in plasma. (a) Model for A-tissue. Three separate compartments for tissue FDOPA, tissue FDA (and its metabolites), and tissue 3-OMFD. (b) Model for B-tissue which is the same as (a), except that there is no compartment for FDA. (c) Model for plasma FDOPA to 3-OMFD in the periphery of one compartment for plasma 3-OMFD and one for the extra-vascular pool.

Time (A-Tissue)

$$\begin{cases} \dfrac{dC_1}{dt} = K_1 C_{fd}(t-\tau)\theta(t-\tau) - (k_2 + k_3)C_1, \\[2mm] \dfrac{dC_2}{dt} = k_3 C_1 - k_4 C_2, \\[2mm] \dfrac{dC_3}{dt} = K_5 C_{omfd}(t-\tau)\theta(t-\tau) - k_6 C_3, \end{cases} \tag{1}$$

Time (B-Tissue)

$$\begin{cases} \dfrac{dC_{p1}}{dt} = K_{p1} C_{fd}(t-\tau)\theta(t-\tau) - k_{p2}C_{p1}, \\[2mm] \dfrac{dC_{p2}}{dt} = K_{p5} C_{omfd}(t-\tau)\theta(t-\tau) - k_{p6}C_{p2}, \end{cases} \tag{2}$$

Time (C-Blood (Plasma))

$$\begin{cases} \dfrac{dC_{omfd}}{dt} = k_{b12}C_{fd} - k_{b2}C_{omfd} + k_{b3}C_x, \\[3mm] \dfrac{dC_x}{dt} = k_{b2}C_{omfd} - k_{b3}C_x. \end{cases} \tag{3}$$

In the compartmental model (A), (B), and (C), every one of the initial values is assumed to be zero because of non-existence of the radio-tracers and their metabolites at starting time $t = 0$. However, there exists the time delay τ of the observed blood curve (C) relative to tissue measurements (A) and (B). That is, τ designates a difference between the starting times of (A), (B), and (C). This effect leads to the terms $C_{fd}(t - \tau)\theta(t - \tau)$ and $C_{omfd}(t - \tau)\theta(t - \tau)$, where $\theta(t)$ is the unit step function defined as follows:

$$\theta(t) = \begin{cases} 0 & (t < 0), \\ 1 & (t > 0). \end{cases}$$

The differential equations describing the A- and B-tissues and the C-blood kinetics models can be changed into the following equations over the Laplace domain:

Laplace (A)

$$\begin{cases} sL[C_1] = K_1 e^{-s\tau} L[C_{fd}] - (k_2 + k_3)L[C_1], \\ sL[C_2] = k_3 L[C_1] - k_4 L[C_2], \\ sL[C_3] = K_5 e^{-s\tau} L[C_{omfd}] - k_6 L[C_3], \end{cases} \tag{4}$$

Laplace (B)

$$\begin{cases} sL[C_{p1}] = K_{p1} e^{-s\tau} L[C_{fd}] - k_{p2} L[C_{p1}], \\ sL[C_{p2}] = K_{p5} e^{-s\tau} L[C_{omfd}] - k_{p6} L[C_{p2}], \end{cases} \tag{5}$$

Laplace (C)

$$\begin{cases} sL[C_{omfd}] = k_{b12} L[C_{fd}] - k_{b2} L[C_{omfd}] + k_{b3} L[C_x], \\ sL[C_x] = k_{b2} L[C_{omfd}] - k_{b3} L[C_x], \end{cases} \tag{6}$$

where $L[f]$ denotes the Laplace transformation of f. Thus, a system of differential equations is transformed into a system of corresponding algebraic equations.

2.3   Rigorous Solution

In the approach without blood sampling, the data observed using PET scanning are limited to the total radioactivities: $Cs(t) = C_1(t) + C_2(t) + C_3(t)$ and $Cc(t) = C_{p1}(t) + C_{p2}(t)$. Let $Cs(s)$ and $Cc(s)$ denote the Laplace transformations of $Cs(t)$ and $Cc(t)$, respectively. Then, the solution to the system of algebraic equations of Laplace (A),

(B), and (C) has been obtained, leading to a rigorous and simple relationship between $Cs(s)$ and $Cc(s)$ as follows:

$$\frac{Cs(s)}{Cc(s)} =$$

$$\frac{(s + k_{p2})(s + k_{p6})}{(s + k_2 + k_3)(s + k_4)(s + k_6)} \times \tag{7}$$
$$\frac{K_5 k_{b12}(s + k_2 + k_3)(s + k_4)(s + k_6)(s + k_{b2} + k_{b3}) + sK_1(s + k_3 + k_4)(s + k_6)(s + k_{b2} + k_{b3})}{K_{p5} k_{b12}(s + k_{b3})(s + k_{p2}) + sK_{p1}(s + k_{b2} + k_{b3})(s + k_{p6})}.$$

Thus, $Cs(s)/Cc(s)$ is a rational function in s to which symbolic methods such as Gröbner base computations can be applied, resulting in exact and efficient parameter determination.

2.4   Procedure to Determine the Kinetic Constants

Procedure overview. Fig. 2 shows an overview of the present procedure for determining the kinetic constants from radio-tracer activity data. The procedure is composed of two parts. First, we fit the observed radioactivity curves by a series of exponentials, and then the fitted series of exponentials are transformed into the corresponding algebraic equations by the Laplace transformation. Second, the kinetic constants in the rigorous equation (7) are determined using an algebraic approach. The details of the above procedure are described below.

Laplace transformation of the observed data. We need a Laplace transformation of the observed data because we perform parameter determination over the Laplace domain. Let $Cso(t)$ and $Cco(t)$ denote the observed data in A- and B-tissues, respectively. By using non-linear regression, $Cso(t)$ and $Cco(t)$ are expressed in terms of a series of exponentials according to [3] as follows:

$$\begin{cases} Cso(t) = a_1 \exp(-m_1 t) + a_2 \exp(-m_2 t) - (a_1 + a_2 + aa) \exp(-m_3 t) + aa, \\ Cco(t) = b_1 \exp(-l_1 t) + b_2 \exp(-l_2 t) - (b_1 + b_2 + bb) \exp(-l_3 t) + bb, \end{cases} \tag{8}$$

where the initial values are assumed to be zero, namely $Cso(0) = 0$ and $Cco(0) = 0$ because of non-existence of radio-tracer at $t = 0$ as mentioned in §2.2. However, this assumption has an inference on regression of the parameters: a_i, b_i, aa, bb, m_i and l_i owing to inaccuracy or noise in the observed data that leads to $Cso(0) \neq 0$ or $Cco(0) \neq 0$. To avoid this inference, we have adopted an additional value: η. We have firstly fitted the observed data with $Cso(t - \eta)$ and $Cco(t - \eta)$, and then have substituted η with 0, that is, η has been ignored. $Cso(t)$ and $Cco(t)$ thus fitted are changed into the Laplace-transformed data as follows:

$$\begin{cases} L[Cso(t)] = \dfrac{a_1}{s + m_1} + \dfrac{a_2}{s + m_2} - \dfrac{a_1 + a_2 + aa}{s + m_3} + \dfrac{aa}{s}, \\ L[Cco(t)] = \dfrac{b_1}{s + l_1} + \dfrac{b_2}{s + l_2} - \dfrac{b_1 + b_2 + bb}{s + l_3} + \dfrac{bb}{s}, \end{cases} \tag{9}$$

where L denotes the Laplace transformation.

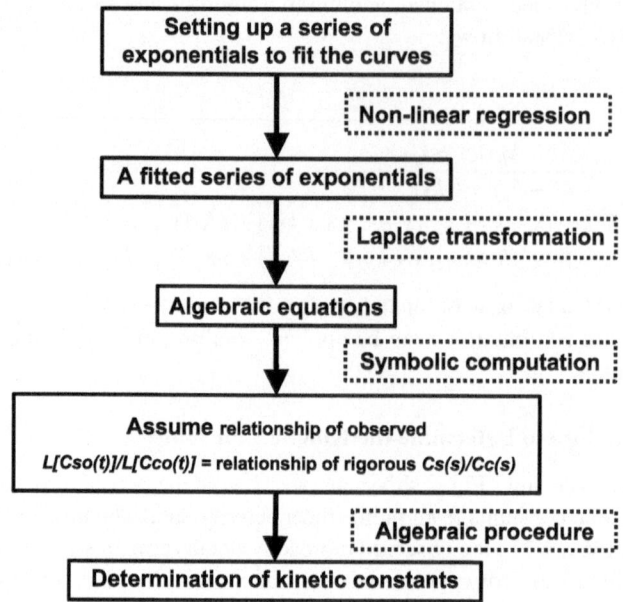

Fig. 2. Overview of the procedure for determining kinetic constants

Algebraic procedure. Without any error, the transformed function of A-tissue data, $L[Cso(t)](s)$, and that of B-tissue data, $L[Cco(t)]$, would be identical with $Cs(s)$ and $Cc(s)$, respectively. This fact has led us to the following procedure to determine the parameters over the Laplace domain.

1. $L[Cso(t)]/L[Cco(t)]$ can be transformed into the form: $F(s)/G(s)$, where $F(s)$ and $G(s)$ are both fifth-order polynomials in s. It follows from Eq. (7) that $-(k_2 + k_3)$, $-k_4$, and $-k_6$ are three of the real roots of $G(s)$. Likewise, $-k_{p2}$ and $-k_{p6}$ are two of the real roots of $F(s)$. It can be proved that both $F(s)$ and $G(s)$ have five real negative roots in PET experiments as mentioned in Appendix A.

2. Let $-r_i$ $(1 \leq i \leq 5)$ and $-t_i$ $(1 \leq i \leq 5)$ denote the real roots of $F(s)$ and $G(s)$, respectively. From (1), $k_2 + k_3, k_4$, and k_6 are three of t_i, e.g., t_1, t_2, and t_3. Likewise, k_{p2} and k_{p6} are, e.g., r_1 and r_2. The number of assignments of the parameters $k_2 + k_3, k_4, k_6, k_{p2}$, and k_{p6} to r_i and t_i is 1200. We apply these 1200 assignments to the two procedures below.

3. The remaining parameters, $K_1/K_{p1}, K_5k_{b12}/K_{p1}, K_{p5}k_{b12}/K_{p1}, k_2, k_3, k_{b2}$, and k_{b3}, are calculated by solving the following system of algebraic equations:

$$H(-r_3) = H(-r_4) = H(-r_5) = I(-t_4) = I(-t_5) = 0, \quad k_2 + k_3 = t_1,$$
$$K_1/K_{p1} = HC(F(s))/HC(G(s)),$$

where $H(s) = K_5k_{b12}(s+k_2+k_3)(s+k_4)(s+k_{b3}) + sK_1(s+k_3+k_4)(s+k_6)(s+k_{b2}+k_{b3})$, $I(s) = K_{p5}k_{b12}(s + k_{b3})(s + k_{p2}) + sK_{p1}(s + k_{b2} + k_{b3})(s + k_{p6})$, and HC denotes the head coefficient. To solve the system of algebraic equations above, we have derived the triangular form with the aid of Gröbner base computations. The third-order polynomial as the elimination ideal with respect to k_{b3} is shown in Appendix B.

4. Because the numerator and denominator of $Cs(s)/Cc(s)$ (Eq. (7)) are both sixth-order polynomials in s while $F(s)$ and $G(s)$ are both fifth-order, the similarity between $Cs(s)/Cc(s)$ and $F(s)/G(s)$ can be calculated by the difference between the roots of the numerator and denominator of $Cs(s)/Cc(s)$ that do not appear as the roots of $F(s)$ or $G(s)$. These two roots are calculated by coefficient comparison as follows:

$$k_3+k_4+k_6+k_{b2}+k_{b3}+K_5 k_{b12}/K_1-(r_3+r_4+r_5),\ k_{b2}+k_{b3}+k_{p6}+K_{p5}k_{b12}/K_{p1}-(t_4+t_5).$$

We record the difference between the above two roots as *diff*. Notice that the roots of $F(s)$ and $G(s)$ correspond to the reciprocals of time constants of PET experiments and that they are distinct from one another.

5. The parameter sets determined above are arranged in ascending order by *diff*. Furthermore, we remove parameter sets that violate an empirical or physiological law. In this paper, we have adopted the following law:

$$k_{p2} < k_{p6} \text{ and } k_3 < 1. \tag{10}$$

The first inequality, $k_{p2} < k_{p6}$, designates a different permeability of FDOPA and 3-OMFD, which cross the BBB (blood-brain barrier) [5, 7]. The second inequality, $k_3 < 1$, is the empirical law.

6. The result of the procedure above is the first parameter set among the sets in ascending order by *diff*.

Using the procedure described above, we can immediately and effectively determine the parameters such that all of them are consistent with PET experiments.

3   Results

We have extracted the observed data of A- and B-tissues from Cumming and Gjedde [4, p.52, Fig. 4], where A- and B-tissues correspond to the caudate and the cerebral (occipital) cortex, respectively. First, we have fitted the ^{18}F radioactivity data in A- and B- tissues of the normal control subject and the patient with Parkinson's disease (PD) as a series of exponentials according to $Cso(t - \eta_A)$ and $Cco(t - \eta_B)$ in Eq. (8). The parameters obtained, $a_i, b_i, aa, bb, m_i, l_i$, are represented in Table 1. Figure 3 shows the fitting of the observed data. As seen in Fig. 3, the estimated curves are fitted from control and PD patient samples.

By using the parameters in Table 1, we have obtained kinetic constants according to the algebraic procedure in §2.4. This calculation needed 15 seconds CPU time and 5.5 MBytes memory via Mathematica 5.2 (Wolfram Research, Inc.) with Intel(R) Xeon(R) CPU 2.33GHz. Table 2 shows almost all kinetic constants of the control and PD patient,

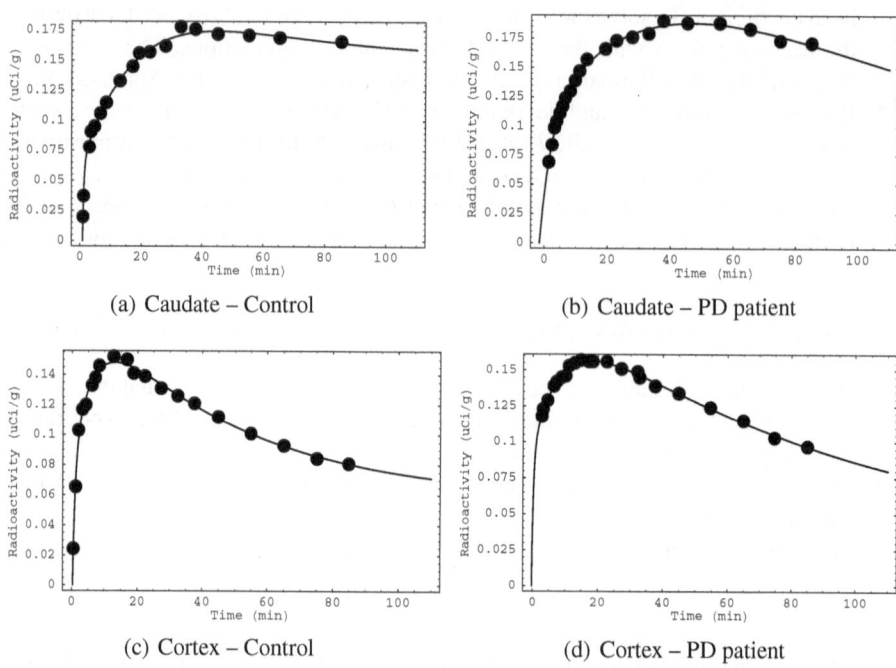

Fig. 3. Time–radioactivity curves in the occipital cortex and caudate of a patient with Parkinson's disease and a normal control subject during 90 min after administration of $[^{18}\text{F}]$fluorodopa. The circles are the observed data that have been extracted from [4, p.52, Fig. 4], and the solid curves are fits by a series of exponentials. (a) Radioactivity in caudate of a normal control subject. (b) Caudate of a patient with Parkinson's disease. (c) Occipital cortex of the control. (d) Occipital cortex of the patient.

in comparison with those in the previous estimation [4]. First, we have determined almost all values of kinetic constants, while the previous work only partially estimated the constants, using graphical analysis. Second, the orders of the kinetic constants obtained by our method are similar to those found by the graphical analysis, for both the control and PD patient. Interestingly, one constant, k_4, which is one of the measures for Parkinson's disease, was slightly different but consistent in our analysis compared with that in the previous study. The difference/consistency of the constants in the two studies will be judged from future work where many samples are analysed. At any rate, we have successfully determined almost all constants without blood-sampling data via our method.

4 Discussion

We have derived the equation Eq. (7), which enables us to determine rigorously almost all of the kinetic constants in the FDOPA model. In contrast, graphical analysis can only approximately determine the kinetic constants around the equilibrium of the

Table 1. The obtained parameters by non–linear regression

A-tissue ($Cso(t - \eta_A)$)	a_1	a_2	aa	m_1	m_2	m_3	η_A
Control	−0.0660074	−4.66107	0.153617	2.31191	0.0399169	0.038911	0.893464
PD patient	−0.10455	−3.02326	0.0588441	0.275661	0.0183807	0.0166495	−1.49546
B-tissue ($Cco(t - \eta_B)$)	b_1	b_2	bb	l_1	l_2	l_3	η_B
Control	−0.0826722	−0.11201	0.058033	1.31972	0.16269	0.021298	0.298255
PD patient	−0.0928105	−0.107709	0.0252521	2.05784	0.113904	0.0105398	−0.267787

Table 2. Kinetic constants of control and PD patient in FDOPA model by the procedure in §2.4 *without* blood–sampling data. The figures in the square bracket denote k_4 by Patlak analysis *with* blood–sampling data [4].

Kinetic constants	k_2	k_3	k_4	k_6	k_{p2}	k_{p6}	k_{b2}	k_{b3}
Control	0.0968	0.220	0.00674 [0.011 ± 0.003]	0.0389	0.0213	1.32	0.0525	0.00122
PD Patient	0.000818	0.0176	0.0166 [0.016 ± 0.004]	0.276	0.00231	0.0646	0.0276	0.112

(Continue)	K_1/K_{p1}	$K_5 k_{b12}/K_{p1}$	$K_{p5} k_{b12}/K_{p1}$
Control	1.29	0.0466	0.979
PD Patient	0.165	0.291	0.120

system under various assumptions and ignorance [6]. For instance, even the striatum (corresponding to A-tissue in this paper) was modelled as a single-tissue compartment [9]. Moreover, replacement with averaged values and ignorance of error terms as a small value are required for the solution to the equation over time domain because of its complicated form. Thus, the present method by the algebraic approach has successfully overcome the difficulties of graphical analysis.

Apart from graphical analyses, Cobelli et al. [2, 3] have studied the relationship between the observational parameters and the unknown model parameters over the Laplace domain. The aim of these works was determination of a model in which, on the assumption that any noise does not exist, it is determined whether the parameters can be determined uniquely or non-uniquely. In contrast, in this paper, we have determined parameters from the observed data with noise via the algebraic procedure as mentioned in §2.4. One of the other procedures to determine parameters from noisy data is the least squares method. We have attempted the least squares method using the following equation:

$$\int_{ls}^{us} (Cs(s)L[Cco(t)](s) - Cc(s)L[Cso(t)](s))^2 \, \mathbf{ds}.$$

Although the selection of interval between ls and us is somewhat ambiguous and it takes about 2.7 hours for each simulation with AMD Opteron(tm) Processor 2.412GHz, this method usually brings us the same results as the algebraic procedure and might be suitable for the equation where blood vessels in tissues are taken into account.

The solution over the Laplace domain is an algebraic equation to which Gröbner base computations can be applied, resulting in a much simpler form and efficient parameter determination (about 15 seconds with Intel(R) Xeon(R) CPU 2.33GHz). In fact, the equivalent equation to Eq. (7) can be described over time domain as follows:

$$Cs(t) = Cc(t) \otimes Y_1(t) \otimes Y_2(t),$$

$$\text{with } Y_1(t) = \frac{(k_2 + k_3 - k_{p2})(k_2 + k_3 - k_{p6})}{(k_2 + k_3 - k_4)(k_2 + k_3 - k_6)} e^{-(k_2+k_3)t} \tag{11}$$

$$- \frac{(k_4 - k_{p2})(k_4 - k_{p6})}{(k_2 + k_3 - k_4)(k_4 - k_6)} e^{-k_4 t} - \frac{(k_6 - k_{p2})(k_6 - k_{p6})}{(k_2 + k_3 - k_6)(k_6 - k_4)} e^{-k_6 t},$$

$$Y_2(t) = \text{Extremely complicated formula over time domain}$$

$$\text{(shown in Supplementary material)},$$

where $\otimes$ denotes the mathematical operation of convolution. The point is that we have solved the system of differential equations over the Laplace domain. In general, the solution including any external force over time domain (in this paper, C_{fd} in the C-Blood model is the external force) leads to the mathematical operation of 'convolution.' Instead, convolution over time domain corresponds to a simpler form of multiplication over the Laplace domain.

Lastly, we note that the present approach can be applied to more complex compartmental models. In compartmental models, the Laplace transformation of differential equations into algebraic equations and the following symbolic computation will reveal a rigorous relationship between kinetic constants. Furthermore, the algebraic procedure seems useful for determining constants from data.

5 Conclusion

We have derived a rigorous relationship for the kinetic constants of compartmental models for FDOPA metabolism, by symbolic computations with the aid of Gröbner bases. The algebraic procedure has successfully determined almost all constants from the observed radioactivity curves. In particular, the rigorous and simple form of a solution for the constants relationship brings us efficient parameter determination *without* blood-sampling data and *only from* PET scanning data that are dozens of minutes short of the equilibrium leading to the considerable reduction of PET scanning periods required for diagnosis.

Acknowledgment

We wish to express our gratitude to Ms. Atsuko Sono and Mr. Shigeo Orii for their supports. H. Y. and K. H. were partly supported by a Grant-in-Aid for Scientific Research on Priority Areas "Systems Genomics" (grant 18016008) from the Ministry of Education, Culture, Sports, Science and Technology of Japan. This study was supported in part by the New Energy and Industrial Technology Development Organization (NEDO) of Japan, and by The Kyushu University Research Superstar Program (SSP) from Special Coordination Funds for Promoting Science and Technology of Japan Science and Technology Agency (JST).

References

[1] Boyes, B.E., Cumming, P., Martin, W.R.W., Macgeer, E.G.: Determination of plasma [^{18}F]-6-fluorodopa during positron emission tomography: elimination and metabolism in carbidopa treated subjects. Life Sci. 39, 2243–2252 (1986)

[2] Cobelli, C., Foster, D., Toffolo, G.: Tracer Kinetics in Biomedical research: From data to model, Kluwer Academic/Plenum Publishers (2000)

[3] Cobelli, C., Toffolo, G.: Theoretical aspects and practical strategies for the identification of unidentifiable compartmental systems. In: chapter 8, pp. 85–91. Pergamon Press, Oxford (1987)

[4] Cumming, P., Gjedde, A.: Compartmental Analysis of Dopa Decarboxylation in living brain from dynamic positron emission tomograms. Synapse 29, 37–61 (1998)

[5] Deep, P., Kuwabara, H., Gjedde, A., Cumming, P.: The kinetic behaviour of [^{3}H]DOPA in living rat brain investigated by compartmental modelling of static autoradiograms. J. Neurosci. Methods 78, 157–168 (1997)

[6] Gunn, R.N., Gunn, S.R., Cunningham, V.J.: Positron emission tomography compartmental models. J. Cereb. Blood Flow Metab. 21, 635–652 (2001)

[7] Huang, S.C., Yu, D.C., Barrio, J.R., Grafton, S., Melega, W.P., Hoffman, J.M., Satyamurthy, N., Mazziotta, J.C., Phelps, M.E.: Kinetics and Modeling of L-6-[^{18}F]Fluoro-DOPA in Human Positron Emission Tomographic Studies. J. Cereb. Blood Flow Metab. 11, 898–913 (1991)

[8] Ichise, M., Ballinger, J.R., Golan, H., Vines, D., Luong, A., Tsai, S., Kung, H.F.: SPECT imaging of dopamine D2 receptors in humans with iodine 123–IBF: a practical approach to quantification not requiring blood sampling. J. Nucl. Med. 36, 11 (1995)

[9] Kawatsu, S., Kato, T., N.-Saito, A., Hatano, K., Ito, K., Ishigaki, T.: New insight into the analysis of 6-[^{18}F]fluoro-L-DOPA PET dynamic data in brain tissue without an irreversible compartment: comparative study of the Patlak and Logan Analyses. Radiation medicine 21, 47–54 (2003)

[10] Kumakura, Y., Danielsen, E.H., Reilhac, A., Gjedde, A., Cumming, P.: Levodopa effect on [^{18}F]fluorodopa influx to brain: normal volunteers and patients with Parkinson's disease. Acta Neurol. Scand. 110, 188–195 (2004), doi:10.1111/j.1600-0404.2004.00299.x

[11] Kumakura, Y., Gjedde, A., Danielsen, E.H., Christensen, S., Cumming, P.: Dopamine storage capacity in caudate and putamen of patients with early Parkinson's disease: correlation with asymmetry of motor symptoms. J. Cereb. Blood Flow Metab. 26, 358–370 (2006), doi:10.1038/sj.jcbfm.9600202

[12] Kumakura, Y., Vernaleken, I., Gründer, G., Bartenstein, P., Gjedde, A., Cumming, P.: PET studies of net blood–brain clearance of FDOPA to human brain; age–dependent decline of [^{18}F]fluorodopamine storage capacity. J. Cereb. Blood Flow Metab. 25, 807–819 (2005), doi:10.1038/sj.jcbfm.9600079

[13] Lammertsma, A.A., Bench, C.J., Hume, S.P., Osman, S., Gunn, K., Brooks, D.J., Frackowiak, R.S.J.: Comparision of methods for analysis of clinical [^{11}C]Raclopride studies. J. Cereb. Blood Flow Metab. 16, 42–52 (1996)

[14] Logan, J., Fowler, J.S., Volkow, N.D., Wang, G.-J., Ding, Y.-S., Alexoff, D.L.: Distribution Volume Ratios without blood sampling from graphical analysis of PET Data. J. Cereb. Blood Flow Metab. 16, 834–840 (1996), doi:10.1097/00004647-199609000-00008

[15] Logan, J., Fowler, J.S., Volkow, N.D., Wolf, A.P., Dewey, S.L., Schlyer, D.J., MacGregor, R.R., Hitzemann, R., Bendriem, B., Gatley, S.J., Christman, D.R.: Graphical analysis of reversible radioligand binding from time–activity measurements applied to [N-^{11}C-methyl]-(–)-Cocaine PET studies in human subjects. J. Cereb. Blood Flow Metab. 10, 740–747 (1990)

[16] Martin, W.R.W., Palmer, M.R., Patlak, C.S., Calne, D.B.: Nigrostriatal Function in Humans Studied with Positron Emission Tomography. Ann. Neurol. 26, 535–542 (1989)

[17] Melega, W.P., Hoffman, J.M., Luxen, A., Nissenson, C.H., Phelps, M.E., Barrio, J.R.: The effects of carbidopa on the metabolism of 6-[^{18}F]fluorodopa-L-DOPA in rats, monkeys and humans. Life Sci. 47, 149–157 (1990)

[18] Patlak, C.S., Blasberg, R.G.: Graphical Evaluation of Blood-to-Brain Transfer Constants from Multiple-Time uptake data. Generalizations. J. Cereb. Blood Flow Metab. 5, 584–590 (1985)

[19] Patlak, C.S., Blasberg, R.G., Fenstermacher, J.D.: Graphical evaluation of blood–to–brain transfer constants from multiple–time uptake data. J. Cereb. Blood Flow Metab. 3, 1–7 (1983)

[20] Rousset, O.G., Deep, P., Kuwabara, H., Evans, A.C., Gjedde, A.H., Cumming, P.: Effect of partial volume correction on estimates of the influx and cerebral metabolism of 6-[^{18}F]fluoro-L-dopa Studied with PET in normal control and Parkinson's Disease subjects. Synapse 37, 81–89 (2000)

Supplementary Material

Hereby, we show $Y_2(t)$, which is not shown in Eq. (11), at the following URL:
  http://www.math.kyushu-u.ac.jp/~phiroshi/pet/Y2.pdf,
where, for instance, $Root[k_1\#1 + k_2\#1^2 + k_3\#1^3, 1]$ denotes the minimum real root of the equation $[k_1 x + k_2 x^2 + k_3 x^3 = 0$ in x and DiracDelta[t] denotes Dirac delta function $\delta(t)$.

Appendix A: Proof of the Existence of Five Real Negative Roots

We shall prove that both $F(s)$ and $G(s)$ have five real negative roots. From Eq. (9):

$$F(s) = (s + l_1)(s + l_2)(s + l_3)F_1(s),$$
$$G(s) = (s + m_1)(s + m_2)(s + m_3)G_1(s),$$

where

$$F_1(s) = aa\, m_3(s + m_1)(s + m_2) + s(-a_2(m_2 - m_3)(s + m_1) - a_1(m_1 - m_3)(s + m_2)),$$
$$G_1(s) = bb\, l_3(s + l_1)(s + l_2) + s(-b_2(l_2 - l_3)(s + l_1) - b_1(l_1 - l_3)(s + l_2)).$$

In PET experiments, we can reasonably postulate $l_1 > l_2 > l_3 > 0$, and, $m_1 > m_2 > m_3 > 0$ because the radioactivity eventually approaches an equilibrium (the finite value). With respect to $F_1(s)$, we can see the following relationships:

$$F_1(0) = aa\, m_1 m_2 m_3,$$

$$F_1(-m_1) = a_1 m_1(m_2 - m_1)(m_1 - m_3), \quad F_1(-m_3) = -(a_1 + a_2 + aa)m_3(m_1 - m_3)(m_3 - m_2).$$

As seen in Fig. 3, $aa > 0$ because the radioactivity is never negative even when $t \to \infty$. Furthermore, the largest and the smallest time constants: $1/m_3$ and $1/m_1$ correspond to the sampling data near the equilibrium and the initial stage, respectively, leading to the coefficient relations of the exponentials, $\exp(-m_3 t)$ and $\exp(-m_1 t)$: $-(a_1 + a_2 + aa) > 0$ and $a_1 < 0$, respectively. These facts lead to $F_1(0) > 0$, $F_1(-m_3) < 0$ and $F_1(-m_1) > 0$, showing that $F_1(s)$ has two real negative distinct roots, and then $F(s)$ has five real negative roots. Likewise, $G(s)$ has five real negative roots.   □

Appendix B: The Third-Order Polynomial in k_{b3}

In §2.4, we have derived the third-order polynomial by calculating the elimination ideal w.r.t. k_{b3}. This calculation needed 35.4 hours CPU time and 220 MBytes memory via Mathematica 5.2 (Wolfram Research, Inc.) with Intel(R) Xeon(R) CPU 2.33GHz. The calculated polynomial is as follows:

$$
\begin{aligned}
&(-r_3+r_4)(r_3-r_5)(r_4-r_5)(t_4-t_5)(t_1 k_4(r_4-k_6)(r_5-k_6)k_{b3}(t_4^2(t_5-k_{b3})(t_5-k_{p2})+t_4(r_4(-(k_{b3}k_{p2})+t_5(k_{b3}+k_{p2}-k_{p6}))-(-(k_{b3}k_{p2})+t_5(k_{b3}+k_{p2}))(t_5- \\
&k_{p6}))-(r_4-t_5)k_{b3}k_{p2}(t_5-k_{p6}))(t_4^2(t_5-k_{b3})(t_5-k_{p2})+t_4(r_5(-(k_{b3}k_{p2})+t_5(k_{b3}+k_{p2}-k_{p6}))-(-(k_{b3}k_{p2})+t_5(k_{b3}+k_{p2}))(t_5-k_{p6}))-(r_5-t_5)k_{b3} \\
&k_{p2}(t_5-k_{p6}))-r_3^2(t_4(-(k_{b3}k_{p2})+t_5(k_{b3}+k_{p2}-k_{p6}))+k_{b3}k_{p2}(-t_5+k_{p6}))(-(t_1 k_4(r_5-k_6)k_{b3}(t_4^2(t_5-k_{b3})(t_5-k_{p2})+t_4(r_5(-(k_{b3}k_{p2})+t_5(k_{b3}+k_{p2} \\
&-k_{p6}))-(-(k_{b3}k_{p2})+t_5(k_{b3}+k_{p2}))(t_5-k_{p6}))-(r_5-t_5)k_{b3}k_{p2}(t_5-k_{p6})))+r_4^2(t_1 k_4 k_{b3}(t_4(k_{b3}k_{p2}-t_5(k_{b3}+k_{p2}-k_{p6}))+k_{b3}k_{p2}(t_5-k_{p6}))+r_5^2 \\
&(-(t_4^2(t_5-k_{b3})(t_5-k_{p2}))-(t_5-t_1-k_4+k_6-k_{b3})k_{b3}k_{p2}(t_5-k_{p6})+t_4(t_5^2(k_{b3}+k_{p2})+k_{b3}k_{p2}(t_1+k_4-k_6+k_{b3}+k_{p6})-t_5(-(k_6 k_{b3})+k_{b3}^2-k_6 k_{p2}+2k_{b3} \\
&k_{p2}+t_1(k_{b3}+k_{p2}-k_{p6})+k_4(k_{b3}+k_{p2}-k_{p6})+k_6 k_{p6}+k_{p2}k_{p6})))+r_5(t_4^2 k_6(t_5-k_{b3})(t_5-k_{p2})-k_{b3}(-(t_5 k_6)+k_4 k_{b3}+t_1(k_4+k_{b3}))k_{p2}(t_5-k_{p6})+t_4(- \\
&(t_5^2 k_6(k_{b3}+k_{p2}))-k_{b3}k_{p2}(k_4 k_{b3}+t_1(k_4+k_{b3})+k_6 k_{p6})+t_5(k_4 k_{b3}(k_{b3}+k_{p2}-k_{p6})+t_1(k_4+k_{b3})(k_{b3}+k_{p2}-k_{p6})+k_6(k_{p2}k_{p6}+k_{b3}(k_{p2}+k_{p6}))))) \\
&))+r_4(t_1 k_4 k_{b3}(-(t_4^2(t_5-k_{b3})(t_5-k_{p2}))-(t_5+k_6)k_{b3}k_{p2}(t_5-k_{p6})+t_4(t_5^2(k_{b3}+k_{p2})+k_{b3}k_{p2}(-k_6+k_{p6})-t_5(-(k_6(k_{b3}+k_{p2}-k_{p6}))+k_{p2}k_{p6}+k_{b3} \\
&(k_{p2}+k_{p6}))))+r_5^2(t_4^2 k_6(t_5-k_{b3})(t_5-k_{p2})-k_{b3}(-(t_5 k_6)+k_4 k_{b3}+t_1(k_4+k_{b3}))k_{p2}(t_5-k_{p6})+t_4(-(t_5^2 k_6(k_{b3}+k_{p2}))-k_{b3}k_{p2}(k_4 k_{b3}+t_1(k_4+k_{b3})+k_6 k_{p6}) \\
&+t_5(k_4 k_{b3}(k_{b3}+k_{p2}-k_{p6})+t_1(k_4+k_{b3})(k_{b3}+k_{p2}-k_{p6})+k_6(k_{p2}k_{p6}+k_{b3}(k_{p2}+k_{p6})))))+r_5(t_4^2(t_5-k_{b3})(-(k_6 k_{b3})+k_4(-k_6+k_{b3})+t_1(k_4-k_6+ \\
&k_{b3}))(t_5-k_{p2})+k_{b3}(k_4 k_6 k_{b3}+t_5(-(k_6 k_{b3})+k_4(-k_6+k_{b3})+t_1(k_4-k_6+k_{b3}))+t_1(k_6 k_{b3}+k_4(k_6+k_{b3}))k_{p2}(t_5-k_{p6})+t_4(t_5^2(k_4(k_6-k_{b3})+k_6 k_{b3}-t_1(k_4-k_6 \\
&+k_{b3}))(k_{b3}+k_{p2})+k_{b3}k_{p2}(k_4 k_6 k_{b3}+k_4 k_6 k_{p6}-k_4 k_{b3}k_{p6}+k_6 k_{b3}k_{p6}+t_1(k_4(k_6+k_{b3}-k_{p6})-k_{b3}k_{p6}+k_6(k_{b3}+k_{p6})))-t_5(k_6 k_{b3}(k_{p2}k_{p6}+k_{b3} \\
&(k_{p2}+k_{p6}))+k_4(k_6(k_{b3}^2+2k_{b3}k_{p2}+k_{p2}k_{p6})-k_{b3}(k_{p2}k_{p6}+k_{b3}(k_{p2}+k_{p6})))+t_1(k_4(k_{b3}^2+k_6(k_{b3}+k_{p2}-k_{p6})-2k_{b3}k_{p6}-k_{p2}k_{p6})+k_6(k_{b3}^2+2 \\
&k_{b3}k_{p2}+k_{p2}k_{p6})-k_{b3}(k_{p2}k_{p6}+k_{b3}(k_{p2}+k_{p6}))))))-r_3(t_1 k_4(r_5-k_6)k_{b3}(t_4^2(t_5-k_{b3})(t_5-k_{p2})+t_4(r_5(-(k_{b3}k_{p2})+t_5(k_{b3}+k_{p2}-k_{p6}))-(-(k_{b3} \\
&k_{p2})+t_5(k_{b3}+k_{p2}))(t_5-k_{p6}))-(r_5-t_5)k_{b3}k_{p2}(t_5-k_{p6}))(-(t_4^2(t_5-k_{b3})(t_5-k_{p2}))-(t_5+k_6)k_{b3}k_{p2}(t_5-k_{p6})+t_4(t_5^2(k_{b3}+k_{p2})+k_{b3}k_{p2}(-k_6+k_{p6} \\
&)-t_5(-(k_6(k_{b3}+k_{p2}-k_{p6}))+k_{p2}k_{p6}+k_{b3}(k_{p2}+k_{p6})))+r_4^2(t_4(-(k_{b3}k_{p2})+t_5(k_{b3}+k_{p2}-k_{p6}))+k_{b3}k_{p2}(-t_5+k_{p6}))(t_1 k_4 k_{b3}(-(t_4^2(t_5-k_{b3})(t_5 \\
&-k_{p2}))-(t_5+k_6)k_{b3}k_{p2}(t_5-k_{p6})+t_4(t_5^2(k_{b3}+k_{p2})+k_{b3}k_{p2}(-k_6+k_{p6})-t_5(-(k_6(k_{b3}+k_{p2}-k_{p6}))+k_{p2}k_{p6}+k_{b3}(k_{p2}+k_{p6}))))+r_5^2(t_4^2 k_6(t_5-k_{b3} \\
&)(t_5-k_{p2})-k_{b3}(-(t_5 k_6)+k_4 k_{b3}+t_1(k_4+k_{b3}))k_{p2}(t_5-k_{p6})+t_4(-(t_5^2 k_6(k_{b3}+k_{p2}))-k_{b3}k_{p2}(k_4 k_{b3}+t_1(k_4+k_{b3})+k_6 k_{p6})+t_5(k_4 k_{b3}(k_{b3}+k_{p2}-k_{p6} \\
&)+t_1(k_4+k_{b3})(k_{b3}+k_{p2}-k_{p6})+k_6(k_{p2}k_{p6}+k_{b3}(k_{p2}+k_{p6})))))+r_5(t_4^2(t_5-k_{b3})(-(k_6 k_{b3})+k_4(-k_6+k_{b3})+t_1(k_4-k_6+k_{b3}))(t_5-k_{p2})+k_{b3}(k_4 k_6 k_{b3}+t_5 \\
&(-(k_6 k_{b3})+k_4(-k_6+k_{b3})+t_1(k_4-k_6+k_{b3}))+t_1(k_6 k_{b3}+k_4(k_6+k_{b3})))k_{p2}(t_5-k_{p6})+t_4(t_5^2(k_4(k_6-k_{b3})+k_6 k_{b3}-t_1(k_4-k_6+k_{b3}))(k_{b3}+k_{p2})+k_{b3}k_{p2}(k_4 k_6 \\
&k_{b3}+k_4 k_6 k_{p6}-k_4 k_{b3}k_{p6}+k_6 k_{b3}k_{p6}+t_1(k_4(k_6+k_{b3}-k_{p6})-k_{b3}k_{p6}+k_6(k_{b3}+k_{p6})))-t_5(k_6 k_{b3}(k_{p2}k_{p6}+k_{b3}(k_{p2}+k_{p6}))+k_4(k_6(k_{b3}^2+2k_{b3} \\
&k_{p2}+k_{p2}k_{p6})-k_{b3}(k_{p2}k_{p6}+k_{b3}(k_{p2}+k_{p6})))+t_1(k_4(k_{b3}^2+k_6(k_{b3}+k_{p2}-k_{p6})-2k_{b3}k_{p6}-k_{p2}k_{p6})+k_6(k_{b3}^2+2k_{b3}k_{p2}+k_{p2}k_{p6})-k_{b3}(k_{p2} \\
&k_{p6}+k_{b3}(k_{p2}+k_{p6}))))))))-r_3(t_1 k_4(r_5-k_6)k_{b3}(t_4^2(t_5-k_{b3})(t_5-k_{p2})+(t_5+k_6)k_{b3}k_{p2}(t_5-k_{p6})+t_4(-(t_5^2(k_{b3}+k_{p2}))+k_{b3}k_{p2}(k_6-k_{p6})+t_5(-(k_6(k_{b3}+k_{p2} \\
&-k_{p6}))+k_{p2}k_{p6}+k_{b3}(k_{p2}+k_{p6}))))^2)+r_5^2(t_4(-(k_{b3}k_{p2})+t_5(k_{b3}+k_{p2}-k_{p6}))+k_{b3}k_{p2}(-t_5+k_{p6})(t_4^2(t_5-k_{b3})(-(k_6 k_{b3})+k_4(-k_6+k_{b3})+t_1(k_4-k_6+ \\
&k_{b3}))(t_5-k_{p2})+k_{b3}(k_4 k_6 k_{b3}+t_5(-(k_6 k_{b3})+k_4(-k_6+k_{b3})+t_1(k_4-k_6+k_{b3}))+t_1(k_6 k_{b3}+k_4(k_6+k_{b3})))k_{p2}(t_5-k_{p6})+t_4(t_5^2(k_4(k_6-k_{b3})+k_6 k_{b3}-t_1(k_4- \\
&k_6+k_{b3}))(k_{b3}+k_{p2})+k_{b3}k_{p2}(k_4 k_6 k_{b3}+k_4 k_6 k_{p6}-k_4 k_{b3}k_{p6}+k_6 k_{b3}k_{p6}+t_1(k_4(k_6+k_{b3}-k_{p6})-k_{b3}k_{p6}+k_6(k_{b3}+k_{p6})))-t_5(k_6 k_{b3}(k_{p2}k_{p6}+k_{b3} \\
&(k_{p2}+k_{p6}))+k_4(k_6(k_{b3}^2+2k_{b3}k_{p2}+k_{p2}k_{p6})-k_{b3}(k_{p2}k_{p6}+k_{b3}(k_{p2}+k_{p6})))+t_1(k_4(k_{b3}^2+k_6(k_{b3}+k_{p2}-k_{p6})-2k_{b3}k_{p6}-k_{p2}k_{p6})+k_6(k_{b3}^2+ \\
&2k_{b3}k_{p2}+k_{p2}k_{p6})-k_{b3}(k_{p2}k_{p6}+k_{b3}(k_{p2}+k_{p6})))))))+r_5(t_4^4(t_5-k_{b3})^2((-k_4+k_6)(k_6-k_{b3})+t_1(k_4-k_6+k_{b3}))(t_5-k_{p2})^2+k_{b3}^2(t_5^2((-k_4+k_6)(k_6-k_{b3} \\
&)+t_1(k_4-k_6+k_{b3}))+k_6(k_4 k_6 k_{b3}+t_1(k_4 k_6+2k_4 k_{b3}+k_6 k_{b3}))+t_5(-(k_6(k_4(k_6-k_{b3})+k_6 k_{b3}))+t_1(k_6(-k_6+k_{b3})+k_4(k_6+2k_{b3}))))k_{p2}^2(t_5-k_{p6})^2-t_4^3(t_5-k_{b3} \\
&)(t_5-k_{p2})(2t_5^2((-k_4+k_6)(k_6-k_{b3})+t_1(k_4-k_6+k_{b3}))(k_{b3}+k_{p2})+k_{b3}k_{p2}(t_1((k_6-k_{b3})(k_6-2k_{p6})-k_4(k_6+2k_{b3}-2k_{p6}))+k_4(k_6-k_{b3})(k_6-2k_{p6})+k_6(k_6 k_{b3} \\
&+2k_6 k_{p6}-2k_{b3}k_{p6}))+t_5(-(k_4(k_6-k_{b3})(k_6(k_{b3}+k_{p2}-k_{p6})-2(k_{p2}k_{p6}+k_{b3}(k_{p2}+k_{p6})))-k_6(-2k_{b3}(k_{p2}k_{p6}+k_{b3}(k_{p2}+k_{p6}))+k_6(k_{b3}^2+2k_{b3}k_{p2}k_{p6} \\
&+k_{b3}(3k_{p2}+k_{p6}))+t_1(k_4(k_6 k_{b3}+k_{p2}-k_{p6})+2(k_{b3}^2-2k_{b3}k_{p6}-k_{p2}k_{p6}))-(k_6-k_{b3})(k_6(k_{b3}+k_{p2}-k_{p6})-2(k_{p2}k_{p6}+k_{b3}(k_{p2}+k_{p6})))))+t_4 k_{b3} \\
&k_{p2}(t_5-k_{p6})(-2t_5^3((-k_4+k_6)(k_6-k_{b3})+t_1(k_4-k_6+k_{b3}))(k_{b3}+k_{p2})+k_{b3}k_{p2}(k_6(k_6 k_{b3}k_{p6}+k_4(2k_6 k_{b3}+k_6 k_{p6}-k_{b3}k_{p6}))+t_1(k_4(k_6+2k_{b3})(2k_6-k_{p6} \\
&)+k_6(2k_{b3}k_6+k_6 k_{p6}-k_{b3}k_{p6})))+t_5(-(k_6 k_{b3}(-2k_{b3}k_{p2}k_{p6}+k_6(3k_{p2}k_{p6}+k_{b3}(2k_{p2}+k_{p6})))+k_4(-2k_{b3}^2 k_{p2}k_{p6}-k_6^2(2k_{b3}^2+4k_{b3}k_{p2}-k_{b3}k_{p6} \\
&+k_{p2}k_{p6})+k_6 k_{b3}(3k_{p2}k_{p6}+k_{b3}(2k_{p2}+k_{p6})))+t_1(-2k_{b3}^2 k_{p2}k_{p6}-k_6^2(2k_{b3}^2+4k_{b3}k_{p2}-k_{b3}k_{p6}+k_{p2}k_{p6})+k_6 k_{b3}(2k_{b3}k_{p2}+k_{b3}k_{p6}+3 \\
&k_{p2}k_{p6})+k_4(-2k_{b3}^2(k_{b3}+k_{p2}-k_{p6})+2k_{b3}^2(2k_{p2}+k_{p6})+k_6(-4k_{b3}^2-2k_{b3}k_{p2}+5k_{b3}k_{p6}+k_{p2}k_{p6})))-t_5^2((k_4(k_6-k_{b3})(k_6(2k_{b3}+2k_{p2}-k_{p6})-2(k_{p2} \\
&k_{p6}+k_{b3}(k_{p2}+k_{p6})))-k_6(k_6(k_{b3}+2k_{p2}(2k_{b3}+k_{p6})-2k_{b3}(k_{p2}k_{p6}+k_{b3}(k_{p2}+k_{p6}))+t_1(k_4(2(2k_{b3}+k_{p2})(k_{b3}-k_{p6})+k_6(k_{b3}+2k_{p2}-k_{p6}))-(k_6 \\
&-k_{b3})(k_6(2k_{b3}+2k_{p2}-k_{p6})-2(k_{p2}k_{p6}+k_{b3}(k_{p2}+k_{p6}))))))+r_4^2(t_4^4((-k_4+k_6)(k_6-k_{b3})+t_1(k_4-k_6+k_{b3}))(k_{b3}^2+4k_{b3}k_{p2}+k_{p2}^2)+k_{b3}^2 k_{p2}^2(k_6 k_{p6}
\end{aligned}
$$

$$
\begin{aligned}
&(-(k_{b3}k_{p6}) + k_6(2k_{b3} + k_{p6})) + k_4(k_{b3}k_{p6}^2 - k_6k_{p6}(2k_{b3} + k_{p6}) + k_6^2(k_{b3} + 2k_{p6})) + t_1(k_{b3}k_{p6}^2 - k_6k_{p6}(2k_{b3} + k_{p6}) + k_6^2(k_{b3} + 2k_{p6}) + k_4(k_6^2 + 2k_6 \\
&(k_{b3} - k_{p6}) + k_{p6}(-4k_{b3} + k_{p6}))) - t_5k_{b3}k_{p2}(k_4(2k_{b3}k_{p6}(k_{p2}k_{p6} + k_{b3}(2k_{p2} + k_{p6})) + k_6^2(2k_{b3}^2 + (3k_{p2} - k_{p6})k_{p6} + k_{b3}(4k_{p2} + k_{p6})) - k_6(2k_{p2} \\
&k_{p6}^2 + k_{b3}k_{p6}(7k_{p2} + k_{p6}) + k_{b3}^2(2k_{p2} + 3k_{p6}))) + k_6(-2k_{b3}k_{p6}(k_{p2}k_{p6} + k_{b3}(2k_{p2} + k_{p6})) + k_6(2k_{p2}k_{p6}^2 + k_{b3}k_{p6}(7k_{p2} + k_{p6}) + k_{b3}^2(2k_{p2} + 3k_{p6} \\
&))) + t_1(2k_{b3}k_{p6}(k_{p2}k_{p6} + k_{b3}(2k_{p2} + k_{p6})) + k_6^2(2k_{b3}^2 + (3k_{p2} - k_{p6})k_{p6} + k_{b3}(4k_{p2} + k_{p6})) - k_6(2k_{p2}k_{p6}^2 + k_{b3}k_{p6}(7k_{p2} + k_{p6}) + k_{b3}^2(2k_{p2} + 3 \\
&k_{p6})) + k_4(2k_6^2(k_{b3} + k_{p2} - k_{p6}) - 2k_{b3}(k_{p2} - 2k_{p6})k_{p6} + 2k_{p2}k_{p6}^2 - 2k_{b3}^2(2k_{p2} + 3k_{p6}) + k_6(4k_{b3}^2 + 2k_{b3}k_{p2} - 7k_{b3}k_{p6} - 3k_{p2}k_{p6} + k_{p6}^2)))) + t_5^3(-(\\
&k_4(k_6 - k_{b3})(k_6(k_{b3}^3 + 3k_{b3}k_{p2} + k_{p2}^2 - k_{b3}k_{p6} - k_{p2}k_{p6}) - 2(k_{p2}^2k_{p6} + k_{b3}^2(2k_{p2} + k_{p6}) + k_{b3}k_{p2}(2k_{p2} + 3k_{p6}))) - k_6(k_6(k_{b3}^3 + 2k_{p2}^2k_{p6} + 5k_{b3} \\
&k_{p2}(k_{p2} + k_{p6}) + k_{b3}^2(7k_{p2} + k_{p6})) - 2k_{b3}(k_{p2}^2k_{p6} + k_{b3}^2(2k_{p2} + k_{p6}) + k_{b3}k_{p2}(2k_{p2} + 3k_{p6}))) + t_1((-k_6 + k_{b3})(k_6(k_{b3}^2 + 3k_{b3}k_{p2} + k_{p2}^2 - k_{b3} \\
&k_{p6} - k_{p2}k_{p6}) - 2(k_{p2}^2k_{p6} + k_{b3}^2(2k_{p2} + k_{p6}) + k_{b3}k_{p2}(2k_{p2} + 3k_{p6}))) + k_4(k_6(k_{b3}^3 + 3k_{b3}k_{p2} + k_{p2}^2 - k_{b3}k_{p6} - k_{p2}k_{p6}) + 2(k_{b3}^3 + k_{b3}^2(k_{p2} - 2 \\
&k_{p6}) - k_{p2}^2k_{p6} - k_{b3}k_{p2}(k_{p2} + 4k_{p6})))) + t_5^2(k_6(k_6(k_{p2}^2k_{p6} + k_{b3}^3(3k_{p2} + k_{p6}) + k_{b3}k_{p2}k_{p6}(7k_{p2} + k_{p6}) + 2k_{b3}^2k_{p2}(3k_{p2} + 4k_{p6})) - k_{b3}(k_{p2}^2 \\
&k_{p6}^2 + 2k_{b3}k_{p2}k_{p6}(3k_{p2} + k_{p6}) + k_{b3}^3(3k_{p2}^2 + 6k_{p2}k_{p6} + k_{p6}^2))) + k_4(k_6^3(k_{b3}^3 + 4k_{b3}k_{p2}^2 + k_{b3}^2(5k_{p2} - k_{p6}) + k_{p2}(k_{p2} - k_{p6})k_{p6}) - k_6(k_{p2}^2 \\
&k_{p6}^2 + k_{b3}^3(3k_{p2} + k_{p6}) + k_{b3}k_{p2}k_{p6}(7k_{p2} + k_{p6}) + 2k_{b3}^2k_{p2}(3k_{p2} + 4k_{p6})) + k_{b3}(k_{p2}^2k_{p6}^2 + 2k_{b3}k_{p2}k_{p6}(3k_{p2} + k_{p6}) + k_{b3}^3(3k_{p2}^2 + 6k_{p2} \\
&k_{p6} + k_{p6}^2))) + t_1(k_6^2(k_{b3}^3 + 4k_{b3}k_{p2}^2 + k_{b3}^2(5k_{p2} - k_{p6}) + k_{p2}(k_{p2} - k_{p6})k_{p6}) - k_6(k_{p2}^2k_{p6}^2 + k_{b3}^3(3k_{p2} + k_{p6}) + k_{b3}k_{p2}k_{p6}(7k_{p2} + k_{p6}) + \\
&2k_{b3}^2k_{p2}(3k_{p2} + 4k_{p6})) + k_{b3}(k_{p2}^2k_{p6}^2 + 2k_{b3}k_{p2}k_{p6}(3k_{p2} + k_{p6}) + k_{b3}^3(3k_{p2}^2 + 6k_{p2}k_{p6} + k_{p6}^2))) + k_4(k_6^2(k_{b3} + k_{p2} - k_{p6})^2 + k_{p2}^2k_{p6}^2 + 4 \\
&k_{b3}k_{p2}k_{p6}(k_{p2} + k_{p6}) - 2k_{b3}^3(3k_{p2} + k_{p6}) + k_{b3}^2(-3k_{p2}^2 + 2k_{p2}k_{p6} + 3k_{p6}^2) + k_6(2k_{b3}^3 + k_{b3}^2(k_{p2} - 5k_{p6}) + k_{p2}k_{p6}(-k_{p2} + k_{p6}) - k_{b3}(k_{p2}^2 + 6 \\
&k_{p2}k_{p6} - 3k_{p6}^2))))))))))K_1/K_{p1}.
\end{aligned}
$$

Efficient Haplotype Inference
with Pseudo-boolean Optimization

Ana Graça[1], João Marques-Silva[2], Inês Lynce[1], and Arlindo L. Oliveira[1]

[1] IST/INESC-ID, Technical University of Lisbon, Portugal
{assg,ines}@sat.inesc-id.pt, aml@inesc-id.pt
[2] School of Electronics and Computer Science, University of Southampton, UK
jpms@ecs.soton.ac.uk

Abstract. Haplotype inference from genotype data is a key computational problem in bioinformatics, since retrieving directly haplotype information from DNA samples is not feasible using existing technology. One of the methods for solving this problem uses the pure parsimony criterion, an approach known as Haplotype Inference by Pure Parsimony (HIPP). Initial work in this area was based on a number of different Integer Linear Programming (ILP) models and branch and bound algorithms. Recent work has shown that the utilization of a Boolean Satisfiability (SAT) formulation and state of the art SAT solvers represents the most efficient approach for solving the HIPP problem.

Motivated by the promising results obtained using SAT techniques, this paper investigates the utilization of modern Pseudo-Boolean Optimization (PBO) algorithms for solving the HIPP problem. The paper starts by applying PBO to existing ILP models. The results are promising, and motivate the development of a new PBO model (RPoly) for the HIPP problem, which has a compact representation and eliminates key symmetries. Experimental results indicate that RPoly outperforms the SAT-based approach on most problem instances, being, in general, significantly more efficient.

Keywords: haplotype inference, pure parsimony, pseudo-Boolean optimization.

1   Introduction

The causes of many common human diseases remain, to this day, largely unknown. Since genetic inheritance is one of the major risk factors for the large majority of diseases, the study of genetic variation in human populations represents one of the critical steps towards a better understanding of the mechanisms of disease.

Although a number of heritable disorders that depend on the variation of one single location in one single gene are known, common diseases usually depend on the combined effects of many different factors, in a number of different genes.

The study of the effects of particular variations of genes is simplified by the fact that, in many cases, there exists a strong correlation between the allele

H. Anai, K. Horimoto, and T. Kutsia (Eds.): AB 2007, LNCS 4545, pp. 125–139, 2007.

present in a particular single nucleotide polymorphism (SNP) and other nearby sites. A given combination of alleles in one chromosome is termed a haplotype, and the deviation from independence that exists between alleles is known as linkage disequilibrium (LD).

For genetic inheritable diseases that are due to a combination of allele values in nearby loci, identifying common haplotypes in the population represents a key first step towards the understanding of the pathogenesis of disease. However, current genotyping methods do not provide haplotype information, which is essential for detailed analysis of the mechanisms of disease.

At a given position for which an individual is heterozygous (i.e., inherited different alleles at a given locus), it is technologically not feasible, in general, to identify the particular chromosome that contains each allele. Additional information can be obtained by genotyping the parents, but significant uncertainty remains. Efficient methods for haplotype inference that can handle large volumes of data are therefore crucial, in order to make adequate use of the results of ongoing efforts like the HapMap project [17], an effort that aims at making available genotype and haplotype information of a significant sample of the human population.

Although a number of different methods has been proposed for the problem of haplotype inference, the Pure-Parsimony criterion [6,10,7] represents a well known approach. Haplotype Inference by Pure-Parsimony (HIPP) aims at finding a solution to the problem that minimizes the total number of distinct haplotypes required. The problem of finding such a solution is APX-hard (and, therefore, NP-hard) [10]. Experimental results [6,18] have shown that the accuracy of the HIPP approach is comparable with the one obtained with other approaches. However, until recently, HIPP inference methods were severely limited on the size of the problems they could handle. Recently, a SAT based approach for this problem, SHIPs [11,12], has shown that the use of effective constraint satisfaction methods leads to an efficient solution of this problem.

Motivated by these results, this paper explores an alternative approach. Existing ILP models only have Boolean variables and, therefore, can be solved with Pseudo-Boolean Optimization (PBO) solvers [5,13]. Hence, this paper starts by considering the utilization of PBO solvers instead of standard ILP solvers. The results are very promising, being competitive with SHIPs. These results motivate the development of a new PBO model (RPoly) for the HIPP problem, which is based on the PolyIP model [1,8] and, in addition, breaks key symmetries and yields a significantly more compact representation. The results show that RPoly is, in general, more efficient than SHIPs, and capable of solving more problem instances in a given time limit.

This paper is organized as follows. First we introduce the haplotype inference by pure parsimony problem. Afterwards, we describe the two main contributions of the paper: (1) how to solve HIPP ILP models using PBO and (2) how to optimize the existing polynomial model. Finally, we conclude and suggest future research work.

Class all with 1183 instances

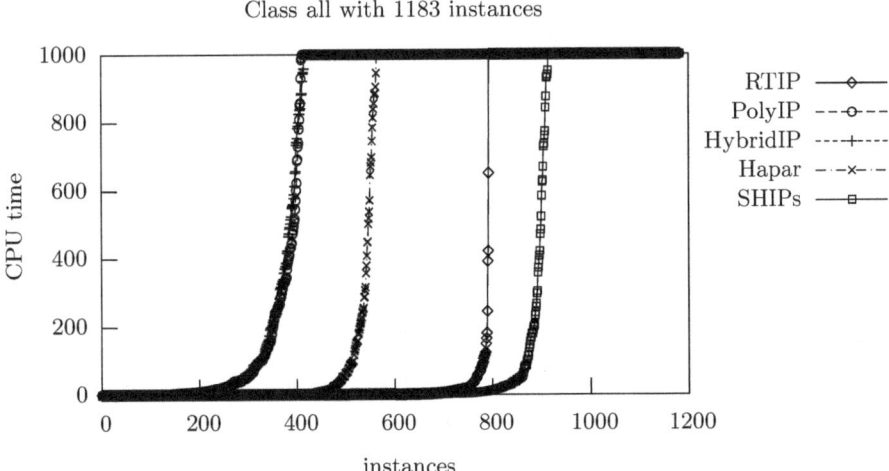

Fig. 1. Relative performance of HIPP solvers

2   Haplotype Inference by Pure Parsimony

A *haplotype* is the genetic constitution of an individual chromosome. The underlying data that forms a haplotype is generally viewed as the set of SNPs in a given region of a chromosome. Normal cells of diploid organisms contain two haplotypes, one inherited from each parent. The *genotype* represents the conflated data of the two haplotypes. The value of a particular SNP may be A, B or A/B, depending on whether the organism is homozygous with allele A, homozygous with allele B or heterozygous.

Starting from a set of genotypes, the haplotype inference by pure parsimony problem consists in finding a minimum set of haplotypes that can be used to derive, by pairwise combinations, the given set of genotypes.

Given a set $\mathcal{G}$ of n genotypes, each of length m, the haplotype inference problem consists in finding a set $\mathcal{H}$ of $2 \cdot n$ haplotypes, not necessarily different, such that for each genotype $g_i \in \mathcal{G}$ there is at least one pair of haplotypes (h_j, h_k), with h_j and $h_k \in \mathcal{H}$ such that the pair (h_j, h_k) explains g_i. The variable n denotes the number of individuals in the sample, and m denotes the number of SNP sites. g_i denotes a specific genotype, with $1 \leq i \leq n$. Furthermore, g_{ij} denotes a specific site j in genotype g_i, with $1 \leq j \leq m$.

Without loss of generality, we may assume that the values of the two possible alleles of each SNP are always 0 or 1. Value 0 represents the wild type and value 1 represents the mutant. A haplotype is then a string over the alphabet $\{0,1\}$. Moreover, genotypes may be represented by extending the alphabet used for representing haplotypes to $\{0,1,2\}$. Homozygous sites are represented by the values 0 or 1, depending on whether both haplotypes have value 0 or 1 at that site, respectively. Heterozygous sites are represented by value 2.

Table 1. Classes of instances used: number of SNPs and genotypes

Class	# Instances	minSNPs	maxSNPs	minGENs	maxGENs
ms	380	4	57	9	94
phasing	329	14	188	34	90
hapmap	24	4	29	5	68
biological	450	4	77	4	49
Total	1183	4	188	4	94

The HIPP problem is to find a minimum-size set $\mathcal{H}$ of haplotypes that explain all genotypes in $\mathcal{G}$. For example, consider the set of genotypes: 2120, 2120 and 1221. There are solutions for this example that use six distinct haplotypes, but solution 0100/1110, 0100/1101, 1011/1101 uses only four distinct haplotypes.

Two strings (denoting genotypes or haplotypes) are *incompatible* if and only if the strings have at least one site where one string has value 1 and the other string has value 0. Otherwise the strings are said to be *compatible*.

A comparison of the performance of alternative approaches to the HIPP problem is summarized in Figure 1. A universe of 1183 problem instances is used, from which 854 instances were taken from [12] and the remaining (harder) instances are described by Schaffner [15] and correspond to the SU-100kb, SU1, SU2 and SU3 classes available from http://www.stats.ox.ac.uk/~marchini/phaseoff.html. All problem instances were simplified in a preprocessing step, according to what has been suggested in [2]: duplicated genotypes and sites were removed, as well as complemented sites. For each class, Table 1 gives the number of instances, and the minimum and maximum number of SNPs and genotypes, respectively, after removing duplicated genotypes and duplicated and complemented sites. The *ms* class includes the uniform and nonuniform classes of instances that have been used in [2] but extended with additional, more complex, problem instances. The *phasing* instances correspond to the instances described in [15] which were generated to evaluate phasing algorithms. The *hapmap* class of instances is also the one used in [2]. Finally, the instances for the *biological* class were generated from publicly available data (e.g. [14,4,3,9]).

The HIPP solvers RTIP [6], PolyIP [1], HybridIP [2], Hapar [18] and SHIPs [12] were considered[1]. The run times for each solver were sorted and plotted, the cutoff point being 1000 seconds. All results shown were obtained on a 1.9 GHz AMD Athlon XP with 1GB of RAM running RedHat Linux. For the ILP-based HIPP solvers, the ILP package used was CPLEX version 7.5. As can be concluded, SHIPs is the HIPP tool capable of solving the largest number of problem instances. SHIPs aborts 268 problem instances out of 1183 instances, whereas RTIP aborts 389 instances, Hapar aborts 619 instances, HybridIP aborts 767 instances and PolyIP aborts 771 instances. Nonetheless, we should note that 95% of the problem instances aborted by RTIP were aborted due to memory exhaustion. Hence, RTIP may be competitive for solving some problem instances but it is not a robust solver.

---

[1] All results were obtained with the tools provided by the authors.

3   Solving ILP HIPP Models with PBO

This section reviews existing ILP models for the HIPP problem [7,10]. In addition, the section includes results using a modern Pseudo-Boolean Optimization (PBO) solver instead of a standard ILP solver.

In a pseudo-Boolean formula, variables have Boolean domains and constraints are linear inequalities with integer coefficients,

$$\sum c_i x_i \geq n \quad c_i, n \in \mathbb{Z}, x_i \in \{0, 1\}. \tag{1}$$

For example, $x + 2y - z \geq 2$ is a pseudo-Boolean constraint (also denoted as PB-constraint). From an ILP point of view, PB-constraints can be seen as a specialization of ILP where all variables are Boolean. This problem formulation is also known as *0-1 integer programming*. From a SAT point of view, PB-constraints can be seen as a generalization of clauses. Furthermore, a pseudo-Boolean formula can be extended with an optimization function.

3.1   Exponential-Size ILP Models

The first ILP model proposed for the HIPP problem, *RTIP* [6], has linear space complexity on the number of possible haplotypes and, therefore, it is exponential on the number of given genotypes.

A Boolean variable $y_{i,u}$ is associated with each pair u of haplotypes that can explain a given genotype g_i, and denotes whether this pair of haplotypes is used for explaining g_i. A cardinality constraint,

$$\sum_u y_{i,u} = 1, \tag{2}$$

requires that exactly one pair of haplotypes must be used for explaining each genotype, among all pairs that can explain the genotype. Each candidate haplotype is associated with a dedicated variable x_v, such that $x_v = 1$ if the haplotype is used. The utilization of a specific pair of haplotypes for explaining a genotype (i.e. $y_{i,u} = 1$) implies the respective x_v variable,

$$y_{i,u} \rightarrow x_v, \tag{3}$$

for each haplotype in the pair. The cost function is used to minimize the number of haplotypes used,

$$\text{minimize} \sum x_v. \tag{4}$$

This model corresponds to the *TIP* model [6]. The *RTIP* (Reduced TIP) model introduces one essential simplification. If the pair of haplotypes for a variable $y_{i,u}$ is such that they are not part of any other pair of haplotypes, then the $y_{i,u}$ variable and the related x_v variables can be removed from the formulation. A key drawback of the RTIP model is that the number of candidate haplotypes grows exponentially with the number of heterozygous sites. Hence, RTIP does not scale for large problem instances.

The RTIP model inspired a branch-and-bound algorithm to the HIPP problem, known as *Hapar* [18].

3.2 Polynomial-Size ILP Models

A more recent ILP model, *PolyIP* [1], is polynomial in the number of sites m and population size n, with a number of constraints and variables, respectively, in $\Theta(n^2 m)$ and $\Theta(n^2 + nm)$. The PolyIP model represents the $2 \cdot n$ candidate haplotypes as sequences of Boolean variables, and then establishes conditions for the haplotypes to explain the corresponding genotypes, such that the total number of distinct haplotypes is minimized. Haplotypes are represented with Boolean variables y_{ij}, $1 \leq i \leq 2n$ and $1 \leq j \leq m$, i.e. m variables for each of the $2 \cdot n$ candidate haplotypes.

First, the PolyIP model defines conditions on the sites, with $1 \leq i \leq n$ and $1 \leq j \leq m$,

$$\begin{aligned}
y_{2i-1\,j} &= 0 \text{ and } y_{2i\,j} = 0, \text{ if } g_{ij} = 0, \\
y_{2i-1\,j} &= 1 \text{ and } y_{2i\,j} = 1, \text{ if } g_{ij} = 1, \\
y_{2i-1\,j} &+ y_{2i\,j} = 1 \text{ if } g_{ij} = 2,
\end{aligned} \tag{5}$$

where $g_{ij} \in \{0, 1, 2\}$ denotes the possible values at each site. Second, the PolyIP model defines conditions for identifying different haplotypes, with $1 \leq l \leq i \leq 2n$ and $1 \leq j \leq m$. Boolean variable $d_{l\,i}$ is defined such that $d_{l\,i} = 1$ if $h_i \neq h_l$. The resulting conditions become

$$\begin{aligned}
y_{ij} - y_{l\,j} &\leq d_{l\,i}, \\
y_{l\,j} - y_{ij} &\leq d_{l\,i}.
\end{aligned} \tag{6}$$

If at least one site of h_i and h_l differs, then $d_{l\,i}$ needs to be assigned value 1.

Third, the model introduces the x_i variables, denoting whether h_i is different from all previous haplotypes h_l, where $1 \leq l < i$, and defines conditions on these variables. Each Boolean variable x_i is defined such that $x_i = 1$ if h_i is unique with respect to the previous haplotypes. Thus, if h_i is unique, then $\sum_{l=1}^{i-1} d_{l\,i} = i - 1$; otherwise $\sum_{l=1}^{i-1} d_{l\,i} < i - 1$. As a result, the condition on variable x_i becomes

$$x_i \geq 2 - i + \sum_{l=1}^{i-1} d_{l\,i}. \tag{7}$$

Finally, the cost function minimizes the number of different haplotypes,

$$\text{minimize} \sum_{i=1}^{2n} x_i. \tag{8}$$

A number of optimizations have been proposed to the basic PolyIP model [1], with the purpose of improving the quality of the LP relaxation step of standard ILP solvers, and therefore pruning the search space to be handled by the ILP solver.

More recently, the same authors introduced a new polynomial-size formulation, *HybridIP* [2], representing a hybrid of the RTIP and PolyIP formulations. Nevertheless, existing experimental results (see Figure 1) suggest that the performance of the two polynomial models does not differ significantly.

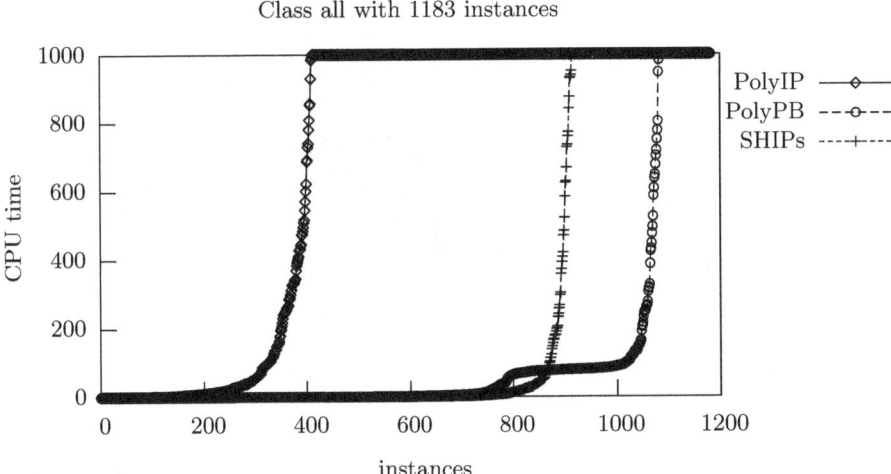

Fig. 2. Relative performance of PolyIP, PolyPB and SHIPs

3.3   ILP vs. PBO Solvers

As is clear from the description of the ILP models, all variables are Boolean and all coefficients are integer. Hence, the HIPP ILP models are also PBO models, and so PBO solvers can be considered. The results summarized in Figure 1 indicate that the performances of the PolyIP and HybridIP models are similar. Moreover, the RTIP model is known to be inadequate for larger problem instances, due to the exponential growth of the model in the number of heterozygous sites per genotype. As a result, this section only evaluates the performance of the PolyIP model using a PBO solver (hereafter referred to as PolyPB). The PBO solver MiniSAT+ [5] is used on all reported PBO results. Although other PBO solvers analyzed in [13] were considered, MiniSAT+ was by far the most efficient.

MiniSAT+ handles PB-constraints through translation to SAT without modifying the SAT procedure itself. In addition, the objective function is satisfied by iteratively calling the SAT solver where for each new iteration the objective function is updated until the problem is unsatisfiable. For example, given a minimization problem with an objective function $f(x)$, MiniSAT+ first runs the solver on the set of constraints (without considering the objective function) to get an initial solution $f(x_0) = k$. Then it adds the constraint $f(x) < k$ and runs the solver again. If the problem is unsatisfiable, then k is the optimum solution. If not, the process is repeated with the new smaller solution. Observe that translating to SAT results in an approach that is particularly suited for problems that are almost pure SAT. Indeed, this is the case for the HIPP problem. Hence, one may expect to get a faster procedure with MiniSAT+ than by applying a native PBO solver, not optimized towards propositional SAT.

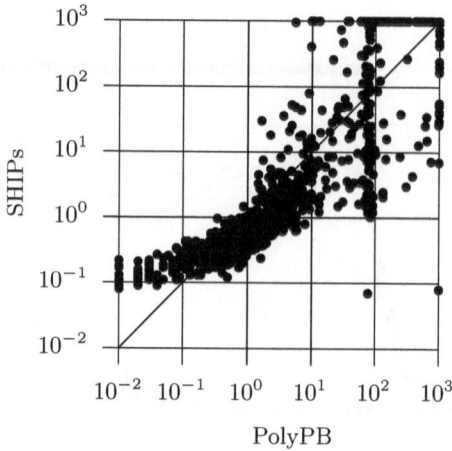

Fig. 3. Run times for PolyPB and SHIPs

Figure 2 compares the PolyIP model using the CPLEX solver, the PolyPB model using the PBO solver MiniSAT+ and the SHIPs solver on the 1183 problem instances described in Table 1 for a timeout of 1000 seconds. Clearly, PolyPB outperforms SHIPs in terms of the number of instances solved. Although both solvers are able to solve the majority of the 1183 problem instances within 1000 seconds, PolyPB only aborts 100 instances whereas SHIPs aborts 268 instances. Observe that PolyIP is significantly worse, aborting 771 out of 1183 instances.

In addition, Figure 3 provides a scatter plot with the run time for PolyPB and SHIPs on each of the problem instances with a timeout of 1000 seconds. For most problem instances SHIPs is faster than PolyPB; PolyPB is faster than SHIPs on 454 out of 1183 instances, with many of these instances being solved in less than one second. Nonetheless, this group of instances for which PolyPB is faster than SHIPs also includes 184 instances that PolyPB is able to solve and SHIPs aborts. On the other hand that there are only 16 instances that SHIPs is able to solve and PolyPB aborts. As a result, we can conclude that PolyPB is more robust than SHIPs. Finally, there are still 84 instances that both solvers are unable to solve within 1000 seconds.

4   RPoly: An Optimized PolyPB Model

Although the results shown in the previous section are promising, it is possible to further optimize the PolyPB model. Indeed, SHIPs is still showing a better performance in a large number of problem instances, which motivates the incorporation of some of the SHIPs model features into the PolyPB model. This section addresses optimizations to the PolyPB model with the main goal of reducing the run times.

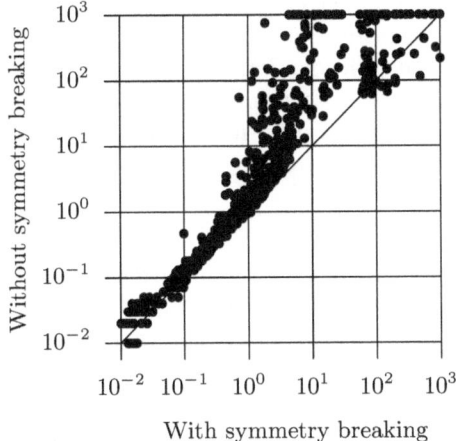

Fig. 4. Run times for PolyPB with and without symmetry breaking

These optimizations are two-fold: (1) the elimination of key symmetries and (2) the reduction of the size of the model. It is well-know that the SHIPs model would not be competitive if it was not for some specific optimizations, which include breaking key symmetries. Symmetries are broken by adding constraints to the model. We have also observed that the PBO instances generated with the PolyPB model are significantly larger than the SAT instances generated with the SHIPs model. The number of constraints in the PBO model can be up to an order of magnitude larger than the number of constraints in the SAT model, whereas the number of variables in the PBO model can be up to a factor of 3 larger than the number of variables in the SAT model.

The resulting model is referred to as *Reduced Poly model (RPoly)*.

4.1   Eliminating Key Symmetries

A key technique for pruning the search space is motivated by observing the existence of symmetry in the problem formulation. Clearly, given a solution to a HIPP problem were a genotype g_i is explained by the pair of haplotypes (h_{2i-1}, h_{2i}), the same genotype g_i may also be explained by the pair of haplotypes (h_{2i}, h_{2i-1}). Eliminating this symmetry significantly reduces the number of solutions and consequently reduces the search space.

In practice, this kind of symmetry is eliminated by adding additional constraints to the model, which guarantee that the elements in a pair of haplotypes are lexicographically ordered. Hence, for each site g_{ij} in a genotype g_i we must force the following:

- If $g_{ij} = 2$ and $g_{ij'} \neq 2$ $(\forall j' : j' < j)$, then $y_{2i-1\,j} - y_{2i\,j} < 0$.

Figure 4 compares the performance of the PolyPB model with and without symmetry breaking constraints. Clearly, with a few exceptions (72 out of 1183 instances), eliminating symmetries accelerates the performance of the PBO solver. The new model is faster than the PolyPB model for 90% of the instances and up to 2 orders of magnitude. This result comes as no surprise, given the success of the same technique when implemented in the SHIPs model. This result is indeed significant, as the new model only aborts 47 instances, whereas the PolyPB model aborts 100 instances.

4.2   Reducing the Model

The organization of RPoly follows the organization of PolyIP: two haplotypes are associated with each genotype, and conditions are defined which capture when a different haplotype is used for explaining a given genotype. However, RPoly has a few key differences. First, the set of variables is different. Instead of associating a variable with each site of each haplotype, RPoly only associates variables with heterozygous sites (since the value of haplotypes in the other sites is known beforehand, and so can be implicitly assumed). In addition, each used variable describes the possible pairs of values for the corresponding heterozygous site.

In practice, the model associates two haplotypes, h_i^a and h_i^b, with each genotype g_i, and these haplotypes are required to explain g_i. Moreover, the model associates a variable t_{ij} with each heterozygous site (i, j) (i.e. with $g_{ij} = 2$). Hence, $t_{ij} = 1$ indicates that $h_{ij}^a = 1$ and $h_{ij}^b = 0$, whereas $t_{ij} = 0$ indicates that $h_{ij}^a = 0$ and $h_{ij}^b = 1$ [2]. The value of h_i^a and h_i^b at homozygous sites j is implicitly assumed.

This alternative definition of the variables associated with the sites of genotypes reduces the number of variables by a factor of 2. In addition, the model only creates variables for heterozygous sites, and, therefore, the number of variables associated with sites equals the total number of heterozygous sites. As a result, the conditions provided by expression (5) are eliminated. It should also be mentioned that this definition of the variables associated with sites follows the SHIPs model [11,12].

Finally, another key modification is that the candidate haplotypes for each genotype are related with candidate haplotypes for other genotypes only if the two genotypes are *compatible*. Clearly, incompatible genotypes are guaranteed not to be explained by the same haplotype.

The proposed modification implies the use of two additional sets of variables. Variable $x_{i_1 i_2}^{p\,q}$, with $p, q \in \{a, b\}$ and $1 \le i_2 < i_1 \le n$, is 1 if the p haplotype of genotype i_1 and the q haplotype of genotype i_2 are incompatible. Clearly, if genotypes i_1 and i_2 are incompatible, then the value of $x_{i_1 i_2}^{p\,q}$ is 1 for the four possible combinations of p and q. Moreover, two genotypes i_1 and i_2 are related only with respect to sites j such that either g_{i_1} or g_{i_2} is heterozygous at that site. In addition, the model uses variables to denote when one of the haplotypes associated with a given genotype is different from all previous haplotypes. Hence, u_i^p, with $p \in \{a, b\}$ and $1 \le i \le n$, is 1 if haplotype p of genotype i is different from all previous haplotypes.

---

[2] Hence, the symmetry in a pair of haplotypes is broken by considering that $t_{ij} = 0$ for the first heterozygous site g_{ij} of each genotype g_i.

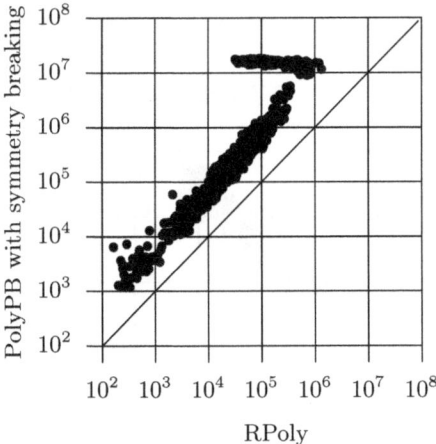

Fig. 5. Number of terms for PolyPB and RPoly

The conditions on the u_i^p variables are based on the conditions for the x_i variables for the PolyIP model,

$$\bigwedge_{1 \leq k < i} (x_{ik}^{p\,a} \wedge x_{ik}^{p\,b}) \rightarrow u_i^p. \tag{9}$$

The conditions on the $x_{i_1 i_2}^{p\,q}$ variables are all of the following form, for all $1 \leq j \leq m$:

$$\neg(R \leftrightarrow S) \rightarrow x_{i_1 i_2}^{p\,q}, \tag{10}$$

where the predicates R and S depend on the values of the sites (i_1, j) and (i_2, j), and on which of the haplotypes is considered, i.e., either a or b. Observe that $1 \leq i_2 < i_1 \leq n$, $1 \leq j \leq m$, and $p, q \in \{a, b\}$. Accordingly, the R and S predicates are defined as follows:

- If $g_{i_1 j} \neq 2$, then $R = \neg(g_{i_1 j} \leftrightarrow (q \leftrightarrow a))$ and $S = t_{i_2 j}$.
- If $g_{i_2 j} \neq 2$, then $R = \neg(g_{i_2 j} \leftrightarrow (q \leftrightarrow b))$ and $S = t_{i_1 j}$.
- If $g_{i_1 j} = 2 \wedge g_{i_2 j} = 2$, then $R = \neg(p \leftrightarrow q)$ and $S = \neg(t_{i_1 j} \leftrightarrow t_{i_2 j})$.

Finally, the cost function is given by

$$\text{minimize} \sum_{i=1}^{n} (u_i^a + u_i^b). \tag{11}$$

The proposed modifications result in significantly smaller PBO problem instances. Figure 5 compares the number of terms for the PolyPB and the RPoly models. The results are consistent and show that the number of terms in RPoly is a factor of 5 to 10 smaller than in PolyPB. Albeit not shown, the number of variables in RPoly can be up to a factor of 3 smaller than the number of

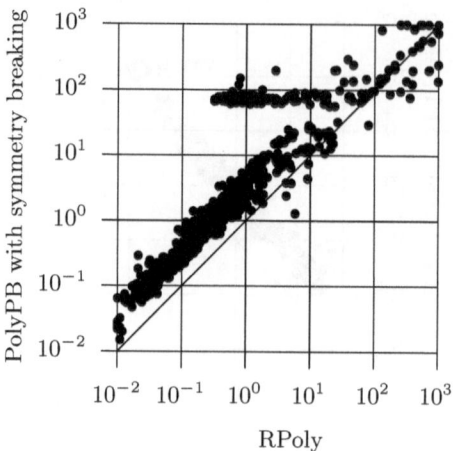

Fig. 6. Run times for PolyPB with symmetry breaking and RPoly

variables in PolyPB. We should note that the *phasing* class of instances exhibits a different behavior: most of these instances have around 10^7 terms in the PBO model with symmetry breaking. The number of terms in RPoly is not reduced for a constant factor, as it is for the other classes of instances. These instances have a higher number of incompatible genotypes when compared with the other classes of instances. Hence, the impact of the reduced model is much more significant. For the same reason, the impact on the run times is also more significant (see Figure 6 where the run time for the *phasing* instances using the PBO model with symmetry breaking is around 10^2 seconds). As a result, for these instances RPoly can outperform PolyPB by two orders of magnitude.

Finally we evaluate the effect of the reductions described above with respect to the run times. Figure 6 compares the PolyPB model extended with symmetry breaking constraints and the RPoly model, both using the PBO solver MiniSAT+, on the set of 1183 problem instances and with a timeout of 1000 seconds. With a few exceptions (28 out of 1183 instances), RPoly is consistently faster than PolyPB, and the speedup can reach 2 orders of magnitude. The few exceptions where RPoly is slower are explained by the branching heuristics used by MiniSAT+, which, in some cases, may not select the most adequate variables to branch on.

4.3   RPoly vs. SHIPs

In this section we measure the progress made with this work, by comparing the SHIPs model [12] with the RPoly model. The RPoly model is based on the PolyIP model but uses a PBO solver, MiniSAT+, and introduces key optimizations: the elimination of symmetries between the elements within a pair of haplotypes and the reduction on the size of the model.

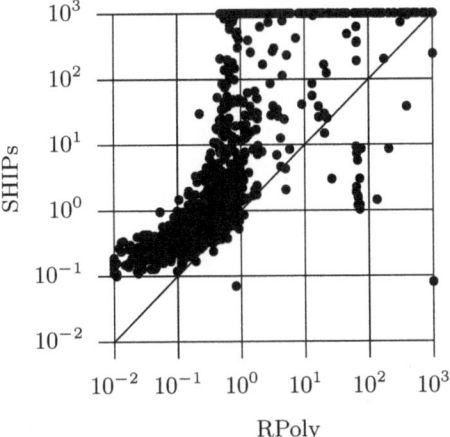

Fig. 7. Run times for RPoly and SHIPs

Although both RPoly and SHIPs use SAT-based technology, the two approaches differ. Whereas SHIPs considers an increasing number of haplotypes until a solution is found, RPoly considers $2 \cdot n$ haplotypes, where n is the number of genotypes, and iteratively reduces the number of different haplotypes until a solution with a minimum number of different haplotypes is found.

Figure 7 compares the RPoly model using the PBO solver MiniSAT+ and the SHIPs solver. For a small number of problem instances (52 out of 1183) SHIPs is faster than RPoly, and the speedup can reach 2 orders of magnitude. However, for most problem instances (1089 out of 1183), RPoly is faster than SHIPs. It should be observed that SHIPs is, in general, slower on very easy problem instances, essentially due to the initial setup time [11]. Nevertheless, the results also clearly show that RPoly is significantly more robust than SHIPs. RPoly aborts on a significantly smaller number of instances, being able to solve more than 96% of the problem instances. Finally, observe that only two instances aborted by RPoly can be solved by SHIPs.

5   Conclusions and Future Work

This paper studies the application of modern PBO solvers to the HIPP problem. By replacing the CPLEX ILP solver with the PBO solver MiniSAT+ [5], the existing PolyIP model [1] is shown to be competitive with the state-of-the-art method, SHIPs [12], being in general more robust. These results motivated the development of a new ILP model for the HIPP problem, RPoly, which entails a number of improvements to the basic PolyIP model inspired by SHIPs. The results for RPoly are significantly more promising than for PolyIP: RPoly is most often faster than SHIPs and is also significantly more robust, aborting only on a small number of problem instances (observe that, with two exceptions, SHIPs also aborts all of these instances).

The results indirectly suggest that the performance improvements obtained with SHIPs [11,12] are to a large extent explained by the efficiency of modern SAT solvers. Indeed, SAT-inspired PBO solvers obtain extremely good results with PolyIP and with RPoly, which are PBO models that differ significantly from the SHIPs SAT-based approach. In addition, the different PBO models provide a new, relevant, and essentially endless, source of challenging real problem instances for PBO solvers.

Despite the promising results obtained using MiniSAT+ with the RPoly model, several challenges remain. A number of problem instances cannot be solved by any HIPP solver. In addition, larger HIPP instances are expected to be significantly more challenging.

Finally, we should mention that having a competitive HIPP solver allows us to extend the pure parsimony approach with some ideas which are on the basis of other haplotype inference approaches. This will enable us to develop parsimony-based methods that explicitly incorporate genetic models (e.g. as in Phase [16]), with the objective of improving the accuracy of the reconstructed haplotypes.

Acknowledgments. This work is partially supported by Fundação para a Ciência e Tecnologia under research project POSC/EIA/61852/2004 and PhD grant SFRH/BD/28599/2006, and by INESC-ID under research project SHIPs.

References

1. Brown, D., Harrower, I.: A new integer programming formulation for the pure parsimony problem in haplotype analysis. In: Jonassen, I., Kim, J. (eds.) WABI 2004. LNCS (LNBI), vol. 3240, pp. 254–265. Springer, Heidelberg (2004)
2. Brown, D., Harrower, I.: Integer programming approaches to haplotype inference by pure parsimony. IEEE/ACM Transactions on Computational Biology and Bioinformatics 3(2), 141–154 (2006)
3. Daly, M.J., Rioux, J.D., Schaffner, S.F., Hudson, T.J., Lander, E.S.: High-resolution haplotype structure in the human genome. Nature Genetics 29, 229–232 (2001)
4. Drysdale, C.M., McGraw, D.W., Stack, C.B., Stephens, J.C., Judson, R.S., Nandabalan, K., Arnold, K., Ruano, G., Liggett, S.B.: Complex promoter and coding region β_2-adrenergic receptor haplotypes alter receptor expression and predict in vivo responsiveness. In: Proceedings of the National Academy of Sciences of the United States of America 97, pp. 10483–10488 (2000)
5. Eén, N., Sörensson, N.: Translating pseudo-Boolean constraints into SAT. Journal on Satisfiability, Boolean Modeling and Computation 2, 1–26 (2006)
6. Gusfield, D.: Haplotype inference by pure parsimony. In: Baeza-Yates, R.A., Chávez, E., Crochemore, M. (eds.) CPM 2003. LNCS, vol. 2676, pp. 144–155. Springer, Heidelberg (2003)
7. Gusfield, D., Orzach, S. (eds.): Handbook on Computational Molecular Biology. Chapman and Hall/CRC Computer and Information Science Series, chapter Haplotype Inference, vol. 9. CRC Press, Boca Raton (2005)

8. Halldórsson, B., Bafna, V., Edwards, N., Lippert, R., Yooseph, S., Istrail, S.: A survey of computational methods for determining haplotypes. In: Istrail, S., Waterman, M.S., Clark, A. (eds.) Computational Methods for SNPs and Haplotype Inference. LNCS (LNBI), vol. 2983, pp. 26–47. Springer, Heidelberg (2004)

9. Kroetz, D.L., Pauli-Magnus, C., Hodges, L.M., Huang, C.C., Kawamoto, M., Johns, S.J., Stryke, D., Ferrin, T.E., DeYoung, J., Taylor, T., Carlson, E.J., Herskowitz, I., Giacomini, K.M., Clark, A.G.: Sequence diversity and haplotype structure in the human ABCD1 (MDR1, multidrug resistance transporter). Pharmacogenetics 13, 481–494 (2003)

10. Lancia, G., Pinotti, C.M., Rizzi, R.: Haplotyping populations by pure parsimony: complexity of exact and approximation algorithms. INFORMS Journal on Computing 16(4), 348–359 (2004)

11. Lynce, I., Marques-Silva, J.: Efficient haplotype inference with Boolean satisfiability. In: National Conference on Artificial Intelligence (AAAI) (July 2006)

12. Lynce, I., Marques-Silva, J.: SAT in bioinformatics: Making the case with haplotype inference. In: International Conference on Theory and Applications of Satisfiability Testing (SAT), pp. 136–141 (August 2006)

13. Manquinho, V., Roussel, O.: The first evaluation of Pseudo-boolean solvers (PB'05). Journal on Satisfiability, Boolean Modeling and Computation 2, 103–143 (2006)

14. Rieder, M.J., Taylor, S.T., Clark, A.G., Nickerson, D.A.: Sequence variation in the human angiotensin converting enzyme. Nature Genetics 22, 59–62 (1999)

15. Schaffner, S., Foo, C., Gabriel, S., Reich, D., Daly, M., Altshuler, D.: Calibrating a coalescent simulation of human genome sequence variation. Genome Reasearch 15, 1576–1583 (2005)

16. Stephens, M., Smith, N., Donelly, P.: A new statistical method for haplotype reconstruction. American Journal of Human Genetics 68, 978–989 (2001)

17. The International HapMap Consortium. A haplotype map of the human genome. Nature, 437, 1299–1320 (2005)

18. Wang, L., Xu, Y.: Haplotype inference by maximum parsimony. Bioinformatics 19(14), 1773–1780 (2003)

An Algebraic Algorithm for the Identification of Glass Networks with Periodic Orbits Along Cyclic Attractors*

Igor Zinovik, Daniel Kroening, and Yury Chebiryak

Computer Systems Institute, ETH Zurich, 8092 Zurich, Switzerland
{izinovik,daniel.kroening,yury.chebiryak}@inf.ethz.ch

Abstract. Glass piecewise linear ODE models are frequently used for simulation of neural and gene regulatory networks. Efficient computational tools for automatic synthesis of such models are highly desirable. However, the existing algorithms for the identification of desired models are limited to four-dimensional networks, and rely on numerical solutions of eigenvalue problems. We suggest a novel algebraic criterion to detect the type of the phase flow along network cyclic attractors that is based on a corollary of the Perron-Frobenius theorem. We show an application of the criterion to the analysis of bifurcations in the networks. We propose to encode the identification of models with periodic orbits along cyclic attractors as a propositional formula, and solving it using state-of-the-art SAT-based tools for real linear arithmetic. New lower bounds for the number of equivalence classes are calculated for cyclic attractors in six-dimensional networks. Experimental results indicate that the run-time of our algorithm increases slower than the size of the search space of the problem.

1 Introduction

Many biological models can be formulated as hybrid systems in which the switch-like behavior of genes is approximated by discontinuous step functions, while the other state variables still change continuously in time. Piecewise-linear differential equations (PLDE) were proposed by Glass and Kaufmann as an approximation for systems in the context of gene regulation [1,2]. These equations are applied to the analysis of gene regulatory networks [3,4,5,6] and neural networks [7,8]. The piecewise linear approach for describing complex nonlinear dynamics is actively studied and utilized in control theory, design of electric and electronic circuits, and embedded software.

The main distinction of biological phenomena is that the interactions are characterized by very localized coupling of the state variables, unlike complex couplings in the context of control and electronic circuit problems. In the resulting model, interactions between genes are present only in the piecewise constant

---

* This research is supported in part by an award from IBM Research and by ETH Research Grant TH-19 06-3.

H. Anai, K. Horimoto, and T. Kutsia (Eds.): AB 2007, LNCS 4545, pp. 140–154, 2007.

terms of the PLDE [9]. Let n denote the number of genes and x_i denote the concentration of the product of gene i. The vector of the x_i-s is denoted by $\mathbf{x}$. The equations can be written in the form

$$\dot{x}_i = -g_i(\mathbf{x}) - \gamma_i x_i \quad \text{for } 1 \leq i \leq n,$$

where $\gamma_i > 0$ is the degradation rate of x_i. The function $g_i : \mathbb{R}^n_{\geq 0} \to \mathbb{R}_{\geq 0}$ describes the coupling of the variables and is defined as

$$g_i(\mathbf{x}) = \sum_{l \in L} k_{il} b_{il}(\mathbf{x})$$

where $k_{il} \geq 0$ is a rate parameter, L is a set of indexes, and $b_{il} : \mathbb{R}^n_{\geq 0} \to \{0,1\}$ is a composition of step functions with the steps located at the prescribed threshold concentrations $x_i = \theta_{il}$. The function b_{il} expresses the conditions under which the gene causes production of the protein at a rate k_{il}. The constant θ_{il} denotes the l-th threshold concentration of the protein encoded by gene i. The thresholds induce a partitioning of the phase space into a set of n-dimensional boxes. In each box, the protein concentrations are described by ODEs with a constant production term μ_i and a rate parameter γ_i:

$$\dot{x}_i = \mu_i - \gamma_i x_i \quad \text{for } 1 \leq i \leq n$$

The global behavior of PLDE with several thresholds for every continuous variable are actively studied in the context of modeling of gene regulatory networks [10] and the qualitative theory of differential equations [11]. If the model of the gene activity is restricted to on/off expressions and the decay rates are identical for all reactions, the PLDE system is reduced to a Glass model [12]. The general form of a Glass network is

$$\dot{x}_i = G_i(\tilde{x}_1, \ldots, \tilde{x}_n) - \alpha x_i \quad \text{for } 1 \leq i \leq n \text{ and } \alpha > 0.$$

The protein production rates are defined via the interaction functions G_i, where $\tilde{x}_i = a$ if $x_i < \theta_i$, and $x_i = b$ if $x_i > \theta_i$ with real constants $a < b$. Using appropriate scaling of the variables, the PLDE can be transformed into the system

$$\dot{y}_i = F_i(\tilde{y}_1, \ldots, \tilde{y}_n) - y_i \quad \text{for } 1 \leq i \leq n,$$

where $\tilde{y}_i = 0$ if $y_i < 0$, and $\tilde{y}_i = 1$ if $y_i > 0$ [12]. The equations describe a network with all thresholds equal 0 and unit decay rate. The equations can be easily integrated, and the trajectories are straight lines in every orthant[1] $\mathcal{O}_k$, $k \in \{1, 2, 3, \ldots, 2^n\}$, of the phase space. The phase flow in each orthant $\mathcal{O}_k$ is defined by its focal point $\boldsymbol{f}^k = (f_1^k, f_2^k, \ldots, f_n^k) \in \mathbb{R}^n$ where $f_i^k = F_i(\tilde{y}_1, \tilde{y}_2, \ldots, \tilde{y}_n)|_{\mathcal{O}_k}$. Thus, the Glass network can be specified by a choice of a set of focal points $\{\boldsymbol{f}^{(k)}\}$, $k \in \{1, 2, 3, \ldots, 2^n\}$.

The phase flow in Glass networks is studied using a *state transition diagram*, which is represented by an n-cube with directed edges. Each orthant of the phase

---

[1] Generalization of a quadrant to the n-dimensional Euclidean space.

space is associated with a vertex of the n-cube, and each common boundary of the orthants corresponds to an edge of the cube. The edge is directed according to the direction of the phase flow across the boundary [13]. Figure 1 illustrates a phase flow with two trajectories of a two-dimensional Glass network. The state transition diagram for a 3-dimensional Glass network is shown in Fig. 2. The vertices of the n-cube are labeled by tuples of n binary variables $(\tilde{y}_1, \tilde{y}_2, \ldots, \tilde{y}_n)$, which define a valuation of the network interaction functions F_i. Periodic trajectories of the networks correspond to closed cycles in the transition graphs (e.g., see thick line in Fig. 2).

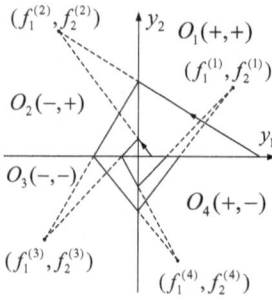

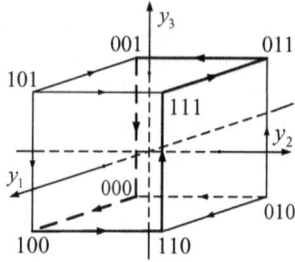

Fig. 1. 2-d phase flow   **Fig. 2.** 3-d transition diagram

The global phase flow in Glass networks can be quite complex. Oscillations towards equilibrium states, cycles and limit cycles may occur when linear parts of the trajectories are connected continuously over sequences of orthants [13,14,12,15]. Numerical simulations [13,16] indicate that for dimensions greater than 4, Glass networks may exhibit aperiodic and chaotic behavior. Studies of the periodic solutions for Glass models show that there are networks that possess a special type of stable limit cycles: the flow between the orthants along these cycles is unambiguous, i.e., for each orthant along the cycle, all trajectories must go to the same successor. In other words, the basin of attraction of the periodic trajectory is composed of all orthants spanned by the trajectory. Networks with such stable cycles are called networks with *cyclic attractors* [13].

Definition 1 (Cyclic Attractor). *A cycle in the state diagram is called a cyclic attractor if a) it is a chord-free simple cycle in the n-cube[2], and b) all edges adjacent to the cycle have to be directed towards the cycle nodes.*

As example, the cycle shown in Fig. 2 is a cyclic attractor.

Models for gene regulatory networks with equilibrium states and stable limit cycles are of special interest in Systems Biology because the models serve for simulation of cell differentiation processes and variability of cell types [17,18]. The classification of the cyclic attractors with respect to symmetry transformation of the n-cube up to dimension 5 has been completed [13]. Numerical studies of the

---

[2] Every edge in the graph that joins two vertices of the cycle is an edge of this cycle.

3-dimensional cyclic attractor identified unique stable oscillations for the value of the bifurcation parameter greater than the Hopf bifurcation point [19]. The network with cyclic attractors was integrated numerically for the 4-dimensional state space to simulate a neural network [20]. Three stable periodic trajectories were found by the parametric study of PLDE models, and the period of each cycle was computed for a set of thresholds θ_i.

To summarize, the classification of the transition diagrams has been obtained for Glass networks up to dimension five. Analytical results on phase flow are presented for three- and four-dimensional networks. The analysis relies on the integration of PLDE and numerical solutions of eigenvalue problems for the matrix associated with the Poincaré return map. Models for the simulation of gene regulatory and neural networks utilize Glass networks with cyclic attractors. Phase flow along cyclic attractors was proven to admit either a stable periodic orbit or to converge to the origin. If the focal coordinates $\{f_i^{(k)}\}$ for the system with cyclic attractor equal ± 1, the flow always is attracted by the unique 1-period trajectory.

The determination of the parameters of gene regulation models based on experimental observations is known to be highly desirable [17], and is a computationally difficult problem [21]. A solution of the inverse problem of 4-dimensional Glass network reconstruction based on a partial information about the transition diagram and the signs of focal coordinates is shown in [3]. The objective of this paper is to suggest an efficient method for the identification of networks with cyclic attractors that exhibit phase flow of a prescribed type for a given set of focal point coordinates.

The problem is stated as follows: based on a given sequence of absolute values of focal point coordinates $\{|f_i^{(k)}|\}$ and the desired flow type, identify a Glass network with an attractor that admits the prescribed flow.

As there are straight-forward upper bounds for the length of cyclic attractors for a given dimension, we propose to use an encoding into propositional satisfiability (SAT) for the search. There are two contributions we present:

1. We propose an algebraic method for analysis of structural stability of phase flow for Glass networks with cyclic attractors. Our method utilizes a corollary of the Perron-Frobenius theorem and gives a criterion for the identification of the flow type along the cyclic attractors.
2. We propose a scalable SAT-based algorithm for identification of the networks with cyclic attractors.

Outline. The paper is organized as follows. In Section 2, we extend a sufficient condition for the identification of networks with cyclic attractors [13] to an algebraic criterion (i.e., necessary and sufficient condition), which is derived from properties of the state transition diagram of the networks. We also present an application of the criterion to the analysis of structural stability of the phase flow as an example, which is useful later on for the construction of the algorithm for network identification. In Section 3, we introduce an algorithm based on SAT for the search for cyclic attractors in the state transition diagram. In Section 4, we integrate the proposed criterion as a part of the algorithm for the identification of Glass networks with periodic orbits along cyclic attractors. We also present experimental results that indicate that the algorithm scales well in the network dimension.

2   Algebraic Criterion for Flow Identification

The flow along cyclic attractors is known either to converge to the origin or to admit a unique stable 1-period orbit. The type of the flow is identified by analyzing a Poincaré plane: the flow with periodic orbit has a unique fixed point, while the Poincaré map for flows converging to the origin does not have fixed points. Thus, a criterion for flow identification should distinguish between Poincaré maps with and without fixed points.

Poincaré return maps of Glass networks can be represented by the composition of *fractional linear maps* $M^{(k)} : \mathbb{R}^n \to \mathbb{R}^n$ [13]. Following the notation in [12], the mapping can be presented as:

$$\boldsymbol{y}^{(k+1)} = M^{(k)}\boldsymbol{y}^{(k)} = B^{(k)}\boldsymbol{y}^{(k)}/(1 + \langle \psi^{(k)}, \boldsymbol{y}^{(k)} \rangle), \tag{1}$$

where $\boldsymbol{y}^{(k)}$ is the coordinate vector on the k-th orthant boundary crossed by the trajectory, and the matrix $B^{(k)}$ is calculated as

$$B^{(k)} = I - (\boldsymbol{f}^{(k)}\boldsymbol{e}_j^T)/f_j^{(k)} \tag{2}$$

where I is the identity matrix. The focal point $\boldsymbol{f}^{(k)}$ associated with the orthant being entered is assumed not to lie in that orthant, $\boldsymbol{e}_j$ denotes the standard basis vector in $\mathbb{R}^n$, and the vector $\psi^{(k)}$ is defined to equal $-\boldsymbol{e}_j/f_j^{(k)}$. The angular brackets denote the Euclidean inner product. Thus, the return map for a cycle restricted to the orthant boundary $y_i = 0$ can be written as

$$M\boldsymbol{y} = A\boldsymbol{y}/(1 + \langle \phi, \boldsymbol{y} \rangle), \tag{3}$$

where $A = ||a_{mp}||$ is the $(n-1) \times (n-1)$ matrix obtained by deleting the i-th column and row of the composition of $B^{(k)}$, and ϕ is the same reduction of the composition of $\psi^{(k)}$ [12].

The values of the matrix elements a_{mp} depend on the choice of the initial orthant boundary as well as on the order of enumeration of the variables, and the prescribed orientation of the basis vectors along the axes. The same N-node cycle in the state diagram may be represented by $N \cdot n! \cdot 2^n$ different matrices. In case of a cyclic attractor, the matrix can be obtained in such a way that all its elements are positive [13]. Subsequently, the Perron-Frobenius theorem guarantees that the flow admits a stable periodic orbit if the dominant eigenvalue r of the positive matrix A is greater than one, and converges to the origin otherwise. Therefore, if a cyclic attractor is represented by a positive matrix, the identification of the flow type does not require the calculation of eigenvalues, but only reasoning about satisfiability of the inequality $r > 1$ for positive matrices. For this purpose, we suggest to utilize a corollary to the Perron-Frobenius theorem. The corollary asserts [22]:

Corollary 1. *A real number λ is greater then the maximal characteristic value r of the (non-negative) matrix A if and only if for this value λ all the successive principal minors of the characteristic matrix $\lambda I - A$ are positive.*

If we are only interested in testing $r > 1$, we need to check the signs of the determinants of the k by k upper left matrices of $A - I$ being $(-1)^k$. Thus, the following algebraic criterion for the identification of flows in cyclic attractors can be used:

Criterion 1. *The flow of an n-dimensional cyclic attractor converges to the origin if and only if the signs of the determinants of the k by k upper left matrices of $A - I$ are $(-1)^k$ for $k = 1, 2, \ldots, n - 2$ and the sign of $\det(A - I)$ is $(-1)^{n-1}$ or $\det(A - I) = 0$, where A is the positive matrix that defines the return map of the attractor by means of Eq. 3. Otherwise, the phase flow along the cyclic attractor admits a unique stable 1-period orbit.*

The analysis of the generic ways in which stable attractors undergo bifurcations in Glass networks is an open question listed in [15]. As a simple example of an application of Criterion 1 to bifurcation analysis, we can consider the structural stability of phase flow along a cyclic attractor for the 3-dimensional Boolean Glass network shown in Fig. 2.

First, we have to define the focal point coordinates of the network. Two conditions are assumed throughout the paper: focal points lie inside orthants and none of them on the orthant boundaries, and the i-th state variable does not change in sign when crossing an orthant boundary in direction i. The conditions ensure that the flow is unambiguous [12,11][3]. In this case, the focal point for every orthant of the cycle lies inside the next cycle orthant.

Example 1. The attractor in Fig. 2 is represented by the orthant sequence

$$(111) \rightarrow (011) \rightarrow (001) \rightarrow (000) \rightarrow (100) \rightarrow (110)$$

Thus, the sequence of focal points is obtained by applying a one-step cyclic shift to the sequence of orthants, and replacing all 0-s by -1, and is written as:

$$(-1, 1, 1) \rightarrow (-1, -1, 1) \rightarrow (-1, -1, -1) \rightarrow (1, -1, -1) \rightarrow (1, 1, -1) \rightarrow (1, 1, 1)$$

Let us consider the perturbations of the first focal point when it remains inside the same orthant (011). The focal point sequence undergoing the perturbations has the form:

$$(-\epsilon_1, \epsilon_2, \epsilon_3) \rightarrow (-1, -1, 1) \rightarrow (-1, -1, -1) \rightarrow (1, -1, -1) \rightarrow (1, 1, -1) \rightarrow (1, 1, 1),$$

where $\epsilon_1 > 0$, $\epsilon_2 > 0$, and $\epsilon_3 > 0$ are free parameters of the network. The matrix $A = \|a_{mp}\|$ for the return map is calculated using Equations (1-3):

$$\begin{pmatrix} \frac{8\epsilon_2}{\epsilon_1} + \frac{5\epsilon_3}{\epsilon_1} & 8 \\ \frac{5\epsilon_2}{\epsilon_1} + \frac{3\epsilon_3}{\epsilon_1} & 4 \end{pmatrix}.$$

---

[3] The conditions can be relaxed using set-valued Filippov solutions. The application of differential inclusions to PLDE is still a current research topic [10], and is not considered in this paper.

All elements of the matrix are positive due to the definition of the perturbation via positive ϵ-s. Thus, Criterion 1 is applicable to the matrix above, and it asserts that the cyclic attractor admits the flow converging to the origin if and only if

$$a_{11} < 0 \wedge (a_{11}a_{22} - a_{21}a_{12} > 0 \vee a_{11}a_{22} - a_{21}a_{12} = 0). \tag{4}$$

The corresponding systems of inequalities are written as

$$\begin{cases} \frac{-\epsilon_1 + 8\epsilon_2 + 5\epsilon_3}{\epsilon_1} < 0 \\ \frac{4(\epsilon_1 + 2\epsilon_2 + \epsilon_3)}{\epsilon_1} > 0 \end{cases} \quad \text{or} \quad \begin{cases} \frac{-\epsilon_1 + 8\epsilon_2 + 5\epsilon_3}{\epsilon_1} < 0 \\ \frac{4(\epsilon_1 + 2\epsilon_2 + \epsilon_3)}{\epsilon_1} = 0 \end{cases}$$

Both systems are inconsistent, and therefore, the flow admits a stable periodic orbit, i.e., the network is stable under any perturbations of the first focal point that leave the point inside the orthant $y_1 < 0, y_2 > 0, y_3 > 0$.

The perturbations of any single focal point within the orthants have been found to preserve the flow type along 3-and 4-dimensional cyclic attractors for Boolean Glass networks. In contrast, simultaneous perturbation of two coordinates of different focal points may change the flow from "periodic" to "converging to the origin". As an example, consider perturbations of the second coordinate of the fifth focal point and the third coordinate of the sixth focal point.

Example 2. Let us consider the sequence of the focal points which is written as

$$(-1, 1, 1) \rightarrow (-1, -1, 1) \rightarrow (-1, -1, -1) \rightarrow (1, -1, -1) \rightarrow (1, \epsilon_1, -1) \rightarrow (1, 1, \epsilon_2)$$

The corresponding positive matrix A is

$$\begin{pmatrix} \frac{5}{\epsilon_1} + \frac{3}{\epsilon_2} + \frac{5}{\epsilon_1 \epsilon_2} & \frac{3}{\epsilon_1} + \frac{2}{\epsilon_2} + \frac{3}{\epsilon_1 \epsilon_2} \\ \frac{3}{\epsilon_2} + \frac{5}{\epsilon_1 \epsilon_2} & \frac{2}{\epsilon_2} + \frac{3}{\epsilon_1 \epsilon_2} \end{pmatrix}$$

The system (4) that represents the criterion is simplified by cylindrical decomposition implemented in *Mathematica*. The sufficient condition for converging flow along the cyclic attractor and the bifurcation condition are written as:

$$\begin{cases} \epsilon_1 > 5 \\ \epsilon_2 > \frac{7 + 5\epsilon_1}{-5 + \epsilon_1} \end{cases} \quad \text{or} \quad \begin{cases} \epsilon_1 > 5 \\ \epsilon_2 = \frac{7 + 5\epsilon_1}{-5 + \epsilon_1} \end{cases}$$

Any solution of this system defines a network with the flow converging to the origin. As an example of one parametric bifurcation diagram we consider a solution of the second system with $\epsilon_1 = 6$ and $\epsilon_2 = 37$. In this case, the dominant eigenvalue r is 1 and the phase flow converges to the origin. The sequence of focal points with the bifurcation parameter μ is written as:

$$(-1, 1, 1) \rightarrow (-1, -1, 1) \rightarrow (-1, -1, -1)$$
$$\rightarrow (1, -1, -1) \rightarrow (1, 6, -1) \rightarrow (1, 1, 37 - \mu)$$

If $\mu \leq 0$, the cyclic attractor admits the flow converging to the origin, and if $\mu > 0$, the location of the fixed point $y^*(\mu)$ on the Poincaré plane $(y_1 > 0, y_2 > 0, y_3 = 0)$ is computed as [12]:

$$y^*(\mu) = \frac{(r-1)v}{\langle \phi, v \rangle},$$

where v is the eigenvector corresponding to the dominant eigenvalue r. The characteristic polynomial for the matrix A is quadratic, and thus, a bifurcation diagram that represents the fixed point coordinate $y^*(\mu)$ can be obtained in a closed analytical form. The bifurcation diagram was found to be similar to a Hopf supercritical bifurcation for non-linear ODE (see Fig. 5 in [23]).

Criterion 1 relies on the condition that matrix A is positive, and thus, the first step of any application of the criterion is to find the sequence of nodes in the n-cube that determines the cyclic attractors with a positive matrix. Such sequences have to satisfy condition (a) of Def. 1, and are called *induced cycles*. The problem of finding longest induced paths in graphs is known to be *NP*-complete [24], and the problem of detecting longest induced cycles in n-cubes is open for dimensions greater than 7 [25]. We propose to encode the search for induced cycles into a satisfiablity (SAT) problem for propositional logic. Thus, the computationally demanding calculations can be handled by the state-of-the-art SAT solvers, which are known to be very efficient for problems with large, tightly constrained search spaces.

3   Computing Induced Cycles

The search for an induced cycle in the network state transition diagram relies on the identification of a cycle with desired properties on n-cubes. The length of the cycle N and the dimension n serve as input parameters. We propose to apply propositional SAT to the search for the attractors.

A state corresponds to a coordinate vector labeling the nodes on the n-cube, i.e., an n-tuple of Boolean variables. Let $s_{i,j}$ with $i \in \{1, \ldots, N\}$, $j \in \{1, \ldots, n\}$ denote the value of bit j in step i. The transitions on the n-cube correspond to sequences of states that satisfy a Gray code condition: the Hamming distance between two neighboring states equals one. We write $H_{k,l}^\alpha$ if the Hamming distance between the states s_k and s_l is α. The Gray code condition is then written as the following conjunction:

$$\Psi^{gray} = \bigwedge_{i=1}^{N-1} H_{i,i+1}^1 \wedge H_{1,N}^1$$

The constraints that eliminate the chords from the paths are represented by the following formula:

$$\Psi^{cycle} = \bigwedge_i \bigwedge_j [H_{i,j}^1 \Leftrightarrow (H_{i-1,j}^0 \vee H_{i+1,j}^0)]$$

The constraints guarantee that the Hamming distance for two of the cycle nodes equals one if and only if one of the nodes is either the previous or the next in the cycle with respect to the other one. A satisfying assignment to

$$\Psi^{ind} = \Psi^{gray} \wedge \Psi^{cycle} \tag{5}$$

identifies an induced cycle of an n-cube. The set of all attractors is represented by the set of all satisfying assignments of formula (5).

Due to the symmetries in the n-cube, the set of cycles that corresponds to the solutions of (5) is highly redundant. Glass proposes equivalence classes that are defined as sets of induced cycles such that all cycles in every set can be obtained via n-cube symmetry transformations of any cycle in the set [2]. The classification for 5-dimensional networks was obtained by Glass [2] using an enumeration approach. We utilize (5) to extend the classification to 6-dimensional networks.

The computation of the equivalence classes utilizes *coordinate* and *interval sequences* for Gray codes and paths on n-cubes. The *coordinate sequence* is a listing of the coordinates that change as the cycle is traversed. The *interval sequence* of a coordinate is a tuple giving the number of coordinates intervening between each successive appearance of the coordinate in the coordinate sequence. A necessary but not sufficient condition that two induced cycles are equivalent is that the set of interval sequences for one cycle are in a one-to-one correspondence with the set of interval sequences of the second cycle, where the interval sequence for any one coordinate can be cyclically permuted [2]. We apply the condition to compute lower bounds for the number of equivalence classes as follows:

1. We obtain the set of induced cycles of a given length N in the n-cube by computing all satisfying assignments of (5). This is an *all-SAT* problem.
2. We construct the set of equivalence classes as follows: every satisfying assignment is decoded back to the coordinates of the induced cycle on n-cube that it represents; if this induced cycle does not belong to any of the computed classes, it is added to the set as the representative.

The pseudocode for the computation is shown in [23]. The all-SAT problem is solved using the *blocking clause* algorithm [26] for the MiniSAT SAT-solver [27]. The algorithm computes a satisfying assignment of the given formula, saves it, and constructs a clause that eliminates the assignment. The clause is added to the formula as an additional constraint, and the previous step is repeated until no satisfying assignment can be found. There are more efficient algorithms available for the all-SAT problem, but these techniques are beyond the scope of this paper.

We applied the algorithm to 5- and 6-dimensional cubes (see the results in Table 1 in [23]). The lower bound for the total number of equivalence classes for six dimensions has been found to increase from 17 to 3007. The computed bound for 5-dimensional networks differs from the exact number of classes [2] by just one class. To the best of our knowledge, these lower bounds for the number of equivalence classes for dimension 6 are presented for the first time.

The network identification may require the evaluation of all induced cycles, even if they belong to the same equivalence class (see the example in [23]). The results of the all-SAT computation indicate that the number of induced cycles increases rapidly with the network dimension: the total number of cycles is 238 and 706336 for 5- and 6-dimensional networks, respectively. Thus, the search over the set of cycles becomes computationally demanding with increasing network dimension.

The size of the search space grows in the order of $2^{n \cdot N}$, i.e., exponentially in the network dimension and the length of the induced cycle. On the other hand, the number of induced cycles *decreases* when the cycle length approaches its maximum value. These opposite trends compromise the efficiency of any algorithm if it identifies Glass networks by enumerating cyclic attractors and applying Criterion 1 subsequently. In the next section, we propose combining the criterion for the flow detection and the search of induced cycles into an identification algorithm that efficiently scales with the network dimension.

4   Algorithm for Network Identification

4.1   Implementation Using SMT

An algorithm that simultaneously detects the flow type and identifies the cyclic attractors is required to conduct a search over both the continuous and discrete parts of the problem. We propose to utilize solvers for *Satisfiability Modulo Theories* for this problem. Sate-of-the-art solvers for Satisfiability Modulo Theories (SMT) decide logical satisfiability (or dually, validity) with respect to a background theory expressed in classical first-order logic with equality. These theories include real or integer arithmetic, which makes SMT solvers a successful tool for the analysis of problems that include linear inequalities over reals [28]. We propose to encode the identification of networks by adding the inequalities that represent the criterion for flow detection to the propositional formula (5). The Boolean structure of the inequality system for the three-dimensional network is defined by formula (4). In case of an arbitrary dimension n, the formula is written as

$$\Psi^{con} = \Psi^{suf} \vee \Psi^{bif},$$

where a sufficient condition for the converging flow is defined by

$$\Psi^{con} = (det(A - I)^{(1)} < 0) \wedge (det(A - I)^{(2)} > 0) \wedge (det(A - I)^{(3)} < 0) \wedge \ldots$$
$$\wedge (det(A - I)^{(n-1)} \gtrless 0),$$

and the condition for the bifurcation point is

$$\Psi^{bif} = (det(A - I)^{(1)} < 0) \wedge (det(A - I)^{(2)} > 0) \wedge (det(A - I)^{(3)} < 0) \wedge \ldots$$
$$\wedge (det(A - I)^{(n-2)} \gtrless 0) \wedge (det(A - I)^{(n-1)} = 0).$$

Here, $det(A - I)^{(k)}$ denotes the determinant of the upper left $k \times k$ matrix of $A - I$ and $\gtrless$ changes accordingly with the sign of $(-1)^k$.

The criterion is applicable if $A = ||a_{mp}||$ is a positive matrix. The following condition guarantees that the cyclic attractor induces a matrix with positive entries:

$$\Psi^{pos} = \bigwedge_m \bigwedge_p (a_{mp} > 0).$$

The matrix elements a_{mp} are calculated using (1-3) based on the prescribed sequence of absolute values $\{|f_i^k|\}$ for the focal point coordinates and a satisfying assignment $s^*_{k,i}$ of the propositional formula (5). The assignment defines the signs in the sequence of focal coordinates $\{|f^k_i|\}$ such that the focal point for every orthant of the induced cycle is located inside the next orthant along the cycle:

$$f^{(k)}_i = \begin{cases} -|f^{(k)}_i| : \text{if } \neg s^*_{k-1,i} \\ |f^{(k)}_i| : \text{otherwise} \end{cases}$$

A Glass network with converging flow along the cyclic attractor is identified from a satisfying assignment for the propositional formula

$$\Psi^{ind} \wedge \Psi^{pos} \wedge \Psi^{con}, \tag{6}$$

and a network with stable periodic orbit is specified by a satisfying assignment for the formula

$$\Psi^{ind} \wedge \Psi^{pos} \wedge \neg\Psi^{con}. \tag{7}$$

An assignment for (6) or (7) solves the corresponding identification problem if all coordinates $\{|f^{(k)}_i|\}$ are given as a sequence of positive real numbers. On the other hand, the formulae allow for an analysis of the structural stability of the network if one coordinate of the sequence is a positive parameter ϵ that undergoes the perturbation. The periodic flow along the cyclic attractor is structurally unstable if there is a satisfying assignment for the formula

$$\Psi^{ind} \wedge \Psi^{pos} \wedge \Psi^{con} \wedge (\epsilon > 0). \tag{8}$$

The perturbation of two or more focal coordinates causes the polynomial inequalities in the criterion to appear (see the examples in Section 2). Non-linear inequalities are not supported by any of the existing SMT solvers, and thus, the calculations are restricted to the case of one parameter. Additional constraints may be added to limit the analysis of structural stability to a particular equivalence class of the n-cube.

4.2   Experiments

We evaluated the Yices and CVCL SMT solvers [29,30]. Yices won the Satisfiability Modulo Theories competition[4] in 2006 in the relevant category. As first step of the experimental evaluation, we compare the run-time of the search for a single induced cycle, i.e., checking satisfiability of the purely propositional formula (5) for various instances. The test cases include the search for the induced cycles of different length in the networks of dimension 4, 5, and 6 (see Table 2 in [23]). A PC with a 1.4 GHz processor and 2 GB RAM was used for the evaluation. We also recorded the run-time of MiniSAT on the same instance as a reference point.

   The difference between MiniSAT and the SMT solvers is that MiniSAT accepts conjunctive normal form (CNF) as input directly, while the SMT solvers use rich

---

[4] Computer-Aided Verification Conference, SMT-COMP, http://www.csl.sri.com/users/demoura/smt-comp/

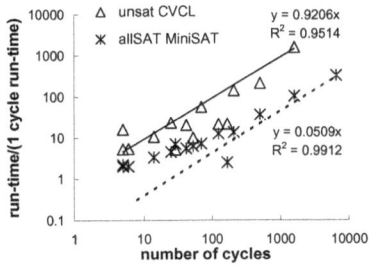

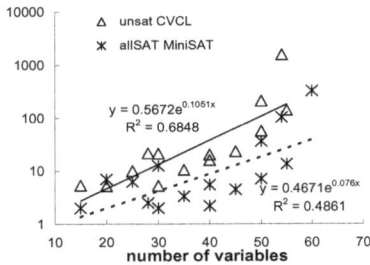

Fig. 3.   **Fig. 4.**

input languages that include all Boolean logic operators. The results indicate that a hand-tuned translation of the propositional formula (5) into CNF may increase the speed of the SMT solver up to the efficiency of MiniSAT, which is currently one of the fastest tools for satisfiability analysis of propositional formulae (see Fig. 6 in [23]).

We chose CVCL over Yices for the evaluation of the identification algorithm because we found the CVCL language more convenient than that of Yices for arithmetic expressions that define the elements of matrix A. The benchmark problem is to identify a Glass model with converging phase flow along cyclic attractors for the same networks as presented above. The satisfiability of formula (6) was evaluated for the input sequence of focal points that were specified to contain only unit coordinates. Such an input restricts the search to Boolean Glass networks that are known not to have flow converging to the origin. Thus, the instance is unsatisfiable, i.e., there is no Glass model satisfying the problem specifications. In this case, the solution requires the evaluation of all induced cycles of the instance and the run-time provides a conservative estimate of the efficiency of the algorithm.

The run-time of CVCL for the benchmark increases linearly with the number of induced cycles in the network (see the solid line in Fig. 3). A linear increase was also observed when MiniSAT was used to solve the corresponding *all-SAT* problems (dashed line in Fig. 3). The same trend indicates that the proposed network identification algorithm scales in the number of network cycles, just as MiniSAT scales well for the problem of computing all induced cycles of this network.

The scalability of the algorithm in the size of search space is estimated using the least square interpolation of the run-time as a function of the number L of Boolean variables in the instance. The interpolation using exponential trend lines is depicted in Fig. 4 by a solid and a dashed line for CVCL and MiniSAT, respectively. The run-time increases approximately as $e^{0.1L}$, while the growth of the size of the search space of a set of L Boolean variables is proportional to $2^L = e^{ln(2)L} \approx e^{0.7L}$. Thus, the experimental results indicate that the run-time of our algorithm increases about 7 times more slowly than the volume of the problem search space.

5   Conclusion

The proposed algorithm belongs to the methods that utilize propositional logic for reasoning about properties of ordinary differential equations. Such methods are widely applied to the analysis of biological networks and hybrid systems. The existing computational tools, developed for the propositional analysis of biological networks, approximate the ODE trajectories using the numerical Runge-Kutta procedure [31], Taylor series [32], or an approximate partitioning of phase space of continuous variables [4]. The computation of the reachable states for hybrid systems also relies on approximations of the PLDE solution [33,34]. We show that an exact algebraic algorithm can be applied for reasoning about the phase flow in a subclass of PLDE that is utilized in the Glass model.

The algorithm is applicable in the case when the PLDE system is near a bifurcation point, where the approximate methods may be inconclusive. We conducted an analysis of the structural stability of the phase flow for Glass networks with cyclic attractors. Cylindrical decomposition has been used for the evaluation of the criterion for identification of the phase flow. The flow for Boolean Glass models has been shown to be stable under the perturbations of any single focal point along the cyclic attractor. The cylindrical decomposition is known to be a powerful tool for evaluating the structural stability of partial and ordinary differential equations [35,36,37]. To the best of our knowledge, the presented stability analysis is a first attempt to apply cylindrical decomposition for the identification of bifurcations in Glass networks.

The proposed algorithm has been found to benefit from the scalability of Bounded Model Checking: new lower bounds for the number of equivalence classes are calculated for cyclic attractors in 6-dimensional networks. Our experimental results also indicate that the run-time of our algorithm increases slower than the size of the search space of the problem.

Acknowledgements. We would like to thank Felix Friedrich for the help with *Mathematica* calculations, and Clark Barrett for kind and detailed answers about CVCL.

References

1. Glass, L., Kaufmann, S.: The logical analysis of continuous non-linear biochemical control networks. J. Theor. Biol. 39, 103–129 (1973)
2. Glass, L.: Combinatorial aspects of dynamics in biological systems. In: Stat. Mech Stat. Methods in Theory and Application, pp. 585–611. Plenum Press, New York (1976)
3. Edwards, R.: Symbolic dynamics and computation in model gene networks. Chaos 11, 160–169 (2001)
4. Ghosh, R., Tiwari, A., Tomlin, C.: Automated symbolic reachability analysis; with application to delta-notch signalic automata. In: Maler, O., Pnueli, A. (eds.) HSCC 2003. LNCS, vol. 2623, pp. 233–248. Springer, Heidelberg (2003)
5. Mason, J., Linsay, P., Collins, J., Glass, L.: Evolving complex dynamics in electronic models of genetic networks. Chaos 14, 707–715 (2004)

6. Batt, G., Ropers, D., de Jong, H., Geiselmann, J., Mateescu, R., Page, M., Schneider, D.: Analysis and verification of qualitative models of genetic regulatory networks: A model-checking approach. In: 19th Int Joint Conference on Artificial Intelligence, pp. 370–375 (2005)
7. Gedeon, T.: Global dynamics of neural nets with infinite gain. Physica D: Nonlinear Phenomena 146, 200–212 (2000)
8. Gedeon, T.: Attractors in continuous time switching networks. Communications on Pure and Applied Analysis 2, 187–209 (2003)
9. de Jong, H.: Modeling and simulation of genetic regulatory systems: a literature review. J. Comp. Biol. 9, 67–103 (2002)
10. Casey, R., de Jong, H., Gouze, J.L.: Piecewise-liner models of genetic regulatory networks: equilibria and their stability. J. Math. Biol. 52, 27–56 (2006)
11. Farcot, E.: Geometric properties of a class of piecewise affine biological network models. J. Math. Biol. 52, 373–418 (2006)
12. Edwards, R.: Analysis of continuous-time switching networks. Physica D 146, 165–199 (2000)
13. Glass, L., Pasternack, J.: Stable oscillations in mathematical models of biological control systems. J. Math. Biol. 6, 207–223 (1978)
14. Mestl, T., Plahte, E., Omholt, S.: Periodic solutions in systems of piecewise-linear differential equations. Dynam. Stabil. Syst. 10, 179–193 (1995)
15. Edwards, R., Glass, L.: Combinatorial explosion in model gene networks. Chaos 10, 691–704 (2000)
16. Mestl, T., Lemay, C., Glass, L.: Chaos in high-dimensional neural and gene networks. Physica D 98, 33–52 (1996)
17. Thomas, R., Kaufman, M.: Multistationarity, the basis of cell differentiation and memory. Chaos 11, 170–195 (2001)
18. Kauffman, S.: A proposal for using the ensemble approach to understand genetic regulatory networks. Theor. Biol. 230, 581–590 (2004)
19. Glass, L.: Global analysis of nonlinear chemical kinetics. Statistical mechanics, part B: time dependent processes, 311–349 (1977)
20. Glass, L., Pasternack, J.: Prediction of limit cycles in mathematical models of biological oscillations. Bull. Math. Biol. 40, 27–44 (1978)
21. Laubenbacher, R., Stigler, B.: A computational algebra approach to the reverse engineering of gene regulatory networks. J. Theor. Biol. 229, 523–537 (2004)
22. Gantmacher, F.: The Theory of Matrices. vol. 2. Chelsea (1974)
23. Zinovik, I., Kroening, D., Chebiryak, Y.: An algebraic algorithm for the identification of Glass networks with periodic orbits along cyclic attractors. Technical Report 557, ETH Zurich, Computer Science Department (2007)
24. Rajan, D., Shende, A.: Maximal and reversible snakes in hypercubes. In: 24th Annual Australasian Conference on Combinatorial Mathematics and Combinatorial Computing (1999)
25. Casella, W.P.D.: Using evolutionary techniques to hunt for the snakes and coils. In: The 2005 IEEE Congress on Evolutionary Computation, vol. 3, pp. 2499–2505. IEEE Press, NJ (2005)
26. Plaisted, D., Biere, A., Zhu, Y.: A satisfiability tester for quantified boolean formulae. J. Discrete Appl. Math. 130, 291–328 (2003)
27. Eén, N., Sörensson, N.: An extendable SAT-solver. Theory and Applications of Satisfiability Testing 2919, 502–518 (2004)
28. Sheini, H., Sakallah, K.: From propositional satisfiability to satisfiability modulo theories. In: Biere, A., Gomes, C.P. (eds.) SAT 2006. LNCS, vol. 4121, pp. 1–9. Springer, Heidelberg (2006)

29. Stump, A., Barrett, C., Dill, D.: CVC: a cooperating validity checker. In: 14th Int. Conf. on Computer-Aided Verification (CAV), pp. 87–105. Springer, Heidelberg (2002)
30. Dutertre, B., de Moura, L.: A fast linear-arithmetic solver for DPLL(T). In: Ball, T., Jones, R.B. (eds.) CAV 2006. LNCS, vol. 4144, pp. 81–94. Springer, Heidelberg (2006)
31. Calzone, L., Chabrier-Rivier, N., Fages, F., Soliman, S.: Machine learning biochemical networks from temporal logic properties. In: Priami, C., Plotkin, G. (eds.) Transactions on Computational Systems Biology VI. LNCS (LNBI), vol. 4220, pp. 68–94. Springer, Heidelberg (2006)
32. Piazza, C., Antoniotti, M., Mysore, V., Policriti, A., Winkler, F., Mishra, B.: Algorithmic algebraic model checking I: Challenges from systems biology. In: Etessami, K., Rajamani, S.K. (eds.) CAV 2005. LNCS, vol. 3576, pp. 5–19. Springer, Heidelberg (2005)
33. Henzinger, T., Preussig, J., Wong-Toi, H.: Some lessons from the HyTech experience. In: Proc of the 40th Annual Conference on Decision and Control (CDC), pp. 2887–2892. IEEE Press, NJ (2001)
34. Frehse, G.: Phaver: Algorithmic verification of hybrid systems past HyTech. In: Morari, M., Thiele, L. (eds.) HSCC 2005. LNCS, vol. 3414, pp. 258–273. Springer, Heidelberg (2005)
35. Hong, H., Liska, R., Steinberg, S.: Testing stability by quantifier elimination. J. Symb. Comp. 11, 1–26 (1996)
36. Wang, D.: Elimination theory, methods, and practice. In: Mathematics, and Mathematics-Mech, pp. 91–137. Shandong Education Publishing House, Jinan (2001)
37. Wang, D., Xia, B.: Stability analysis of biological systems with real solution classification. In: ISSAC'05, vol. 3414, pp. 354–361. ACM, New York (2005)

Analyzing Pathways Using SAT-Based Approaches*

Ashish Tiwari, Carolyn Talcott, Merrill Knapp, Patrick Lincoln,
and Keith Laderoute

SRI International, Menlo Park, CA 94025

Abstract. A network of reactions is a commonly used paradigm for representing knowledge about a biological process. How does one understand such generic networks and answer queries using them? In this paper, we present a novel approach based on translation of generic reaction networks to Boolean *weighted MaxSAT*. The Boolean weighted MaxSAT instance is generated by encoding the equilibrium configurations of a reaction network by weighted boolean clauses. The important feature of this translation is that it uses reactions, rather than the species, as the boolean variables. Existing weighted MaxSAT solvers are used to solve the generated instances and find equilibrium configurations. This method of analyzing reaction networks is generic, flexible and scales to large models of reaction networks. We present a few case studies to validate our claims.

1 Introduction

A network of reactions is a convenient way to represent knowledge about a biological process. Each reaction converts some reactants into products in the presence of certain other molecules. There is no single universal meaning, or a single formal semantics, that can be ascribed to the various reaction networks and pathways in the literature. Consequently, it is unclear how to build computational support for understanding and reasoning about large reaction networks.

A reaction network can be interpreted in various ways. They are often mapped onto a continuous dynamical system, where the dynamics are given by ordinary differential equations. These differential equations can be generated using different kinetic laws, such as Mass Action and Michaelis-Menten. However, it is not easy to experimentally determine, especially for biochemical reactions, the rate constants required to build the continuous dynamical system. As a result, fully specified and experimentally validated continuous dynamical system models are rarely available. Moreover, it has also been argued that the assumptions used to arrive at the differential equations may not be valid inside a biological compartment, where certain molecules may be few.

---

* This work was supported in part by Public Health Service grant GM068146-03 from the National Institute of General Medical Sciences and by the National Science Foundation under grants IIS-0513857 and CCR-0326540.

H. Anai, K. Horimoto, and T. Kutsia (Eds.): AB 2007, LNCS 4545, pp. 155–169, 2007.

A reaction network can also be interpreted as a dynamical system over a discrete state space. In this case, the state space consists of mappings from the set of species to the natural numbers that specifies the number of molecules of each species. The dynamics over this state space can be defined either in continuous time (using a stochastic model) as a Chemical Master Equation, or in discrete time as a (standard or stochastic) Petri net. While such models are considered to be more accurate, they are difficult to analyze because of the horrendously huge state space. For example, when analyzing systems containing just 100 total molecules of 4 different species, the state space size is 4^{100}.

All continuous time models require reaction rates in some form. To overcome this requirement, discrete time models are considered that abstract time to a before-after relationship. When considered over the discrete state space mentioned above, a reaction network simply maps to a Petri net. Analyzing Petri nets is not easy. For instance, while Petri net reachability is decidable, there is no known upper-bound.

To overcome the state space problem, the discrete-time discrete-space models are further simplified. For instance, boolean models abstract species to being either *present* or *absent*. Other qualitative abstractions, such as *absent, present in low quantities*, and *present in large quantities* are also possible. In the absence of accurate detailed models, these abstract models have been found to be highly useful for representing and understanding biological knowledge.

In this paper, we present a new scalable approach for analyzing large reaction networks interpreted in the discrete-time and abstract discrete-space domain. There are three main features in our approach. First, it is based on qualitatively abstracting the *reactions* into two states–*on* and *off*. This is dual to the more conventional approach where the presence or absence of molecular species, and not *reactions*, is used to define the state of the system [4,8,9]. Second, it uses a boolean MaxSat as its backend engine. There is a generic translation from reaction networks to boolean MaxSat instances. Third, it is flexible. Clauses and their weights can be adjusted for reaction networks encoding specific aspects, such as signaling pathways, or transcriptional regulation.

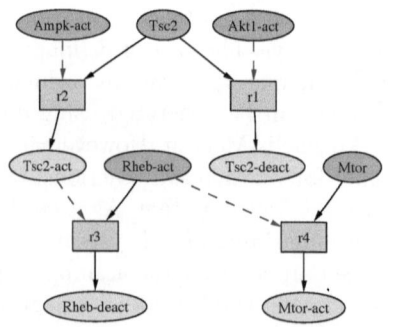

Consider, for example, the very simple network shown here (not necessarily biologically accurate). This network consists of 4 reactions:

$$r_1: \quad \text{Tsc2} \xrightarrow{\text{Akt1-act}} \text{Tsc2-deact}$$

$$r_2: \quad \text{Tsc2} \xrightarrow{\text{Ampk-act}} \text{Tsc2-act}$$

$$r_3: \quad \text{Rheb-act} \xrightarrow{\text{Tsc2-act}} \text{Rheb-deact}$$

$$r_4: \quad \text{Mtor} \xrightarrow{\text{Rheb-act}} \text{Mtor-act}$$

We will use this network as a running example in the paper.

It is not immediately obvious how to understand even this simple network. Using the approach described in this paper, all possible "steady-state" behaviors of the above network can be computed. For this example, the tool computes two possible behaviors. Either Akt1-act is present, deactivating Tsc2, while Mtor gets activated by Rheb-act (Reactions 1 and 4 are "on"); or, Ampk-act is present, activating Tsc2, which in turn deactivates Rheb-act (Reactions 2 and 3 are "on"). The important point here is that the steady-state behavior is thought of a subset of reactions that can be consistently "on", as opposed to the traditional viewpoint where steady-state refers to species reaching some equilibrium concentrations.

As mentioned earlier, our approach is flexible and additional constraints can be added to specialize the search for certain steady-state configurations. We can specify an initial dish consisting of some of the species and search for most likely steady-state configurations resulting from the given initial dish. In the above example, if the initial dish only contains Tsc2, Ampk-act, Rheb-act and Mtor, then our tool identifies that the second and third reactions can be "on", and that the other option, where reactions 1 and 4 are "on" is less likely. Similarly, target species can be specified, and the tool will generate paths (scenarios) that produce the target species. Each such scenario will be assigned a weight indicating its relative likelihood.

1.1 Motivation

The definition of "steady-state" behavior we use in this paper is nonstandard. Traditionally, a steady-state refers to all species in the network being at their equilibrium concentrations. In this paper, a steady-state refers to a subset of reactions that can be consistently "on". This new definition is motivated by the observation that *signaling pathways* are best understood this way. More than the individual species concentrations, it is the chain of reactions that captures how information flows from the cell membrane to effect downstream activities in a cell. This chain of reactions corresponds directly to the notion of a steady-state in our approach.

The different reactions in the steady-state chain of reactions will, in reality, be temporally separated. While certain phosphorylation activity may occur in a few minutes after a cell is hit by ligands, other downstream activities may occur much later. In our approach, we identify the whole chain as one possible steady-state behavior of the reaction network. The complete chain of reactions may never simultaneously be "on" in reality. However, they are still useful in understanding the function of a given complex reaction network.

The approach based on translation to MaxSat is motivated by the need for flexibility. Reaction networks have slightly different meaning in different contexts. Metabolic pathways, signaling pathways, and transcriptional regulation networks work on different notions of species and reactions. Our basic semantics attempts to capture the minimal common meaning that can be ascribed to any such network. The weights on the MaxSat instance give flexibility in making certain constraints harder than others in different contexts.

Finally, it should be mentioned that the technology for solving SAT and MaxSAT problems has made significant advances in recent years and problems with thousands of boolean variables and even more clauses are routinely solved in a few seconds. We have used our tool on the HumanCyc database of metabolic pathways (containing over a thousand reactions) and we can answer queries in a few seconds.

2   Reaction Networks

In this section, we formalize our terminology. A *species* is a generic name used to denote any entity, such as a molecule, ion, protein, enzyme, ligand, receptor, complex, or a postranscriptionally modified form of a protein. We do not differentiate between these different roles and just formally identify a species with a unique name. The set of all species will be denoted by S. A *reaction* consists of a set of *reactants*, a set of *modifiers*, and a set of *products*. Thus, a reaction r is a 3-tuple $\langle R, M, P \rangle$, where R, M, P are pairwise disjoint subsets of S. Given a reaction r, we denote its set of reactants, modifiers, and products by $R(r)$, $M(r)$, and $P(r)$ respectively. Given a species s, the set of reactions in which s occurs as a reactant (modifier, product) is denoted by $R^{-1}(s)$ (respectively, $M^{-1}(s)$, $P^{-1}(s)$).

A *network* $\mathcal{N}$ is a collection of reactions. A *network instance* is a network together with an optional set of *input* species, a set of *forbidden* species, and a set of *target* species.

A *pathway* is a special kind of network. Informally, a pathway contains a *related* set of reactions that can be consistently switched "on". The following sections will formally define the constraints we impose to identify pathways.

2.1   Semantics of Reaction Networks

As mentioned in the introduction, motivated by the need to handle unknown model parameters while maintaining computational feasibility of analysis, we use a discrete-time abstract discrete-state semantics of reaction networks. The key aspect of our semantics is that we introduce a boolean variable for each reaction (and not for each species). Thus, the semantics of a biochemical network $\mathcal{N} = \{r_1, r_2, \ldots, r_n\}$ with n reactions is given as a state transition system defined over n boolean variables $b_1, \ldots, b_n$, where the i-th boolean variable b_i represents whether the i-th reaction r_i is "on" or "off".

Let $present4i(s, i)$ denote the formula $\bigvee_{r_j \in P^{-1}(s)} b_j \ \wedge \ \bigwedge_{r_j \in R^{-1}(s), j \neq i} \neg b_j$, which means some reaction that produces s is "on" and every reaction other than r_i that consumes s is "off". Intuitively, $present4i(s, i)$ represents the availability of species s for reaction r_i. The transitions of the state transition system are given by nondeterministically applying one of the following $2n$ guarded commands:

$$\neg b_i \wedge \bigwedge_{s \in R(r_i) \cup M(r_i)} present4i(s, i) \longrightarrow b_i' := true$$

$$b_i \wedge \bigvee_{s \in R(r_i) \cup M(r_i)} \neg present4i(s, i) \longrightarrow b_i' := false$$

The first guarded command says that if a reaction r_i is "off", but each of its reactants and modifiers is "present" (for r_i), then it can be turned "on". The second guarded command says that if a reaction r_i is "on", but one of its reactants or modifiers is not present (for r_i), then it can be turned "off".

3   Biochemical Networks to Boolean SAT

In this section, we describe the procedure that generates a set of boolean constraints from a network. The boolean constraints represent the equilibrium configurations of the network in the semantics given above. Later in this section, we describe the additional constraints that are generated from a network instance.

An equilibrium state is defined as a state in which none of the $2n$ guarded transitions are enabled. Hence, if a state $(b_1, \ldots, b_n)$ is an equilibrium state of the above state transition system, then it should be the case that, for all i,

$$\neg(\neg b_i \wedge \bigwedge_{s \in R(r_i) \cup M(r_i)} present4i(s,i)) \wedge \neg(b_i \wedge \bigvee_{s \in R(r_i) \cup M(r_i)} \neg present4i(s,i))$$

This is equivalent to saying that for all i,

$$b_i \Leftrightarrow \bigwedge_{s \in R(r_i) \cup M(r_i)} present4i(s,i) \tag{1}$$

Any boolean assignment that satisfies these constraints is an equilibrium state of the given reaction network. In the implementation (Section 5), we break up the constraint in Formula 1 into the following constraints to enable the MaxSat solver to partially satisfy these constraints.

$$b_i \Rightarrow \bigwedge_{s \in R(r_i) \cup M(r_i)} \bigvee_{r_j \in P^{-1}(s)} b_j \tag{2}$$

$$b_i \Rightarrow \bigwedge_{s \in R(r_i)} \bigwedge_{r_j \in R^{-1}(s), j \neq i} \neg b_j \tag{3}$$

$$b_i \Rightarrow \bigwedge_{s \in M(r_i)} \bigwedge_{r_j \in R^{-1}(s), j \neq i} \neg b_j \tag{4}$$

$$\neg b_i \Rightarrow \neg \bigwedge_{s \in R(r_i) \cup M(r_i)} present4i(s,i) \tag{5}$$

Formula 2 captures the rule that if a reaction is "on", then each of its reactants and modifiers is produced by some "on" reaction. Formula 3 encodes the inhibitory effect that a reaction may have on another that shares a reactant with it by saying that if a reaction is "on", then none of its reactants is consumed (used as a reactant) by any *other* reaction. Formula 4 encodes the competitive inhibition between reactions through a species that is a reactant in one reaction and a modifier in another. Note that if two reactions share a modifier, then they do not inhibit each other. Finally, Formula 5 encodes that if all reactants and modifiers of a reaction are present, then it should be "on".

3.1   Completing the Network

Biological databases of biochemical networks are often incomplete. They often use species that are not created by any reaction in the network. In the running example, Tsc2, Akt1-act, Ampk-act, Rheb-act, and Mtor are all species with no producers. The presence of such species is a problem for our encoding since, to be "on", a reaction requires all of its reactants (and modifiers) to be produced by some other reaction. If there are no producers of certain species, then reactions using that species can never be turned on.

We solve this problem by adding dummy reactions that create species that have no producers. Specifically, for each species s such that $P^{-1}(s) = \emptyset$, we add a new reaction $r = \langle R, M, P \rangle$, where $R = \emptyset$, $M = \emptyset$, and $P = \{s\}$. We perform this step as a preprocessing step. As a result, these additional dummy reactions are taken into account when the constraints given in Formula 1 are generated.

We also encode the fact that these dummy reactions are different from other reactions by adding boolean constraints that force these dummy reactions to be "off". For each dummy reaction r, if b is the corresponding boolean variable, then we add the following clause

$$\neg b \tag{6}$$

This constraint says that the dummy reaction, and hence the corresponding species, should *preferably* not be used. In Section 4, we will discuss how this preference is effected by means of weights.

In the running example, for each of the 5 species that have no producers, we add one new dummy reaction. Thus, we have new dummy reactions $r_5, \ldots, r_9$ that respectively produce Tsc2, Akt1-act, Ampk-act, Rheb-act, and Mtor. Thus the complete network has 9 reactions, and hence, the boolean encoding will be over 9 boolean variables $b_1, \ldots, b_9$. The constraints given by Formula 1 will be:

$$b_1 \Leftrightarrow (b_5 \wedge \neg b_2) \wedge (b_6) \qquad b_3 \Leftrightarrow (b_8) \wedge (b_2)$$
$$b_2 \Leftrightarrow (b_5 \wedge \neg b_1) \wedge (b_7) \qquad b_4 \Leftrightarrow (b_9) \wedge (b_8 \wedge \neg b_3)$$

Additionally, we will also get boolean constraints $\neg b_5, \neg b_6, \ldots, \neg b_9$ coming from Formula 6. Note that Reaction r_3 requires the modifier Tsc2-act, which is produced by Reaction r_2. This gets reflected as $b_3 \Rightarrow b_2$ above. As an example of competitive inhibition, note that Reaction r_1 and Reaction r_2 share a common reactant, namely Tsc2. This shows up as $b_1 \Rightarrow \neg b_2$ and $b_2 \Rightarrow \neg b_1$. Similarly, Reaction r_3 and Reaction r_4 compete for Rheb-act—Reaction r_3 uses it as a reactant, whereas Reaction r_4 requires it as a modifier. This generates the constraint $b_4 \Rightarrow \neg b_3$.

3.2   Optional Clauses

In case of analyzing a network *instance*, we may optionally have additional information about the input species, forbidden species, and target species. We now show how these are incorporated into the constraints.

Initial Species. The set of species specified as initial are assumed to be present. If a set of initial species is specified, then the preprocessor adds a dummy reaction that produces all the initial species. Specifically, if S_{init} is the set of initial species, then the preprocessor will add a dummy reaction $r = \langle R, M, P \rangle$, where $R = M = \emptyset$ and $P = S_{init}$. Furthermore, the boolean variable b corresponding to this reaction is forced to be "on" by simply adding a clause b in the generated set of boolean constraints. If some initial species are specified, then the initial dummy reaction is added to the network *before* the network is completed (Section 3.1). Hence, fewer dummy reactions get added in the network completion phase if some of the species with no producers in the network are assumed to be in the initial soup.

Target Species. The set of target species is a list of species that should be *present* in the equilibrium configurations generated by the tool. If a set of target species is specified, then the boolean constraint generator adds additional constraints that say that for each target species, there is at least one producer of it turned "on".

For each species s in the set of target species, we add the constraint,

$$\bigvee_{r_i \in P^{-1}(s)} b_i \tag{7}$$

Forbidden Species. The set of forbidden species specifies the set of species that should not be used in any equilibrium configuration generated by the system. If this set is provided, then the following additional boolean constraint is generated for each species s in this forbidden set,

$$\bigwedge_{r_i \in P^{-1}(s)} \neg b_i \tag{8}$$

3.3 Mode Based Constraints

Given the above constraints, we can try to turn "on" as many reactions as possible, or turn "on" as few reactions as possible. These two possibilities are encoded as two different sets of constraints.

If we wish to turn "on" as many reactions as possible, then, for each reaction $r_i \in \mathcal{N}$, we add the clause

$$b_i \tag{9}$$

to the set of constraints. This clause simply says that reaction r_i is "on".

If we wish to turn "on" as few reactions as possible (say to find minimal pathways), then, for each reaction $r_i \in \mathcal{N}$, we add the clause

$$\neg b_i \tag{10}$$

to the set of constraints.

4   Biochemical Pathway to Boolean Max-SAT

The constraints outlined above are not all equally important. This is captured by adding a weight (number) to each constraint that indicates its relative importance.

In particular, constraints obtained by instantiating Formula 2, Formula 3, and Formula 5 are each given a very large weight W. In the current implementation, W is equal to the total number of reactions in the completed network. The constraint represented in Formula 4 is given weight equal to $W/2$ since competitive inhibition between reactions via a species that is a modifier in one reaction and a reactant in another is intuitively weaker than the inhibition via shared reactants. The constraint saying that species with no producers should not be used (Formula 6) is given intermediate weight (approximately $W/(k+1)$, where k is the total number of species with no producers). Whenever present, the constraint for creation of target species (Formula 7) is given weight W. The constraints that specify the hints (Formula 9 and Formula 10) are given weight 1.

The choice of weights for each constraint gives additional flexibility that can be used, in the future, to encode other biologically relevant information that is not generic to all biochemical processes.

4.1   Weighted MaxSAT

A *solution* is a mapping from the boolean variables to $\{true, false\}$. In our context, a solution maps reactions to either "on" or "off". Under a given solution, constraints also evaluate to either *true* or *false*.

Each solution can be associated with a weight: the sum of the weights of all the constraints that are made *true* by that solution. A *weighted MaxSAT solver* finds a solution that has the maximum weight.

In our running example, using the above rule for assigning weights (we do not break Formula 1 into smaller parts and assign it a weight $W = 9$ for simplicity here), we get the following weighted basic constraints:

c_1 : $b_1 \Leftrightarrow (b_5 \wedge \neg b_2) \wedge (b_6)\ w_1 = 9$ c_2 : $b_3 \Leftrightarrow (b_8) \wedge (b_2)$ $w_2 = 9$
c_3 : $b_2 \Leftrightarrow (b_5 \wedge \neg b_1) \wedge (b_7)\ w_3 = 9$ c_4 : $b_4 \Leftrightarrow (b_9) \wedge (b_8 \wedge \neg b_3)\ w_4 = 9$
c_5 : $\neg b_5$ $w_5 = 1$ c_6 : $\neg b_6$ $w_6 = 1$
c_7 : $\neg b_7$ $w_7 = 1$ c_8 : $\neg b_8$ $w_8 = 1$
c_9 : $\neg b_9$ $w_9 = 1$ c_{10} : b_1 $w_{10} = 1$
c_{11} : b_2 $w_{11} = 1$ c_{12} : b_3 $w_{12} = 1$
c_{13} : b_4 $w_{13} = 1$

Note that the 5 constraints, $c_5, \dots, c_9$, encode the fact that the five species with no producers can be used by paying a small penalty; and the last 4 constraints say that each reaction should preferably be turned "on".

For this set of constraints, the solution in which all b_i are *false* has a weight 41 (since only $c_{10}, \dots, c_{13}$ are violated). The solution $b_1 = b_4 = b_5 = b_6 = b_8 = b_9 = true$ (and the rest *false*) has weight 39; and the solution $b_2 = b_3 = b_5 = b_7 =$

$b_8 = true$ has weight 40. These three solutions are the top three maximum weight solutions. The latter two correspond exactly to the two scenarios described in Section 1. The first solution captures the scenario where no reaction is "on", which can be eliminated by using a nonempty initial set of species that includes (some of) the 5 species with no producers.

5   Implementation and Case Studies

We have implemented a tool based on the technique described in this paper. As a backend MaxSAT solver, we use Yices [14,3], which is a more general *satisfiability modulo theory* solver. The input format for our tool is a network or network instance described in a very simple intermediate language. We also have several front-ends that convert from other formats to our intermediate language format. For example, we have front-ends for Pathway Logic [12,11] and BioCyc [5,7].

In this section, we describe the results obtained using this tool on some specific networks.

5.1   Sporulation Initiation in *B. Subtilis*

Bacillus subtilis is considered a model organism for Gram-positive bacteria and has been extensively studied in the laboratory. It is an endospore-forming bacteria most commonly found in the soil. Endospore formation is initiated when nutrients become limiting and is an adaptive response of the bacteria to their environment.

Sporulation is a one-way decision and once the decision is made, the cell undergoes changes which take 6 to 8 hours in most organisms. If conditions improve in the meantime, then the cell will be at a disadvantage. Hence the decision to initiate sporulation is important to the organism and is subject to a variety of control.

The formation of spores in Bacillus subtilis is a developmental process under genetic control. The decision to either grow vegetatively or sporulate is regulated by the state of phosphorylation of the SpoOA transcription factor [10,6]. SpoOA obtains its phosphate through a phosphorylation pathway (see Figure 1), the so-called *phosphorelay*, in which at least three histidine protein kinases transfer phosphate to the relay protein, SpoOF, then to SpoOB, and finally to SpoOA (represented by ReactionIDs $r17$, $r19$, and $r20$ in Table 1). In addition, the phosphorylation state of SpoOA is modulated by specific phosphatases, such as SpoOE, which dephosphorylates SpoOA-P, and RapA, which dephosphorylates SpoOF-P (ReactionIDs $r18$, $r21$).

The *SinI* and *SinR* pair is a regulatory operon in the sporulation initiation network. While SinR is a transcriptional regulator that represses *spoOA* transcription, SinI disrupts the SinR tetramer through the formation of a SinI-SinR heterodimer. This aspect, along with the logic regulating SinI transcription, is encoded in ReactionIDs $r1$, $r2$, $r3$, and $r4$.

The activity of protein RapA is modulated by quorum sensing, the process of sensing activity in neighboring cells and reacting in a cell-density-specific

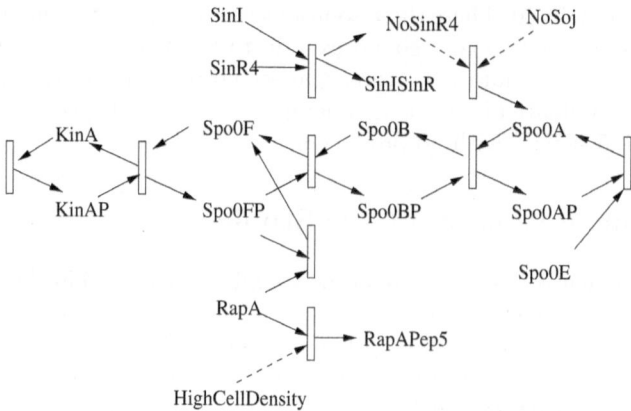

Fig. 1. Selected reactions from the sporulation initiation network of *B. Subtilis*. The reactions are represented using standard Petri net notation and show the main phosphorelay.

fashion. Under high population density, RapA is inhibited by PhrA pentapeptide (not modeled in the reactions). These aspects are captured in ReactionIDs $r13$, $r15$. The protein kinase KinA is a sensor that initiates the phosphorelay and is modeled here by ReactionIDs $r16$, $r17$. Most of the remaining reactions encode transcriptional regulation logic for different proteins.

On this simplified model of sporulation initiation, the tool implementing the approach described in this paper can find possible stable behaviors of the network. These behaviors are found as subsets of reactions in the network that can be consistently "on". The tool finds 3 different possibilities for the model above.

- SinI is produced, and it binds to SinR, thus preventing it from repressing *spo0A*. RapA is converted to RapAPep5, thus preventing it from dephosphorylating Spo0A-P. In the presence of stress signals, KipI is prevented from inhibiting KinA from self-kinasing. The self-kinasing of KinA triggers the phosphorelay, which leads to production of Spo0A-P, a precursor for sporulation.
- In the second stable state scenario, RapA dephosphorylates Spo0F-P, thus breaking the phosphorelay chain. Thus, there is no production of Spo0A-P.
- The third stable state scenario is similar to the first, except that Spo0E dephosphorylates the produced Spo0A-P, thus using up the produced Spo0A-P.

The three stable scenarios each make different assumptions about the environment. In our case, the environment consists of the species that are not created by any of the reactions in the network. In the network above, HighCellDensity, and NoFood, are two examples of *input species*.

The tool can also be used in the mode in which a desired target set of species is specified (for example, Spo0A-P). In this case, the tool will generate the first stable scenario above to show how Spo0A-P could be produced.

Table 1. The list of reactions modeling the sporulation initiation network

ID	Reactants	+Modifiers	⟶Products
r1		+(Spo0AP, NoSinR4)	⟶SinI
r2		+(Spo0AP, NoAbrB6, NoHpr)	⟶SinI
r3	SinI, SinR4	+	⟶SinISinR, NoSinR4
r4	SinR	+	⟶SinR4
r5		+(NoSinR4, sigmaH, NoSoj)	⟶Spo0A
r6		+(NoAbrB6)	⟶Spo0E
r7	AbrB, AbrB6	+(Spo0AP)	⟶NoAbrB6
r8		+(NoSpo0AP)	⟶AbrB
r9		+(NoAbrB6)	⟶AbrB
r10	AbrB, NoAbrB6	+	⟶AbrB6
r11	NoHpr	+(AbrB6)	⟶Hpr
r12	Hpr	+(NoAbrB6)	⟶NoHpr
r13		+(ComAP)	⟶RapA
r14	RapA	+(Spo0AP, Hpr)	⟶
r15	RapA	+(HighCellDensity)	⟶RapAPep5
r16	KinA	+(NoKipI)	⟶KinAP
r17	KinAP, Spo0F	+	⟶Spo0FP, KinA
r18	Spo0FP, RapA	+	⟶Spo0F
r19	Spo0FP, Spo0B	+	⟶Spo0BP, Spo0F
r20	Spo0A, Spo0BP	+(NoSoj)	⟶Spo0AP, Spo0B
r21	Spo0AP, Spo0E	+	⟶Spo0A, NoSpo0AP
r22		+(sigmaH, sigmaA)	⟶Spo0F
r23		+(sigmaA)	⟶Spo0B
r24	KipI	+(NoFood, NoNitrogen)	⟶NoKipI

5.2   MAPK Signaling Network

The Mitogen-Activated Protein kinase (MAPK) network regulates several cellular processes, including the cell cycle machinery. The MAPK cascade communicates signals from growth factors that bind receptor kinases to transcription and other cellular processes [2]. A simplified model of this network, taken from [2], can be encoded in our notation as shown in Table 2. The tool finds two stable sets of behavior for this network.

- The positive feedback loop is active. In this case, either Grb2, Sos1, or PKC* turns on Ras. This causes, in steps, the phosphorylation of Raf, MEK, and Erk. Activated Erk causes production of AA*, which stimulates PKC.
- The negative feedback loops are active. In this case, protein phosphatase 2A (PP2A) dephosphorylates both Raf* and Mek*, and MKP dephosphorylates Erk*. MKP is created by transcription of *MKP* gene, and this is promoted by Erk*.

The two stable solutions clearly identify the positive cycle and the multiple negative cycles that break the positive cycle. The overall system behavior is seen to be a result of the close interaction between the positive and negative cycles.

Table 2. The list of reactions modeling the MAPK signaling network

ID	Reactants+Modifiers		$\longrightarrow$Products
r1	Ras	+(Grb2, Sos1)	$\longrightarrow$Ras*
r2	Ras	+(PKC*)	$\longrightarrow$Ras*
r3	Raf	+(Ras*)	$\longrightarrow$Raf*
r4	Raf*	+(PP2A)	$\longrightarrow$Raf
r5	Mek	+(Raf*)	$\longrightarrow$Mek*
r6	Mek*	+(PP2A)	$\longrightarrow$Mek
r7	Erk	+(Mek*)	$\longrightarrow$Erk*
r8	Erk*	+(MKP)	$\longrightarrow$Erk
r9		+(Erk*, MKPgene)	$\longrightarrow$MKP
r10	AA	+(Erk*, Ca)	$\longrightarrow$AA*
r11	PKC	+(DAG, Ca, AA*)	$\longrightarrow$PKC*

We also used the detailed model of the MAPK signaling network from [1]. The total running time on the full network is of the order of a few seconds.

5.3   EGF Stimulation Network

In the Pathway Logic project [12,11], a model of Egf stimulation is being developed by curating a network of biochemical reactions involved in mammalian cell signaling from the literature. When a cell is stimulated by Egf, certain species are experimentally observed to be present in the cell after its initial stimulation. These observations can be used to validate the model by checking whether the model predicts the observations. To carry out the validation, we started with a network of about 400 reactions and created a network instance by adding initial and target species. Specifically, we started with a set of about 250 *initial* species and 62 *target* species that are experimentally observed in response to EGF stimulation.

When this network instance is analyzed by our tool, our tool attempts to find a set of reactions that will create each of the target species using the initial species and the reactions in the network. A "–no-assume" option tells the tool to not assume any species not already specified in the initial set. (Recall that, by default, species that have no producers can be assumed, with a moderate penalty.)

The output of the tool indicated that it was not possible to find a solution without violating one Type 3 and one Type 4 *competitive inhibition* constraints. Specifically, the species (Frap1:Lst8)-CLc [1] is a reactant in two different reactions that are *both* required to be "on" to create the target species. This causes a Type 3 constraint to be violated. The Type 4 constraint that is violated is caused by the species Src-CLi, which is used as a reactant in a reaction to create Src-act-CLi, and it is also used as a modifier in the reaction that creates Cbl-Yphos-CLi. This violation pointed out a typing error in specifying the reaction

---

[1] A complex containing Frap1 and Lst8 located in the cytoplasm, CLc.

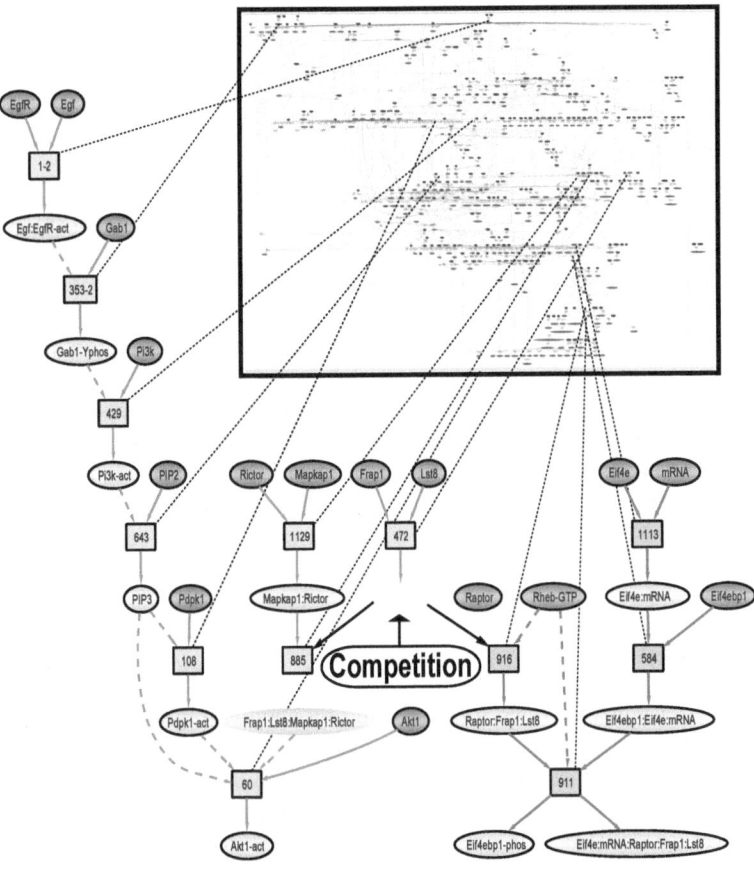

Fig. 2. A simple network with competing rules

rules which has been corrected. Figure 2 shows the pathways competing for (Frap1:Lst8)-CLc in the context of the larger network.

Using our tool provided two valuable forms of feedback to the model developer. One was a form of meta analysis or type-checking that detected syntactic problems with the model. (The first pass detected a number of inconsistencies that were easily repaired.) The second was the identification of the point of competition. Using the Pathway Logic Assistant [13] one can check whether a given set of observations is predicted, singly or jointly. However if a prediction fails there is no feedback as to the cause of failure. Using MaxSAT, candidate conflicting constraints can be identified to guide the modeler.

Starting with the discovered Type 3 violation and studying the subnetwork connected to this reaction lead to two hypotheses: (1) (Frap1:Lst8)-CLc splits into two *populations* one for each of the two competing reactions; (2) there is a feedback loop that can reset the state of (Frap1:Lst8)-CLc and the system oscillates between the two pathways. Experiments are ongoing to test these hypotheses.

6   Related Work

We compare here with work that is closer in spirit to our work, and do not mention all the literature devoted to building various kinds of models and improving understanding of specific biological phenomena, such as sporulation and MAPK signaling.

Senachak et al. [8] give a generic interpretation to a reaction network by translating it to a graph. Strongly-connected components of the graph are related to the pathways. The construction of the graph has some unusual steps, such as *cascading*, that arise primarily because the authors use species as defining the nodes of the graph. The main difference in our approach is that, in our approach, the boolean variables correspond to reactions in the network. We believe this leads to a much simpler and natural encoding of the "cascading"-style constraints of [8].

7   Conclusion

We presented a new approach for analyzing biochemical reaction networks using MaxSAT. The novelty here is that we make reactions central to the notion of a steady-state behavior. A steady-state behavior is a subset of reactions that can be mutually and consistently "on".

The attractiveness of our approach is that it is generic and applies to networks coming from different kinds of biological networks. Additionally, it is also flexible and allows encoding of knowledge specific to certain kinds of networks via suitable manipulation of the weights on the generic constraints.

The analysis approach is promising. Even for the largest networks we have studied, the analysis takes at most a few seconds to compute answers.

Possible future work include studying quantitative variants of the boolean constraints. Fortunately, our backend tool, Yices, supports reasoning over linear arithmetic constraints. We can replace the use of boolean MaxSAT with MaxSAT over arbitrary combination of boolean and linear arithmetic constraints.

Acknowledgments. We thank the referees for helpful suggestions.

References

1. Bhalla, U.S., Iyengar, R.: Robustness of the bistable behavior of a biological signalling feedback loop. Chaos, 11(1) (2001)
2. Bhalla, U.S., Ram, P.T., Iyengar, R.: MAP kinase phosphatase as a locus of flexibility in a Mitogen-Activated Protein kinase signaling network. Science, 297 (2002)

3. Dutertre, B., de Moura, L.M.: A fast linear-arithmetic solver for dpll(t). In: Ball, T., Jones, R.B. (eds.) CAV 2006. LNCS, vol. 4144, pp. 81–94. Springer, Heidelberg (2006)

4. Fages, F., Soliman, S., Chabrier-Rivier, N.: Modelling and querying interaction networks in the biochemical abstract machine BIOCHAM. Journal of Biological Physics and Chemistry 4(2), 64–73 (2004)

5. Keseler, I.M., Collado-Vides, J., Gama-Castro, S., Ingraham, J., Paley, S., Paulsen, I.T., Peralta-Gil, M., Karp, P.D.: EcoCyc: A comprehensive database resource for Escherichia coli. Nucleic Acids Research 33, D334–D347 (2005)

6. Prescott, L.M., Klein, D.A., Harley, J.P.: Microbiology. McGraw-Hill, New York (2002)

7. Romero, P., Wagg, J., Green, M.L., Kaiser, D., Krummenacker, M., Karp, P.D.: Computational prediction of human metabolic pathways from the complete human genome. Genome Biology 6(R2), 1–17 (2004)

8. Senachak, J., Vestergaard, M., Vestergaard, R.: Rewriting game theory and protein signalling in MAPK cascades. In: Proc. CMSB (2006)

9. Shankland, C., Tran, N., Baral, C., Kolch, W.: Reasoning about the ERK signal transduction pathway using BioSigNet-RR. In: Plotkin, G. (ed.) Proceedings of the Third International Conference on Computational Methods in System Biology (2005)

10. Stragier, P., Losick, R.: Molecular genetics of sporulation in bacillus subtilis. Annu. Rev. Genet. 30, 297–341 (1996)

11. Talcott, C., Eker, S., Knapp, M., Lincoln, P., Laderoute, K.: Pathway logic modeling of protein functional domains in signal transduction. In: Proceedings of the Pacific Symposium on Biocomputing (January 2004)

12. Talcott, C.: Symbolic modeling of signal transduction in pathway logic. In: Perrone, L.F., Wieland, F.P., Liu, J., Lawson, B.G., Nicol, D.M., Fujimoto, R.M. (eds.) 2006 Winter Simulation Conference (2006)

13. Talcott, C., Dill, D.L.: Multiple representations of biological processes. Transactions on Computational Systems Biology VI 4220, 221–245 (2006)

14. Yices home page, http://yices.csl.sri.com/

Algorithmic Algebraic Model Checking IV: Characterization of Metabolic Networks[*]

Venkatesh Mysore[1,2] and Bud Mishra[2,3]

[1] D.E. Shaw Research, New York, NY, U.S.A.
[2] Courant Institute, New York University, New York, NY, U.S.A.
[3] NYU School of Medicine, New York University, New York, NY, U.S.A.
venkatesh.mysore@deshaw.com, mishra@nyu.edu

Abstract. A series of papers, all under the title of Algorithmic Algebraic Model Checking (AAMC), has sought to combine techniques from algorithmic algebra, model checking and dynamical systems to examine how a biochemical hybrid dynamical system can be made amenable to temporal analysis, even when the initial conditions and unknown parameters may only be treated as symbolic variables. This paper examines how to specialize this framework to metabolic control analysis (MCA) involving many reactions operating at many dissimilar time-scales. In the earlier AAMC papers, it has been shown that the dynamics of various biochemical semi-algebraic hybrid automata could be unraveled using powerful techniques from computational real algebraic geometry. More specifically, the resulting algebraic model checking techniques were found to be suitable for biochemical networks modeled using general mass action (GMA) based ODEs. This paper scrutinizes how the special properties of metabolic networks– a subclass of the biochemical networks previously handled–can be exploited to gain improvement in computational efficiency. The paper introduces a general framework for performing symbolic temporal reasoning over metabolic network hybrid automata that handles both GMA-based equilibrium estimation and flux balance analysis (FBA). While algebraic polynomial equations over $\mathbb{Q}[x_1, \ldots, x_n]$ can be symbolically solved using Gröbner bases or Wu-Ritt characteristic sets, the FBA-based estimation can be performed symbolically by rephrasing the algebraic optimization problem as a quantifier elimination problem. Effectively, an approximate hybrid automaton that simulates the metabolic network is derived, and is thus amenable to manipulation by the algebraic model checking techniques previously described in the AAMC papers.

1 Introduction

Recently, several biologists have convincingly argued for a systems level analysis, as opposed to the traditional reductionist approach of molecular biology [13,7,12]. When aimed at understanding the holistic properties of the dynamics of biochemical networks, this approach could not only lead to giant leaps in our elucidation of the basic science of biology, but could also contribute more directly to many practical applications, e.g., the drug and vaccine discovery process, diagnosis, agricultural and manufacturing technologies, and synthetic biology of the future. Algebraic analysis may hold the

[*] The work reported in this paper was supported by two grants from NSF ITR program.

H. Anai, K. Horimoto, and T. Kutsia (Eds.): AB 2007, LNCS 4545, pp. 170–184, 2007.
© Springer-Verlag Berlin Heidelberg 2007

key to success of this venture, as it enables obtaining richer answers to deeper questions, even when both initial conditions and rate parameters can only be presented as symbolic variables. It is hoped that this methodology will expose important algebraic functional relationships among the emergent phenomena, the kinetic parameters and the initial conditions, thus revealing many fundamental unifying principles of biology. The starting point for this approach is the fundamental general mass action (GMA) law of chemical kinetics, which supplies a system of ordinary differential equations (ODEs) governing the rate of change of the concentrations of interacting biochemicals. Let k_is denote the rate constants, n_is the number of molecules that appear in the reactions, and W_js their concentrations. Then, the continuous dynamics within each state may be described through the GMA-based ODEs [11,28,58]:

$$\dot{W}_h = +\{\Sigma_{j \in h_+} n_j k_j \Pi_i^j W_i\} - \{\Sigma_{j \in h_-} n_j k_j \Pi_i^j W_i\}. \tag{1}$$

Each equation above is an algebraic differential equation consisting of two affine summation terms: a positive term representing synthesis (all processes producing W_h) and a negative term representing degradation (all processes consuming W_h). The number of W_js (an integer) multiplied in each term is equal to the number of molecules of reactants (and similarly, products) in that reaction (e.g., higher order terms like $100kW_i^3W_j^{10}W_k^5$ are possible[1]).

For tools analyzing GMA ODEs, their ability to handle unknown parameters or uncertainties in their estimates becomes crucial as kinetic parameters are seldom measured under ideal conditions [14]. In response to this challenge, we have extended the GMA models to the algebraic domain, by developing decidable and approximable techniques for symbolic temporal analysis within our Algorithmic Algebraic Model Checking (AAMC) framework, described in a series of publications [46,42,41,6,40,39]. In this framework, the process of numerically integrating the differential equations and extracting a simpler examinable representation is substituted with an algebraic procedure (based on Computational Real Algebraic Geometry [37,15,38]) that can answer complex queries about the symbolic states of the system.

In this paper, we specialize this approach to *metabolism*, which is comprised of the complex enzyme-catalyzed pathways (excluding signal transduction and genetic regulation) that produce and consume the "metabolites" in any living cell. The system of ODEs for metabolic networks lends itself to simplification and efficient analysis because of three key properties [17]: (1) A subset of the metabolites interact with each other through reactions much faster than the rest of the system; (2) These fast reactions always reach a quasi-equilibrium state, which is local (involving only this subset of metabolites) and momentary (it is modulated by the slower reactions in the rest of the system); (3) Mass is conserved during such equilibrium recomputation, and the equilibrium configuration is completely determined by the total concentration of the metabolites. Powerful computational methods have emerged to exploit this structure of metabolic networks; in these methods, only the dynamic GMA simulation of the slow reactions are performed, while, under the assumption that the fast reactions respond quickly, the equilibria of the fast reactions are recomputed at each time-step. A

---

[1] Though negative and real exponents can be indirectly handled, we restrict our analyses to terms with non-negative integer exponents.

list of the most prominent methods that venture in this direction would include the following: tendency modeling [57], dynamic flux balance analysis [33], hybrid static + dynamic simulation [65], intrinsic low dimensional manifolds [52] and singular perturbation analysis [19,32]. This paper builds upon several concepts from the earlier studies to arrive at a trichotomic characterization of the metabolites – the slow irreversible dynamic reactants X, the fast reversible quasi-equilibrium reactants Y and the interface reactants Z. Their properties are elaborated below:

1. *Dynamic Reactants*: All the reactions involving these metabolites (denoted by X) are modeled using detailed general mass action-based differential equations. Typically, these reactions are understood to be slow and irreversible, with dynamics of the form: $\dot{X} = F(X,Z,K)$, where K are the symbolic (rate) parameters.
2. *Quasi-Equilibrium Reactants*: All the reactions involving these metabolites (denoted by Y) are modeled in terms of their dynamic equilibria alone. They always participate in at least one reaction as a substrate and in at least one reaction as a product. Typically, these reactions are understood to be fast and reversible, with dynamics of the form: $\dot{Y} = G(Y,Z,K)$.
3. *Interface Reactants*: These reactants (denoted by Z) interact with both the dynamic reactants and the quasi-equilibrium reactants. Thus, their general mass action based flow equations (from slow reactions) are modified because of the fast reactions with the quasi-equilibrium reactants, giving rise to dynamics of the form: $\dot{Z} = D(X,Z,K) + P(Y,Z,K)$.

Example 1. Consider a simple metabolic network composed of just two reactions: a slow irreversible reaction $A + B \xrightarrow{k_s} R + S$, and a fast reversible reaction $E + S \underset{k_r}{\overset{k_f}{\rightleftharpoons}} C$. This reaction could describe an enzymatic process involving, say, an enzyme (E) and substrate (S) interacting to produce the enzyme substrate complex (C). We wish to study how an external slow reaction producing the substrate can control the equilibrium configuration. Let us denote the metabolite concentrations, $[E]$, $[S]$, $[C]$, by the letters e, s, and c, respectively. The dynamic reactants X are A, B and R. The quasi-equilibrium reactants Y are E and C. The interface reactant Z is S. Their flow equations are: $\dot{a} = \dot{b} = -\dot{r} = -k_s ab$, $\dot{e} = -\dot{c} = k_r c - k_f es$ and $\dot{s} = k_s ab + k_r c - k_f es$. The dynamics are often rephrased using the flux variables $U_1 = k_s ab$, $U_2 = k_r c$ and $U_3 = k_f es$.   □

The existing tools [59] for metabolic networks are all structured primarily on analyses that use numerical simulation, numerical perturbation, random sampling and parameter sweeping techniques. A list of tools in this category includes: Gepasi [35], Systems Biology Workbench [23], E-Cell [54] and BioSpice [29]. Conclusions about the behavior of the network are often made by alternating between (a) tracing specific trajectories over a suitable time frame and then (b) verifying temporal logic properties such as reachability or safety [56,8,2]. The slow reactions in metabolic networks are typically modeled and analyzed as per this approach. The fast reaction systems are typically subject only to a quasi-equilibrium characterization with minimal dynamic characterization. Some of the popular techniques following this strategy are: Metabolic Control Analysis (MCA) [21], Metabolic Flux Analysis (MFA) [31], Flux Balance Analysis (FBA) [27], Cybernetic approaches [43] and Metabolic Pathway Analysis (MPA)

[49]. While the algebraic estimation of the equilibrium concentrations has been studied extensively [8,36,64,3], in contrast, a directed effort to handle both GMA-based simulation and direct equilibrium estimates (via GMA or FBA) algebraically seems conspicuously absent.

Rather than pursuing the traditional numerical simulation based analysis, this paper suggests an entirely symbolic algorithmic algebraic framework for the unified analysis of metabolic networks. It proceeds by first mathematically characterizing the hybrid dynamical system to which metabolic networks correspond, and then integrating general mass action [11] and flux balance analysis [27] based equilibrium estimation. Next the paper shows that the algebraic equilibrium description is decidable, both using GMA and FBA. Our proof of the decidability of the algebraic approach are based on the well-established Gröbner basis and characteristic set techniques [5,47,62,18] for solving polynomial equations, and the decidability of semi-algebraic[2] optimization using real quantifier elimination [55]. The paper then examines how to move from the equilibrium description to its derivative (rate of change), which can then be combined with the ODEs of the slow reactants to complete an algebraic description of the metabolic network. These steps directly lead to efficient algebraic model-checking, since, at this point, they have ensured that all the interactions operate at roughly the same time scale. Hence a bigger time-step suitable for the slow interactions is sufficient (as opposed to the smaller time-step that would have been necessary for the fast reactions), be it for simulation or algebraic temporal logic analysis, based on the techniques described in the earlier AAMC papers.

2   Preliminaries: Algebraic Analysis of a Biochemical Hybrid System

Biochemical systems are conveniently approximated as hybrid automata operating in one of many *discrete states* (or modes). In each state, the *continuous evolution* of different chemicals, reactions, assumptions and ODEs predominate, with *discrete transitions* to other states possible under certain *guard* conditions, leading to the variables being reassigned as per the *reset* relations. Within each state, the temporal properties of the network of interacting biochemicals are captured algebraically by the *flow* relation (from GMA-based ODEs) that relates two neighboring system-states at time instants t and $t + h$, and the biochemical interactions (synthesis, degradation, multimerization, etc.) that occur in that short time interval h.

In the *Algorithmic Algebraic Model Checking* approach [46,42,41,6,40,39], it was shown how most temporal logic query-answering can be expressed as a series of quantifier elimination problems over the reals. The resulting mathematical problem has been known to be decidable [55] and elementarily computable (e.g., using Qepcad [22] or Redlog [16]), though computationally expensive – time complexity, unfortunately, still remains doubly exponential in the number of variables. For such analyses to be possible, each discrete state should have only polynomial ODEs, with the guard, reset and

---

[2] Unquantified first-order formulæ over the theory of reals (i.e., over $(\mathbb{R}, +, \times, =, <)$); see [37,15,38] for details.

invariant relations also being semi-algebraic (Boolean combinations of polynomial equations and inequalities), thus yielding a new class of hybrid systems, as defined below:

Definition 1. Semi-Algebraic Hybrid Automata: [46,42] *A k-dimensional* hybrid automaton *is a septuple, H = (W, V, E, Init, Inv, Flow, Jump), consisting of the following components:*

- *$W = \{W_1, \ldots, W_k\}$ and $W' = \{W'_1, \ldots, W'_k\}$ are two finite sets of variables ranging over the reals $\mathbb{R}$;*
- *(V,E) is a directed graph of discrete states and transitions;*
- *Each discrete state $v \in V$ is labeled by "Init"(initial), "Inv"(invariant) and "Flow" labels of the form $Init_v[W]$, $Inv_v[W]$, and $Flow_v[W,W',t,h]$*
- *Each edge $e \in E$ is labeled by a "Jump" condition of the form $Jump_e[W,W'] \equiv Guard_e(W) \land Reset_e(W,W')$*
- *Init, Inv, Flow, and Jump are all semi-algebraic.*   □

Within each state of a biochemical hybrid dynamical system, the network of interacting biochemicals is modeled using variables that represent their concentrations (see *Eqn.* 1). The semi-algebraic hybrid automaton structure requires that the continuous dynamics of each discrete state v be captured in the *flow* relation $Flow_v[W,W',t,h]$ that connects the symbolic state W of the system at time t with the symbolic state W' at time $t+h$. To derive an approximate *flow* relation, the polynomial differential equations describing the continuous evolution are integrated using one of the symbolic schemes (e.g., the *Taylor* series, the linear *Euler* or the higher degree *Runge-Kutta*). The error is controlled by an upper bound on the time spent in one continuous step, as we aim for over- or under-approximating the flow equations (also see [30]). Thus, we can write the flow equations for the biochemical dynamical system as shown here[3]:

$$Flow_v[W,W',t,h] \equiv \{W' = W + h\dot{W}(W,K)\}.$$

Here, W represents the vector of concentrations at time t, $\dot{W}$ is the vector of first temporal derivatives (from the GMA-based ODEs) expressed as a polynomial in W and the rate constants K (and t, if necessary as with many time-variant systems), and W' is the approximate value of $W(t+h)$ (with $O(h^2)$ error, in the case of the Euler forward integration). Note that the incompleteness that results from following the biochemical traces using a fixed time step (chosen based on the desired integration error bound) that plagues numerical methods is not alleviated in the algebraic procedure detailed here.

Since the guard, resets and invariants are also restricted to be Boolean combinations of polynomial equations and inequalities, the complete transition relation (see *Defn. 3 – Semantics of Hybrid Automata* in [42]) of the biochemical hybrid dynamical system can be written in terms of a semi-algebraic expression. Once such a relation is derived, temporal logic analysis can be performed to algebraically characterize global and emergent dynamical properties of the biochemical network (for example, see the analysis of the Delta-Notch pathway using Timed Computation Tree Logic in the tool Tolque [42]).

---

[3] Without loss of generality, in this paper, we will adopt the Euler forward symbolic integration scheme [42] to compute the trajectories of the metabolic reactions.

3 Algebraic Analysis of Metabolic Hybrid Systems

The basic outline of our algebraic procedure is as follows:

1. Start with a complete general mass action based hybrid automaton model of the entire metabolic network, with symbolic variables (parameters) substituted in place of unknowns.
2. Within each discrete state:
    (a) Identify sub-networks of reversible fast reactions (using information from biochemistry literature).
    (b) Compute the dynamic equilibrium concentrations and fluxes of the fast subnetworks. This step can be performed accurately over the GMA model using the Gröbner basis and Wu-Ritt characteristic set techniques (see *Sec.* 3.1). Similar analysis can also be obtained from the FBA approach, using algebraic optimization (see *Sec.* 3.2). Irrespective of which algorithm is used, we formulate an algebraic description of the equilibrium state of the reactants participating in fast reactions. (In some cases, this equilibrium description might yield differential equations – see *Defn.* 5 and *Note* 1)
3. Now the entire hybrid system is ready to be simulated or analyzed using a timestep appropriate for the slow biochemical reactions, with the fast reactants in each discrete state updated as determined by the equilibrium relations (or in some cases, the new differential equations) derived in *Step* 2(*b*).

Steps 1, 2(*a*) and 3 are part of the standard procedure [57,33,65], and there is no need for a new algebraic version. This paper provides the necessary mathematical details for *Step* 2(*b*), where we wish to symbolically characterize the momentary quasi-equilibria that the fast variables (interface and quasi-equilibrium metabolites) reach in response to a change in the slow interactions (dynamic reactants) at each time-step. We first formally capture the dynamical system to which the subclass of metabolic networks corresponds, as constrained by our assumptions (see *Sec.* 1 for details).

Definition 2. *Metabolic Dynamics: A metabolic network comprises the slow irreversible dynamic reactants X, the fast reversible quasi-equilibrium reactants Y, the interface reactants Z that participate in both slow and fast reactions, and symbolic (rate) parameters K, such that the following differential algebraic equations hold:*

$$\dot{X} = F(X,Z,K) \, , \ \dot{Y} = G(Y,Z,K) \, , \ \dot{Z} = D(X,Z,K) + P(Y,Z,K). \quad \Box$$

As before, let X, Y and Z be the concentrations of the dynamic, quasi-equilibrium and interface metabolites respectively, at time t — the start of the integration step. The goal is to derive the $Flow(\{X,Y,Z\},\{X',Y',Z'\},h,K)$ relation (in each discrete state[4] of the semi-algebraic hybrid automaton of the metabolic network) that expresses the algebraic values of the concentrations X', Y' and Z' at time $t+h$ in terms of their concentrations at time t, the small time-step h and the rate parameters K (and time t, if required to capture

---

[4] The subscript v denoting the discrete state and the explicit time variable t are dropped for clarity from the $Flow_v$ notation.

some other external aspects of the dynamics). The flow equations of the dynamic reactants X do not involve any simplification and are directly given by the Euler forward approximation as $X' = X + hF(X,Z,K)$. Thus,

$$Flow(\{X,Z,Y\}, \{X',Z',Y'\}, h, K) \equiv$$
$$\{ X' = X + hF(X,Z,K) \ \wedge \ Flow(\{Z,Y\}, \{Z',Y'\}, h, K).$$

Thus, the essence of the problem is the expression of $Flow(\{Z,Y\}, \{Z',Y'\}, h, K)$ — the flow of the quasi-equilibrium and interface reactants, *algebraically*. As a result of the way we have formulated the problem, the complete set of constraints, which must be true to achieve quasi-equilibrium are given by:

Definition 3. *Quasi-Equilibrium Relation:*

$$\mathscr{E}(Z,Y,K) \equiv \{P(Z,Y,K) = 0 \wedge G(Y,Z,K) = 0\} \qquad \square$$

GMA follows the straightforward approach of solving the quasi-equilibrium equations to obtain the exact *concentrations*. FBA instead guesses what the equilibrium *fluxes* must be by optimizing some function, without using the kinetic parameters K; the exact concentrations are then obtained by substituting the concentration terms for the flux variables. The algebraic versions of the two procedures and their mathematical details are further elaborated below.

3.1   General Mass Action Based Approximation

Since the quasi-equilibrium characterization (see *Defn.* 3) involves only equalities, the relation $\mathscr{E}$ is effectively just a system of polynomial equations, which needs to be solved for Z and Y. The issue of simultaneous solution of polynomial equations, especially in the context of biochemical networks, has been addressed before [4,8,36]. The well-established methods for solving such systems of simultaneous multivariate polynomial equations with symbolic parameters are to be found in the Gröbner Basis algorithm [5] and the Wu-Ritt characteristic set [47,62] algorithm. Their many implemented forms include PoSSo [10], CoCoA [9] and Macaulay-2 [20].

In the case of metabolic dynamical systems, the system of polynomial equations can be solved more easily by exploiting the fact that the concentration of each chemical form of a metabolite at pseudo-equilibrium is dictated by the total concentration of its different chemical forms. In other words, each substrate of each reaction involving at least one interface metabolite (Z) also as a substrate, is associated with a mass conservation equation. As suggested in the literature [57,33,65], the total concentration of these substrate metabolites in their many chemical forms at equilibrium is captured using *equilibrium pool* variables T. The mass conservation equations $T = \mathscr{M}(Z,Y)$ have the form: $T_i = \Sigma_{j \in Pool_i} W_j$, where $Pool_i$ represents the set of the different chemical forms W_j, in which the i-th substrate metabolite exists. Effectively, as a result of the structure of metabolic pathways, the equilibrium concentrations of Z and Y are expressible in terms of the equilibrium pool concentrations T. The simplified GMA equilibrium relation may thus be expressed as:

Definition 4. *GMA Equilibrium Relation:*

$$\mathscr{E}_{GMA}(Z,Y,T,K) \equiv \{Z = E_Z(T,K) \wedge Y = E_Y(T,K)\},$$

where E_Z and E_Y represent the solutions obtained using the Gröbner basis technique over $\{P(Z,Y,K) = 0 \wedge G(Y,Z,K) = 0 \wedge T = \mathscr{M}(Z,Y)\}$.   □

We are now ready to construct $Flow(\{X,Y,Z,T\},\{X',Y',Z',T'\},h,K)$ – the continuous flow expression, which connects the state of the system $\{X,Y,Z,T\}$ at time t and the state of the system $\{X',Y',Z',T'\}$ at time $t + h$. During the quasi-instantaneous recomputation of the equilibrium point, the total concentrations of the pool variables T can be assumed to remain unchanged. This assumption is justifiable because the time required for re-establishing the equilibrium is negligible compared to the time-step used for simulating the slow reactions. Consequently, the change in the concentrations attributed to the slow reactions is negligible compared to the effect of the equilibrium recomputation (which only redistributes the metabolites among the different chemical forms being added in each equilibrium pool variable). Thus, the ODEs for Z and Y can be directly approximated from E_Z and E_Y by differential calculus:

$$\dot{Z} \approx \frac{dE_Z(T,K)}{dt} = \frac{\partial E_Z(T,K)}{\partial T} \cdot \frac{dT}{dt} = \frac{\partial E_Z}{\partial T} H,$$

where $\dot{T}_i = \Sigma_{j \in Pool_i} \dot{W}_j = H_i(X,Y,Z,K)$. The same applies to Y as well. Thus we have our final result:

Definition 5. *GMA-Approximated Metabolic Dynamics:*

$$Flow_{GMA}(\{X,Y,Z,T\},\{X',Y',Z',T'\},h,K) \equiv$$
$$\{(X' = X + hF(X,Z,K)) \wedge (T' = T + hH(X,Y,Z,K) \wedge$$
$$(Z' = Z + h\dot{Z}(T,X,Y,Z,K) \wedge (Y' = Y + h\dot{Y}(T,X,Y,Z,K))\},$$
$$\text{where}: \dot{X} = F(X,Z,K) \quad , \quad \dot{T} = H(X,Y,Z,K),$$
$$\dot{Z}(T,X,Y,Z,K) \approx \frac{\partial E_Z}{\partial T} H \quad \& \quad \dot{Y}(T,X,Y,Z,K) \approx \frac{\partial E_Y}{\partial T} H \qquad \square$$

3.2 Flux Balance Analysis Based Approximation

Flux Balance Analysis [27] aims to estimate the steady-state flux distribution using the stoichiometric matrix and the input and output fluxes of the system to constrain the solution space, without relying on any kinetic parameters. Since the number of fluxes is always greater than the number of metabolites, the system of linear flux equations is under-determined. FBA overcomes this hurdle by assuming that the biochemical network would have so evolved as to optimize certain physiologically important functions such as growth. Thus, the essence of flux balance analysis is optimizing a function under the set of equilibrium and other external constraints.

The general optimization (in its maximization formulation) problem can be rephrased as follows: *for all values U that differ from the optimal value Ǔ and still satisfy the constraints $\mathscr{C}(U,V)$ involving parametric variables V (not being optimized), the value of the function $\mathscr{F}(U,V)$ is, by definition, less than $\mathscr{F}(\check{U},V)$.* This step immediately leads to the following characterization of $\{\check{U},V\}$:

Definition 6. Optimization Relation: $Optimize(\check{U}, \mathscr{F}(U,V), \mathscr{C}(U,V)) \equiv \mathscr{C}(\check{U},V)$
$\bigwedge\{\forall U, (U \neq \check{U} \wedge \mathscr{C}(U,V)) \Rightarrow (\mathscr{F}(U,V) < \mathscr{F}(\check{U},V))\}$ □

If $\mathscr{C}$ is semi-algebraic and $\mathscr{F}$ is polynomial, then $Optimize(\check{U}, \mathscr{F}(U,V), \mathscr{C}(U,V))$ is a quantified semi-algebraic set. Gröbner bases or characteristic sets cannot be used to solve this optimization problem as they can handle only equations and not inequality relations. Instead, the general technique of real quantifier elimination [55] has to be employed to perform the algebraic optimization [1]. In addition to quantifier elimination tools like Qepcad [22] and Redlog [16], specialized systems such as the Maple-based Symbolic-Numeric toolbox for Real Algebraic Constraints (SyNRAC) [63] could also be exploited for performing algebraic optimization.

Unlike the GMA-based approach which uses concentrations to describe the dynamics, FBA uses the *flux* variables: $U_j \equiv n_j k_j \Pi_i^j W_i$. For metabolic networks, the flux variables may be divided into $U_{zx} \equiv U_z \cup U_x \cup U_{z\wedge x}$ and $U_{zy} \equiv U_{z\wedge y} \cup U_y$ based on whether the reactions are fast or slow[5]: reactions in which only Z_is and/or X_is participate contribute the slow flux terms U_{zx}; reactions in which Z_is and Y_is interact and those in which only Y_is interact contribute the fast flux terms U_{zy}. Thus the metabolic dynamics (see *Defn. 2*) may be rephrased as:

$$\dot{X} = F_U(U_{zx}) \,, \ \dot{Y} = G_U(U_{zy}) \,, \ \dot{Z} = D_U(U_{zx}) + P_U(U_{zy}).$$

Let $\mathscr{C}(U_{zy}, U_{zx})$ represent the semi-algebraic constraints on the kinetic parameters, rates of change, bounds on parameters, energy balance equations, etc. Let $\mathscr{F}(U_{zy}, U_{zx})$ represent the function that the metabolic network is assumed to be optimizing. Thus, the complete set of equations and inequalities that needs to be true at the equilibrium predicted by FBA may be represented thus:

Definition 7. FBA Equilibrium Relation:

$$\mathscr{E}_{FBA}(\check{U}_{zy}, U_{zx}) \equiv \{ Optimize(\check{U}_{zy}, \mathscr{F}(U_{zy}, U_{zx}), \mathscr{C}(U_{zy}, U_{zx})) \wedge$$
$$G_U(\check{U}_{zy}) = P_U(\check{U}_{zy}) = 0 \}.   □$$

Consistent with the static optimization based dynamic flux balance analysis approach [33], it is assumed that at the beginning of each small time interval h, the fast reactions optimize growth (or some other physiological function) by re-establishing equilibrium (U_{zy}) based on the current concentrations of the fast and dynamic reactants (U_{zx}). The slow reactions are then integrated assuming that these fluxes stay constant over that time period h. Thus, the FBA-based dynamics can now be characterized algebraically as:

Definition 8. FBA-Approximated Metabolic Dynamics:

$$Flow_{FBA}(\{X,Y,Z\}, \{X',Y',Z'\}, h, K) \equiv$$
$$\{\mathscr{E}_{FBA}(U_{z''y'}, U_{zx}) \wedge X' = X + hF(X, Z'', K) \wedge Z' = Z'' + hD(X, Z'', K)\}$$

where $U_i \equiv n_i k_i \Pi_j^i W_j$.   □

---

[5] Note that since X and Y do not interact, there are no $U_{x\wedge z}$ terms.

Remark 1. Alternatively, one could perform FBA using the concentration variables themselves. Let $\mathscr{C}(Z',Y',Z,Y,K)$ represent the semi-algebraic constraints on the kinetic parameters, rates of change, bounds on parameters, energy balance equations, etc. Let $\mathscr{O}(Z',Y',Z,Y)$ represent the function[6] that the metabolic network is assumed to be optimizing. Since FBA assumes that the kinetic parameters K are unavailable, the effective set of constraints over which the optimization must be performed *may be* obtained by eliminating K from the accurate equilibrium relation $\mathscr{E}$ (see *Defn. 3*). Note that if K is not eliminated, the equilibrium is exactly defined by the relation $\mathscr{E}$; hence there is no room for optimization. Further, the existential quantifier captures the assumption that there exist some kinetic parameters (involved in the genetic variation, and discovered during evolution via natural selection) for which the network optimizes the physiologically relevant function (i.e., its "fitness" function). Thus, the dynamics may be approximated thus:

$$\mathscr{O}(Z',Y',Z'',Y) \equiv \exists K, \{\mathscr{C}(Z',Y',Z'',Y,K) \wedge \mathscr{E}(Z',Y',K)\},$$

$$\mathscr{E}_{FBA}(Z'',Y,Z',Y') \equiv Optimize(\{Z',Y'\}, \mathscr{F}(Z',Y',Z'',Y), \mathscr{O}(Z',Y',Z'',Y)) \ \&$$

$$Flow_{FBA}(\{X,Y,Z\}, \{X',Y',Z'\}, h, K) \equiv$$
$$\{(X' = X + hF(X,Z,K)) \wedge (Z'' = Z + hD(X,Z,K)) \wedge \mathscr{E}_{FBA}(Z'',Y,Z',Y')\}.$$

The validity and utility of this approach need to be investigated further.

Note 1. In some cases, the solution after optimization and substitution with the concentration variables might be a set of polynomial equations, which can then be solved (by Gröbner basis like methods, say) to yield the general solution

$$Flow_{FBA}(\{Y,Z\}, \{Y',Z'\}, h, K) \equiv \{Z' = E_Z(Z,Y,T,K) \wedge Y' = E_Y(Z,Y,T,K)\}.$$

Then, we can write:

$$\dot{Z} = \frac{\partial E_Z}{\partial Z}\dot{Z} + \frac{\partial E_Z}{\partial Y}\dot{Y} + \frac{\partial E_Z}{\partial T}\dot{T} \ , \ \dot{Y} = \frac{\partial E_Y}{\partial Z}\dot{Z} + \frac{\partial E_Y}{\partial Y}\dot{Y} + \frac{\partial E_Y}{\partial T}\dot{T}.$$

By solving these two equations, one can obtain the general solution:

$$\dot{Y} = \frac{\frac{\frac{\partial E_Y}{\partial Z}\frac{\partial E_Z}{\partial T}}{1 - \frac{\partial E_Z}{\partial Z}} + \frac{\partial E_Y}{\partial T}}{1 - \frac{\frac{\partial E_Y}{\partial Z}\frac{\partial E_Z}{\partial Y}}{1 - \frac{\partial E_Z}{\partial Z}} - \frac{\partial E_Y}{\partial Y}}\dot{T} \ , \ \dot{Z} = \frac{\frac{\partial E_Z}{\partial Y}\dot{Y} + \frac{\partial E_Z}{\partial T}\dot{T}}{1 - \frac{\partial E_Z}{\partial Z}}.$$

Also note that $\{\dot{Z} = \frac{\partial E_Z}{\partial T}\dot{T} \ , \ \dot{Y} = \frac{\partial E_Y}{\partial T}\dot{T}\}$ derived in the GMA based approximation is just a special case where $\frac{\partial E}{\partial Y} = \frac{\partial E}{\partial Z} = 0$.

4   Example

Our approach is now illustrated on the *Example* 1 introduced earlier. Recall that $X = \{A,B,R\}$, $Y = \{E,C\}$ and $Z = \{S\}$.

---

[6] The primed variables may be necessary to capture relations involving the rate of change of concentrations.

4.1 GMA-Based Approximation

The only reaction with an interface metabolite as a substrate is $E + S \underset{k_r}{\overset{k_f}{\rightleftharpoons}} C$. The mass-conservation equations can be written for the two substrates E and S as $e_T = e + c$ and $s_T = s + c$, where e_T and s_T are the new equilibrium pool variables. At equilibrium, $\mathscr{E}(\{s\}, \{e,c\}, K) \equiv \{(k_f es - k_r c = 0)\}$, i.e., $k_f es = k_r c$. Rewriting in terms of the equilibrium pool variables, we get $k_f(e_T - c)(s_T - c) = k_r c$. Let $k = k_f/k_r$. In the general case, we would solve these equations using the Gröbner basis technique. Here, these quadratic equations can be solved directly, under the constraint that all concentrations are non-negative, leading to the solution:

$$\mathscr{E}_{GMA} \equiv \{c = \frac{(s_T + e_T + 1/k) - \sqrt{(s_T + e_T + 1/k)^2 - 4(s_T + e_T)}}{2} \ \wedge$$

$$e = e_T - c \ \wedge \ s = s_T - c\}.$$

Observe that $\dot{T}$ is: $\{\dot{s}_T = \dot{s} + \dot{c} = k_s ab, \ \dot{e}_T = \dot{e} + \dot{c} = 0\}$. Thus,

$$\hat{c} = \frac{\partial c}{\partial T}\dot{T} = \frac{\partial c}{\partial s_T}\dot{s}_T = \frac{1}{2}(1 - \frac{2(s_T + e_T + 1/k) - 4}{2\sqrt{(s_T + e_T + 1/k)^2 - 4(s_T + e_T)}})k_s ab$$

$$\hat{e} = -\dot{c}, \ \hat{s} = k_s ab - \dot{c}. \ \textit{Thus, we get:}$$

$$\textit{Flow}_{GMA} \quad (\{\{a,b,r\}, \{s\}, \{e,c\}, \{e_T, s_T\}\},$$
$$\{\{a',b',r'\}, \{s'\}, \{e',c'\}, \{e'_T, s'_T\}\}, h, k) \equiv$$
$$\{(a' = a + h\dot{a}) \ \wedge \ (b' = b + h\dot{b}) \ \wedge \ (r' = r + h\dot{r}) \ \wedge \ (s' = s + h\hat{s}) \ \wedge$$
$$(e' = e + h\hat{e}) \ \wedge \ (c' = c + h\hat{c}) \wedge (e'_T = e_T + h\dot{e}_T) \wedge (s'_T = s_T + h\dot{s}_T)\}.$$

4.2 FBA-Based Approximation

Observe that $U_{zx} = U_{z \wedge x} = \{U_1\}$ and $U_{zy} = U_{z \wedge y} = \{U_2, U_3\}$. Let $\mathscr{C}(\{U_2, U_3\}, \{U_1\})$ represent the external constraints under which the network is assumed to be optimizing the function $\mathscr{F}(\{U_2, U_3\}, \{U_1\})$. Thus, the equilibrium may be characterized as:

$$\mathscr{E}_{FBA}(\{\check{U}_2, \check{U}_3\}, \{U_1\}) = \mathscr{C}(\{\check{U}_2, \check{U}_3\}, \{U_1\}) \ \wedge$$
$$\{\forall U_2, U_3, (U_2 \neq \check{U}_2 \ \vee \ U_3 \neq \check{U}_3) \ \wedge \ \mathscr{C}(\{U_2, U_3\}, \{U_1\})$$
$$\Rightarrow \ \mathscr{F}(\{U_2, U_3\}, \{U_1\}) < \mathscr{F}(\{\check{U}_2, \check{U}_3\}, \{U_1\})\}$$
$$\wedge \ \check{U}_2 = \check{U}_3.$$

This leads to the complete flow characterization:

$$\textit{Flow}_{FBA} \quad (\{\{a,b,r\}, \{s\}, \{e,c\}\},$$
$$\{\{a',b',r'\}, \{s'\}, \{e',c'\}\}, h, \{k_s, k_f, k_r\}) \equiv$$
$$\{\mathscr{E}_{FBA}(\{U_2, U_3\}, \{U_1\}) \ \wedge$$
$$a' = a + h\dot{a} \ \wedge \ b' = b + h\dot{b} \ \wedge \ r' = r + h\dot{r} \ \wedge \ s' = s'' + h\dot{s}\},$$

where $U_1 = k_s ab$, $U_2 = k_f e's''$, $U_3 = k_r c'$, $\dot{a} = \dot{b} = -\dot{r} = -k_s ab$, and $\dot{s} = k_s ab + k_r c' - k_f e's''$.

5   Discussion

Several extensions of the mathematical theory [37,15] are necessary for the approach to be more practical and useful. To improve computational complexity, it is necessary to develop more efficient, albeit less general, techniques for equilibrium estimation: for instance, applications of the Wu-Ritt characteristic set algorithm [34], resultant computation followed by eigen decomposition [60], and heuristics for choosing among them [45]. To reduce the computational overload due to the algebraic optimization involved in FBA, some less universal quantifier elimination approaches may be used [61,26]. More recent efforts at efficient optimization include the following: constraint logic programming with first-order constraints $CLP(RL)$ [53] based on Redlog [16], systems theoretic algebraic optimization [25], and semidefinite programming [44].

In additional to the purely algebraic research described previously, several Systems Biology extensions and applications also necessitate further investigation. For instance, the relative merit of flux-based and concentration-based characterization of dynamics (see *Remark* 1) has to be further investigated, in terms of both the complexity gain and the repertoire of constraints that can be handled (such as Minimization of Metabolic Adjustment (MOMA) [50] and Regulatory On-Off Minimization (ROOM) [51]). Similarly, an integration with singular perturbation analysis-like methods [52,19,32] can potentially help automate the classification of the metabolic system interactions as fast and slow, and the decomposition into sub-modules of a large network. Other approximate methods to estimate the equilibrium fluxes (e.g., cybernetic modeling [43]) may also become more powerful when extended into the algebraic domain. Another important perspective comes from the mathematically rigorous approaches being developed in non-linear Control Theory [24,48]. A related thorny problem that remains to be properly addressed is the semi-automatic (approximate) translation of a one-state biochemical dynamical system into a multi-state hybrid system.

In summary, we have exploited techniques from the AAMC approach to enable efficient analysis of metabolic networks. This paper shows how the numerical procedure for exploiting the inherent multi-time-scale quasi-equilibrium structure of metabolic networks could be extended to the algebraic domain, using techniques from Computational Real Algebraic Geometry: namely, real quantifier elimination, Gröbner bases, Wu-Ritt Characteristic sets, and algebraic optimization. Our approach is thus an algebraic generalization of numerical approaches, as typified by tendency modeling [57], dynamic flux balance analysis [33] and hybrid static + dynamic simulation [65]. The more general mathematical approaches [52,19,32] make fewer assumptions about the structure of metabolic networks, and can be incorporated into the proposed framework. Further, the paper provides a uniform algebraic framework to handle two distinct approaches for equilibrium estimation: (i) solving the general mass action-based polynomial equations and (ii) optimizing the flux distribution using flux balance analysis. Thus, the paper demonstrates how a standard biochemistry problem description can be automatically transformed into an entirely algebraic dynamical system specification. This algebraic framework can potentially elicit a powerful symbolic functional description of the dynamical behavior of the metabolic network, in terms of the quasi-equilibrium states of its fast reversible sub-networks. This algebraic approach is to be contrasted with the conventional analysis, which involves performing numerical integration of the ordinary differential equations (ODEs), time-course data assimilation, visualization and model-checking of concentration-traces.

In conclusion, we note the success of Algorithmic Algebraic Model Checking project, which was initiated to integrate relevant theory in dynamical systems, model checking, hybrid automata and systems biology, in an effort to establish a sound and rigorous procedure for symbolic temporal reasoning over biochemical networks. While, in terms of building a suitable theoretical foundation, it has been successful, it has also pointed to newer theoretical and pragmatic problems that were unforeseen at the outset. One apparent shortcoming of our approach is its computational complexity; but it is hoped that this hurdle could be overcome, when the different avenues of extending these ideas are explored in the theoretical and practical realms.

References

1. Anai, H.: On solving semidefinite programming by quantifier elimination. In: Proceedings of the American Control Conference (June 1998)
2. Antoniotti, M., Policriti, A., Ugel, N., Mishra, B.: Reasoning about Biochemical Processes. Cell Biochemistry and Biophysics 38, 271–286 (2003)
3. Barnett, M.P.: Computer algebra in the life sciences. SIGSAM Bull. 36(4), 5–32 (2002)
4. Barnett, M.P., Capitani, J.F., Gathen, J., Gerhard, J.: Symbolic calculation in chemistry: Selected examples. International Journal of Quantum Chemistry 100, 80–104 (2004)
5. Buchberger, B.: Grobner bases: An algorithmic method in polynomial ideal theory. Recent Trends in Multidimensional Systems Theory, pp. 184–232 (1985)
6. Casagrande, A., Mysore, V., Piazza, C., Mishra, B.: Independent dynamics hybrid automata in systems biology. In: First International Conference on Algebraic Biology (2005)
7. Cascante, M., Boros, L.G., Comin-Anduix, B., de Atauri, P., Centelles, J.J., Lee, P.W.-N.: Metabolic control analysis in drug discovery and design. Nature Biotechnology 20, 243–249 (2002)
8. Celik, E., Bayram, M.: Application of grobner basis techniques to enzyme kinetics. Applied Mathematics and Computation 153, 97–109 (2004)
9. CoCoATeam. CoCoA: a system for doing Computations in Commutative Algebra. (2005), Available at http://cocoa.dima.unige.it
10. European Commission. Posso: Polynomial system solving research project. (1996), http://posso.dm.unipi.it
11. Cornish-Bowden, A.: Fundamentals of Enzyme Kinetics, 3rd edn. Portland Press, London (2004)
12. Cornish-Bowden, A., Cardenas, M.L.: Metabolic analysis in drug design. C. R. Biologies 326, 509–515 (2003)
13. Cornish-Bowden, A., Cardenas, M.L.: Systems biology may work when we learn to understand the parts in terms of the whole. Biochemical Society Transactions, 33(3) (2005)
14. Cornish-Bowden, A., Hofmeyr, J.-H.S.: Enzymes in context: Kinetic characterization of enzymes for systems biology. The Biochemist 27, 11–14 (2005)
15. Cox, D.A., Little, J.B., O'Shea, D.: Ideals, Varieties, and Algorithms: An Introduction to Computational Algebraic Geometry and Commutative Algebra. Springer, Heidelberg (1996)
16. Dolzmann, A., Sturm, T.: REDLOG: Computer algebra meets computer logic. SIGSAM Bulletin 31(2), 2–9 (1997)
17. Fell, D.A.: Understanding the Control of Metabolism. Portland Press, London (1997)
18. Gallo, G., Mishra, B.: Wu-ritt characteristic sets and their complexity. DIMACS series in Discrete Mathematics and Theoretical Computer Science 6, 111–136 (1991)

19. Gerdtzen, Z.P., Daoutidis, P., Hu, W.S.: Non-linear reduction for kinetic models of metabolic reaction networks. Metab. Eng. 6(2), 140–154 (2004)
20. Grayson, D.R., Stillman, M.E.: Macaulay 2, a software system for research in algebraic geometry. Available at http://www.math.uiuc.edu/Macaulay2/
21. Hofmeyr, J.-H.S.: Metabolic control analysis in a nutshell. In: Proceedings of the Second International Conference on Systems Biology, pp. 291–300 (2001)
22. Hong, H.: Quantifier elimination in elementary algebra and geometry by partial cylindrical algebraic decomposition, version 13. (1995), WWW site www.eecis.udel.edu/~saclib
23. Hucka, M., Finney, A., Sauro, H.M., Bolouri, H., Doyle, J., Kitano, H.: The erato systems biology workbench: Enabling interaction and exchange between software tools for computational biology. In: Proceedings of the Pacific Symposium on Biocomputing (2002)
24. Ingalls, B.P.: A control theoretic interpretation of metabolic control analysis (submitted) (2005), http://www.math.uwaterloo.ca/~bingalls/Pubs/con.pdf
25. Jibetean, D.: Algebraic optimization with applications to system theory. PhD Thesis, Department of Mathematics, Vrije University, Amsterdam (2003)
26. Jirstrand, M.: Nonlinear control system design by quantifier elimination. J. Symbolic Computation 24, 137–152 (1997)
27. Kauffman, K.J., Prakash, P., Edwards, J.S.: Advances in flux balance analysis. Curr. Opin. Biotechnol. 14, 491–496 (2003)
28. Keener, J.P., Sneyd, J.: Mathematical Physiology. Springer, New York (1998)
29. Kumar, S.P., Feidler, J.C.: Biospice: A computational infrastructure for integrative biology. OMICS: A Journal of Integrative Biology 7(3), 225–225 (2003)
30. Lanotte, R., Tini, S.: Taylor Approximation for Hybrid Systems. In: Morari, M., Thiele, L. (eds.) HSCC 2005. LNCS, vol. 3414, pp. 402–416. Springer, Heidelberg (2005)
31. Lee, D.Y., Yun, H., Park, S., Lee, S.Y.: Metafluxnet: the management of metabolic reaction information and quantitative metabolic flux analysis. Bioinformatics 19(16), 2144–2146 (2003)
32. Litcanu, G., Velazquez, J.J.L.: Singular perturbation analysis of camp signalling in dictyostelium discoideum aggregates. J. of Mathematical Biology 52(5), 682–718 (2006)
33. Mahadevan, R., Edwards, J.S., Doyle-III, F.J.: Dynamic flux balance analysis of diauxic growth in escherichia coli. Biophysical Journal 83, 1331–1340 (2002)
34. Manocha, D., Canny, J.F.: Multipolynomial resultant algorithms. J. Symbolic Computation 15, 99–122 (1993)
35. Mendes, P.: Biochemistry by numbers: simulation of biochemical pathways with gepasi 3. Trends in Biochemical Sciences 22, 361–363 (1997)
36. Minimair, M., Barnett, M.P.: Solving polynomial equations for chemical problems using Gröbner bases. Molecular Physics 102(23–24), 2521–2535 (2004)
37. Mishra, B.: Algorithmic Algebra. In: Texts and Monographs in Computer Science, Springer, New York (1993)
38. Mishra, B.: Computational Real Algebraic Geometry, pp. 740–764. CRC Press, Boca Raton, FL (2004)
39. Mysore, V.: Algorithmic Algebraic Model Checking: Hybrid Automata and Systems Biology. Ph.D. Thesis, New York University, New York, USA (2006)
40. Mysore, V., Casagrande, A., Piazza, C., Mishra, B.: Tolque – A Tool for Algorithmic Algebraic Model Checking. In: HSCC'06 Poster Session (March 2006)
41. Mysore, V., Mishra, B.: Algorithmic Algebraic Model Checking III: Approximate Methods. In: Infinity'05 ENTCS 149(1), 61–77 (2006)
42. Mysore, V., Piazza, C., Mishra, B.: Algorithmic Algebraic Model Checking II: Decidability of Semi-Algebraic Model Checking and its Applications to Systems Biology. In: Peled, D.A., Tsay, Y.-K. (eds.) ATVA 2005. LNCS, vol. 3707, pp. 217–233. Springer, Heidelberg (2005)

43. Namjoshi, A.A., Doraiswami, R.: A cybernetic modeling framework for analysis of metabolic systems. Computers & chemical engineering 29(3), 487–498 (2005)
44. Parrilo, P., Lall, S.: Semidefinite programming relaxations and algebraic optimization in control. European Journal of Control 9(2–3), 307–321 (2003)
45. Petitjean, S.: Algebraic geometry and computer vision: Polynomial systems, real and complex roots. Journal of Mathematical Imaging and Vision 10, 191–220 (1999)
46. Piazza, C., Antoniotti, M., Mysore, V., Policriti, A., Winkler, F., Mishra, B.: Algorithmic Algebraic Model Checking I: The Case of Biochemical Systems and their Reachability Analysis. In: Etessami, K., Rajamani, S.K. (eds.) CAV 2005. LNCS, vol. 3576, pp. 5–19. Springer, Heidelberg (2005)
47. Ritt, J.F.: Differential Algebra, vol. XXXII. AMS Colloquium Publications, New York (1950)
48. Sauro, H.M.: The computational versatility of proteomic signaling networks. Current Proteomics 1, 67–81 (2004)
49. Schilling, C.H., Schuster, S., Palsson, B.O., Heinrich, R.: Metabolic pathway analysis: Basic concepts and scientific applications in the post-genomic era. Biotechnol. Prog. 15, 296–303 (1999)
50. Segre, D., Vitkup, D., Church, G.M.: Analysis of optimality in natural and perturbed metabolic networks. PNAS 99(23), 15112–15117 (2002)
51. Shlomi, T., Berkman, O., Ruppin, E.: Constraint-based modelling of perturbed organisms: A room for improvement. In: ISMB (2004)
52. Singh, S., Powers, J.M., Paolucci, S.: On slow manifolds of chemically reactive systems. The Journal of Chemical Physics 117(4), 1482–1496 (2002)
53. Sturm, T.: Quantifier elimination-based constraint logic programming. Technical Report MIP-0202, Fakultät für Mathematik und Informatik, Universität Passau (2002)
54. Takahashi, K., Kaizu, K., Hu, B., Tomita, M.: A multi-algorithm, multi-timescale method for cell simulation. Bioinformatics 20(4), 538–546 (2004)
55. Tarski, A., Decision, A.: Method for Elementary Algebra and Geometry. University of California Press, 2 edn. (1948)
56. Tiwari, A., Khanna, G.: Series of Abstraction for Hybrid Automata. In: Tomlin, C.J., Greenstreet, M.R. (eds.) HSCC 2002. LNCS, vol. 2289, pp. 465–478. Springer, Heidelberg (2002)
57. Visser, D., van der Heijden, R., Mauch, K., Reuss, M., Heijnen, S.: Tendency modeling: A new approach to obtain simplified kinetic models of metabolism applied to s. cerevisiae. Metabolic Engineering 2, 252–275 (2000)
58. Voit, E.O.: Computational Analysis of Biochemical Systems. A Pratical Guide for Biochemists and Molecular Biologists. Cambridge University Press, Cambridge (2000)
59. Voit, E.O.: The dawn of a new era of metabolic systems analysis. Drug Discovery Today: BioSilico 2(5), 182–189 (2004)
60. Wallack, A., Emiris, I.Z., Manocha, D., MARS: A MAPLE/MATLAB/c resultant-based solver. In: Intl. Symposium on Symbolic and Alg. Computation, pp. 244–251 (1998)
61. Weispfenning, V.: Simulation and optimization by quantifier elimination. J. Symb. Comput. 24(2), 189–208 (1997)
62. Wu, W.-T.: On the decision problem and the mechanization of theorem proving in elementary geometry. Scientia Sinica 21(2), 159–172 (1978)
63. Yanami, H., Anai, H.: Development of SyNRAC. In: Computer Algebra Systems and Applications, CASA (2005)
64. Yildirim, N.: Use of symbolic and numeric computation techniques in analysis of biochemical reaction networks. International Journal of Quantum Chemistry (2005)
65. Yugi, K., Nakayama, Y., Kinoshita, A., Tomita, M.: Hybrid dynamics/static method for large-scale simulation of metabolism. T. Biology and Medical Modelling, 2(42) (2005)

Cascaded Games

Jittisak Senachak, Mun'delanji Vestergaard, and René Vestergaard*

JAIST, Nomi, Ishikawa, Japan
vester@jaist.ac.jp
http://www.jaist.ac.jp/~vester/

Abstract. We introduce a novel model construction, *cascaded games*, that is intended to allow us to study the notion of *steady states* algebraically and structurally. The model construction is inspired by the chemical underpinning and the prevailing conceptualisation of *mitogen-activated protein kinase (MAPK) cascades*. To analyse the models, we use the recent notion of *change-of-mind equilibria*. We exemplify our proposal with gene regulation and MAPK cascades, capturing basic as well as advanced issues such as *prophage induction* and *tauopathy* causation.

1 Introduction

The core contribution of this paper is conceptual and formal support for feasible, i.e., sub-exponential, *steady-state* analysis. Steady-state analysis is based on the idea that an autonomous system is most likely to be found in certain configurations (that we shall attempt to characterise abstractly) and not in others, and that an analysis therefore may serve to predict (emergent) behaviour. In the biological sciences, the idealised form of this idea is that life itself is a reflection of the possible steady states of the involved system constituents, along with provoked transitions between the steady states. One of the best known qualitative steady-state analysis is due to Kauffman [17] and Thomas [37]. The concrete insight behind their analysis is that understanding whether genes influence each other's *expression* positively or negatively suffices to identify the expression configurations that are characteristic of the organism's main functionality. This may be as concrete as the ways in which an organism may replicate, see Section 4.

In earlier work [5], we showed that Kauffman/Thomas steady-state analysis is a concrete use of a recent notion of dynamic Nash equilibria, called change-of-mind equilibria [32]. Nash equilibria are known for making reliable real-life predictions based on mathematical models in a range of situations [15]. Change-of-mind equilibria are seemingly the first adaptation of the technology that allows us to address dynamic stability, e.g., in the form of *homeostasis* [4]. As implied by the name, the concepts behind *cascaded games* come from MAPK cascades and game theory. Cascaded games are not inherently about either but both types of considerations are directly identifiable at the technical level of our construction, e.g., by incentives having first-class status and, indeed, being pivotal.

---

* Corresponding author.

H. Anai, K. Horimoto, and T. Kutsia (Eds.): AB 2007, LNCS 4545, pp. 185–201, 2007.
© Springer-Verlag Berlin Heidelberg 2007

One widely understood problem with Kauffman/Thomas analysis is that it involves the construction of an exponential-sized state-space graph and that the analysis therefore does not scale. Cascaded games are explicitly constructed to contain only as many nodes as appear to be warranted by the complexity (in a sense we make precise) of the considered system and will typically have a polynomial upper-bound; in fact, the models appear to be much smaller than their upper-bounds in most situations. Conceptually speaking, we pursue a structural line of thinking by identifying *points of interaction,* and not merely influences, between the objects and let the involved *catalysts* determine what possibilities/graph edges to consider. We focus on catalysts because they alter the *affinity* (i.e., the chemical incentives) between the involved compounds, typically leading to an increase in reaction kinetics by a factor of 10^6 to 10^{12} [40]. Pursuant to the construction, the identified equilibria are different in nature than those of Kauffman/Thomas analysis (structural vs functional) and the fact that they appear to be directly comparable is of independent interest [13], see Conclusion.

Following preliminaries in Section 2, we define cascaded games and the abstract formalism they apply to, viz *auto-regulating systems,* in Section 3. In Section 4, we show how to apply the technology to gene-regulation analysis and in Section 5 we go into details with the subtle and not-so-subtle issues that cascaded games allow us to address in a complete account of mammalian MAPK.

The Cascaded Game tool is available through the corresponding author's homepage or directly at `http://cascade.jaist.ac.jp/`.

2   Preliminaries

In this section, we briefly review basic game theory, the theory behind change-of-mind equilibria, and Kauffman/Thomas gene-regulation analysis.

2.1   Abstract Nash Equilibria

A Nash equilibrium is a game situation in which no agent who can move away wants to do so. The notion makes sense in many different concrete classes of games. Abstractly speaking, the notion of Nash equilibrium is definable using only the following four concepts, aka conversion/preference (C/P) games.

Definition 1 (C/P Games [32]). G^{cp} *are 4-tuples* $\langle \mathcal{A}, \mathcal{S}, (\succ_{\!\!a})_{a \in \mathcal{A}}, (\lhd_a)_{a \in \mathcal{A}} \rangle$:

- $\mathcal{A}$ *is a non-empty set of* agents.
- $\mathcal{S}$ *is a non-empty set of* synopses *(i.e., game situations).*
- *Each* $\succ_{\!\!a}$ *is a binary* conversion *relation on* $\mathcal{S}$.
- *Each* $\lhd_a$ *is a binary* preference *relation on* $\mathcal{S}$.

A *strategic-form game* [24,26] is a C/P game where $\mathcal{S}$ is an $\mathcal{A}$-indexed Cartesian product and each $\succ_{\!\!a}$ allow for free movement in the a-dimension of a synopsis.

Definition 2 ([32]). *Synopsis* s *is an* (abstract) Nash equilibrium *for* G^{cp} *if*

$$\mathrm{Eq}^{\mathrm{aN}}_{\mathrm{G}^{\mathrm{cp}}}(s) \quad \triangleq \quad \forall a \in \mathcal{A}, s' \in \mathcal{S} . s \succ_{\!\!a} s' \;\Rightarrow\; \neg(s \lhd_a s')$$

Nash equilibria can also be viewed in direct graph-theoretic terms.

Definition 3 ([32]). *The* (free) change-of-mind *relation for agent* a *is* $\to_a$ $\triangleq \succ_a \cap \lhd_a$. *Let* $\to \triangleq \bigcup_{a \in \mathcal{A}} \to_a$ *and let* $\to^*$ *be the reflexive, transitive closure of* $\to$.

Nash equilibria are exactly the terminal nodes of $\to$ and thus, modulo reflexivity:

$$\text{Eq}_{G^{cp}}^{aN}(s) \quad \Leftrightarrow \quad \forall s' \in \mathcal{S} \,.\, s \to^* s' \Leftrightarrow s = s' \tag{1}$$

2.2   Compromises

Games cannot in general be guaranteed to have Nash equilibria. Exceptions (with a guarantee) include i) extensive-form games with perfect information [19,39] and ii) C/P games with $\mathcal{S}$ a non-empty, convex, compact subset of Euclidian space and with preference relations that allow for a continuous *synchronised update* function [24,26]. Nash's Theorem uses the fact that the latter class includes the probabilised version of any finite strategic-form game with preferences induced from real-valued payoffs. Our alternative starting point is the observation that (1) can be naturally relaxed to address the absence of *unexpected* updates.

Definition 4 ([32]). *Let* $\xrightarrow{S} \triangleq \to \cap (S \times S)$, *viz the sub-graph induced by synopses S. For* G^{cp} *and non-empty* S, $\xrightarrow{S}$ *is a* change-of-mind equilibrium *if*

$$\text{Eq}_{G^{cp}}^{com}(\xrightarrow{S}) \quad \triangleq \quad \forall s \in S, s' \in \mathcal{S} \,.\, s \to^* s' \Leftrightarrow s' \in S$$

Note that a *pure* Nash equilibrium is a static change-of-mind equilibrium, and vice versa. The main theorem and the supporting theory reads as follows.

Theorem 5 ([32]). *In any finite* G^{cp} *there is a (non-empty)* $S \subseteq \mathcal{S}$ *such that* $\text{Eq}^{com}(\xrightarrow{S})$. *More, for given finite* G^{cp}, *all such* S *can be found in* $|\mathcal{S}|^2$.

Lemma 6 ([32]). *For any* G^{cp} *and S, the following are equivalent.*[1]

1. $\text{Eq}_{G^{cp}}^{com}(\xrightarrow{S})$
2. $\text{Eq}_{\lfloor G^{cp} \rfloor}^{aN}(S)$, *where* $\lfloor G^{cp} \rfloor$ *is* G^{cp}*'s shrunken game (i.e., with change-of-mind given as the shrunken graph over the strongly connected components of* $\to$*).*
3. S *is a least, non-empty fixed point of* $\mathcal{U}(S) \triangleq \bigcup_{s \in S}\{s' \mid s \to^* s'\}$.

Following Nash [24], the lemma says that our compromises, 1., are Nash equilibria in a derived game 2., as well as fixed points of an update function, 3. Crucially, our approach admits i) a direct characterisation of the identified compromises, through the notion of change-of-mind equilibrium, which is what makes our equilibria dynamic in nature and ii) an algorithm for computing all equilibria as the terminal strongly connected components of the change-of-mind graph in linear time in the number of nodes plus the number of edges [35]. Nash's probabilistic and our dynamic compromises can seemingly not be quantitatively distinguished, e.g., in terms of size, expected/average payoff, or even in terms of what parts of games can be involved in that the two can be identical, disjoint, subsume each other, and can overlap non-trivially [32].

---

[1] We shall revisit the three in Section 5 and see what they mean for Systems Biology.

2.3   Kauffman/Thomas Gene-Regulation Analysis

The starting point of Kauffman/Thomas gene-regulation analysis is a so-called *influence graph,* indicating regulatory influences between considered objects. The analyses of Kauffman and Thomas differ in two main regards. One is that Kauffman assumes that objects are either being expressed or not, while Thomas allows for genes to be expressed at several levels. For example, in case of Thomas:

The graph shows how the cI and cro genes are influenced by each other and the context in *bacteriophage lambda,* whether positively: $\rightarrow$, or negatively: $\dashv$ [36,5]. The gene cI is able to assume two states, say $0, 1$, and cro is able to assume three states, say $0, 1, 2$. An influence (i.e., an arrow out of a gene) may take place when the gene is on (i.e., in state 1, or above), unless annotated differently. For example, cro auto-represses in state 2, while it represses cI in states 1 and 2. The influences are translated into the associated state space by evaluating them against each state. Thomas considers the likely move of each gene out of each state, see below, while Kauffman lets all genes make a *synchronous* move.

$$\langle cI_0, cro_0 \rangle$$

$$\langle cI_0, cro_1 \rangle \qquad \langle cI_1, cro_0 \rangle$$

$$\langle cI_1, cro_1 \rangle$$

$$\langle cI_0, cro_2 \rangle \longleftarrow \langle cI_1, cro_2 \rangle$$

The translation is not unambiguous and so-called K functions are typically used to resolve joint positive and negative influences into one polarity. We shall return to this point in Section 4. For now, and from our perspective, we note that the above is a change-of-mind graph (that is underpinned by natural conversion and preference relations for agents that in general can be chosen variably) [5]. The considered game therefore has two change-of-mind equilibria: (the static) $\langle cI_1, cro_0 \rangle$ and the (dynamic) cycle between $\langle cI_0, cro_1 \rangle$ and $\langle cI_0, cro_2 \rangle$. This is interesting because the former is characteristic of λ-phage's *lysogenic* way of getting replicated by a bacteria host while the latter is its *lytic* way [41].

3   Auto-regulating Systems and Cascaded Games

We are interested in the likely observable behaviours of what we call *auto-regulating systems,* i.e., arbitrary 4-ary relations over sets of some set of objects.

Definition 7. *For set $\mathcal{O}$, let* $\mathrm{ARS}_\mathcal{O} \triangleq 2^{2^\mathcal{O} \times 2^\mathcal{O} \times 2^\mathcal{O} \times 2^\mathcal{O}}$.

The intended semantics is that, for a specific relationship, the first component, the *substrates,* can turn into the second component, the *products,* when the third component, the *catalysts,* is present and the fourth component, the *inhibitors,* is absent.

Definition 8. *For* r $\in$ rs $\in$ ARS, *with* r $= Ss \xrightarrow[Is]{Cs} Ps$, *let*

$$\pi_s(r) \triangleq Ss \qquad \pi_p(r) \triangleq Ps \qquad \pi_c(r) \triangleq Cs \qquad \pi_i(r) \triangleq Is$$

Let $\pi_x[rs] \triangleq \bigcup_{r \in rs} \pi_x(r)$, *for* $x \in \{s, p, c, i\}$.

The word 'auto-regulating' refers to the fact that all four components come from the same $\mathcal{O}$, thereby allowing for co-regulation between objects.

3.1 MEK, ERK Cascade

An ARS may express chemistry when each of the considered relationships obey *stoichiometric laws*, i.e., when the relationships indicate chemical reactions.

$$MEK.P + ATP \xrightarrow{c-Raf^*} MEK.PP + ADP \qquad MEK.PP + H_2O \xrightarrow{Pase_2} MEK.P + Pi$$
$$ERK + ATP \xrightarrow{MEK.PP} ERK.P + ADP \qquad ERK.P + H_2O \xrightarrow{Pase_3} ERK + Pi$$
$$ERK.P + ATP \xrightarrow{MEK.PP} ERK.PP + ADP \qquad ERK.PP + H_2O \xrightarrow{Pase_3} ERK.P + Pi$$

The above reactions are "triggered" by c-Raf* in a *kinase*-role, i.e., as an enzyme that *phosphorylates* a protein, i.e., that catalyses the affixation of a phosphate group, P, from the "energy molecule" ATP. With two phosphate groups affixed, also the targeted MEK becomes a kinase (aka is *activated*) and may double-phosphorylate ERK. The right column of reactions are deactivations, aka de-phosphorylations catalysed by *phosphatases*. This kind of "rolling" activation of proteins is what is referred to as (kinase) *cascading*.

3.2 State-Space Analysis

In terms of the Kauffman/Thomas technology discussed in Section 2.3, the above reactions amount to the following influence graph over the potentially regulating/regulatable objects, i.e., over the proteins. (As part of our modelling abstraction, we assume that non-proteins are freely available, meaning in particular that we are assuming that we are considering a cell with normal metabolism.)

$$c\text{-Raf}^* \longrightarrow MEK^{\{P,PP\}}_\sharp \xrightarrow{PP} ERK^{\{\sqcup,P,PP\}}_\sharp$$
$$\underset{Pase_2}{\top} \qquad \underset{Pase_3}{\top}$$

The possible states (and their ordering) of each protein is superscripted. For simplicity, we consider only one implicit (active) state for each of c-Raf*, Pase$_2$, Pase$_3$. In order to apply Kauffman/Thomas to the current non DNA-bound (read: free-flowing) situation, as frequently claimed possible, we have introduced a special state, $\sharp$ or 'absent', for MEK and ERK, which we subscript to indicate that it is non-regulatable. The induced state space has 12 nodes (3 MEK-states, 4 ERK-states) and 16 edges. The resulting change-of-mind equilibria are as follows.

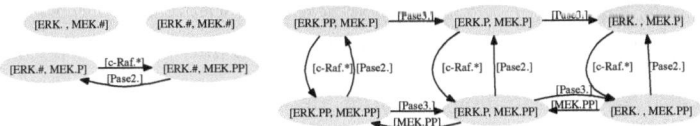

Note the stand-alone MEK.P, MEK.PP loop with absence of ERK (to cascade onto), and the singleton ERK corresponding to the absence of (catalysing) MEK.PP. Note, also, the repeated use of the MEK.P, MEK.PP loop inside the loops for ERK. Finally, note that without ♯ we would observe just one equilibrium, namely the one with 6 nodes. (Following Kauffman, we could also have constructed the state space with, e.g., MEK.P and MEK.PP taken as different objects and states 'present', 'absent'. The state-space graph is bigger than above in this case, 32 nodes and 78 edges, but the equilibria turn out to be exactly the same).

3.3 Cascaded Games

Cascaded games are based primarily on the second issue noted above. More precisely, it is our observation that the exponential granularity of state spaces typically is not taken advantage of. In our first definition of cascaded games, we shall, for simplicity, consider only ARSs of the form $2^{\mathcal{O} \times \mathcal{O} \times \mathcal{O}_\perp}$, with $\mathcal{O}_\perp$ meaning one or none. (But, we retain set notation.) Such relationships involve exactly one substrate, one product, at most one catalyst, and no inhibitor. To start, note that the example (with non-proteins suppressed) amounts to a labelled graph.

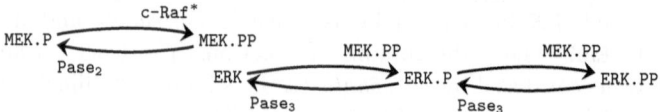

To avoid the inherent exponential blow-up in a state-space graph, the cascaded game for this example will be constructed roughly by collapsing MEK.PP and ERK.

Formally, we first identify the reactions that can be cascaded onto as follows.

Definition 9. *For* rs $\in 2^{\mathcal{O} \times \mathcal{O} \times \mathcal{O}_\perp}$, r $\in$ rs *is a* cascadee *if*

$$\mathrm{Casc}^{\mathrm{dee}}(r) \quad \triangleq \quad \pi_c(r) \cap \pi_p[\mathrm{rs}] \neq \emptyset$$
$$\wedge \, (\forall r' . \, r' \neq r \Rightarrow \neg(\pi_p(r') = \pi_s(r) \wedge \pi_c(r') = \pi_c(r)))$$

The lower conjunct says that the MEK.PP-reaction from ERK.P to ERK.PP is not a cascadee because the MEK.PP-reaction from ERK to ERK.P is an invalidating r'.

Observation 10. *We do not identify either B-reaction below as a cascadee.*

$$A \xrightarrow{\perp} B \qquad C \underset{B}{\overset{B}{\rightleftarrows}} D$$

We could, of course, make Definition 9 more complex but we shall not need to address any situations with cyclic catalysis by one and the same object.

As seen, catalysts that are not produced anywhere do not have their regulative role regulated by another object and we will assume they are freely available.

Definition 11 (Cascaded Players). *For* $r \in rs \in 2^{\mathcal{O} \times \mathcal{O} \times \mathcal{O}_\perp}$, *let*

$$\mathcal{A}_{rs} \; \triangleq \; \bigcup \pi_c[rs]_\perp \qquad\qquad \pi_c(r)_\perp \; \triangleq \; \pi_c(r) \cap \pi_p[rs]$$

The definition implies that our equilibrium analysis will be over the catalytic/regulatory effects (read: incentives) of those objects (read: players, $\mathcal{A}_{rs}$) that may themselves be regulated upon. We define the nodes of the graph to be analysed in a similar manner, namely as the potential *points of interaction* without explicitly listing all the contexts that a considered interaction may take place in. If a reaction is a cascadee, we create a node containing the substrate(s) of the reaction and any catalyst that is involved in cascading. If the reaction is not a cascadee, we create a node with just the substrate(s). In particular, if a reaction is not a cascadee for the reason that the MEK.PP-reaction from ERK.P to ERK.PP is not a cascadee above, we shall consider the catalyst(s) to be only implicitly present.

Definition 12 (Cascaded Synopses). *For* $rs \in 2^{\mathcal{O} \times \mathcal{O} \times \mathcal{O}_\perp}$, *let*

$$\mathcal{S}_{rs}^{poi} \; \triangleq \; \bigcup_{r \in rs} \{\pi_s(r) \cup \pi_c(r)_\perp \mid \mathrm{Casc}^{dee}(r)\} \cup \bigcup_{r \in rs} \{\pi_s(r) \mid \neg\mathrm{Casc}^{dee}(r)\}$$

$$\mathcal{S}_{rs} \; \triangleq \; \mathcal{S}_{rs}^{poi} \cup \bigcup_{r \in rs} \{\pi_p(r) \mid \forall n \in \mathcal{S}_{rs}^{poi} . \pi_p(r) \not\subseteq n\}$$

The full set of situations we are interested in consists of all points-of-interaction, as well as (singleton) nodes for products that are not involved in any interaction. The latter set of objects are candidates for deadlocked situations or, more technically speaking, may be static change-of-mind equilibria, aka Nash equilibria. (An example is λ-phage's lysogenic state, see Sections 2.3 and 4).

When it comes to the conversion and preference parts of cascaded games, we note that their intersection, i.e., the change-of-mind relation, is already the de facto object of study in chemistry. The natural conversion relation is simply that of stoichiometric laws. In particular, the conversion relation is shared by all objects and is reversible (in principle). Catalysts alter the involved affinities, typically resulting in an increase in the observable difference of the two directions of a reaction by a factor of 10^6 to 10^{12} [40], which gives rise to the relevant notion of preference for the catalysts. In other words, when we see an oriented chemical reaction with a catalyst annotated, we are actually seeing the induced change-of-mind relation. We follow suit and treat change-of-mind as primitive.

Having identified the relevant points-of-interaction for a given ARS, inserting edges/change-of-mind is almost as straightforward as going from substrate and (non-suppressed) catalyst to product and (non-suppressed) catalyst. First, though, we note that the catalyst will be consumed in case of self-catalysis.

Definition 13. *For* $r \in rs \in 2^{\mathcal{O} \times \mathcal{O} \times \mathcal{O}_\perp}$, *the perfect-match start, end would be:*

$$\mathbb{S}(r) \; \triangleq \; \pi_s(r) \cup \pi_c(r)_\perp \qquad\qquad \mathbb{E}(r) \; \triangleq \; \pi_p(r) \cup (\pi_c(r)_\perp \setminus \pi_s(r))$$

Secondly, we note that we have not explicitly created a product-catalyst node for each reaction, and substrate-catalyst nodes have only been created for cascadees. Perfect-match nodes may have been created by another reaction and, so, edge insertions go between perfect nodes if they exist and, otherwise, between any nodes that contain the substrate and product. Exceptionally, though, we note that a suppressed catalyst amounts to it being freely available. Also, we only allow reflexive ARS-steps to result in reflexive edges (note, though, that reflexive steps may have resulted in nodes that might not otherwise have been created).

Definition 14 (Cascaded Change-of-Mind). *For* rs $\in 2^{\mathcal{O}\times\mathcal{O}\times\mathcal{O}_\perp\times 2^{\mathcal{O}}}$, *let*[2]

$$\to_{\text{rs}} \triangleq \bigcup_{r\in\text{rs}} \bigcup_{s,e\in\mathcal{S}_{\text{rs}}} \{\langle s,e,\pi_c(r)\perp\rangle \mid (\pi_c(r)\perp \neq \emptyset \wedge \mathbb{S}(r)\in\mathcal{S}_{\text{rs}}$$
$$\Rightarrow s=\mathbb{S}(r) \parallel s\supseteq\pi_s(r))$$
$$\wedge (\pi_c(r)\perp \neq \emptyset \wedge \mathbb{E}(r)\in\mathcal{S}_{\text{rs}}$$
$$\Rightarrow e=\mathbb{E}(r) \parallel e\supseteq\pi_p(r))$$
$$\wedge \pi_i(r)\cap s=\pi_i(r)\cap e=\emptyset$$
$$\wedge \pi_s(r)=\pi_p(r)\Rightarrow s=e\}$$

We write $P\Rightarrow Q\parallel R$ *to mean* $(P\Rightarrow Q)\wedge(\neg P\Rightarrow R)$.

Definition 15 (Cascaded Games). *are defined, for given* rs $\in 2^{\mathcal{O}\times\mathcal{O}\times\mathcal{O}_\perp}$, *as*

$$G_{\text{rs}}^{\text{casc}} \triangleq \langle\mathcal{A}_{\text{rs}},\mathcal{S}_{\text{rs}},\to_{\text{rs}}\rangle$$

See Definitions 1, 3, 11, 12, 14 for more details.

As defined, cascaded games have at most $|\mathcal{O}|^2$ many nodes, seeing that no node contains more than two objects, and a naive implementation and equilibrium analysis will take time $|\mathcal{O}|^9$ ($|\mathcal{O}|^3 \times (|\mathcal{O}|^2)^3$), see Definition 14 and Theorem 5.[3]

3.4   The General Case

Cascaded games can be applied to all ARSs as it stands, not just to $2^{\mathcal{O}\times\mathcal{O}\times\mathcal{O}_\perp}$, but the required discussions and analyses to address the appropriateness of doing so are sufficiently subtle and open-ended to warrant maturation of the technology before being undertaken. Some conditions may need to be stated in ways that are equivalent for $2^{\mathcal{O}\times\mathcal{O}\times\mathcal{O}_\perp}$ but behave differently elsewhere. We are confident that cascaded games will scale to multiple catalysts, substrates, and products but extensive experimentation is needed to understand how to deal with first-class inhibition. The main challenge is to identify what nodes to construct. The naive approach is to construct a collapsed node involving every object that is not an inhibitor but, as we shall see in Section 4, this is merely one particular possibility that, in fact, is not guaranteed to be the right thing to do in all circumstances. While the trade-off may appear to exclusively be between (the asymptotic) number of nodes and the ability to distinguish different situations, we will make two possibly surprising observations in Section 4. Firstly, not making certain distinctions can bring out issues that are not brought out, e.g., in the exponential Kauffman/Thomas models, see Section 4.3. Secondly, for certain types of inhibition, the right thing to do is seemingly to not have the inhibitors impact the nodes we construct, see Section 4.1. See also the Conclusion.

---

[2] The explicit treatment of inhibition at this point is intentional, see Section 3.4.

[3] Our actual implementation avoids replication of work and has an $|\mathcal{O}|^8$ upper-bound in general, with an upper-bound of $|\mathcal{O}|^5$ for all but exotic ARSs.

3.5   Cascaded MEK, ERK

The cascaded game, Definition 15, for the MEK, ERK cascade ARS in Section 3.1 (without non-proteins) has one change-of-mind equilibrium, see Theorem 5.[4]

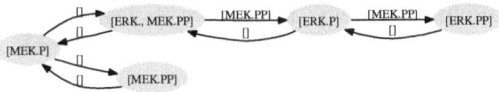

We see that the MEK.P, MEK.PP loop corresponding to the absence of ERK is retained, alongside the cascaded-onto double-phosphorylation of ERK by MEK.PP. In this analysis, there is no duplication, in part at the cost of ERK existing only in a collapsed configuration with the triggering MEK.PP. The graph makes it clear that a double-phosphorylating phosphate group arriving to MEK.P either initiates double-phosphorylation for an ERK in sufficiently close physical proximity or, in the alternate case, is simply released in short order by a phosphatase. We shall see in Section 5 how this process is used by cells to eliminate the need for proteins carrying a "signal" to a cell to be allowed to penetrate the cell membrane. Indeed, MAPK cascades, of which MEK, ERK are part, serve to transduce such signals into a phosphate-group form that *autopoietically* [22] can target the nucleus.

4   Cascaded Gene-Regulation Analysis

As implied by our discussions, influence graphs (for gene expression) may be translated into ARS-form. We considered the reverse translation in Section 3.2, and we shall now formalise the forward direction. The translation will be exemplified with bacteriophage lambda, and we will compare the information that can be gained with our and the Kauffman/Thomas approaches.

4.1   Influence Graphs as Auto-regulating Systems

The arrows in an influence graph are different from but related to the arrows in an ARS. Take, for example, this situation from Section 2.3.

$$cI \vdash\!\!\!\longrightarrow cro \qquad\qquad cI.1 \xrightarrow{cro.1} cI.0 \qquad\qquad cI.1 \xrightarrow{cro.2} cI.0$$

The influence graph, left, says that *cro* in states 1, 2 represses *cI*, which amounts to the ARS-steps on the right. While the steps do not obey stoichiometric laws they are nonetheless expressing (high-level) chemical changes, namely in terms of how the protein that is synthesised, e.g., from *cro*.1 regulates the transcription of *cI*. When there is no ambiguity between positive and negative influences, this translation is straightforward. In case of conflicts, we use inhibition with the stronger influence to make the weaker influence assume a secondary role. According to the disambiguating K functions for bacteriophage lambda [5], for example, the above repression of *cI* by *cro* is stronger than *cI*'s auto-activation.

---

[4] The complete model construction and equilibrium analysis (excluding file I/O) takes around 0.05 seconds, using a naive Java implementation running on a laptop.

$$\begin{array}{cc} \circlearrowleft cI \vdash \!\!\!\!\!\!\!\!\!\!\!-\!\!\!\!\!-\!\!\!\!\!- cro & \qquad cI.1 \xrightarrow[cro.1,cro.2]{cI.1} cI.1 \end{array}$$

The chemical justification for prioritising the influences is that each of them is associated with a kinetics and what concerns us in our discrete presentation is the net effect of combining them. The above description of our translation of influence graphs into ARSs is essentially complete, with only a few caveats.

- Influences may have identical priority, with neither inhibiting the other.
- Influences without an influencer are not annotated with a catalyst.
- Multiple catalysts would be needed, e.g., if gene g is influenced by g_1, g_2, g_3, with g_1, g_2 positive and g_3 negative, and with g_3's influence stronger than g_1, g_2 separately but weaker when both positive influences are on jointly.
- Cascaded games treat catalysts and inhibitors conjunctively: each catalyst must be present (modulo free availability) and each inhibitor must be absent. Disjunction is via multiple steps, i.e., ARSs are disjunctive normal forms.
- We use reflexive steps for both activation and repression, e.g., $cI.1 \xrightarrow{cro.1} cI.1$.
- We do not allow multiple states of an object to co-exist in one node, e.g., $cro.1 \xrightarrow{cro.2} cro.0$ is suppressed (but $cro.1 \longrightarrow cro.2$, $cro.1 \xrightarrow[cro.2]{} cro.0$ are not).
- We suppress inhibitor states that occur in the product (unless they also occur in the substrate). For example, the $cro.2$-inhibition of $cro.1 \xrightarrow[cI.1,cro.2]{} cro.2$ is suppressed (i.e., we consider $cro.1 \xrightarrow[cI.1]{} cro.2$ instead), in order to allow for the temporal delay in the creation of the inhibitor in the product.

Proposition 16. *Lambda-phage's influence graph, see Section 2.3, is this ARS.*

$$\begin{array}{llll} cI.1 \xrightarrow{cro.1} cI.0 & cI.0 \xrightarrow[cro.1,cro.2]{} cI.1 & cro.2 \xrightarrow{cI.1} cro.1 & cro.0 \xrightarrow[cI.1]{} cro.1 \\[4pt] cI.0 \xrightarrow{cro.1} cI.0 & cI.1 \xrightarrow[cro.1,cro.2]{cI.1} cI.1 & cro.1 \xrightarrow{cI.1} cro.0 & cro.1 \xrightarrow[cI.1]{} cro.2 \\[4pt] cI.1 \xrightarrow{cro.2} cI.0 & & cro.0 \xrightarrow{cI.1} cro.0 & \\[4pt] cI.0 \xrightarrow{cro.2} cI.0 & & cro.2 \xrightarrow{cro.2} cro.1 & \end{array}$$

(We note that the K functions could also be transliterated into ARSs. K functions indicate the likely next state of each gene for each combination of influences and transliteration would annotate all possible influences on a particular gene to its ARS-steps. The influencer of an influence that is on would go into the catalysts, with the others becoming inhibitors. We refer to Sections 3.4, 4.3, and the Conclusion for discussions of why we do not pursue transliteration here.)

4.2   "Catalysed State-Space" for Bacteriophage Lambda

Using a construction in the style of Definition 14 on the ARS in Proposition 16 and a state-space set of nodes results in the following (non-cascaded) graph.

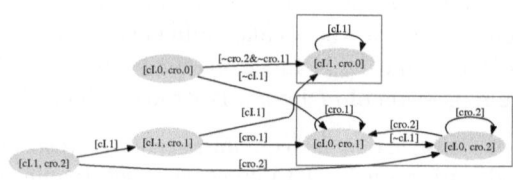

Other than the reflexive steps, the annotations on the arrows, and the boxes indicating the change-of-mind equilibria/steady states, it is identical to the Thomas graph for lambda phage in Section 2.3. The arrows in a Thomas graph are implicitly annotated with the gene whose state-change the arrow captures. Our arrows are annotated with the influences that facilitate the change, along with ~-prefixed annotations of the inhibitors. The node for $cI.0$, $cro.0$ makes explicit the retrospectively obvious fact that re-activation is driven by the context when both genes are off, seeing that neither out-going arrow has a catalyst annotation.

4.3   Cascaded Bacteriophage-Lambda Analysis

The cascaded game built from lambda-phage's ARS, see Proposition 16, looks as follows.[5] The solid boxes are the change-of-mind equilibria; they coincide with the established Thomas steady states (that correspond to lambda-phage's lysogenic and lytic states). The dashed boxes are strongly connected components that are not terminal but have only inhibited arrows out of them, i.e., they are pre-equilibria, whose collapse are preventable. (Hollow heads indicate multi-arrows, with comma-separations of their annotations.)

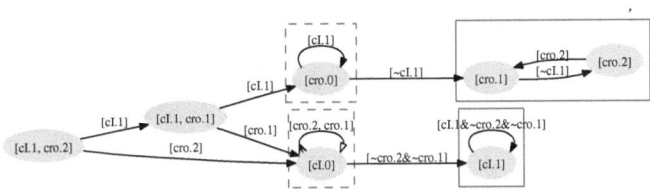

Our graph has seemingly more nodes than the (Kauffman/Thomas) state space. However, this is due to the context influences being explicit for us but implicit in Kauffman/Thomas' K functions. The full state space would have 24 nodes. More subtly, we note that the lysogenic state (lower solid box) is characterised by gene cI being on and the reflexive arrow makes it clear that this can be sustained as long as gene cro is not expressed. The top dashed box is the dual view on this situation, saying that cI can keep cro off as long as the cI-expressed protein is not depleted. In case the protein is depleted, our analysis predicts that the observed lysogenic state (top dashed box) would collapse to lambda-phage's lytic cycle (top solid box). In the Thomas graph in Section 4.2, this information is implicit in the $cI.0$, $cro.0$ node, along with what is for us the analogous situation for the lower boxes. The Thomas graph and the cascaded game have similar layouts and our dashed boxes are superficially in the same positions as the Thomas steady-states in the state space. In the first instance this means that we in principle could have done the analysis with just four nodes. On the other hand, the outgoing arrows from the dashed boxes are undeniable chemical possibilities.

Prophage Induction. The arrow out of the top dashed box is readily observable. It is referred to as *prophage* (or *lysogen*) *induction* and serves as a sort

---

[5] The complete model construction and equilibrium analysis (excluding file I/O) takes around 0.05 seconds, using a naive Java implementation running on a laptop.

of panic button for the bacteria-infecting lambda-phage virus. Lysogenic replication is relatively slow. Conversely, lytic replication is relatively fast but leads
to *lysis*, i.e., bursting of the host and, with it, the release of approximately 100
lambda phages [2]. Prophage induction (out of slow lysogenic and into fast but
destructive lytic replication) has been observed in lambda-phages under attack
by the host, e.g., with *cI* being stimulated into *auto-proteolysis* by the host DNA
repair protein RecA responding to UV-radiation damage to the DNA [11].

Integrated View on Bacteriophage-Lambda Gene Regulation. Contrary
to prophage induction, the arrow out of the lower dashed box does not appear to
have been observed in nature. While it may in theory be possible for either *cI* or
cro to activate when both are off, the present integrated analysis suggests that
cro always is faster than *cI*. We therefore propose to suppress $cI.0 \xrightarrow{cro.1, cro.2} cI.1$ in
the ARS for lambda-phage, see Proposition 16. The similarly-adjusted influence
graph reads as follows (with the Thomas-steady states remaining unchanged.)

$$\left(cI \rightleftharpoons cro \right) 2$$

The cascaded game analysis for the adjusted lambda-phage ARS is as follows.[6]

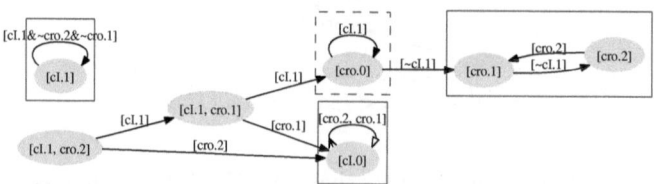

This version predicts lysogenic replication (stand-alone box), prophage induction
(dashed box), and dual sustainability of lytic replication through i) alternating
state-change for *cro* (right box) and ii) unwavering repression of *cI* (lower box).
It is our contention that the absence of a location in the state-space graph from
which lysogenic induction may explicitly be seen to take place is a shortcoming.
Further to Definitions 12 and 14, the problem is related to the functional/non-
structural view on the *cI*.0, *cro*.0 gene configutation in the state-space approach.

5   MAPK Cascaded Signal Transduction

This section returns to the eponymous application of cascaded games, namely
MAPK cascades, and addresses in some detail the game-theoretic foundation of
the technology. In particular, we shall consider separately the three alternative
views on change-of-mind equilibria in Lemma 6 and highlight what they mean
in terms of biochemistry and systems biology. In order, the three amount to
sustainability, **inevitability**, and **atomicity** of the equilibria [32].

---

[6] The complete model construction and equilibrium analysis (excluding file I/O) takes
around 0.05 seconds, using a naive Java implementation running on a laptop.

5.1   Sustainability, Signal Transduction in the Chemistry of MAPK

An elaborate system of proteins, from trans-membrane receptor proteins via cytosolic proteins to target proteins in the nucleus, enable the cell to respond to a particular signal in a specific manner. Responses include cell growth, survival, apoptosis, differentiation and proliferation [1,27]. Intracellular proteins include kinases, phosphatases and GTP-binding proteins (GTPases). Studies have shown that cells respond to external stimuli using clearly defined *signalling pathways* [1]. These encompass all the biochemical phenomena that start with perception of an extracellular signalling molecule (aka ligand) to the response of the cell. MAPK signal transducted pathways are among the most widespread in eukaryotes [18]. In mammalian systems, five distinguishable MAPK pathways have been identified so far: extracellular signal-regulated kinases 1 and 2 (ERKs 1/2), c-Jun N-terminal kinases 1,2 and 3 (JNK 1/2/3), p38 ($\alpha/\beta/\gamma/\delta$), ERKs 3/4 and ERK 5 [31]. The most widely studied, in vertebrates, are ERKs 1/2, JNK and p38 [21]. ERKs 1/2, preferentially regulate cell growth and differentiation whilst JNK and p38 are strongly activated by stress and inflammatory cytokines [31,6]. MAPK cascaded signal-transduction systems have remained unchanged during the course of evolution and currently exist in virtually identical form in a wide range of species. Chemically, they are triggered by a receptor protein on the cell membrane (without penetration) and transduces the received signal by the transfer of phosphate groups. We have created an ARS compendium of all chemical reactions said to be involved in MAPK cascades in [3,7,16,18,25,29,30,38], see [34]. The compendium contains 109 reactions and 21 proteins, each with 2 or 3 states, for a total of 53 ($= 3 \times 11 + 2 \times 10$) distinct objects/protein states and thus an upper-bound of 2809 ($= 53^2$) cascaded nodes. The resulting cascaded game has:[7]

- 15 cascaded players/co-regulating enzymes, see Definition 11,
- 71 cascaded synopses/points-of-interaction, see Definition 12,
- 207 cascaded changes-of-mind/edges (with multiplicity), see Definition 14,
- 2 change-of-mind equilibria, covering the whole cascaded game.

Further to the last point, we see that the chemical underpinning of MAPK cascades is **sustainable**, i.e., the cascades can keep running with no other support than a functioning metabolism: they are *autopoietic* [22]. In other words, MAPK cascades are ideal building blocks for a fundamental biological process that is expected to operate within the confines of a membrane-protected part of a living organism, which is what signal transduction systems do. One change-of-mind equilibrium is as follows; it is the ERK pathway [33]. (See also Section 5.3).

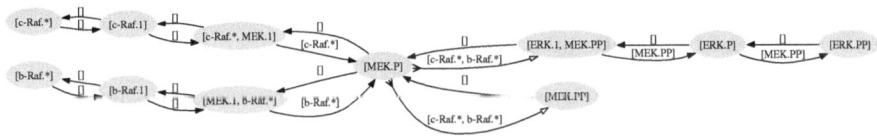

---

[7] The complete model construction and equilibrium analysis (excluding file I/O) takes around 0.4 seconds, using a naive Java implementation running on a laptop.

Although all MAPK-transducted pathways have their own unique properties, they share a number of characteristics. They, e.g., comprise three kinases: MAP-KKK (MAPK kinase kinase), MAPKK (MAPK kinase), MAPK. These can be recognised (in order) in the equilibrium: Rafs, MEK, ERK, with the latter sitting close to the nucleus physically, conceptually, and graphically [33].

5.2 Inevitability and Pathologies in the Chemistry of MAPK

The MAPKs we consider multi-phosphorylate tau as target protein at serine and threonine sites, with phosphorylated tau binding together other proteins. If tau becomes fully saturated the result is a condition called *hyperphosphorylation* in which insoluble *neurofibrillary tangles* of the tau-bound proteins are created [14,8]. This phenomenon has been implicated in several diseases, the most common of which are *tauopathies* such as Alzheimer's disease (AD), Pick's disease (PiD), and Parkinson's disease (PD) [10,12]. One of the enzymes that dephosphorylates ERK.PP and other active MAPKs is PP2A. Inhibition of PP2A seems to be minimally counter-acted by other Pase$_3$s [28]. Re-analysing our MAPK compendium with Pase$_3$ knocked-out results in the following six change-of-mind equilibria.[8]

All six are activated MAPKs and the fact that our analysis predicts that they are the **inevitable** outcomes of Pase$_3$-inhibited MAPK suggests that PP2A inhibition could cause tau hyperphosphorylation. This prediction is validated by recent work [28]. Indeed, transgenic mice exhibit increased expression of JNK and p38 in AD and PiD (i.e., tauopathic) individuals [9].

5.3 Atomicity vs. Cross-Talk in the Chemistry of MAPK

The change-of-mind equilibrium that we did not discuss in Section 5.1 consists of 60 cascaded synopses/points-of-interaction. As seen in Section 5.2, the involved MAPK proteins are JNK and different isoforms of p38. In other words, and as can be checked on closer inspection, the equilibrium comprises the JNK- and p38-pathways [34]. The **atomicity** property of change-of-mind equilibria therefore correctly predicts that there is *cross-talk* between these pathways. Although we do not discuss it here, it is possible and feasible to (semi-automatically) run our analysis several times with different proteins suppressed to identify the cross-talking points-of-interactions, in order to understand the nature of the cross-talk and possibly how to regulate it [34].

5.4 State-Space Analysis

We briefly attempted to analyse our whole MAPK compendium using a Kauffman/Thomas state-space graph. Further to Sections 3.2 and 3.5, our expectation

---

[8] The complete model construction and equilibrium analysis (excluding file I/O) takes around 0.4 seconds, using a naive Java implementation running on a laptop.

is that the result will be compatible with the above analyses. However, even an SGI Altix 3700 with 768GB physical memory failed to build the required graph within a 24 hour period. Of course, this is not surprising as the state space has around 250 billion $(= (3 + 1)^{11} \times (2 + 1)^{10})$ nodes in the considered case.[9]

6 Conclusion

Based mainly on game-theoretic considerations, we have introduced the novel notion of *cascaded games* that can be used to analyse the potential steady states of what we call *auto-regulating systems*. In analogy with Kauffman/Thomas analysis, steady states are captured formally as *change-of-mind equilibria*. Unlike Kauffman/Thomas analysis, however, our analysis is typically of polynomial complexity and, by virtue of our model construction using *points-of-interaction* between *co-regulating* objects, is *structural* in nature. Although seemingly closely related to Kauffman/Thomas steady states, our results may be substantially different. One difference in the analysis of bacteriophage lambda is our explicit identification of *prophage induction*. In the case of MAPK cascades, our analysis, e.g., proves the cascades to be *sustainable*, captures known causes of tauopathies, and avoids large-scale duplication to the tune of using a 71-node graph to analyse what Kauffman/Thomas analysis would need 250 billion nodes for. Indeed, all the cascaded-game analyses we undertake here take less than half a second. Current work is addressing, e.g., ARSs with multiple objects in each reaction and is looking at large-scale data sets. Theoretically, one of the most interesting issues that our work has opened up is the mathematical comparison of our structural equilibria (aka autopoiesis) with the functional equilibria (aka homeostasis) of Kauffman/Thomas. As far as we can tell, this issue is emerging as a major future challenge, seeing that "the form that provides better functionality is likely to be selected [in evolution]" [13]. More to the point, "[i]f form follows function, then we should be able to infer systems function and evolution, as well as their interplay, from the architecture of complex biochemical networks" [13] and, in addition to mathematical results, we are therefore also looking at conceptually clearer ways of reading the various equilibria and of relating them to real-life situations.

Acknowledgements. This work has benefitted from discussions with Jérôme Puisségur on specificity of MAPK cascades.

References

1. Alberts, B., Johnson, A., Julian, L., Raff, M., Roberts, K., Walter, P.: Molecular Biology of the Cell. In: Garland Science, 4th edn. Taylor and Francis Group, London (2002)

---

[9] While a matrix implementation for the state-space approach may help in this particular case, the exponential size is an undeniable problem. By contrast, cascaded games scale and analysing, e.g., 7,000 influences over 3,000 objects takes 20 minutes.

2. Atsumi, S., Little, J.W.: Role of the lytic repressor in prophage induction of phage lambda as analyzed by a module-replacement approach. PNAS, 103 (2006)
3. BioCarta. BioCarta, http://www.biocarta.com
4. Bradford Cannon, W.: The Wisdom of the Body. W.W. Norton, New York (1932)
5. Chettaoui, C., Delaplace, F., Lescanne, P., Vestergaard, M., Vestergaard, R.: Rewriting game theory as a found. for state-based models of gene regulation. In: Priami, C. (ed.) CMSB 2006. LNCS (LNBI), vol. 4210, Springer, Heidelberg (2006)
6. Comperts, B.D., Kramer, I.M., Tatham, P.E.R. (eds.): Signal Transduction. Elservier, Amsterdam (2004)
7. Chen, Z., et al.: MAP kinase. Chemical Review 101, 2449–2476 (2001)
8. Feijoo, C., Campbell, D.G., Jakes, R., Goedert, M., Cuenda, A.: Evidence that phosphorylation of the microtubule-associated protein tau by SAPK4/p38d at Thr50 promotes microtubule assembly. J. Cell Sc., 118 (2005)
9. Ferrer, I.: Stress kinases involved in tau phosphorylation in Alzheimer's disease, tauopathies and APP transgenic mice. Neurotoxicity Research, 6 (2004)
10. Ferrer, I., Gomez-Isla, T., Puig, B., Freixes, M., Ribe, E., Dalfo, E., Avila, J.: Current advances on different kinases involved in tau phosphorylation, and implications in Alzheimer's disease and tauopathies. Curr. Alz's R., 2 (2005)
11. Gilman, A., Arkin, A.P.: Genetic "code": Representations and dynamical models of genetic components and networks. Annu Rev Genomics Hum Genet. 3 (2002)
12. Gong, C.-X., Liu, F., Grundke-Iqbal, I., Iqbal, K.: Dysregulation of protein phosphorylation/dephosphorylation in Alzheimer's disease. J of Biomed, Biotech (2006)
13. Guimerà, R., Sales-Pardo, M.: Form follows function: the architecture of complex networks. Molecular Systems Biology, 2(42) (2006)
14. Haddad, J.J.: Mitogen-activated protein kinases and the evolution of Alzheimer's: a revolutionary neurogenetic axis for therapeutic intervention. Progress in Neurobiology, 73 (2004)
15. Harsanyi, J.C., Nash, J.F., Selten, R.: The Bank of Sweden Prize in Economic Sciences in Memory of Alfred Nobel, for their pioneering analysis of equilibria in the theory of non-cooperative games (1994)
16. Kanehisa, M., et al.: KEGG, MAP signaling pathway - homo sapiens (human), http://www.genome.jp/kegg/
17. Kauffman, S.A.: The Origins of Order: Self-Organization and Selection in Evol. OUP, Oxford (1993)
18. Kiriakis, J.M., Avruch, J.: Mammalian mitogen-activated protein kinase signal transduction pathways activated by stress and inflammation. Physiological Reviews, 81 (2001)
19. Kuhn, H.W.: Extensive games and the problem of information. Contributions to the Theory of Games II, Reprinted in [20] (1953)
20. Kuhn, H.W. (ed.): Classics in Game Theory. Princeton Uni. Press, Princeton (1997)
21. Manning, A.M., Davis, R.J.: Targeting JNK for therapeutic benefit: From junk to gold? Nature Reviews Drug Discovery 2, 554–565 (2003)
22. Maturana, H.R., Varela, F.J.: Autopoiesis and Cognition: The Realization of the Living. D. Reidel Publishing, Dordrecht (1980)
23. Nash, J.F.: Non-Cooperative Games PhD thesis. Princeton University, Princeton (1950)
24. Nash, J.F.: Non-cooperative games. Annals of Mathematics, 54, Reprinted in [20]; published version of [23] (1951)
25. Oda, K., Matsuoka, Y., Funahashi, A., Kitano, H.: A comprehensive pathway map of epidermal growth factor receptor signalling. Mol. Sys. Bio. (2005)

26. Osborne, M.J., Rubinstein, A.: A Course in Game Theory. The MIT Press, Cambridge (1994)
27. Pearson, G., Robinson, F., Gibson, T.B., Xu, B., Karandikar, M., Berman, K., Cobb, M.H.: Mitogen-activated protein (MAP) kinase pathways: Regulation and physiological fynctions. Endocrine Reviews 22, 153–183 (2001)
28. Pei, J.-J., Gong, C.-X., An, W.-L., Winblad, B., Cowburn, R.F., Grundke-Iqbal, I., Iqbal, K.: Okadaic-acid-induced inhibition of protein phosphatase 2A produces activation of mitogen-activated protein kinases ERK1/2, MEK1/2, and p70S6, similar to that in Alzheimer's disease. Am. J. Path, 163 (2003)
29. ProteinLounge. Proteinlounge databases, http://www.proteinlounge.com
30. Qi, M., Elion, E.A.: MAP kinase pathways. Journal of Cell Science 116, 3569–3572 (2003)
31. Rous, P.P., Blenis, J.: ERK and p38 MAPK-activated protein kinases: a family of protein kinases with diverse biologcal functions. Microbio and MBR 68, 320–344 (2004)
32. Le Roux, S., Lescanne, P., Vestergaard, R.: A discrete Nash theorem with quadratic complexity and dynamic equilibria. Technical Report IS-RR-2006-006, JAIST (May 2006)
33. Schoeberl, B., Eichler-Jonsson, C., Gilles, E.D., Muller, G.: Computational modeling of the dynamics of the MAP kinase cascade activated by surface and internalized EGF receptors. Nature Biotechnology, 20 (2002)
34. Senachak, J., Vestergaard, M., Vestergaard, R.: Rewriting game theory applied to protein signalling in MAPK cascades TechReport IS-RR-2006-007, JAIST, May 2006 (Unpublished)
35. Tarjan, R.E.: Depth first search and linear graph algorithms. SIAM JoC, 146–160 (1972)
36. Thieffry, D., Thomas, R.: Dynamical behaviour of biological regulatory networks–II. immunity control in bacteriophage lambda. Bull Math Biol. 57(2) (1995)
37. Thomas, R., Kaufman, M.: Multistationarity, the basis of cell differentiation and memory I: Structural conditions of multistationarity and other nontrivial behavior. Chaos, 11(1) (2001)
38. Upstate. Cell signaling, MAPK pathway, http://www.cellsignaling.com
39. Vestergaard, R.: A constructive approach to sequential Nash equilibria. IPL 97, 46–51 (2006)
40. Voet, D., Voet, J.G.: Biochemistry. John Wiley and Sons, Inc. West Sussex (1995)
41. Watson, J.D., Baker, T.A., Bell, S.P., Gann, A., Levine, M., Losick, R.: Molecular Biology of the Gene, 5th edn. The Benjamin/Cummings Publishing Company (2004)

On Differential Algebraic Decision Methods for the Estimation of Anaerobic Digestion Models

Elena Chorukova[1], Sette Diop[2], and Ivan Simeonov[1]

[1] Institute of Microbiology,
Bulgarian Academy of Sciences,
Acad. G. Bonchev St., Block 26, Sofia 1113, Bulgaria
{elena,issim}@microbio.bas.bg
[2] Labratoire des Signaux et Systèmes,
CNRS – Supélec – Univ. Paris Sud,
Plateau de Moulon, 91192 Gif sur Yvette cedex, France
diop@lss.supelec.fr

Abstract. Monitoring and control of anaerobic digestion of organic wastes by microorganisms are parts of actual world efforts to preserve environment. The anaerobic digestion is a biochemical process in which microorganisms (or bacteria) biodegrade organic matters into biogas (methane and carbon dioxide). Given the complexity of biochemical processes going on in such a bioreactor, control models are almost exclusively written in terms of mass balances of various species of interest. Such models are highly nonlinear and may contain many parameters which need to be identified. But the most challenging part of this estimation work concerns the online estimation of a key variable named the *specific growth rates* of microorganisms. It is invoked in most mass balance models. There is no devices to measure it so as techniques of estimation are very welcome in this field. The communication presents how differential algebraic decision methods can help find partial answers to the problem of online estimation of biomass specific growth rates based upon easily available measurements.

Keywords: Differential algebra, Differential algebraic decision methods, Characteristic set, Observability, Software sensors, Dynamic systems.

1 Introduction

The anaerobic digestion is a biochemical process in which microorganisms (or bacteria) biodegrade organic matters into biogas (methane and carbon dioxide). This process is interesting in 2 aspects: depollution (the biodegradation of some organic wastes) and production of biogas from domestic, agriculture and industrial wastes. In order to efficiently control this process the first task is modeling. Given the complexity of biochemical processes going on in such a bioreactor, control models are almost exclusively written in terms of mass balances of various species of interest [12,4,6]. Such models are highly nonlinear and may contain many parameters which need to be identified [12,6]. But the most challenging

H. Anai, K. Horimoto, and T. Kutsia (Eds.): AB 2007, LNCS 4545, pp. 202–216, 2007.

part of this estimation work concerns the online estimation of a key variable named the *specific growth rate* of microorganisms. It is invoked in most mass balance models. There is no devices to measure it so as techniques of estimation are very welcome in this field.

The differential algebraic approach of general observation problems is used to investigate the identification and observation of some of the most popular models of these processes.

The so-called one-stage model is first considered. It results from the assumption that the biogas is produced by only one type of microorganism species. Then it is shown that

- the specific growth rate of microorganisms is *not observable* according to the differential algebraic approach;
- but, with the help of differential algebraic decision methods, a special differential equation involving the specific growth rate of microorganisms and the supposedly online measured quantities is obtained and used as a software sensor for the growth rate of such species.

Next, a more elaborate model, namely the three-stage model, is considered. It results from the assumption that the biogas is produced at a third stage by a type of microorganisms which, at a second stage, are given by another type of bacteria which, at first stage, have started the process through a well-known model due to Hill & Barth [8]. It is then shown that the previous two conclusions of the study of estimation problems of the one-stage model are still valid. Turning to even more elaborate models of the anaerobic digestion, a five-stage model is considered. Here the authors were not able to draw the same conclusions as for the previous two simpler models.

The differential algebraic approach of observation problems dates back to late eighties and early nineties with works of [11,5,7,3]. See [2] for a recent survey. The main point of this approach, as first clarified in [3], is that a quantity, say z, of a system is observable with respect to some other one, say w (which is supposed to be available in some time interval), if each component of z is a solution of a (non differential) algebraic equation with coefficients eventually depending on w and finitely many of its time derivatives. The theory applies to models of systems in terms of *differential algebraic equations* only but which may be *implicit* in the variables to be observed.

It is a matter of fact that biotechnological process models are often described in terms of differential algebraic equations. The only non polynomial expressions that enter these models are often rational expressions. But, as argued in [2], the basic differential algebraic approach may handle such rational expressions. In summary, the differential algebraic approach to nonlinear observability can tackle the identifiability and observability questions which arise in biotechnological processes. The main limitations that may be encountered when following this approach are the availability of computation resources which are enough to carry over all the suggested calculations.

The differential algebraic approach is among the rare ones which provide *explicit* tests of observability. There are many such differential algebraic decision methods. The most attractive one uses Joseph F. Ritt's notion of *characteristic set*. The observability of a variable z with respect to a supposedly measured variable w is verified by running characteristic set algorithms through the system equations relatively a ranking which orders w and all its derivatives before z and all its derivatives, and any derivative of these two variables before any potentially remaining variable of the system.

The rest of this communication is organized as follows. The differential algebraic approach which is used in these studies is first presented very briefly in Section 2. In Section 3 the one-stage model is considered, and the observability of the biomass specific growth rate is discussed. In Section 4 the three-stage model is considered and it is shown how the conclusions of the one-stage model studies are ascertained. In Section 5 the five-stage model is considered and it is shown how far the previous conclusions can be carried out.

2   On the Differential Algebraic Approach

The reader is referred to [2] for details and references on this approach. Here are the main lines of the application of this theory to anaerobic digestion models. First note that the theory applies to systems which are described by differential polynomials with coefficients which are meromorphic functions of the time. The theory may be extended to include descriptions involving differential rational fractions. Then one needs to consider differential equations and inequations (the sign $\neq$). In summary, in order to check the observability of a latent variable z with respect to, say u and y, of a system

$$P_i(u, \dot{u}, \ldots, y, \dot{y}, \ldots, z, \dot{z}, \ldots, \xi, \dot{\xi}, \ldots) = 0$$

$(i = 1, 2, \ldots)$ one computes the characteristic set of the previous set of differential polynomials with respect to a ranking

$$\{\{u, y\}, \{z\}, \{\xi\}\}.$$

This notation of rankings is very intuitive. It says that all derivatives of u and y are lower than z, and all derivatives of z are lower than ξ. The characteristic set is merely a set $\mathcal{A}$ of differential polynomials each one being *led* by one (and only one) of the variables (at some derivative order). The testing device then reads as: z is observable with respect to u and y if, and only if, each component of z *leads* (i.e., is the highest variable derivative according to the ranking which appears in $\mathcal{A}$) one differential polynomial in $\mathcal{A}$.

The examples in this communication all have been computed by a REDUCE package called **astb** based upon Kolchin's revisit of Ritt's characteristic set algorithm. **astb** follows the account in [1]. The reader who is familiar with differential algebraic decision methods knows that a practical and complete effective algorithm is still lacking. The package **astb** will fail to yield a characteristic set of systems in many circumstances due to factorization issues for instance. But when `astb` exhibits a differential polynomial then of course the latter is consequence of the system's equations.

3 The One-Stage Model

The following model of the anaerobic digestion process [12]:

$$\begin{cases} \dot{X} = \mu\,X - D\,X\,, \\ \dot{S} = -K_1\,\mu\,X + D\,(S_{\text{in}} - S)\,, \\ Q = K_2\,\mu\,X\,. \end{cases} \tag{1}$$

is first considered. It is a mass-balance model. The first equation describes the growth and changes of the biomass X consuming the appropriate substrate S. The first term in the right hand side reflects the growth of the bacteria and the second one reflects the effluent flow rate of liquid. The mass balance for the substrate is described by the second equation, where the first term reflects consumption by the bacteria, the second term reflects the influent flow rate of liquid with concentration of the inlet diluted organics S_{in}. The last equation in system (1) describes the formation of biogas with flow rate Q. In systems terms the dilution rate D is the control input, the output is the biogas flow rate Q, and S_{in} is a disturbance. The quantities K_1 and K_2 are *constant* parameters.

The specific growth rate of bacteria μ is a quite complex function of the process variables. It is standard in the anaerobic digestion control literature to approximate μ by an empirical function of X and S. The most popular among these growth rate empirical models are the following three nonlinear expressions (respectively known as Monod, Contois and Haldane models) [12,6]:

$$\mu = \frac{\mu_{\text{max}}\,S}{K_S + S}\,, \tag{2}$$

$$\mu = \frac{\mu_0\,S}{K_{\text{m}}\,X + S}\,, \tag{3}$$

$$\mu = \frac{\mu_0\,K_{\text{i}}\,S}{K_S\,K_{\text{i}} + K_{\text{i}}\,S + S^2}\,. \tag{4}$$

In the model (1) and (2) K_1 and K_2 are yield coefficients, and K_S is a kinetic coefficient. In practical applications most of the coefficients are not exactly known.

The choice of such a model however usually is difficult and is done on the basis of an expert's knowledge [12]. That is why μ is preferably assumed to be *unknown* and to be reconstructed via estimation techniques [6].

3.1   Observability of the Specific Growth Rate of Bacteria

The decision on the observability of the biomass specific growth rate with respect to D, Q, S_{in} (and possibly K_1 and K_2) can be made on the basis of a characteristic set of the following set of differential polynomials

$$\begin{cases} \dot{X} - \mu X + DX \,, \\ \dot{S} + K_1 \mu X - D(S_{in} - S) \,, \\ Q - K_2 \mu X \,, \\ \dot{K}_1 \,, \\ \dot{K}_2 \,, \end{cases}$$

with respect to the ranking

$$\{\{D, Q, S_{in}, K_1, K_2\}, \{\mu\}, \{X, S\}\} \,.$$

Using astb the following set of differential polynomials

$$\begin{cases} \dot{K}_1 \,, \\ \dot{K}_2 \,, \\ Q\dot{\mu} + Q\mu^2 - \dot{Q}\mu - DQ\mu \,, \\ -K_2\mu X + Q \,, \\ K_2\dot{S} + K_2 DS - K_2 D S_{in} + K_1 Q \,. \end{cases} \tag{5}$$

is obtained. It indicates that μ *is not observable with respect to* D, Q, S_{in}, K_1 and K_2 since the differential polynomial which introduces μ (namely the third line of (5)) is of *order* 1 (and not 0) in μ.

As the output of astb was not proved to be, in general, a characteristic set of its input, it is left to verify that the μ is not observable with respect to D, Q, S_{in}, K_1 and K_2. Here are the main arguments of the proof.

Observability of μ with respect to D, Q, S_{in}, K_1 and K_2 means the existence of a (nondifferential) polynomial m in one indeterminate and with coefficients depending on D, Q, S_{in}, K_1 and K_2 and such that $m(\mu) = 0$. Given the first equation in (1) that observability of μ would imply that the biomass concentration dynamics depends only on D and S_{in}, and is free of the concentration S of the substrate, which is nonsense. The following lemma is thus obtained.

Lemma 1. *The biomass specific growth rate* μ *of anaerobic digestion when the latter process evolves according to model* (1) *is not observable with respect to* D, Q, S_{in}, K_1 *and* K_2. *Moreover a differential equation (in one indeterminate and with coefficients depending on derivatives of* D, Q, S_{in}, K_1 *and* K_2) *of lowest order satisfied by* μ *is given by*

$$Q\dot{\mu} + Q\mu^2 - \dot{Q}\mu - DQ\mu = 0 \,. \tag{6}$$

A consequence of this lemma is that it is not possible to devise an estimator for μ which is based upon online measurements of D, Q and S_{in} only, and with *freedom* to choose its speed of convergence. The next section provides a closer look on equation (6) yielding an estimator for μ but with a speed of convergence totally depending on D.

Assuming μ to be estimated the way indicated in the next section the new situation then raises the following question: are X and S observable with respect to D, Q, S_{in}, K_1, K_2 and μ?

Concerning X the answer is positive and is directly given by the fourth polynomial in (5). The answer for S is negative as suggested by the last differential polynomial in (5). To prove this it is necessary to refer to [1] or § IV.9 of [10] where it is indicated that in order to ascertain the output of astb factorization must be undertaken. But the last differential polynomial in (5)

$$\dot{S} + DS - D\,S_{\text{in}} + \frac{K_1}{K_2}\,Q$$

is of the form

$$\dot{S} + P_0$$

with $P_0 = DS + P_0^*$ where P_0^* does not involve S. It cannot be factored in the form

$$(\dot{S} + P_1)\,(1 + P_2)$$

with P_1 and P_2 two (nondifferential) polynomials in S (with coefficients depending on derivatives of D, Q, S_{in}, K_1, K_2 and μ); that would contradict the fact that P_0 is of degree 1 in S. This proves the following lemma.

Lemma 2. *The biomass concentration X of anaerobic digestion when the latter process evolves according to model (1) is observable with respect to D, Q, S_{in}, K_1, K_2 and μ. But the substrate concentration S is not observable with respect to D, Q, S_{in}, K_1, K_2 and μ.*

3.2 Estimation of the Specific Growth Rate of Bacteria

The differential polynomial introducing μ is next examined

$$Q\,\dot{\mu} + Q\,\mu^2 - \dot{Q}\,\mu - D\,Q\,\mu - 0\,. \tag{6}$$

Rewriting this equation as follows

$$\dot{Q}\,\mu - Q\,\dot{\mu} = -D\,Q\,\mu + Q\,\mu^2\,,$$

then in time intervals where μ is not identically zero, it may be put in the following form

$$\left(\frac{Q}{\mu}\right)^{\cdot} = -D\left(\frac{Q}{\mu}\right) + Q\,. \tag{7}$$

Given the constant sign of D and Q the quantity

$$z = \frac{Q}{\mu}$$

can thus be estimated thanks to the exponential stability of the previous dynamic equation. The biomass specific growth rate is estimated by merely simulating the following dynamics using online measurements D and Q:

$$\begin{cases} \dot{z} = -D\,z + Q\,, \\ \hat{\mu} = \dfrac{Q}{z}\,, \end{cases} \tag{8}$$

This dynamics is better initialized with $z(t_0) = z_0 = \dfrac{Q(t_0)}{D(t_0)}$. This estimation scheme for μ works as long as none of D and Q vanishes.

4   The Three-Stage Model

The model is as follows

$$\begin{cases} \dot{S}_0 = -DS_0 - \beta X_1 S_0 + DY_{\mathrm{p}} S_{\mathrm{in}}\,, \\ \dot{X}_1 = (\mu_1 - k_1 - D)X_1\,, \\ \dot{S}_1 = -D\,S_1 + \beta\,X_1\,S_0 - \mu_1 \dfrac{X_1}{Y_1}\,, \\ \dot{X}_2 = (\mu_2 - k_2 - D)X_2 \\ \dot{S}_2 = -D\,S_2 + Y_{\mathrm{b}}\,\mu_1\,X_1 - \mu_2 \dfrac{X_2}{Y_2}\,, \\ Q = Y_{\mathrm{g}}\mu_2 X_2\,, \end{cases} \tag{9}$$

where the S are the organic matter concentrations and the X are the microorganisms concentrations at the respective three stages of the process.

4.1   Observability of the Growth Rates

All constant parameters are supposed to be identified by ad-hoc methods, and the estimation of the growth rates is considered with respect to dynamic variables, namely D, Q and S_2.

The decision on the observability of the growth rates μ_1 and μ_2 with respect to D, Q and S_2 is made on the characteristic set of system (9) with respect to the ranking

$$\{\{D, Q, S_{\mathrm{in}}, \beta, k_1, k_2, Y_1, Y_2, Y_{\mathrm{b}}, Y_{\mathrm{g}}, Y_{\mathrm{p}}, S_2\}, \{\mu_1, \mu_2\}, \{X_1, X_2, S_0, S_1\}\}\,.$$

Using astb the following three differential polynomials

$$\begin{aligned} -k_1\,Y_2\,Y_{\mathrm{g}}\,\dot{S}_2\,\mu_1 &- k_1\,Y_2\,Y_{\mathrm{g}}\,D\,S_2\,\mu_1 + Y_2\,Y_{\mathrm{g}}\,\dot{S}_2\,\dot{\mu}_1 + Y_2\,Y_{\mathrm{g}}\,D\,S_2\,\dot{\mu}_1 \\ &+ Y_2\,Y_{\mathrm{g}}\,\dot{S}_2\,\mu_1^2 + Y_2\,Y_{\mathrm{g}}\,S_2\,D\,\mu_1^2 - Y_2\,Y_{\mathrm{g}}\,\ddot{S}_2\,\mu_1 - 2\,Y_2\,Y_{\mathrm{g}}\,D\,\dot{S}_2\,\mu_1 \\ &- Y_2\,Y_{\mathrm{g}}\,\dot{D}\,S_2\,\mu_1 - Y_2\,Y_{\mathrm{g}}\,D^2\,S_2\,\mu_1 - k_1\,Q\,\mu_1 + Q\,\dot{\mu}_1 + Q\,\mu_1^2 \\ &- \dot{Q}\,\mu_1 - D\,Q\,\mu_1 = 0\,, \end{aligned} \tag{10}$$

$$-k_2 D \mu_2 + Q \dot{\mu}_2 + Q \mu_2^2 - \dot{Q} \mu_2 - D Q \mu_2 = 0 \,. \tag{11}$$

$$(Y_2 Y_b Y_g \mu_1) X_1 = Q + Y_2 Y_g D S_2 + Y_2 Y_g \dot{S}_2 \,. \tag{12}$$

Again, `astb` being not a complete algorithm for characteristic set computation it remains here to verify that equations (10) and (11) are the lowest order ones (in μ_1 and μ_2, respectively) that are consequences of system (9). This is still an open problem.

4.2   Estimation of the Growth Rates

Equation (11) is readily recognized as very similar to equation (6). The sole difference is the appearance of the term $-k_2 D \mu_2$ in (11). The same manipulations as those which were done on equation (6) are repeated for equation (11) to obtain the following estimation scheme for μ_2:

$$\begin{cases} \dot{z}_2 = -(D + k_2) z_2 + Q \,, \\ \widehat{\mu}_2 = \dfrac{Q}{z_2} \,. \end{cases} \tag{13}$$

As for equation (10) a careful examination of it shows that it can be rewritten as follows

$$\left(\frac{Q + Y_2 Y_g D S_2 + Y_2 Y_g \dot{S}_2}{\mu_1} \right)^{\!\!\cdot} =$$
$$-(D + k_1) \left(\frac{Q + Y_2 Y_g D S_2 + Y_2 Y_g \dot{S}_2}{\mu_1} \right) + Q + Y_2 Y_g D S_2 + Y_2 Y_g \dot{S}_2 \,. \tag{14}$$

yielding the following estimation scheme for μ_1

$$\begin{cases} \dot{z}_1 = -(D + k_1) z_1 + Q + Y_2 Y_g D S_2 + Y_2 Y_g \widehat{\dot{S}}_2 + Y_2 Y_g D S_2 \,, \\ \widehat{\mu}_1 = \dfrac{Q + Y_2 Y_g D S_2 + Y_2 Y_g \widehat{\dot{S}}_2 + Y_2 Y_g D S_2}{z_1} \,. \end{cases} \tag{15}$$

In order to implement this estimation scheme one needs to estimate $\dot{S}_2$. This is done using regularized numerical differentiation.

Assuming μ_1 and μ_2 thus estimated, X_1 and X_2 can be estimated through equation (12) for X_1 and through the biogas flaw rate equation in system (9).

5   The Five-Stage Model

The following model is a modification of known ones [12,13] with additional two specific biochemical reactions participating in the anaerobic biodegradation, namely the syntrophic acetate oxidation and hydrogenotrophic methanogenesis.

$$
\begin{cases}
\dot{S}_0 = -\dfrac{\beta\,X_1\,S_0}{S_2 + K_{i,\text{acet}}} + Y_e\,D\,S_{0\text{in}} - D\,S_0, \\[2mm]
\dot{X}_1 = \mu_1\,X_1 - D\,X_1, \\[2mm]
\dot{S}_1 = -Y_{\text{glu}/X_1}\mu_1\,X_1 + \dfrac{\beta S_0 X_1}{S_2 + K_{i,\text{acet}}} - D\,S_1, \\[2mm]
\dot{X}_2 = \mu_2 X_2 - D\,X_2, \\[2mm]
\dot{S}_2 = Y_{\text{acet}/X_1}\mu_1\,X_1 - Y_{\text{acet}/X_2}\mu_2 X_2 - Y_{\text{acet}/X_3}\mu_3 X_3 - D\,S_2, \\[2mm]
\dot{X}_3 = \mu_3 X_3 - D\,X_3, \\[2mm]
\dot{S}_3 = Y_{H_2/X_1}\mu_1\,X_1 + Y_{H_2/X_3}\mu_3 X_3 - Y_{H_2/X_4}\mu_4 X_4 - K_{H_2}S_3 - D\,S_3, \\[2mm]
\dot{X}_4 = \mu_4 X_4 - D\,X_4, \\[2mm]
\dot{S}_4 = Y_{CO_2/X_1}\mu_1\,X_1 + Y_{CO_2/X_2}\mu_2 X_2 + Y_{CO_2/X_3}\mu_3 X_3 \\[1mm]
\qquad\quad -Y_{CO_2/X_4}\mu_4 X_4 - K_{CO_2}S_4 - D\,S_4, \\[2mm]
Q = Y_{CH_4/X_2}\mu_2 X_2 + Y_{CH_4/X_4}\mu_4 X_4 + K_{CO_2}S_4.
\end{cases}
\tag{16}
$$

5.1   Observability with Respect to D, Q, S_2, S_3 and S_4

Not only yield coefficients are all assumed constant and *known* but organic substrate concentrations S_2, S_3 and S_4 are also supposed to be *measured online*. The soluble organics concentration, S_0, and the substrate concentration S_1 are *not* assumed measured.

As for the previous simpler models the observability of the biomass specific growth rates μ_1, μ_2, μ_3 and μ_4 with respect to the yield coefficients and S_2, S_3 and S_4 is again decided through the characteristic set of system (16) with respect to a ranking similar to the previously used ones for the one- and three-stage models:

$$
\begin{aligned}
&\{\{K_{H_2}, K_{CO_2}, Y_{\text{acet}/X_1}, Y_{\text{acet}/X_2}, Y_{\text{acet}/X_3}, Y_{H_2/X_1}, Y_{H_2/X_3}, \\
&\quad Y_{H_2/X_4}, Y_{CO_2/X_1}, Y_{CO_2/X_2}, Y_{CO_2/X_3}, Y_{CO_2/X_4}, Y_{CH_4/X_2}, Y_{CH_4/X_4}, Y_e, \beta, \quad (17)\\
&\quad D, Q, S_2, S_3, S_4\}, \{\mu_1, \mu_2, \mu_3, \mu_4\}, \{S_0, S_1, X_1, X_2, X_3, X_4\}\}
\end{aligned}
$$

The resulting differential polynomials for the specific growth rates are of the following form

$$
\dot{\mu}_i + \mu_i^2 + f_i\mu_i = 0 \tag{18}
$$

where the f_i are functions of S_2, S_3, S_4 and their derivatives, D and Q.

This partial result lets think that the specific growth rates are again not observable with respect to measured variables.

Moreover, the fundamental exponential stability property which made equations (6), (10) and (11) very interesting for the estimation of specific growth rates does not appear in (18). In other words, not only the specific growth rates are probably not observable but there does not appear any clue on how to estimate them on the basis of online measurements of D, Q, S_2, S_3 and S_4.

5.2   Estimation with Respect to D, Q, S_2, S_3 and S_4

For the five-stage model (16), in the absence of specific growth rates estimation schemes with respect to easily measured variables, one of the last resorts are standard empirical methods such as Monod, Contois, Haldane models (2), (3), (4). Assuming specific growth rates to be identified as functions of the S_i's and X_i's it is often interesting to answer to the question: can the biomass concentrations be estimated based upon these empirical models?

Momentarily assuming the specific growth rates to be available online it is found that the expressions of the biomass concentrations are as follows

$$a\,\mu_i\,X_i = b_i \quad i = 1, 2, 3 \text{ or } 4 \tag{19}$$

where

$$b_i = c_{i1}Q + c_{i2}D\,S_2 + (c_{i3} + c_{i4}D)S_3 + (c_{i5} + c_{i6}D)S_4 + c_{i7}\dot{S}_2 + c_{i8}\dot{S}_3 + c_{i9}\dot{S}_4 .$$

The quantities a, and the c_{ij}'s are functions of the process constant parameters only. Their expressions are appended to the end of this paper. It is also noticeable that the b_i's depend on the supposedly measured variables: D, Q, S_2, S_3 and S_4, only.

Remark 1. Note that the relations (19) for $i = 2$ and $i = 4$ are valid if

$$\begin{aligned} -Y_{\text{acet}/X_3}\,Y_{\text{CH}_4/X_4}\,Y_{\text{CO}_2/X_2} - Y_{\text{acet}/X_3}\,Y_{\text{CH}_4/X_2}\,Y_{\text{CO}_2/X_4} \\ +Y_{\text{acet}/X_2}\,Y_{\text{CH}_4/X_4}\,Y_{\text{CO}_2/X_3} \neq 0, \end{aligned} \tag{20}$$

and the relation (19) for $i = 3$ assumes

$$Y_{\text{CO}_2/X_4}\,Y_{\text{H}_2/X_3} - Y_{\text{CO}_2/X_3}\,Y_{\text{H}_2/X_4} \neq 0 . \tag{21}$$

to be true.

The most important information revealed by the previous equations is the fact that each biomass concentration X_i depends only on specific growth rate μ_i and not on $\mu_j, j \neq i$.

Now according to [9] the specific growth rates may be empirically identified as follows

$$\begin{cases} \mu_1 = \mu_{1\text{max}} \dfrac{S_1}{K_{S_1} + S_1}, \\[3mm] \mu_2 = \mu_{2\text{max}} \dfrac{K_{i,\text{NH}_4^+}}{K_{i,\text{NH}_4^+} + S_{\text{NH}_4^+}} \dfrac{S_2}{K_m X_2 + S_2}, \\[3mm] \mu_3 = \mu_{3\text{max}} \dfrac{S_2}{K_{S_3} + S_2}, \\[3mm] \mu_4 = \mu_{4\text{max}} \dfrac{S_3}{K_{S_4} + S_3} \dfrac{S_4}{K_{S_4} + S_4}. \end{cases}$$

If the yield coefficients in the preceding empirical models of the specific growth rates are identified through ad-hoc methods then, with the help of formulae (19), the biomass concentrations X_3 and X_4 may be estimated from the online measurements of D, Q, S_2, S_3, and S_4 (and the process parameters). Concerning X_2, here is its expression:

$$\left(\mu_{2\max} \frac{K_{i,\mathrm{NH}_4^+}}{K_{i,\mathrm{NH}_4^+} + S_{\mathrm{NH}_4^+}} a\, S_2 - K_m b_2 \right) X_2 = b_2\, S_2 \,,$$

which holds if (20) is satisfied. Therefore, in all time intervals where the quantity between parentheses in the preceding formula is not too small X_2 may be estimated from the measurements.

References

1. Diop, S.: Differential algebraic decision methods, and some applications to system theory. Theoret. Comput. Sci. 98, 137–161 (1992)
2. Diop, S.: From the geometry to the algebra of nonlinear observability. In: Anzaldo-Meneses, A., Bonnard, B., Gauthier, J.P., Monroy-Perez, F. (eds.) Contemporary Trends in Nonlinear Geometric Control Theory and its Applications, pp. 305–345. World Scientific Publishing Co, Singapore (2002)
3. Diop, S., Fliess, M.: On nonlinear observability. In: Commault, C., Normand-Cyrot, D., Dion, J.M., Dugard, L., Fliess, M., Titli, A., Cohen, G., Benveniste, A., Landau, I.D. (eds.) Proceedings of the European Control Conference, Hermès, Paris, pp. 152–157 (1991)
4. Dochain, D., Vanrolleghem, P.A.V.: Dynamical Modeling and Estimation in Wastewater Treatment Processes. IWA Publishing, London (2001)
5. Fliess, M.: Quelques remarques sur les observateurs non linéaires. In: Proceedings Colloque GRETSI Traitement du Signal et des Images, GRETSI, pp. 169–172 (1987)
6. Galava, H.V., Angelidaki, I., Ahring, B.K.: Kinetics and modeling of anaerobic digestion process. Adv. Biochem. Engrg. Biotechnol. 81, 57–93 (2003)
7. Glad, S.T., Ljung, L.: Model structure identifiability and persistence of excitation. In: Proceedings of the IEEE Conference on Decision and Control, IEEE Press, New York (1990)
8. Hill, D.T., Barth, C.L.: A dynamic model for simulation of animal waste digestion. J. Water Pollution Center Fed. 10, 2129 (1977)
9. Karakashev, D., Simeonov, I., Galabova, D., Stefanova, L., Yordanov, St.: Mathematical model of the anaerobic biodegradation of waste activated sludge from municipal wastewaters treatment plants. J. Ecological Engrg. Environment Protection 3, 58–67 (2004)
10. Kolchin, E.R.: Differential Algebra and Algebraic Groups. Academic Press, New York (1973)
11. Pommaret, J.F.: Géométrie différentielle algébrique et théorie du contrôle. C.R. Acad. Sci. Paris Sér. I 302, 547–550 (1986)
12. Simeonov, I.: Mathematical modeling and parameters estimation of anaerobic fermentation processes. Bioprocess Engrg. 21, 377–381 (1999)
13. Simeonov, I.: Modelling and control of biological anaerobic wastewaters treatment processes. Internat. J. Archives of Control Sciences 9, 53–78 (1999)

Appendix

$$a_1 = -Y_{\text{acet}/X_1} Y_{\text{CH}_4/X_4} Y_{\text{CO}_2/X_2} Y_{\text{H}_2/X_3} - Y_{\text{acet}/X_1} Y_{\text{CH}_4/X_2} Y_{\text{CO}_2/X_4} Y_{\text{H}_2/X_3}$$
$$+ Y_{\text{acet}/X_1} Y_{\text{CH}_4/X_2} Y_{\text{CO}_2/X_3} Y_{\text{H}_2/X_4} - Y_{\text{acet}/X_3} Y_{\text{CH}_4/X_4} Y_{\text{CO}_2/X_2} Y_{\text{H}_2/X_1}$$
$$- Y_{\text{acet}/X_3} Y_{\text{CH}_4/X_2} Y_{\text{CO}_2/X_4} Y_{\text{H}_2/X_1} + Y_{\text{acet}/X_3} Y_{\text{CH}_4/X_2} Y_{\text{CO}_2/X_1} Y_{\text{H}_2/X_4}$$
$$- Y_{\text{acet}/X_2} Y_{\text{CH}_4/X_4} Y_{\text{CO}_2/X_1} Y_{\text{H}_2/X_3} + Y_{\text{acet}/X_2} Y_{\text{CH}_4/X_4} Y_{\text{CO}_2/X_3} Y_{\text{H}_2/X_1}$$

$$c_{11} = -Y_{\text{acet}/X_3} Y_{\text{CO}_2/X_2} Y_{\text{H}_2/X_4} - Y_{\text{acet}/X_2} Y_{\text{CO}_2/X_4} Y_{\text{H}_2/X_3}$$
$$+ Y_{\text{acet}/X_2} Y_{\text{CO}_2/X_3} Y_{\text{H}_2/X_4}$$

$$c_{12} = -Y_{\text{CH}_4/X_4} Y_{\text{CO}_2/X_2} Y_{\text{H}_2/X_3} - Y_{\text{CH}_4/X_2} Y_{\text{CO}_2/X_4} Y_{\text{H}_2/X_3}$$
$$+ Y_{\text{CH}_4/X_2} Y_{\text{CO}_2/X_3} Y_{\text{H}_2/X_4}$$

$$c_{13} = K_{\text{H}_2} \left(-Y_{\text{acet}/X_3} Y_{\text{CH}_4/X_4} Y_{\text{CO}_2/X_2} \right.$$
$$\left. -Y_{\text{acet}/X_3} Y_{\text{CH}_4/X_2} Y_{\text{CO}_2/X_4} + Y_{\text{acet}/X_2} Y_{\text{CH}_4/X_4} Y_{\text{CO}_2/X_3} \right)$$

$$c_{14} = -Y_{\text{acet}/X_3} Y_{\text{CH}_4/X_4} Y_{\text{CO}_2/X_2} - Y_{\text{acet}/X_3} Y_{\text{CH}_4/X_2} Y_{\text{CO}_2/X_4}$$
$$+ Y_{\text{acet}/X_2} Y_{\text{CH}_4/X_4} Y_{\text{CO}_2/X_3}$$

$$c_{15} = K_{\text{CO}_2} \left(Y_{\text{acet}/X_3} Y_{\text{CH}_4/X_2} Y_{\text{H}_2/X_4} + Y_{\text{acet}/X_3} Y_{\text{CO}_2/X_2} Y_{\text{H}_2/X_4} \right.$$
$$-Y_{\text{acet}/X_2} Y_{\text{CH}_4/X_4} Y_{\text{H}_2/X_3} + Y_{\text{acet}/X_2} Y_{\text{CO}_2/X_4} Y_{\text{H}_2/X_3}$$
$$\left. -Y_{\text{acet}/X_2} Y_{\text{CO}_2/X_3} Y_{\text{H}_2/X_4} \right)$$

$$c_{16} = Y_{\text{acet}/X_3} Y_{\text{CH}_4/X_2} Y_{\text{H}_2/X_4} - Y_{\text{acet}/X_2} Y_{\text{CH}_4/X_4} Y_{\text{H}_2/X_3}$$

$$c_{17} = -Y_{\text{CH}_4/X_4} Y_{\text{CO}_2/X_2} Y_{\text{H}_2/X_3} - Y_{\text{CH}_4/X_2} Y_{\text{CO}_2/X_4} Y_{\text{H}_2/X_3}$$
$$+ Y_{\text{CH}_4/X_2} Y_{\text{CO}_2/X_3} Y_{\text{H}_2/X_4}$$

$$c_{18} = -Y_{\text{acet}/X_3} Y_{\text{CH}_4/X_4} Y_{\text{CO}_2/X_2} - Y_{\text{acet}/X_3} Y_{\text{CH}_4/X_2} Y_{\text{CO}_2/X_4}$$
$$+ Y_{\text{acet}/X_2} Y_{\text{CH}_4/X_4} Y_{\text{CO}_2/X_3}$$

$$c_{19} = Y_{\text{acet}/X_3} Y_{\text{CH}_4/X_2} Y_{\text{H}_2/X_4} - Y_{\text{acet}/X_2} Y_{\text{CH}_4/X_4} Y_{\text{H}_2/X_3}$$

$$c_{21} = -Y_{\text{acet}/X_1} Y_{\text{CO}_2/X_4} Y_{\text{H}_2/X_3} + Y_{\text{acet}/X_1} Y_{\text{CO}_2/X_3} Y_{\text{H}_2/X_4}$$
$$- Y_{\text{acet}/X_3} Y_{\text{CO}_2/X_4} Y_{\text{H}_2/X_1} + Y_{\text{acet}/X_3} Y_{\text{CO}_2/X_1} Y_{\text{H}_2/X_4}$$

$$c_{22} = Y_{\text{CH}_4/X_4} \left(Y_{\text{CO}_2/X_1} Y_{\text{H}_2/X_3} - Y_{\text{CO}_2/X_3} Y_{\text{H}_2/X_1} \right)$$

$$c_{23} = K_{\text{H}_2} Y_{\text{CH}_4/X_4} \left(Y_{\text{acet}/X_1} Y_{\text{CO}_2/X_3} + Y_{\text{acet}/X_3} Y_{\text{CO}_2/X_1} \right)$$

$$c_{24} = Y_{\text{CH}_4/X_4} \left(Y_{\text{acet}/X_1} Y_{\text{CO}_2/X_3} + Y_{\text{acet}/X_3} Y_{\text{CO}_2/X_1} \right)$$

$$c_{25} = K_{\text{CO}_2} \left(-Y_{\text{acet}/X_1} Y_{\text{CH}_4/X_4} Y_{\text{H}_2/X_3} + Y_{\text{acet}/X_1} Y_{\text{CO}_2/X_4} Y_{\text{H}_2/X_3} \right.$$
$$-Y_{\text{acet}/X_1} Y_{\text{CO}_2/X_3} Y_{\text{H}_2/X_4} - Y_{\text{acet}/X_3} Y_{\text{CH}_4/X_4} Y_{\text{H}_2/X_1}$$
$$\left. +Y_{\text{acet}/X_3} Y_{\text{CO}_2/X_4} Y_{\text{H}_2/X_1} - Y_{\text{acet}/X_3} Y_{\text{CO}_2/X_1} Y_{\text{H}_2/X_4} \right)$$

$$c_{26} = -Y_{\text{CH}_4/X_4} \left(Y_{\text{acet}/X_1} Y_{\text{H}_2/X_3} + Y_{\text{acet}/X_3} Y_{\text{H}_2/X_1} \right)$$

$$c_{27} = Y_{CH_4/X_4} \left(Y_{CO_2/X_1} Y_{H_2/X_3} - Y_{CO_2/X_3} Y_{H_2/X_1} \right)$$

$$c_{28} = Y_{CH_4/X_4} \left(Y_{acet/X_1} Y_{CO_2/X_3} + Y_{acet/X_3} Y_{CO_2/X_1} \right)$$

$$c_{29} = -Y_{CH_4/X_4} \left(Y_{acet/X_1} Y_{H_2/X_3} + Y_{acet/X_3} Y_{H_2/X_1} \right)$$

$$
\begin{aligned}
c_{31} = & \; Y_{acet/X_1} Y_{acet/X_3} Y_{CH_4/X_4} Y^2_{CO_2/X_2} Y_{H_2/X_4} \\
& + Y_{acet/X_1} Y_{acet/X_3} Y_{CH_4/X_2} Y_{CO_2/X_4} Y_{CO_2/X_2} Y_{H_2/X_4} \\
& - Y_{acet/X_1} Y_{acet/X_2} Y_{CH_4/X_4} Y_{CO_2/X_3} Y_{CO_2/X_2} Y_{H_2/X_4} \\
& - Y_{acet/X_3} Y_{acet/X_2} Y_{CH_4/X_4} Y_{CO_2/X_4} Y_{CO_2/X_2} Y_{H_2/X_1} \\
& + Y_{acet/X_3} Y_{acet/X_2} Y_{CH_4/X_4} Y_{CO_2/X_1} Y_{CO_2/X_2} Y_{H_2/X_4} \\
& - Y_{acet/X_3} Y_{acet/X_2} Y_{CH_4/X_2} Y^2_{CO_2/X_4} Y_{H_2/X_1} \\
& + Y_{acet/X_3} Y_{acet/X_2} Y_{CH_4/X_2} Y_{CO_2/X_4} Y_{CO_2/X_1} Y_{H_2/X_4} \\
& + Y^2_{acet/X_2} Y_{CH_4/X_4} Y_{CO_2/X_4} Y_{CO_2/X_3} Y_{H_2/X_1} \\
& - Y^2_{acet/X_2} Y_{CH_4/X_4} Y_{CO_2/X_1} Y_{CO_2/X_3} Y_{H_2/X_4}
\end{aligned}
$$

$$
\begin{aligned}
c_{32} = & -Y_{acet/X_3} Y^2_{CH_4/X_4} Y^2_{CO_2/X_2} Y_{H_2/X_1} \\
& - 2 Y_{acet/X_3} Y_{CH_4/X_4} Y_{CH_4/X_2} Y_{CO_2/X_4} Y_{CO_2/X_2} Y_{H_2/X_1} \\
& + Y_{acet/X_3} Y_{CH_4/X_4} Y_{CH_4/X_2} Y_{CO_2/X_1} Y_{CO_2/X_2} Y_{H_2/X_4} \\
& - Y_{acet/X_3} Y^2_{CH_4/X_2} Y^2_{CO_2/X_4} Y_{H_2/X_1} \\
& + Y_{acet/X_3} Y^2_{CH_4/X_2} Y_{CO_2/X_4} Y_{CO_2/X_1} Y_{H_2/X_4} \\
& + Y_{acet/X_2} Y^2_{CH_4/X_4} Y_{CO_2/X_3} Y_{CO_2/X_2} Y_{H_2/X_1} \\
& + Y_{acet/X_2} Y_{CH_4/X_4} Y_{CH_4/X_2} Y_{CO_2/X_4} Y_{CO_2/X_3} Y_{H_2/X_1} \\
& - Y_{acet/X_2} Y_{CH_4/X_4} Y_{CH_4/X_2} Y_{CO_2/X_1} Y_{CO_2/X_3} Y_{H_2/X_4}
\end{aligned}
$$

$$
\begin{aligned}
c_{33} = K_{H_2} \Big(& \; Y_{acet/X_1} Y_{acet/X_3} Y^2_{CH_4/X_4} Y^2_{CO_2/X_2} \\
& + 2 Y_{acet/X_1} Y_{acet/X_3} Y_{CH_4/X_4} Y_{CH_4/X_2} Y_{CO_2/X_4} Y_{CO_2/X_2} \\
& + Y_{acet/X_1} Y_{acet/X_3} Y^2_{CH_4/X_2} Y^2_{CO_2/X_4} \\
& - Y_{acet/X_1} Y_{acet/X_2} Y^2_{CH_4/X_4} Y_{CO_2/X_3} Y_{CO_2/X_2} \\
& - Y_{acet/X_1} Y_{acet/X_2} Y_{CH_4/X_4} Y_{CH_4/X_2} Y_{CO_2/X_4} Y_{CO_2/X_3} \\
& + Y_{acet/X_3} Y_{acet/X_2} Y^2_{CH_4/X_4} Y_{CO_2/X_1} Y_{CO_2/X_2} \\
& + Y_{acet/X_3} Y_{acet/X_2} Y_{CH_4/X_4} Y_{CH_4/X_2} Y_{CO_2/X_4} Y_{CO_2/X_1} \\
& - Y^2_{acet/X_2} Y^2_{CH_4/X_4} Y_{CO_2/X_1} Y_{CO_2/X_3} \Big)
\end{aligned}
$$

$$
\begin{aligned}
c_{34} = & \; Y_{acet/X_1} Y_{acet/X_3} Y^2_{CH_4/X_4} Y^2_{CO_2/X_2} \\
& + 2 Y_{acet/X_1} Y_{acet/X_3} Y_{CH_4/X_4} Y_{CH_4/X_2} Y_{CO_2/X_4} Y_{CO_2/X_2} \\
& + Y_{acet/X_1} Y_{acet/X_3} Y^2_{CH_4/X_2} Y^2_{CO_2/X_4} \\
& - Y_{acet/X_1} Y_{acet/X_2} Y^2_{CH_4/X_4} Y_{CO_2/X_3} Y_{CO_2/X_2} \\
& - Y_{acet/X_1} Y_{acet/X_2} Y_{CH_4/X_4} Y_{CH_4/X_2} Y_{CO_2/X_4} Y_{CO_2/X_3} \\
& + Y_{acet/X_3} Y_{acet/X_2} Y^2_{CH_4/X_4} Y_{CO_2/X_1} Y_{CO_2/X_2} \\
& + Y_{acet/X_3} Y_{acet/X_2} Y_{CH_4/X_4} Y_{CH_4/X_2} Y_{CO_2/X_4} Y_{CO_2/X_1} \\
& - Y^2_{acet/X_2} Y^2_{CH_4/X_4} Y_{CO_2/X_1} Y_{CO_2/X_3}
\end{aligned}
$$

$$c_{35} = K_{CO_2} \left(-Y_{\text{acet}/X_1} Y_{\text{acet}/X_3} Y_{CH_4/X_4} Y_{CH_4/X_2} Y_{CO_2/X_2} Y_{H_2/X_4} \right.$$
$$-Y_{\text{acet}/X_1} Y_{\text{acet}/X_3} Y_{CH_4/X_4} Y_{CO_2/X_2}^2 Y_{H_2/X_4}$$
$$-Y_{\text{acet}/X_1} Y_{\text{acet}/X_3} Y_{CH_4/X_2}^2 Y_{CO_2/X_4} Y_{H_2/X_4}$$
$$-Y_{\text{acet}/X_1} Y_{\text{acet}/X_3} Y_{CH_4/X_2} Y_{CO_2/X_4} Y_{CO_2/X_2} Y_{H_2/X_4}$$
$$+Y_{\text{acet}/X_1} Y_{\text{acet}/X_2} Y_{CH_4/X_4} Y_{CH_4/X_2} Y_{CO_2/X_3} Y_{H_2/X_4}$$
$$+Y_{\text{acet}/X_1} Y_{\text{acet}/X_2} Y_{CH_4/X_4} Y_{CO_2/X_3} Y_{CO_2/X_2} Y_{H_2/X_4}$$
$$-Y_{\text{acet}/X_3} Y_{\text{acet}/X_2} Y_{CH_4/X_4}^2 Y_{CO_2/X_2} Y_{H_2/X_1}$$
$$-Y_{\text{acet}/X_3} Y_{\text{acet}/X_2} Y_{CH_4/X_4} Y_{CH_4/X_2} Y_{CO_2/X_4} Y_{H_2/X_1}$$
$$+Y_{\text{acet}/X_3} Y_{\text{acet}/X_2} Y_{CH_4/X_4} Y_{CO_2/X_4} Y_{CO_2/X_2} Y_{H_2/X_1}$$
$$-Y_{\text{acet}/X_3} Y_{\text{acet}/X_2} Y_{CH_4/X_4} Y_{CO_2/X_1} Y_{CO_2/X_2} Y_{H_2/X_4}$$
$$+Y_{\text{acet}/X_3} Y_{\text{acet}/X_2} Y_{CH_4/X_2} Y_{CO_2/X_4}^2 Y_{H_2/X_1}$$
$$-Y_{\text{acet}/X_3} Y_{\text{acet}/X_2} Y_{CH_4/X_2} Y_{CO_2/X_4} Y_{CO_2/X_1} Y_{H_2/X_4}$$
$$+Y_{\text{acet}/X_2}^2 Y_{CH_4/X_4}^2 Y_{CO_2/X_3} Y_{H_2/X_1}$$
$$-Y_{\text{acet}/X_2}^2 Y_{CH_4/X_4} Y_{CO_2/X_4} Y_{CO_2/X_3} Y_{H_2/X_1}$$
$$\left. +Y_{\text{acet}/X_2}^2 Y_{CH_4/X_4} Y_{CO_2/X_1} Y_{CO_2/X_3} Y_{H_2/X_4} \right)$$

$$c_{36} = -Y_{\text{acet}/X_1} Y_{\text{acet}/X_3} Y_{CH_4/X_4} Y_{CH_4/X_2} Y_{CO_2/X_2} Y_{H_2/X_4}$$
$$-Y_{\text{acet}/X_1} Y_{\text{acet}/X_3} Y_{CH_4/X_2}^2 Y_{CO_2/X_4} Y_{H_2/X_4}$$
$$+Y_{\text{acet}/X_1} Y_{\text{acet}/X_2} Y_{CH_4/X_4} Y_{CH_4/X_2} Y_{CO_2/X_3} Y_{H_2/X_4}$$
$$-Y_{\text{acet}/X_3} Y_{\text{acet}/X_2} Y_{CH_4/X_4}^2 Y_{CO_2/X_2} Y_{H_2/X_1}$$
$$-Y_{\text{acet}/X_3} Y_{\text{acet}/X_2} Y_{CH_4/X_4} Y_{CH_4/X_2} Y_{CO_2/X_4} Y_{H_2/X_1}$$
$$+Y_{\text{acet}/X_2}^2 Y_{CH_4/X_4}^2 Y_{CO_2/X_3} Y_{H_2/X_1}$$

$$c_{37} = -Y_{\text{acet}/X_3} Y_{CH_4/X_4}^2 Y_{CO_2/X_2}^2 Y_{H_2/X_1}$$
$$-2 Y_{\text{acet}/X_3} Y_{CH_4/X_4} Y_{CH_4/X_2} Y_{CO_2/X_4} Y_{CO_2/X_2} Y_{H_2/X_1}$$
$$+Y_{\text{acet}/X_3} Y_{CH_4/X_4} Y_{CH_4/X_2} Y_{CO_2/X_1} Y_{CO_2/X_2} Y_{H_2/X_4}$$
$$-Y_{\text{acet}/X_3} Y_{CH_4/X_2}^2 Y_{CO_2/X_4}^2 Y_{H_2/X_1}$$
$$+Y_{\text{acet}/X_3} Y_{CH_4/X_2}^2 Y_{CO_2/X_4} Y_{CO_2/X_1} Y_{H_2/X_4}$$
$$+Y_{\text{acet}/X_2} Y_{CH_4/X_4}^2 Y_{CO_2/X_3} Y_{CO_2/X_2} Y_{H_2/X_1}$$
$$+Y_{\text{acet}/X_2} Y_{CH_4/X_4} Y_{CH_4/X_2} Y_{CO_2/X_4} Y_{CO_2/X_3} Y_{H_2/X_1}$$
$$-Y_{\text{acet}/X_2} Y_{CH_4/X_4} Y_{CH_4/X_2} Y_{CO_2/X_1} Y_{CO_2/X_3} Y_{H_2/X_4}$$

$$c_{38} = Y_{\text{acet}/X_1} Y_{\text{acet}/X_3} Y_{CH_4/X_4}^2 Y_{CO_2/X_2}^2$$
$$+2 Y_{\text{acet}/X_1} Y_{\text{acet}/X_3} Y_{CH_4/X_4} Y_{CH_4/X_2} Y_{CO_2/X_4} Y_{CO_2/X_2}$$
$$+Y_{\text{acet}/X_1} Y_{\text{acet}/X_3} Y_{CH_4/X_2}^2 Y_{CO_2/X_4}^2$$
$$-Y_{\text{acet}/X_1} Y_{\text{acet}/X_2} Y_{CH_4/X_4}^2 Y_{CO_2/X_3} Y_{CO_2/X_2}$$
$$-Y_{\text{acet}/X_1} Y_{\text{acet}/X_2} Y_{CH_4/X_4} Y_{CH_4/X_2} Y_{CO_2/X_4} Y_{CO_2/X_3}$$
$$+Y_{\text{acet}/X_3} Y_{\text{acet}/X_2} Y_{CH_4/X_4}^2 Y_{CO_2/X_1} Y_{CO_2/X_2}$$
$$+Y_{\text{acet}/X_3} Y_{\text{acet}/X_2} Y_{CH_4/X_4} Y_{CH_4/X_2} Y_{CO_2/X_4} Y_{CO_2/X_1}$$
$$-Y_{\text{acet}/X_2}^2 Y_{CH_4/X_4}^2 Y_{CO_2/X_1} Y_{CO_2/X_3}$$

$$c_{39} = -Y_{\text{acet}/X_1} Y_{\text{acet}/X_3} Y_{CH_4/X_4} Y_{CH_4/X_2} Y_{CO_2/X_2} Y_{H_2/X_4}$$
$$-Y_{\text{acet}/X_1} Y_{\text{acet}/X_3} Y^2_{CH_4/X_2} Y_{CO_2/X_4} Y_{H_2/X_4}$$
$$+Y_{\text{acet}/X_1} Y_{\text{acet}/X_2} Y_{CH_4/X_4} Y_{CH_4/X_2} Y_{CO_2/X_3} Y_{H_2/X_4}$$
$$-Y_{\text{acet}/X_3} Y_{\text{acet}/X_2} Y^2_{CH_4/X_4} Y_{CO_2/X_2} Y_{H_2/X_1}$$
$$-Y_{\text{acet}/X_3} Y_{\text{acet}/X_2} Y_{CH_4/X_4} Y_{CH_4/X_2} Y_{CO_2/X_4} Y_{H_2/X_1}$$
$$+Y^2_{\text{acet}/X_2} Y^2_{CH_4/X_4} Y_{CO_2/X_3} Y_{H_2/X_1}$$

$$c_{41} = -Y_{\text{acet}/X_1} Y_{CO_2/X_2} Y_{H_2/X_3} - Y_{\text{acet}/X_3} Y_{CO_2/X_2} Y_{H_2/X_1}$$
$$-Y_{\text{acet}/X_2} Y_{CO_2/X_1} Y_{H_2/X_3} + Y_{\text{acet}/X_2} Y_{CO_2/X_3} Y_{H_2/X_1}$$

$$c_{42} = Y_{CH_4/X_2} \left(-Y_{CO_2/X_1} Y_{H_2/X_3} + Y_{CO_2/X_3} Y_{H_2/X_1} \right)$$

$$c_{43} = -K_{H_2} Y_{CH_4/X_2} \left(Y_{\text{acet}/X_1} Y_{CO_2/X_3} + Y_{\text{acet}/X_3} Y_{CO_2/X_1} \right)$$

$$c_{44} = -Y_{CH_4/X_2} \left(Y_{\text{acet}/X_1} Y_{CO_2/X_3} + Y_{\text{acet}/X_3} Y_{CO_2/X_1} \right)$$

$$c_{45} = K_{CO_2} \left(Y_{\text{acet}/X_1} Y_{CH_4/X_2} Y_{H_2/X_3} + Y_{\text{acet}/X_1} Y_{CO_2/X_2} Y_{H_2/X_3} \right.$$
$$+Y_{\text{acet}/X_3} Y_{CH_4/X_2} Y_{H_2/X_1} + Y_{\text{acet}/X_3} Y_{CO_2/X_2} Y_{H_2/X_1}$$
$$\left. +Y_{\text{acet}/X_2} Y_{CO_2/X_1} Y_{H_2/X_3} - Y_{\text{acet}/X_2} Y_{CO_2/X_3} Y_{H_2/X_1} \right)$$

$$c_{46} = Y_{CH_4/X_2} \left(Y_{\text{acet}/X_1} Y_{H_2/X_3} + Y_{\text{acet}/X_3} Y_{H_2/X_1} \right)$$

$$c_{47} = Y_{CH_4/X_2} \left(-Y_{CO_2/X_1} Y_{H_2/X_3} + Y_{CO_2/X_3} Y_{H_2/X_1} \right)$$

$$c_{48} = -Y_{CH_4/X_2} \left(Y_{\text{acet}/X_1} Y_{CO_2/X_3} + Y_{\text{acet}/X_3} Y_{CO_2/X_1} \right)$$

$$c_{49} = Y_{CH_4/X_2} \left(Y_{\text{acet}/X_1} Y_{H_2/X_3} + Y_{\text{acet}/X_3} Y_{H_2/X_1} \right)$$

Protein Structure Prediction Using Residual Dipolar Couplings

Ioannis Z. Emiris[1] and Sotirios I. Pantos[2]

[1] Department of Informatics and Telecommunications,
National Kapodistrian University of Athens,
Panepistimiopolis 15784, Greece
last_name@di.uoa.gr
[2] Department of Biology,
National Kapodistrian University of Athens,
Panepistimiopolis 15784, Greece
spantos@biol.uoa.gr

Abstract. NMR is important for the determination of protein structures, but the usual NOE distance constraints cannot capture large structures. However, RDC experiments offer global orientation constraints for the H–N backbone vectors. Our first application validates local structure from 3 RDC values, by solving an elliptical equation. Second, we model the protein backbone by drawing upon robot kinematics, and compute the relative orientation of consecutive pairs of peptide planes; we obtain a unique orientation by considering also NOE distances. Third, we present a novel algebraic method for determining the relative orientation of secondary structures, a crucial question in fold classification. The orientation of the magnetic vector relative to the secondary structures is determined using two media, leading to a rotation matrix mapping one molecular frame to the other. A unique solution is obtained from RDC data, with no NOE constraints. Our algorithms use robust algebraic operations and are implemented in MAPLE.

Keywords: Inverse kinematics, MAPLE implementation, polynomial equations, protein fold, protein kinematics, RDC data, Saupe tensor, secondary structure.

1 Introduction

The functional properties of proteins depend upon their three-dimensional structures. The latter arises as sequences of amino acids in polypeptide chains fold to compact domains with specific three-dimensional structure. The folded domains can serve as modules for building up large assemblies or can provide specific catalytic or binding sites. Until 1984 structural information could only be determined by X-ray diffraction techniques.

Nuclear magnetic resonance (NMR) has made it possible to obtain structures in a solution environment that is much closer to that in a living being. The NMR method is based on the variation of the resonance frequencies of nuclear

H. Anai, K. Horimoto, and T. Kutsia (Eds.): AB 2007, LNCS 4545, pp. 217–231, 2007.

spins in an external magnetic field [6]. This interaction depends on the chemical structure, the conformation of the molecule, and the solvent environment. NMR yields a wealth of indirect structural information from which the three-dimensional structure can be obtained by extensive calculations. The method of distance geometry was used first for protein structure calculation. This method finds molecular configurations that satisfy a network of interatomic NOE (Nuclear Overhauser Effect) distance measurements. However, NOE measurements correspond to short-range distance constraints, thus they have limitations in the extended biomolecules.

More recent methods utilize the partial alignment of biomolecules in liquid crystal media [10] where interactions such as the dipole-dipole couplings (Residual Dipolar Coupling - RDC) are not averaged to zero as is the case in isotropic solutions. These interactions can be observed in spectra as splittings or contributions of splittings. RDC depends on the angle between an internuclear vector and the magnetic field. Thus RDC can provide valuable structural constraints on the relative orientation of even spatially remote parts of biomolecules.

One of the usual approaches for the study of protein conformation with RDC is to look for fragments of the protein backbone that have rigidity and known geometry. Such fragments can be a peptide plane, a secondary structure or even a protein domain. We use the RDC orientational constraints for each fragment in order to find their relative orientation. In particular, we treat the molecule or molecular fragment as a rigid body containing a number of dipolar interaction vectors with known geometric relationships between them, e.g. as with the $^1H - ^{15}N$ RDC interaction vectors of protein backbone in an α-helix. An arbitrary coordinate system, fixed within the molecule, allows the description of the vectors' orientation within the molecular frame, e.g. the orientation of $^1H - ^{15}N$ unit vector is described by angle ϕ_i relative to molecular frame axis i, where $i \in x, y, z$, as in Fig. 1.

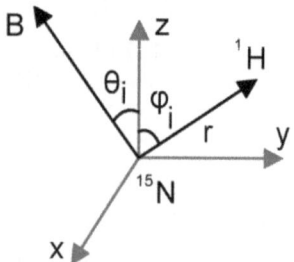

Fig. 1. Molecular frame

The overall orientation of a molecular frame relative to the magnetic unit vector in the laboratory frame is described by Saupe's tensor (S_{ij}), a traceless, symmetric 3×3 matrix with five independent elements. The experimentally measured RDC of an internuclear vector, denoted by D, is described by the following RDC equation:

$$D = D_{max} \sum_{i,j \in x,y,z} S_{ij} \cos \phi_i \cos \phi_j =$$

$$\begin{pmatrix} \cos \phi_x & \cos \phi_y & \cos \phi_z \end{pmatrix} \begin{pmatrix} S_{xx} & S_{xy} & S_{xz} \\ S_{yx} & S_{yy} & S_{yz} \\ S_{zx} & S_{zy} & S_{zz} \end{pmatrix} \begin{pmatrix} \cos \phi_x \\ \cos \phi_y \\ \cos \phi_z \end{pmatrix} = v^T S v . \quad (1)$$

where v is an unit vector relative to the molecular frame. The five independent elements of the Saupe matrix are determined by solving a linear system of RDC equations of the form $Ax = b$. Matrix A contains the direction cosines of unitary internuclear vectors, x is the vector of five independent Saupe elements and b is the vector of RDC measurements [5,1]. Later we explain how we utilize the RDC equation (1) to obtain structural information.

Previous work. In [5], an overdetermined system of linear equations was set up by RDC interaction data within a structurally well defined fragment of a molecule. The system was solved by Singular Value Decomposition and the Saupe order tensor was determined. The method was illustrated on a two-domain fragment of the barley lectin protein and a Sauson-Flamsteed projection map was used for the determination of the relative orientation of the two domains.

The above method was also applied in [8], where RDC data were used for the orientation of helical segments of proteins with PDB codes ACP and NodF. For the determination of the Saupe matrix, they assumed helical segments to be idealized polyalanine α-helices using standard backbone torsional angles. The Sauson-Flamsteed plots depicted qualitatively the 4 relative orientations of each pair of fragments. The 4-fold degeneracy was resolved by the software POSE using a small set of NOE and van der Waals constraints. In [2], a procedure based on Saupe tensors was developed that completely removes the 4-fold relative orientation degeneracy, by combining RDC measurements from two alignment media. The solution was found by a Sauson-Flamsteed map depicting the orientations of molecular fragments of the zinc rubredoxin protein.

Another approach has been implemented in [12], where exact solutions from analytic expressions compute the directions of a NH vector and ϕ, ψ angles for a single residue using only NH RDCs in two media. A backbone structure can be computed by a systematic search to find the best conformation out of a finite number of all possible conformations. The Saupe tensors for secondary structures are computed after refinement from ideal secondary structures built with the average PDB[1] ϕ, ψ angles. The degeneracy of the relative orientations between secondary structures is resolved by using 4 NOE constraints.

Main contributions. We illustrate the power of RDC measurements, and remove the need for NOE distances for most applications we consider, since these are not reliable for extended proteins. We also avoid the use of graphics qualitative procedures, thus making our methods general and robust. Moreover, we

---

[1] Protein Data Bank: www.rcsb.org/pdb/home/home.do

exploit the veracity and rigour of algebraic methods and show that for structural problems of reasonable size, computer algebra offers a very attractive alternative to numerical approximation methods.

First, we apply the RDC equation in a peptide plane which is a rigid structure with known geometry since bond lengths and bond angles are known. In particular, we take a peptide plane from ubiquitin protein which is deposited in PDB. We determine the Saupe tensor by solving a system of linear equations. We confirm the coordinates of $^{13}C - {}^{15}N$ internuclear vector by restraining with the $^{15}N - {}^{1}H$ vector. We solve a polynomial equation and find that one of the 4 solutions is in agreement with the experimentally observed $^{13}C - {}^{15}N$ vector.

Next, we consider two consecutive pairs of peptide planes. This time we assume that we do *not* know the atom coordinates but we know certain ϕ, ψ, and bond angles. From the system of RDC equations of each pair we determine the Saupe tensor. This tensor specifies the orientation of each pair relative to a common laboratory frame. We determine the values of ϕ, ψ angles of the residue that connects these pairs by solving a polynomial system. Next, we examine the effect of NOE distance constraints, and we show that trigonal distance constraints give unique solutions for the ϕ, ψ angles. Our numerical results are sufficiently accurate for biological applications.

Finally, we propose an algebraic method for the computation of the relative orientation between two secondary structures from RDC measurements only. Structural genomics projects do not target all proteins in the genome, but attempt to produce representative structures in each protein fold family. The aim of RDC experiments is to speed up the classification of proteins into structural and operational protein families. Previous work [8] dealt with this problem using a minimal set of NOE distance constraints together with a qualitative graphics method. Our approach removes the need of NOE information and relies on rigorous algebraic operations instead of graphics methods.

We calculate the coordinates of the magnetic unit vector relative to a fixed molecular frame on each secondary structure. Then, we find a rotation matrix that transforms the coordinates of the magnetic unit vector from one molecular frame to the other. We use RDC values from two anisotropic media and formulate a system of 3 polynomial equations with 3 unknowns. The true solution will be the one with the minimum RDC error and consistent with the orientations calculated from a third secondary structure.

The rest of the paper is organized as follows. Section 2 describes an algebraic methodology for the validation of coordinates of a backbone vector of ubiquitin. Section 3 models a protein backbone fragment as a macroscopic robot mechanism with revolute joins. By solving algebraic equations we find the conformation of consecutive peptide pairs. Section 4 formulates an inverse kinematics problem to find the relative orientation of secondary structures. Section 5 summarizes the main ideas and methods we have applied for the determination of protein conformation using RDC data.

2 Structure Validation Using RDC Data

We sample one internuclear vector from a peptide plane of ubiquitin (1D3Z) and confirm its coordinates from RDC data. We determine the Saupe parameters and solve for a vector from its RDC equation and its constraint with another vector. This is also the approach of [13]. But in our approach, measurements of the axial and rhombic Saupe tensor elements are *not* required. We determine the Saupe parameters of RDC equations with an algebraic method, similar to [5,1].

Specifically, we consider the PDB coordinate system of ubiquitin as the molecular frame of peptide plane (residues 24-25). From BMRB[2] (Biological Magnetic Resonance Data Bank) [4], we use the RDC values of $^{13}C^{\alpha}_{24} - {}^{13}C_{24}$, $^{13}C_{24} - {}^{15}N_{25}$, $^{15}N_{25} - {}^{1}H_{25}$ vectors (see Fig. 2) and formulate the linear system of $Ax = b$, where

$$A = \begin{pmatrix} \cos^2\phi^1_y - \cos^2\phi^1_x & \cos^2\phi^1_z - \cos^2\phi^1_x & 2\cos\phi^1_x\cos\phi^1_y & 2\cos\phi^1_x\cos\phi^1_z & 2\cos\phi^1_y\cos\phi^1_z \\ \cos^2\phi^2_y - \cos^2\phi^2_x & \cos^2\phi^2_z - \cos^2\phi^2_x & 2\cos\phi^2_x\cos\phi^2_y & 2\cos\phi^2_x\cos\phi^2_z & 2\cos\phi^2_y\cos\phi^2_z \\ \cos^2\phi^3_y - \cos^2\phi^3_x & \cos^2\phi^3_z - \cos^2\phi^3_x & 2\cos\phi^3_x\cos\phi^3_y & 2\cos\phi^3_x\cos\phi^3_z & 2\cos\phi^3_y\cos\phi^3_z \end{pmatrix},$$

$$x = \begin{pmatrix} S_{yy} \\ S_{zz} \\ S_{xy} \\ S_{xz} \\ S_{yz} \end{pmatrix}, b = \begin{pmatrix} D^{C^{\alpha}C_{24}} \\ D^{CN_{25}} \\ D^{NH_{25}} \end{pmatrix}. \tag{2}$$

We solve the linear system with the method of least squares as implemented on the computer algebra system MAPLE[3] and determine the 5 independent elements (vector x) of the Saupe order matrix. The molecular frame is transformed to the principle order frame (POF) by an orthogonal matrix transformation which diagonalizes the Saupe matrix. Thus, the RDC equation of $^{13}C - {}^{15}N$ vector is converted to

$$D = S_{xx}P_x^2 + S_{yy}P_y^2 + S_{zz}P_z^2. \tag{3}$$

where $P = (P_x, P_y, P_z)$ is the unitary $^{13}C - {}^{15}N$ vector relative to POF (see Fig. 2) and $P_x^2 + P_y^2 + P_z^2 = 1$. By eliminating P_x from (3), the RDC equation is transformed to an elliptical one:

$$\frac{P_y^2}{\frac{D-S_{xx}}{S_{yy}-S_{xx}}} + \frac{P_z^2}{\frac{D-S_{xx}}{S_{zz}-S_{xx}}} = 1. \tag{4}$$

where $a^2 = (D - S_{xx})/(S_{yy} - S_{xx})$, $b^2 = (D - S_{xx})/(S_{zz} - S_{xx})$ are the semi-axes of the ellipse, and (4) is written in parametric form as expression (5), provided that $0 < a^2, b^2 < 1$:

$$P_y = a\cos\omega, \ P_z = b\sin\omega, \ P_x = \pm\sqrt{1 - P_y^2 - P_z^2}. \tag{5}$$

---

[2] http://www.bmrb.wisc.edu/
[3] http://www.maplesoft.com

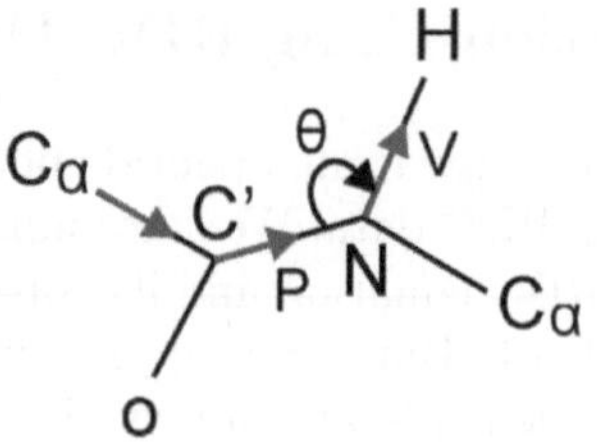

Fig. 2. Peptide plane of residues 24-25

Now, we constrain the solutions of vector P with the angle $\theta = 119.70°$ to the known $^{15}N_{25} - {}^{1}H_{25}$ bond vector, such that $P_x V_x + P_y V_y + P_z V_z = \cos(180 - \theta)$ and solve the trigonometric equation:

$$P_x^2 + P_y^2 + P_z^2 = 1, \tag{6}$$
$$P_y = a \cos\omega,$$
$$P_z = b \sin\omega,$$
$$P_x = (\cos(180 - \theta) - V_y a \cos\omega - V_z b \sin\omega)/V_x.$$

Using the change of variables $\cos\omega = (1 - u^2)/(1 + u^2)$ and $\sin\omega = 2u/(1 + u^2)$ in (6) we solve with MAPLE the following equations:

$$104.33258\, u^4 - 70.8993 u^3 - 31.36879 u^2 + 13.91787 u + 3.69760 = 0,$$
$$u = \tan(\omega/2).$$

The calculated coordinates for the $^{13}C_{24} - {}^{15}N_{25}$ unit vector are:

$$\begin{pmatrix} 0.984843436073204904 \\ 0.138817009304170570 \\ -0.103986754507100265 \end{pmatrix}, \begin{pmatrix} 0.217333787120009752 \\ -0.470787629256483209 \\ 0.855058497069312584 \end{pmatrix},$$

$$\begin{pmatrix} -0.329176754687332318 \\ -0.0208926352387638904 \\ 0.944037163115530320 \end{pmatrix}, \begin{pmatrix} 0.428406582542063984 \\ 0.880300095194688659 \\ -0.203812501026730070 \end{pmatrix}.$$

The second solution agrees up to 4 decimal digits with the experimentally observed $^{13}C_{24} - {}^{15}N_{25}$ vector:

$$\begin{pmatrix} 0.217338329640301086 \\ -0.470774374930192863 \\ 0.855064639449899966 \end{pmatrix}.$$

This is very accurate for biological applications.

The analytic expression used above may be employed to explore all possible orientations of the backbone internuclear vectors relative to POF and, together with other experimental or geometric constraints, we may devise ways to limit the size of our solutions. No other experimental parameters are necessary besides the RDC measurements, while the Saupe parameters are computed by linear algebra.

3 Relative Orientation of Peptide Planes Using RDCs

We use RDC orientational constraints to find the ϕ, ψ angles of consecutive peptide pairs. We model the backbone as a robot mechanism with revolute joints and attach to every atom a right-handed orthogonal coordinate system [11,14] (see Fig. 3). Thus, the problem is reduced to a set of polynomial equations that determine the Saupe matrix and the ϕ, ψ angles. Complementarily to RDC, NOE constraints are added to reduce the set of solutions. Our numerical results are sufficiently accurate for biological applications.

We have seen that the RDC equation (1) relates a measured RDC value with the coordinates of an internuclear vector relative to a fixed molecular frame. In the previous application we placed the molecular frame in the PDB coordinate system and we assumed that the coordinates of the internuclear vectors were known. In this application we go a step further and place the molecular frame on a molecular fragment of known geometry. This allows us to calculate the coordinates of internuclear vectors relative to the fixed molecular frame. Thus, the Saupe matrix can be determined by at least 5 internuclear vectors together with their corresponding RDC values solving a linear system $Ax = b$ where A, x, b are defined in expression (2). The Saupe matrix expresses the orientation of the molecular frame relative to the laboratory frame. The problem is reduced to a system of two polynomial equations.

We demonstrate our method by choosing the well studied protein structure of ubiquitin (1D3Z – 5th model). We take the first two pairs of peptide planes of α-helix and calculate the ϕ, ψ angles of the 26th residue. In particular, we consider as known the bond angles of peptide planes within the residues 24-26 and 26-28, as well as the ϕ, ψ angles of residues 25 and 27. We attach to every atom of the backbone a right handed coordinate system and define the molecular frames MF1, MF2 to be on the ^{15}N atoms of residues 25 and 27 as illustrated in Fig. 3. For each peptide plane pair, we calculate the coordinates of

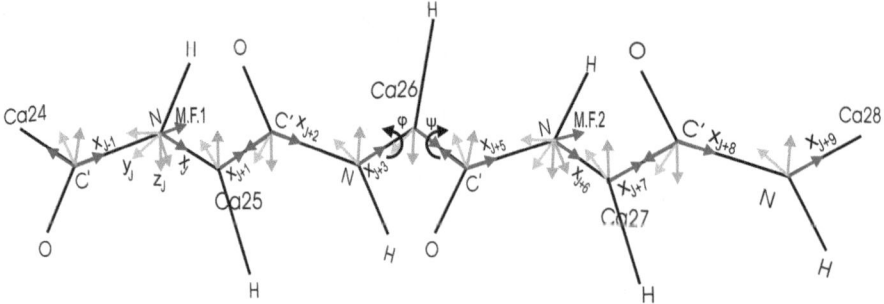

Fig. 3. Peptide planes of 24-28 residues of 1D3Z

^{13}C$^{\alpha}$ – ^{13}C, ^{13}C – ^{15}N, ^{15}N – ^{1}H, ^{13}C$^{\alpha}$ – ^{1}H$^{\alpha}$ internuclear unit vectors relative to the molecular frame as follows:

$$\begin{pmatrix} \cos\phi_x \\ \cos\phi_y \\ \cos\phi_z \end{pmatrix} = {}^{MF}T_J \begin{pmatrix} 1 \\ 0 \\ 0 \end{pmatrix}. \tag{7}$$

where $^{MF}T_J$ transforms local coordinate system J to the molecular frame. Matrix T is the product $T_1 \cdot T_2 \cdots T_n$, each T_i corresponding to a stepwise coordinate transformation along the bond path. The RDC values and the coordinates of the above internuclear vectors form a linear system of 7 equations with 5 unknowns. This system is solved with least squares implemented on MAPLE.

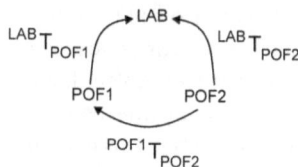

Fig. 4. Transformation matrices of molecular frames

Next, we transform the coordinate system of molecular frame to POF such that the Saupe matrix becomes diagonal. We treat the diagonal Saupe matrix as the orientation of POF to laboratory frame ($^{LAB}T_{POF}$). The transformation of POF2 to the laboratory frame is equal to the product of two transformation matrices, one from POF2 to POF1 and one from POF1 to the laboratory frame as depicted in Fig. 4. Thus, we obtain the following equation:

$$^{LAB}T_{POF1}{}^{POF1}T_{POF2} = {}^{LAB}T_{POF2} . \tag{8}$$

where transformation matrix $^{POF1}T_{POF2}$ depends on the angles ϕ, ψ of the 26th residue. We transform the trigonometric equations of matrix elements (2,2) and (3,3) into polynomial equations and derive 4 solutions for ϕ, ψ angles. However, we have to take into account the orientational degeneracy of the POF axes relative to the magnetic unit vector [9]. We choose one orientation for the first molecular frame and consider the 4 possibilities for the second molecular frame as in Fig. 5 (as seen in [9]).

The expressions that describe the orientational degeneracy of POF2 are:

$$\begin{aligned} ^{LAB}T^1_{POF2} &= {}^{LAB}T_{POF2,} \\ ^{LAB}T^2_{POF2} &= {}^{LAB}T_{POF2}R_x(180^\circ), \\ ^{LAB}T^3_{POF2} &= {}^{LAB}T_{POF2}R_x(180^\circ)R_y(180^\circ), \\ ^{LAB}T^4_{POF2} &= {}^{LAB}T_{POF2}R_y(180^\circ) . \end{aligned} \tag{9}$$

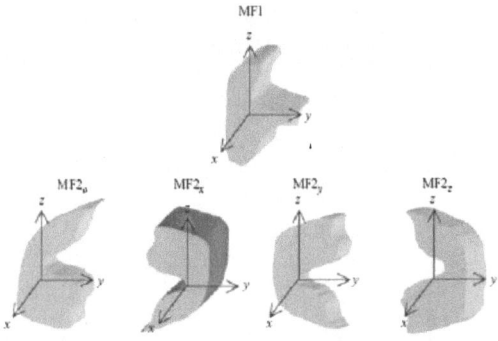

Fig. 5. Orientational degeneracy of molecular frames

We substitute all the transformation matrices of (9) in (8) and obtain 4 sets of solutions. Only $^{\text{LAB}}T^3_{\text{POF2}}$ gives ϕ, ψ values near the experimentally observed ones. The calculated angles are:

$$(\phi, \psi) = ((-76.34232842^o, -38.09716745^o), (-71.21931380^o, 61.72378527^o),$$
$$(26.06896998^o, -62.93116277^o), (37.52245783^o, 31.28649670^o)) \ .$$

The calculated pair $(-76.34232842^o, -38.09716745^o)$ is nearest to the experimentally observed values $(-62.94^o, -43.73^o)$. This accuracy is considered sufficient for all biological applications.

To constrain the number of the above solutions we need to use NOE distance constraints. The question we need to answer is how many NOE distances would be sufficient in order to find a unique solution for ϕ, ψ. We focus on the smallest (proton) distance constraints between ^{1}H atoms in consecutive peptide planes which can be expressed as a function of ϕ, ψ. Let $d(\text{H}^N_i, \text{H}^\alpha_i)$ be the measured distance between nuclei within residue i, and let $d(\text{H}^\alpha_i, \text{H}^N_{i+1})$ be the measured nuclei distance between residues i and $i+1$ (see Fig. 6).

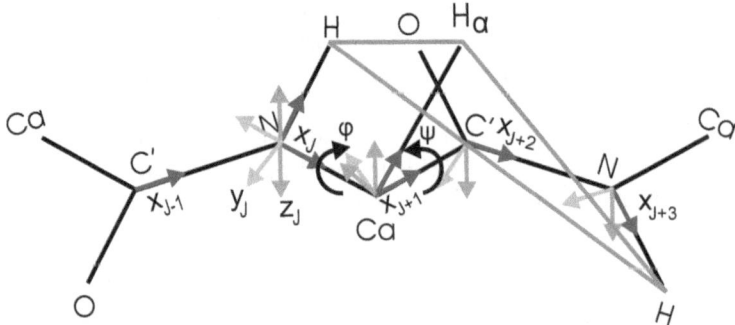

Fig. 6. NOE distance constraints in consecutive peptide planes

The calculated distance of ^{1}H atoms relative to frame J of residue i, denoted by $\|H^N - {}^JT_{J+1}H^\alpha\|$, is a function of angle ϕ (see [7]). Thus, we obtain:

$$\|H^N - {}^JT_{J+1}H^\alpha\|^2 = d^2(H_i^N, H_i^\alpha) \ . \tag{10}$$

Accordingly, the calculated distance of ^{1}H atoms between residues i and $i+1$ relative to frame $J+1$, denoted by $\|H^\alpha - {}^{J+1}T_{J+3}H^N\|$, is a function of angle ψ. Thus, we obtain the equation:

$$\|H^\alpha - {}^{J+1}T_{J+3}H^N\|^2 = d^2(H_i^\alpha, H_{i+1}^N) \ . \tag{11}$$

We use the 26th residue of ubiquitin to demonstrate the above equations. By solving (10) for ϕ we obtain the values $-168.43^o, -62.89^o$ and by solving (11) for ψ we obtain the values $-80.9^o, -43.08^o$. From all possible pairs of ϕ, ψ values, only $(-62.89^o, 43.08^o)$ satisfies the observed distance $d(H_i^N, H_{i+1}^N)$

The above method exploits the known geometry of peptide planes by modeling the backbone structure as a robot kinematic mechanism. The problem has been formulated as a polynomial system, where exact solutions are obtained, without the need of an iterative method.

4 Relative Orientation of Secondary Structures

Relative orientation of secondary structures determines the protein fold which is fundamental for the classification of proteins into structural and operational protein families. Previous works [2,5,8], that deal with the determination of molecular fragments relative orientation, are not sufficiently general. They require either a graphics qualitative procedure or use a (minimal) number of NOE distance constraints, which is hard to obtain for extended proteins.

We propose a new algebraic method that solves a system of polynomial equations using only $^{15}N - {}^1H$ RDC data of a protein dissolved in two different media. Since we know the orientation of the magnetic vector relative to secondary structures, the problem is reduced to an inverse kinematics problem. We find a rotation matrix which maps the coordinates of the magnetic vector from one molecular frame to the other. In the following paragraphs we present an analytic expression that computes a number of orientations of the unitary magnetic vector, relative to a molecular frame, and we show an algorithm for restricting their number.

According to (1), we can choose a global molecular frame that makes the Saupe tensor diagonal and we call this frame Principal Alignment System (PAS) [9]. Then, the RDC equation is expressed as

$$D = u^T S u \ . \tag{12}$$

where u is an internuclear unit vector relative to PAS and S is the diagonalized molecular alignment tensor. Without loss of generality, we may assume that the magnetic vector is collinear with the Z principle axis of PAS, cf. [3]. In each submolecular structure (e.g. secondary structures, domains) we can assign a submolecular coordinate system (POF) which will make the corresponding Saupe tensors diagonal. We may find a transformation matrix where its columns are the unitary vectors that define the x, y, z axes of POF to the X, Y, Z axes of PAS such that $u = (^{\text{PAS}}T_{\text{POF}})v$ where v are the internuclear unit vectors relative to POF. Thus, (12) becomes

$$D = u^{\text{T}} \cdot S \cdot u = v^{\text{T}} (^{\text{POF}}T_{\text{PAS}} \cdot S \cdot {}^{\text{PAS}}T_{\text{POF}})v \ . \tag{13}$$

where the tensor $s = {}^{\text{POF}}T_{\text{PAS}} \cdot S \cdot {}^{\text{PAS}}T_{\text{POF}}$ can be computed in the same way as in Sect. 2 (see also [5]). Thus, we obtain the following expression

$$^{\text{POF}}T_{\text{PAS}}^{\text{T}}(s)^{\text{POF}}T_{\text{PAS}} = S \ . \tag{14}$$

where the S_{ZZ} element of molecular alignment tensor S is given by

$$- (\mu_{\text{o}}h/16\pi^3)\mathbf{S}\gamma_{\text{N}}\gamma_{\text{H}}r_{\text{NH}}^{-3}S_{\text{ZZ}} = D_{\text{a}} \ . \tag{15}$$

where D_{a} is the experimental value of the tensorial axial component, γ the gyromagnetic ratio, r_{NH} the distance between nuclei, and $\mathbf{S}$ the generalized order parameter which falls between 0.85 to 0.95, cf. [15]. Thus, from (14) we obtain the following

$$
\begin{aligned}
S_{\text{ZZ}} &= x^2 s_{\text{xx}} + y^2 s_{\text{yy}} + z^2 s_{\text{zz}}, \\
S'_{\text{ZZ}} &= x'^2 s'_{\text{xx}} + y'^2 s'_{\text{yy}} + z'^2 s'_{\text{zz}}, \\
1 &= x^2 + y^2 + z^2 \ .
\end{aligned}
\tag{16}
$$

where $B = (x, y, z)$, $B' = (x', y', z')$ are the coordinates of the magnetic unit vector relative to the POF and POF' in the first and second alignment medium respectively. We set $B' = {}^{\text{POF}'}T_{\text{POF}}B$ where $^{\text{POF}'}T_{\text{POF}} = {}^{\text{MF}}T_{\text{POF}'}^{-1}\,{}^{\text{MF}}T_{\text{POF}}$ and obtain from (16) at most 8 magnetic vectors.

Let us focus on the problem of finding the relative orientation between two secondary structures. We attach to each secondary structure a fixed molecular frame in a way that diagonalizes the corresponding Saupe tensors (see Fig. 7).

Since we know the coordinates of B in each molecular frame, we need to specify a matrix which rotates one molecular frame to the other. This is an inverse kinematics problem where we have 3 equations with 9 unknowns. To reduce the number of unknowns we factor the requested rotational matrix as a product of 3 Euler angles and we obtain the following equation:

$$
\begin{pmatrix} x_1 \\ y_1 \\ z_1 \end{pmatrix} = R_{\text{z}}(\alpha)R_{\text{y}}(\beta)R_{\text{z}}(\gamma) \begin{pmatrix} x_2 \\ y_2 \\ z_2 \end{pmatrix} ,
\tag{17}
$$

where $^{\text{POF1}}B = (x_1, y_1, z_1)$ and $^{\text{POF2}}B = (x_2, y_2, z_2)$ are the coordinates of the unitary magnetic vector relative to the POFs of two secondary structures and α, β, γ are the unknown Euler angles.

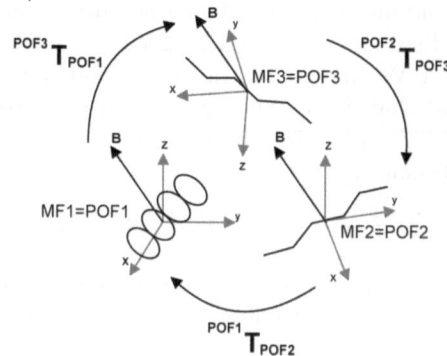

Fig. 7. The magnetic unit vector relative to molecular frames

Next we describe an algorithm for finding the relative orientation between two secondary structures:

1. For each secondary structure assign a fixed molecular frame coincident with the POF. Obtain a second set of POFs from measurements in a second medium.
2. Solve (16) and find one set of magnetic vector coordinates relative to the POF of the first secondary structure and another set of vectors relative to the POF of the second secondary structure.
3. Use all the possible magnetic vector pairs in (17) and solve for the relative orientation between the two POFs.
4. Use the POF of the first secondary structure and calculate a new Saupe matrix for the complex of the two secondary structures using the $^{15}N - {}^{1}H$ RDC vectors and the relative orientations from step 3.
5. Filter the relative orientations of the previous step and obtain only those that have calculated RDC error less than 5%.
6. Use a third secondary structure and apply the steps 1 to 5.
7. The minimum norm of the difference between the relative orientations of the original pair of secondary structures and the relative orientations we obtain via the third one will be the criterion for finding the true solution.

Now we illustrate the above procedure by using the well studied NMR structure of human ubiquitin (1D3Z – 5th model) [4]. We attach a molecular frame to the 29th, 14th, 67th backbone ^{15}N atom of α-helix (residues 25 to 33), second β-strand (residues 11 to 17) and fifth β-strand (residues 65 to 70). Then, with at least 5 $^{15}N - {}^{1}H$ RDC values and their known PDB internuclear vector coordinates we calculate the Saupe matrices for each molecular frame. After Saupe matrix diagonalization, reattach the molecular frames to POFs.

We use the alignment media DMPC/DHPC and DMPC/DHPC/CTAB with D_a values $-9.15\,\text{Hz}$ and $-15.51\,\text{Hz}$ respectively [4,16]. Thus, the S_{ZZ}, S'_{ZZ} values of (16) are $-0.00081036, -0.001373627$ where $\mathbf{S} = 0.9$.

Next, to find the relative orientation between the α-helix and the second β-strand it is sufficient to find the relative orientation of the corresponding POFs ($^{\mathrm{POF}_1}T_{\mathrm{POF}_2}$) (see Fig. 7). We apply steps 1 to 5 of the above algorithm and we obtain 2 orientations with RDC error less than 5%. Next, calculate the orientation of the second and fifth β–strand ($^{\mathrm{POF}_2}T_{\mathrm{POF}_3}$) and obtain 4 orientations with RDC error less than 5%. Eventually, calculate the orientation between the first and fifth β–strands ($^{\mathrm{POF}_3}T_{\mathrm{POF}_1}$)and obtain 4 orientations with RDC error less than 7%.

Obtaining the 32 possible orientations of the three secondary structures we select the one which satisfies the minimum norm ($min \parallel {}^2T_1 - {}^2T_3{}^3T_1 \parallel$), where ^jT_i stands for $^{\mathrm{POF}_j}T_{\mathrm{POF}_i}$. The best solution (norm 0.15) is also the true solution in comparison with the observed relative orientations below

$$
{}^1T_2^{\mathrm{Calc}} = \begin{pmatrix} -.8324 & .5501 & .067 \\ -.5542 & -.8263 & -.1006 \\ 0 & -.1209 & .9927 \end{pmatrix}, {}^1T_2^{\mathrm{Obs}} = \begin{pmatrix} -.8605 & .5090 & -.0194 \\ -.5071 & -.8597 & -.0617 \\ -.0481 & -.0433 & .9979 \end{pmatrix}.
$$

$$
{}^2T_3^{\mathrm{Calc}} = \begin{pmatrix} .9999 & -.0037 & .0001 \\ -.0037 & -.9996 & .029 \\ 0 & -.029 & -.9996 \end{pmatrix}, {}^2T_3^{\mathrm{Obs}} = \begin{pmatrix} .9869 & .1512 & -.0558 \\ .1521 & -.9882 & .0121 \\ -.0533 & -.0205 & -.9984 \end{pmatrix}.
$$

$$
{}^3T_1^{\mathrm{Calc}} = \begin{pmatrix} -.8708 & -.4873 & -.0646 \\ -.4835 & .8728 & -.0662 \\ .0887 & -.0264 & -.9957 \end{pmatrix}, {}^3T_1^{\mathrm{Obs}} = \begin{pmatrix} -.7708 & -.6278 & -.1073 \\ -.6328 & .7742 & .015 \\ .0736 & .0795 & -.9941 \end{pmatrix}.
$$

One additional feature of the above algorithm is that once we know the true orientation in one place of the cycle of Fig. 7, then it propagates the true solution to the neighbor substructures.

The above ideas can be extended as follows. The calculation of Saupe tensors assumed that the protein backbone conformation is known. However, we might calculate an initial Saupe tensor assuming the average ϕ, ψ, and bond angles of secondary structures and then follow an iterative process of fitting the calculated RDCs in with the observed ones as it has been implemented in [12].

To summarize the benefits of our method we have to take into account that we formulate the protein fold problem with analytic algebraic equations. Their special form makes them easy to solve. The runtime of our algorithm was about 7 min for computing the relative orientation of 3 secondary structures, implemented in MAPLE 9 on an Intel Pentium 2.5 GHz PC. Furthermore, with these analytic expressions, we are able to explore all possible solutions and determine the orientation of molecular fragments quantitatively, without a graphics qualitative method [2].

5   Conclusion

In this section we summarize the main ideas for the problems of protein structure validation, peptide planes conformation and protein fold. In all cases, our methods offer sufficient accuracy for biological applications, coupled with the rigor and robustness of algebraic methods.

In our first application (Sect. 2), we showed that an RDC equation of an internuclear vector in a peptide plane can be expressed in an elliptic parametric form. Constraining the solutions with a known geometric relation, we obtain 4 distinct solutions one of which is in agreement with the experimental results. The accurate results of our method imply that a reliable protein structure may be determined very fast using the analytic expressions of RDC equations together with other experimental constraint measurements.

In our second application (Sect. 3) we explored the relative orientation of two consecutive pairs of peptide planes of ubiquitin. Each pair was considered as a rigid body of known geometry; we attached to each a molecular frame. The relative orientation of the two molecular frames should confirm the experimental ϕ, ψ angles of the intermediate residue. We formulate an inverse kinematics problem and obtain 4 sets of solutions. We showed that trigonal precise experimental NOE distances in consecutive peptide pairs lead to a unique solution for ϕ, ψ. The NOE constraint has to be considered as an upper distance bound between protons.

In our last application (Sect. 4), RDCs determine the orientation of the laboratory magnetic unit vector relative to a molecular frame. We find a transformation matrix that maps the magnetic vector from one molecular frame into the other. The method uses only RDC constraints from two media and is applicable even in cases where NOE distances are *not* available. Such cases are proteins with extended molecular fragments where their remote parts are *not* connected with NOE. In this way, we answer a crucial question in the problem of protein fold.

Acknowledgments. Research supported by project 70/4/6452: the project is co-funded by the European Social Fund and National Resources KAPODIS-TRIAS. We thank Professors S. Hamodrakas, B. Donald, G. Cornilescu and W. Peti for their valuable help.

References

1. Andrec, M., Du, P., Levy, R.M.: Protein Structural Motif Recognition via NMR Residual Dipolar Couplings. J. Am. Chem. Soc. 123, 1222–1229 (2001)
2. Al-Hashimi, H.M., Valafar, H., Terrell, M., Zartler, E.R., Eidsness, M.K., Prestegard, J.H.: Variation of Molecular Alignment as a Means of Resolving Orientational Ambiguities in Protein Structures from Dipolar Couplings. J. Magnetic Resonance 143, 402–406 (2000)
3. Canet, D.: Nuclear Magnetic Resonance Concept and Methods. J. Wiley & Sons, West Sussex (1996)
4. Cornilescu, G., Marquardt, J.L., Ottiger, M., Bax, A.: Validation of Protein Structure from Anisotropic Carbonyl Chemical Shifts in a Dilute Liquid Crystalline Phase. J. Am. Chem. Soc. 120, 6836–6837 (1998)
5. Losonczi, J.A., Andrec, M., Fischer, M.W.F., Prestegard, J.H.: Order Matrix Analysis of Residual Dipolar Couplings Using Singular Value Decomposition. J. Magnetic Resonance 138, 334–342 (1999)
6. Levitt, M.H.: Spin Dynamics. J. Wiley & Sons, West Sussex (2001)

7. Emiris, I.Z., Fritzilas, E.D., Manocha, D.: Algebraic algorithms for structure determination in biological chemistry. Intern. J. Quantum Chemistry 106, 190–210 (2006)
8. Fowler, C.A., Tian, F., Al-Hashimi, H.M., Prestegard, J.H.: Rapid Determination of Protein Folds Using Residual Dipolar Couplings. J. Mol. Biol. 304, 447–460 (2000)
9. Prestegard, J.H., Al-Hashimi, H.M., Tolman, J.R.: NMR structures of biomolecules using field oriented media and residual dipolar couplings. Quarterly Reviews of Biophysics 33, 371–424 (2000)
10. Tjandra, N., Bax, A.: Direct Measurement of Distances and Angles in Biomolecules by NMR in a Dilute Liquid Crystalline Medium. Science 278, 1111–1114 (1997)
11. Quine, J.R., Cross, T.A., Chapman, M.S., Bertram, R.: Mathematical Aspects of Protein Structure Determination with NMR Orientational Restraints. Bulletin of Math Biology 66, 1705–1730 (2004)
12. Wang, L., Donald, B.R.: Exact solutions for internuclear vectors and backbone dihedral angles from NH residual dipolar couplings in two media, and their application in a systematic search algorithm for determining protein backbone structure. J. Biomolecular NMR 29, 223–242 (2004)
13. Wedemeyer, W.J., Rohl, C.A., Scheraga, H.A.: Exact solutions for chemical bond orientations from residual dipolar couplings. J. Biomolecular NMR 22, 137–151 (2002)
14. Zhang, M., White, R.A., Wang, L., Goldman, R., Kavraki, L., Hassett, B.: Improving conformational searches by geometric screening. Bioinformatics 21, 624–630 (2005)
15. Clore, G.M., Gronenborn, A.M., Bax, A.: A Robust Method for Determining the Magnitude of the Fully Asymmetric Alignment Tensor of Oriented Macromolecules in the Absence of Structural Information. J. Magnetic Resonance 133, 216–221 (1998)
16. Hus, J., Peti, W., Griesinger, C., Brüschweiler, R.: Self-Consistency Analysis of Dipolar Couplings in Multiple Alignments of Ubiquitin. J. Am. Chem. Soc. 125, 5596–5597 (2003)

A Stochastic Pi Calculus for Concurrent Objects

Céline Kuttler[1], Cédric Lhoussaine[2], and Joachim Niehren[3]

[1] The Microsoft Research - University of Trento
Centre for Computational and Systems Biology, Italy
[2] University of Lille 1, LIFL, Lille, France
[3] INRIA Futurs, Lille, France, Mostrare project

Abstract. We present SPiCO, a new modeling and simulation language for systems biology. SPiCO is based on the stochastic π-calculus. It supports higher level modeling via multi-profile concurrent objects with static inheritance. We present a semantics for SPiCO in terms of continuous time Markov chains, and show how to compile SPiCO back into the biochemical stochastic π-calculus while preserving semantics.

1 Introduction

A central objective of systems biology is the investigation of the dynamics in living cells, that arises from interactions between its molecular components. Modeling and simulation increasingly complement knowledge acquisition through experimentation. Discrete event based approaches are advantageous with respect to detailed cellular control by small numbers of molecular actors. Deterministic approaches offer benefits when modeling large populations.

Regev and co-authors [22] proposed to apply the stochastic π-calculus as a modeling language for systems biology, based on Priami's [20] refinement of the synchronous π-calculus [17] by a notion of time. Expression in the π-calculus then abstract chemical solutions, in which molecules interact concurrently. As for earlier stochastic process algebras [9], stochastic parameters impose exponential distributions of waiting times on reaction. Thus π-calculus expressions give rise to continuous time Markov chains (CTMCs). Their execution yields stochastic simulation, based on Gillespie's algorithm [7].

Both existing simulation engines for the (biochemical) stochastic π-calculus–SPiM [19] and BioSpi [22] – have been applied in case studies of small to medium size [11,15,16]. Alternative modeling languages as BioCham [4] or SBML [10] directly specify systems of chemical reaction rules. This approach is simpler, yet seems less expressive with respect to concurrent control, i.e. intricate conditions for rule application.

From the *modeling* perspective in systems biology, the minimality of the π-calculus is sometimes unfortunate. Concurrent control soon requires sophisticated protocols [11], tricky to both design and understand. Such protocols are at a low level and must be adapted upon model extension. This constitutes a major obstacle to up-scaling models.

H. Anai, K. Horimoto, and T. Kutsia (Eds.): AB 2007, LNCS 4545, pp. 232–246, 2007.
© Springer-Verlag Berlin Heidelberg 2007

In this paper we present SPICO, a new modeling language for systems biology extending on a *stochastic π-calculus for concurrent objects*. SPICO was indeed developed concomitantly with modeling case studies [12,13]. The main insight behind SPICO is that concurrent objects (as in programming languages) appropriately represent interacting molecules for systems biology. Object *interfaces* avoid communication protocols, while object *inheritance* renders models more extensible. The technical contributions can be summarized as follows:

1. We present Core SPICO, a novel stochastic π-calculus with input patterns, that originate from the distributed programming language TYCO [18,25] for typed concurrent objects in the asynchronous π-calculus. SPICO assigns stochastic rates to pairs of channel and function names.
2. We define a *stochastic semantics* assigning CTMCs to SPICO's process expressions, carefully distinguising timed and instantaneous reactions. Technically, this is the most difficult part of the paper. Previous semantics for the stochastic π-calculus do not define CTMCs at all [19,22], or disregard immediate reactions [20] essential for expressiveness and modeling. Experiments with SPiM confirm a correct treatment in implementations nevertheless.
3. We identify multi-profile objects with expressions in Core SPICO. Each profile comes with its own interface, similarly as TYCO's non-uniform objects [23]. Beyond these, multi-profile objects allow choice with mixed input and output on possibly different channels and synchronous communication.
4. We define a notion of inheritance for multi-profile objects that is compiled into the core of SPICO. We present a module system for SPICO providing syntax for definitions of objects with inheritance.
5. We discuss a programming technique to model mutual exclusion of molecular events, as frequently encountered in cellular regulation. Its essence lies in escaping inconsistent intermediate states by immediate reactions, that are applied before timed ones. This solves tedious *atomicity* problems, without introducing transactions [5].
6. We encode SPICO back into the biochemical stochastic π-calculus, so that we can run SPICO programs in SPiM or BioSpi. The main challenge is to encode input patterns, while preserving the stochastic semantics.

In previous work we proposed a first ad hoc abstraction of interacting molecules as objects that switch between discrete states [6]. How to explicitly support objects with multiple profiles in a more conservative language with proper syntax and semantics remained open.

Other languages for systems biology were recently proposed. Beta binders [21] are inspired by the π-calculus, but enable interactions by type coincidence rather than channel name equality. Others [3,24] adress spatial aspects at membranes.

Outline. The core of SPICO is introduced in Section 2 and illustrated for modeling molecular binding at overlapping sites in Section 3. SPICO's multi-profile objects with inheritance are discussed in Section 4. CTMCs for chemical reactions in Section 5 motivate the stochastic semantics of SPICO in Section 6. In Section 7, we show how to encode input patterns by a naming discipline. For space considerations, the proofs are not included here but can be found in [14].

2   A Stochastic Pi-calculus with Input Patterns

The core of SPiCO (Core SPiCO) consists in a novel stochastic π-calculus with *input patterns*, a linguistic feature introduced by Vasconcelos and Tokoro for typed concurrent objects in the asynchronous π-calculus (TYCO) [18,25]. Input patterns are motivated by pattern matching in functional programming languages of the ML family. In TYCO, they are closely tied to communication: objects only receive tuples if they provide a matching input pattern.

Core SPiCO's vocabulary consists in an infinite set of *channel names* $\mathcal{N} = \{x, y, z, \ldots\}$, a set of *process names* A, and a set of *function names* $f \in \mathcal{F}$. Process and function names have fixed arities. We write A/n or f/n for a symbol of arity $n \geq 0$. In order to account for *stochastic rates*, the vocabulary comprises functions $\rho : \mathcal{F} \to]0, \infty]$ to define *stochastic rates* for every channel. If some function ρ is assigned to x then $\rho(f)$ is the rate of the pair (x, f).

Table 1 defines the syntax of Core SPiCO. We write $\tilde{x}$ for finite, possibly empty sequences of channels $x_1, \ldots, x_n$ where $n \geq 0$. When using tuples $f(\tilde{x})$ or terms $A(\tilde{x})$ the number of arguments (the length of $\tilde{x}$) is assumed equal to the respective arity of f or A. Process expressions are ranged over by P. The only atomic expression (not decomposable into others) is the guarded choice of length $n = 0$ that we write as $\mathbf{0}$. Expressions $P_1 | P_2$ denote the parallel composition of processes P_1 and P_2. A term $\mathbf{new}\ x{:}\rho.\ P$ introduces a new channel x scoping over P; the rate function ρ fixes stochastic rates $\rho(f)$ for all pairs (x, f) where $f \in \mathcal{F}$. We can omit rate functions ρ in the declaration of a channel x if all reactions on x are instantaneous, i.e. $\rho(f) = \infty$ for all $f \in \mathcal{F}$. An expression $A(\tilde{x})$ applies the definition of a parametric process A with actual parameters $\tilde{x}$.

A sum of guarded processes $C_1 + \ldots + C_n$ offers a *choice* between $n \geq 0$ communication alternatives $C_1, \ldots, C_n$. A guarded input $x?f(\tilde{y})$ describes a communication act, ready to *receive* over x a tuple constructed by f. The channels $\tilde{y}$ in input guards serve as pattern variables; these bound variables are replaced by the channels received as input. An output guarded process $x!f(\tilde{y}).P$ describes a communication act willing to send tuple $f(\tilde{y})$ over channel x and continue as P.

A definition of a parametric process has the form $A(\tilde{x}) \triangleq P$ where A is a process name with $\tilde{x}$ as formal parameters - that is, a sequence of bound channels. For modeling convenience, we permit free channel names in P besides the parameters

Table 1. Syntax of Core SPiCO

Processes	$P ::= P_1 \mid P_2$	parallel composition
	$\mid\ \mathbf{new}\ x{:}\rho.\ P$	channel creation
	$\mid\ C_1 + \ldots + C_n$	sum $(n \geq 0)$
	$\mid\ A(\tilde{x})$	application
Guarded processes	$C ::= x?f(\tilde{y}).P$	pattern input
	$\mid\ x!f(\tilde{y}).P$	tuple output
Definitions	$D ::= A(\tilde{y}) \triangleq P$	

Table 2. Axioms of the structural congruence

$$(P_1|P_2)|P_3 \equiv P_1|(P_2|P_3) \qquad\qquad\qquad P_1|P_2 \equiv P_2|P_1$$
$$\ldots + C_1 + C_2 + \ldots \equiv \ldots + C_2 + C_1 + \ldots \qquad P|\mathbf{0} \equiv P$$
$$\textbf{new } x{:}\rho.\ (P_1|P_2) \equiv P_1|\ \textbf{new } x{:}\rho.\ P_2 \quad \text{if } x \notin fv(P_1) \qquad P_1 \equiv P_2 \text{ if } P_1 \equiv_\alpha P_2$$
$$\textbf{new } x_1{:}\rho_1.\ \textbf{new } x_2{:}\rho_2.\ P \equiv \textbf{new } x_2{:}\rho_2.\ \textbf{new } x_1{:}\rho_1.\ P \quad \text{if } x_1 \neq x_2$$

Table 3. Reduction relation for a finite set of definitions Δ

Communication, choice, and pattern matching:

$$x!f(\tilde{y}).P_1 + \ldots \mid x?f(\tilde{z}).P_2 + \ldots \quad\rightarrow\quad P_1 \mid P_2[\tilde{z} \mapsto \tilde{y}] \qquad \text{if } \tilde{z} \text{ free for } \tilde{y} \text{ in } P_2$$

Application of definitions:

$$A(\tilde{x}) \quad\rightarrow\quad P[\tilde{y} \mapsto \tilde{x}] \qquad \text{if } A(\tilde{y}) \triangleq P \text{ in } \Delta, \text{ and } \tilde{y} \text{ free for } \tilde{x} \text{ in } P$$

Context and congruence closure:

$$\frac{P \rightarrow P'}{\textbf{new } c{:}\rho.\ P \rightarrow \textbf{new } c{:}\rho.\ P'} \qquad \frac{P \rightarrow P'}{P \mid Q \rightarrow P' \mid Q} \qquad \frac{P \equiv P' \quad P' \rightarrow Q' \quad Q \equiv Q'}{P \rightarrow Q}$$

in $\tilde{x}$. The set of *free channel names* for processes P and guarded processes C are denoted by $fv(P)$ and $fv(C)$ respectively. There are three scope baring constructs: new binder **new** $x{:}\rho.\ P$, input patterns $\_?f(\tilde{x}).P$, and definitions $A(\tilde{x}) \triangleq P$.

We define an (non-stochastic) operational semantics for the π-calculus in terms of a binary relation between expressions, the so called (one step) reduction. We will later refine it to a ternary relation adding stochastic labels. The reduction relation is closed under the usual structural congruence (Table 2) between expressions.

Table 3 defines the *reduction relation*. The first axiom tells how to interpret choices; it comprises channel communication and pattern matching. It applies to two complementary matching alternatives in parallel choices, an output alternative $x!f(\tilde{y}).P_1$ willing to send a term $f(\tilde{y})$ and an input pattern $x?f(\tilde{z}).P_2$ on the same channel x; this pattern matches in that it is built using the same function symbol f. Reduction cancels all other alternatives, substitutes the pattern's variables $\tilde{z}$ by the received channels $\tilde{y}$ in the continuation P_2 of the input, and reduces the result in parallel with the continuation of the output P_1.

Only matching tuples can be received over a channel. Other sending attempts suspend until a suitable input pattern becomes available. This fact proves extremely useful for concurrent modeling. Upon reception, tuples are immediately decomposed, in contrast to the π-calculus with data terms [1].

The application axiom unfolds one of the definitions of the parametric processes in a given set Δ. An application $A(\tilde{y})$ reduces in one step to definition P in which the formal parameters $\tilde{y}$ were replaced by the actual parameters $\tilde{x}$. Parametric definitions may be recursive, e.g. A may occur in P. Reduction can be applied in arbitrary contexts, however not under choices or in definitions.

The syntax of the biochemical stochastic π-calculus is the same as ours except for function names, and our more flexible assignment of stochastic rates. We can express polyadic input and output by using dummy function names UNIT$_i$ for all arities $i \geq 0$ in following shortcuts for all sequences $\tilde{y}$ of channel names of length i:

$$x?(\tilde{y}).P =_{\text{def}} x?(\text{UNIT}_i(\tilde{y})).P \qquad \text{and} \qquad x!(\tilde{y}).P =_{\text{def}} x!(\text{UNIT}_i(\tilde{y})).P$$

3   Molecular Binding at Overlapping Sites

We illustrate the modeling power of SPICO by a frequent control mechanism between molecular interactions, as binding of molecules: mutual exclusion [11,12]. Consider overlapping sites allowing for a unique visitor at a time - i.e. overlapping *semaphores*.

Figure 1(a) illustrates two such overlapping sites s and s'. Each can be either free, bound, or blocked. Only a free site can become bound by a visitor - while blocking the peer. We model sites as multi-profile object with three profiles Site_free, Site_bound, and Site_blocked. Figure 2 presents their definitions in the π-calculus. Beside of its own identity me, a site is parametrized by the identity of the other overlapping site. The defining sums specify interfaces for profiles, i.e. which functions are offered or applied and on which channels. Profile Site_free for instance, offers functions bind and block by which it can become bound or blocked, and can apply function unblock of the other site.

Multi-profile objects yield an elegant solution to express semaphores (sites with at most one visitor). A visitor can only bind to free sites since no other profile offers the bind function. This exploits the clever coupling between pattern matching and synchronization by input patterns.

The most tedious aspect of overlapping sites is to keep states consistent. Whenever a site gets bound, its overlapping peer must **immediately** become blocked, i.e. without any elapse of simulated time. The actor Site_bound(me,other) enforces this by applying function block on its peer. The stochastic rate of this function needs must thus be ∞. This technique works only if immediate transitions have priority over time-consuming ones, and under the assumption that function block is immediate. Highest priority of immediate transitions is guaranteed by the stochastic semantics to come (see rule (SUM) in Table 4). This way, we solve a tedious atomicity problems while avoiding heavier extensions of the π-calculus by transactions [5].

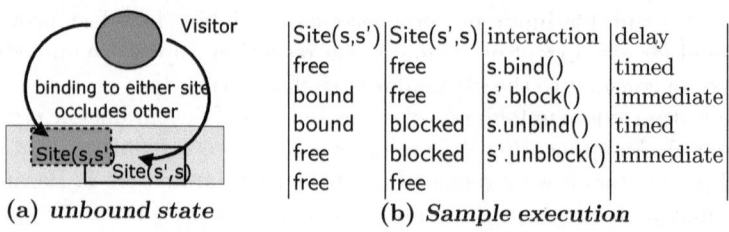

Site(s,s')	Site(s',s)	interaction	delay
free	free	s.bind()	timed
bound	free	s'.block()	immediate
bound	blocked	s.unbind()	timed
free	blocked	s'.unblock()	immediate
free	free		

(a) *unbound state*   (b) *Sample execution*

Fig. 1. Overlapping sites located at s and s'

```
module 'overlapping sites'
export Site with bind/0, unbind/0
define
   Site(me,other) ≜ Site_free(me,other)
   Site_free(me,other) ≜
      me?bind(). Site_bound(me,other)
    + me?block(). Site_blocked(me,other)
    + other!unblock(). Site_free(me,other)
   Site_bound(me,other) ≜
      me?unbind(). Site_free(me,other)
    + other!block(). Site_bound(me,other)
   Site_blocked(me,other) ≜
      me?unblock(). Site_free(me,other)
```

Fig. 2. Overlapping sites module

Table 4. Timed transitions of Core SPICO with respect to a set Δ of definitions in prenex normal form, and a global assignment ϱ of channels to rate functions

Labeled reduction steps

(COM)

$$C_{i_1}^{j_1} = x?f(\tilde{z}).\textbf{new } \widetilde{x_1{:}\rho_1}.\, Q_1 \qquad C_{i_2}^{j_2} = x!f(\tilde{y}).\textbf{new } \widetilde{x_2{:}\rho_2}.\, Q_2$$

$$\Pi_{i=1}^n \sum_{j=1}^{m_i} C_i^j \xrightarrow[i_1,j_1,i_2,j_2]{\varrho(x)(f)} \left\{ \begin{array}{l} \textbf{new } \widetilde{x_1{:}\rho_1}. \textbf{ new } \widetilde{x_2{:}\rho_2}. \\ (Q_1[\tilde{z} \mapsto \tilde{y}] \mid Q_2 \mid \Pi_{i=1,i\neq i_1,i_2}^n \sum_{j=1}^{m_i} C_i^j) \end{array} \right.$$

where Q_1, Q_2 have no top-level **new**-binders and $1 \leq i_1 \neq i_2 \leq n$, $1 \leq j_1 \leq m_{i_1}$, $1 \leq j_2 \leq m_{i_2}$

(APP)

$$P_{i_1} = A(\tilde{y}) \qquad A(\tilde{x}) \triangleq \textbf{new } \widetilde{z{:}\rho}.\, Q \text{ in } \Delta$$

$$\Pi_{i=1}^n P_i \xrightarrow[i_1]{\infty} \textbf{new } \widetilde{z{:}\rho}.\, (Q[\tilde{x} \mapsto \tilde{y}] \mid \Pi_{i=1,i\neq i_1}^n P_i)$$

where Q has no top-level **new**-binders and $1 \leq i_1 \leq n$

(NEW)

$$\frac{P \xrightarrow[w]{s} Q \qquad \varrho(x) = \rho}{\textbf{new } x{:}\rho.\, P \xrightarrow[w]{s} \textbf{new } x{:}\rho.\, Q} \qquad \text{where } s \in \mathbb{R}^+ \cup \{\infty\},\ w \in \mathbb{N} \cup \mathbb{N}^4$$

Time consuming transitions $(r, r' \in \mathbb{R}^+,\ w \in \mathbb{N}^4)$

(SUM)

$$\frac{P \equiv P' \qquad r = \sum_{P' \xrightarrow{r'}_w Q' \equiv Q} r' \neq 0 \qquad \neg\exists R\exists w' \in \mathbb{N} \cup \mathbb{N}^4.\, P' \xrightarrow[w']{\infty} R}{P \xrightarrow{r} Q}$$

Immediate transitions

(COUNT)

$$\frac{P \equiv P' \qquad \begin{array}{l} n = \#\{w \in \mathbb{N} \cup \mathbb{N}^4 \mid P' \xrightarrow[w]{\infty} Q' \equiv Q\} \neq 0 \\ m = \#\{w \in \mathbb{N} \cup \mathbb{N}^4 \mid P' \xrightarrow[w]{\infty} Q''\} \end{array}}{P \xrightarrow{\infty(n/m)} Q}$$

```
module 'visitors for sites s or s' '
public s s'
export Visitor
define
   Visitor() ≜ Visitor_free()
   Visitor_free()≜ s!bind().Visitor_at(s)
                 + s'!bind().Visitor_at(s')
   Visitor_at(site) ≜ site!unbind().Visitor_free()
```

Fig. 3. Visitors module for sites s and s'

One possible sequence of state changes is given in Figure 1(b). Initially, we assume a parallel composition of two free sites and two free visitors. The first parameter of Site_free refers to its identity and the second to its peer's:

$$\begin{aligned}
& \text{Site\_free(s,s')} \mid \text{Site\_free(s',s)} \mid \text{Visitor\_free} \mid \text{Visitor\_free} \\
\rightarrow\ & \text{Site\_bound(s,s')} \mid \text{Site\_free(s',s)} \mid \text{Visitor\_free} \mid \text{Visitor\_at(s)} \\
\nrightarrow\ & \text{Site\_bound(s,s')} \mid \text{Site\_bound(s',s)} \mid \text{Visitor\_at(s)} \mid \text{Visitor\_at(s')}
\end{aligned}$$

The first reduction step is an application of function bind of s by the second Visitor_free defined in Figure 3, which consumes time. Now Site_free(s',s) has a potential choice between a time consuming transition where function bind of s' is applied by the first Visitor_free, and an immediate transition applying function block of s' by Site_bound(s,s'). Priority is given to immediate transitions, so only the latter function can be applied.

$$\xrightarrow{\infty}\ \text{Site\_bound(s,s')} \mid \text{Site\_blocked(s',s)} \mid \text{Visitor\_at(s)} \mid \text{Visitor\_free}$$

Thereby, it becomes impossible to enter into an erroneous configuration in which both Sites are bound.

4   Multi-profile Objects with Inheritance

The full SPICO language features multi-profile objects with static inheritance. In this section, we define these concepts formally and show how to compile them to Core SPICO.

SPICO supports the paradigm of "molecules as concurrent objects" a refinement of the paradigm "molecules as processes" by Regev and Shapiro. *Object classes* correspond to species of molecules. A class of a *multi-profile object* is a set of definitions by sums, each of which defines a profile.

$$\text{Obj\_p}_1(\tilde{x}_1) \triangleq C_1^1 + \ldots + C_{n_1}^1$$
$$\ldots$$
$$\text{Obj\_p}_m(\tilde{x}_m) \triangleq C_1^m + \ldots + C_{n_m}^m$$

A major advantage of object-orientation for biological systems is model extensibility by object inheritance. Numerous examples are elaborated in [12]. A

```
module 'repressible promoter'
    import Site from 'overlapping sites'
export
    Promoter extends Site by initiate/0
define
    Promoter_bound(me, other) extended by
        me? initiate(). Promoter_free(me, other)
```

Fig. 4. Promoters inherit from overlapping sites

simpler case is given in Figure 4. This is a promoter, a DNA region controlling transcription initiation, which overlaps with an operator region. A promoter is thus like an overlapping site, except that it can initiate transcription when bound by a polymerase. This new functionality is added by inheritance. We next define inheritance for multi-profile objects. We extend class Obj to Obj2 as follows:

```
Obj2 extends Obj
```
$$\mathsf{Obj2\_p_1}(\tilde{z}_1) \text{ extended by } C^1_{k_1+1} + \ldots + C^1_{l_1}$$
$$\ldots$$
$$\mathsf{Obj2\_p_n}(\tilde{z}_n) \text{ extended by } C^n_{k_n+1} + \ldots + C^n_{l_n}$$

This specification with inheritance can be compile into definitions of Core SPICO:

$$\mathsf{Obj2\_p_1}(\tilde{z}_1) \triangleq C^1_1 + \ldots + C^1_{l_1} \ [\mathsf{Obj}\mapsto \mathsf{Obj2}]$$
$$\ldots$$
$$\mathsf{Obj2\_p_n}(\tilde{z}_n) \triangleq C^n_1 + \ldots + C^n_{l_n} \ [\mathsf{Obj}\mapsto \mathsf{Obj2}]$$

The substitution renames all recursive calls to profiles $\mathsf{Obj\_p_i}$ into recursive calls to $\mathsf{Obj2\_p_i}$ for $1 \leq i \leq n$.

SPICO provides a module system for grouping sets of definitions together so that they can be extended by multiple inheritance. Modules import definitions from others as usual. Such module dependencies can be resolved statically, as long as they remain acyclic, which SPICO assumes. The details of the module systems are out of the scope of this paper.

5 Markov Chains for Chemical Reactions

The stochastic semantics of our π-calculus is guided by the analogy to continuous time Markov chains (CTMCs) for chemical reactions.

We first recall CTMCs with countably infinite state spaces. We assume a countable set S called the *state space*. A *continuous time stochastic process* with states $q \in S$ is a family $\{X_t \mid t \in \mathbb{R}^+\}$ of random variables with values in S. These define probabilities $Pr(X_t \in S')$ for all subsets $S' \subseteq S$, i.e. the probability that the process is in some state of S' at time t.

A *continuous time Markov chain (CTMC)* is a continuous time stochastic process (CTSP), with memoryless sojourn times for all states. More formally, a

CTMC over S is a CTSP $\{X_t \mid t \in \mathbb{R}^+\}$ with states in S, that satisfies the Markov property, i.e. for all $q_0, \ldots, q_{n+1} \in S$ and all time points $0 \le t_0 < \ldots < t_{n+1}$:

$$Pr(X_{t_{n+1}} = q_{n+1} \mid X_{t_n} = q_n, \ldots, X_{t_0} = q_0) = Pr(X_{t_{n+1}} = q_{n+1} \mid X_{t_n} = q_n)$$

The probabilistic behavior of a CTMC is determined by the distribution of its initial states (at time 0) and its *transition rates*. The transition rate r from state q to state q' is a value that "scales how the (one step) transition probability between q and q' increases with time" [8]. We write $q \xrightarrow{r} q'$ in this case. For simplicity, we consider CTMCs with a single initial state. These can be identified with a Markovian transition system $(S, (\xrightarrow{r})_{r \in \mathbb{R}^+}, q_0)$ where $q_0 \in S$ is the initial state and $\xrightarrow{r} \subseteq S \times S$ are transition relations for all $r \in \mathbb{R}^+$, such that for all $q, q' \in S$ there exists at most one $r \in \mathbb{R}^+$ satisfying $q \xrightarrow{r} q'$.

The stochastic time evolution of a CTMC can be computed by Gillespie's *first reaction method* (1976) [7] if each state permits only a finite number of transitions, as we assume in the sequel. At time 0 the process starts in state q_0. Suppose that the process has moved to state q at time point t and let $q\{\xrightarrow{r_i} q_i\}_i$ be all (finitely many) transitions starting in q. Draw delays $t_i > 0$ for all i from an exponential distribution with rate r_i. Draw with equal probability some j, with minimal t_j. Move to state q_j at time point $t + t_j$.

Gillespie's direct method equivalently determines the stochastic behavior of a CTMC [7]. In state q at time t it first computes the delay until the next transition (called *sojourn time*), by drawing a number from the exponential distribution with rate $\downarrow s =_{\text{def}} \sum_{q \xrightarrow{r_i} q_i} r_i$. Second, the state q_j to go to is drawn with probability

$$Pr(q \to q_j) =_{\text{def}} r_j / \sum_{q \xrightarrow{r'} q'} r' \text{ if } q \xrightarrow{r_j} q_j \text{ and } 0 \text{ otherwise.}$$

We next illustrate CTMCs for systems of chemical reaction rules. We start from a set of chemical species X, Y, Z and a set of chemical reaction rules of the following form, where $r \in \mathbb{R}^+$, reserving the symbol $+$ for choice:

$$X \mid Y \xrightarrow{r} Z_1 \mid \ldots \mid Z_k$$

Chemical solutions P are multisets of species, where each occurrence in the multiset represents a molecule of the species. Chemical rules as above apply as follows to a chemical solution P. Each pair of molecules of species X and Y can interact at rate r, yielding one molecule of each of the species $Z_1, \ldots, Z_k$. The solution obtained is $P - \{\!| X, Y |\!\} \cup \{\!| Z_1, \ldots, Z_k |\!\}$. According to the *Chemical Law of Mass Action*, the speed of a chemical reaction in a solution is proportional to the number of possible interactions of its reactants in the solution. It is distributed exponentially, and defines a CTMC with chemical solutions as states and the following transitions:

$$P \xrightarrow{n \cdot r} \begin{cases} P - \{\!| X, Y |\!\} \\ \cup \{\!| Z_1, \ldots, Z_k |\!\} \end{cases} \quad \text{where} \quad n = \begin{cases} \sharp(X \in P) \times \sharp(Y \in P) & \text{if } X \ne Y \\ \binom{\sharp(X \in P)}{2} & \text{else} \end{cases}$$

The expression $\binom{m}{2} = \frac{1}{2} m (m - 1)$ counts the number of two-element subsets in sets of cardinality m.

6   Stochastic Semantics of Core SpiCO

We define the stochastic semantics of Core SpiCO by associating a π-calculus process with a CTMC. The states of this Markov chain are the (countably infinite) set of congruence classes of π-calculus processes with respect to structural congruence. This differs from [20] where two congruent processes are associated with two different states. Since congruent processes are behaviorally equivalent we believe that their associated stochastic states should not be distinguished neither. Moreover, in [20], the author proposes a *labeled semantics* where labels are so-called *proof terms*, i.e. (possibly long) strings used to localize interacting subterms. Those labels are necessary to properly calculate interaction rates. We instead propose a *reduction semantics*, a style for defining semantics known to be more intuitive and elegant. Still, we temporarily use labels but in a much simpler form: a label is an integer or a tuple of four integers. Finally, and contrary to [20], our semantics takes into account immediate transitions of which we emphasized the importance in the biological example in section 3. Such transitions require specific consideration: we show how they can be removed in order to obtain an equivalent Markovian transition system. The theorem 1 states the correctness of this transformation.

6.1   Transition Relations

We first consider the *fragment* of the π-calculus without proper summation, parametric processes, infinite rates, and **new**-binders. The remaining processes are parallel compositions $C_1 \mid \ldots \mid C_n$. The structural congruence turns them into multisets of guarded processes, i.e. into chemical solutions whose species are guarded processes.

Suppose we know the rate functions $\varrho(x)$ for all channels x. The π-calculus with input patterns then defines the following chemical reaction rule:

$$x?f(\tilde{z}).Q_1 \mid x!f(\tilde{y}).Q_2 \xrightarrow{\varrho(x)(f)} Q_1[\tilde{z} \mapsto \tilde{y}] \mid Q_2$$

This defines a CTMC. For example, assume n molecules of a first species $x!f().P_1$ and m of another different one $x!f().P_2$, which all want to react with a single molecule of a third kind $x?f().P$. The Markovian transitions are:

$$\prod_{i=1}^{n} x!f().P_1 \mid \prod_{i=1}^{m} x!f().P_2 \mid x?f().P \begin{cases} \xrightarrow{n \times \varrho(x)(f)} \prod_{i=1}^{n-1} x!f().P_1 \mid \prod_{i=1}^{m} x!f().P_2 \mid P \\ \xrightarrow{m \times \varrho(x)(f)} \prod_{i=1}^{n} x!f().P_1 \mid \prod_{i=1}^{m-1} x!f().P_2 \mid P \end{cases}$$

We first discuss time consuming transitions $P \xrightarrow{r} P'$ where $r \in \mathbb{R}^+$. These capture everything, except parametric process unfolding and invocation of functions of rate ∞.

We first define labeled reduction steps $P \xrightarrow[w]{s} Q$ where P and Q are in prenex normal form, that is a parallel composition of sums where restrictions have been pushed ahead and in which bound variables are renamed apart. The rate function $\varrho(x)$ is then read off from the quantifier prefix in rule (NEW).

Definition 1. *P is in prenex normal form (pnf for short) iff $P = \mathbf{new}\ \widetilde{x{:}\rho}.\ (P_1 \mid \ldots \mid P_m)$ where each P_i either is an application $A(\tilde{y})$, or a sum $C_1 + \ldots + C_n$ where each C_j is in pnf, or a guarded process $x?f(\tilde{y}).Q$ or $x!f(\tilde{y}).Q$ where Q is in pnf. Moreover, a definition $A(\tilde{y}) \triangleq P$ is in pnf iff P is in pnf.*

What remains from pnfs after removing top-level **new**-binders are multisets of sums and applications. All applications must have been reduced before time consuming transitions can apply, so we have a multiset of sums. Each sum is like a molecule, except that each of its choices offers its own interactions.

In $x?f().0 + x?f().0 \mid x!f().0$ there are two possible interactions with rate $r = \varrho(x)(f)$ leading to the same state. We can think of $x?f().0 + x?f().0$ as a protein with two identical domains, complementary to one domain of some other protein represented by $x!f().0$. The overall rate of the interaction thus doubles:

$$\boxed{\text{x?f() .0}} + x?f().0 \mid \boxed{\text{x!f() .0}} \xrightarrow[1.1.2.1]{r} 0$$

$$\text{and } x?f().0 + \boxed{\text{x?f() .0}} \mid \boxed{\text{x!f() .0}} \xrightarrow[1.2.2.1]{r} 0$$

$$\text{sums up to } \quad x?f().0 + x?f().0 \mid x!f().0 \xrightarrow{2r} 0$$

Rule (COM) defines labeled reductions $P \xrightarrow[i_1,j_1,i_2,j_2]{r} Q$ that distinguish communication actions with identical reactants and results, while using different occurrences of choice alternatives in sums. Those occurrences are identified by *labels* in $\mathbb{N}^4$ that specify the numbers of the reacting sums (i_1, i_2) and the reacting choices (j_1, j_2). Rule (SUM) defines transitions $P \xrightarrow{r} Q$ by summing up all rates of all different interactions leading from P to Q. These reduction rules are defined with care, so that corresponding interactions in structurally congruent processes are not counted twice.

We next turn to *immediate transitions* $P \xrightarrow{\infty(p)} Q$, where $p \in [0, 1]$ is a probability. Rule (SUM) ensures that time consuming transitions apply only after all immediate have been reduced. In this case, all calls $A(\tilde{y})$ on top level must have been reduced before. Note that this order is important for a proper count of the possible interactions. Indeed, if an application hides an interaction on some pattern, the application unfolding changes the rate of the action involving this pattern. Immediate transitions can be licensed by communication (COM), or by applications of parametric process definitions (APP). Their labels are in $\mathbb{N} \cup \mathbb{N}^4$. Note that the labeled reduction is independent of the choice of the pnf.

We merge labeled immediate transitions with rule (COUNT). Although being immediate we want to associate probabilities, which characterize the number of immediate interactions leading to a common state with respect to the total number of enabled immediate interactions. For instance, let $\varrho(x)(f) = \infty$, in $x?f().P + x?f().P \mid x?f().Q \mid x!f().0$, for some $P \not\equiv Q$, the associated probabilities reflect that 2 out of 3 interactions lead to P, and 1 out of 3 to Q:

$$x?f().P + x?f().P \mid x?f().Q \mid x!f().0 \xrightarrow{\infty(2/3)} P$$

$$x?f().P + x?f().P \mid x?f().Q \mid x!f().0 \xrightarrow{\infty(1/3)} Q$$

Table 5. Elimination of immediate transitions and merging timed transitions

$$(\text{ELIM}_1) \quad \frac{P \xrightarrow[w]{\infty} Q \qquad n = \sharp\{w' \in \mathbb{N} \cup \mathbb{N}^4 \mid P \xrightarrow[w']{\infty} Q'\}}{P \xrightarrow[w]{\infty(1/n)} Q} \qquad w \in \mathbb{N} \cup \mathbb{N}^4$$

$$(\text{ELIM}_2) \quad \frac{P \xrightarrow[w]{r} Q \qquad Q \xrightarrow[w_1]{\infty(p_1)} \cdots \xrightarrow[w_n]{\infty(p_n)} Q_n \not\xrightarrow{\infty}}{P \underset{ww_1\ldots w_n}{\overset{rp_1\ldots p_n}{\Longrightarrow}} Q_n} \qquad r \in \mathbb{R}^+$$

$$(\text{ELIM}^{\text{sum}}) \quad \frac{P \equiv P' \qquad r = \sum_{P' \underset{w_1\ldots w_n}{\overset{r'}{\Longrightarrow}} Q' \equiv Q} r'}{P \overset{r}{\Longrightarrow} Q}$$

6.2   CTMCs with Immediate Reactions

In the presence of immediate transitions, the reduction relation $\xrightarrow{r}$ does not define a Markovian transition system (in which all rates are finite). To capture the stochastic dynamics of processes, we instead define the *sojourn time parameters* (i.e. the parameter of an exponentially distributed probability which determine the sojourn time in a given state) and the *probabilities of state changes* for all P, Q as follows[1]:

$$\downarrow P = \begin{cases} \infty & \text{if } P \xrightarrow{\infty(p)} Q, \\ \sum_{P \xrightarrow{r} Q} r & \text{otherwise.} \end{cases} \qquad Pr(P \to Q) = \begin{cases} r/\sum_{P \xrightarrow{r'} Q'} r' & \text{if } P \xrightarrow{r} Q \\ p & \text{if } P \xrightarrow{\infty(p)} Q \\ 0 & \text{otherwise} \end{cases}$$

We are now giving an interpretation of the reduction semantics with immediate transitions in terms of CTMCs for processes that can not exhibit infinite sequences of immediate transitions. The Markovian transition system deriving statements $P \overset{r}{\Longrightarrow} Q$ is defined in Table 5. The idea is quite similar to that of [2]: the transitions are obtained by integrating immediate transitions into time consuming transitions. An example for this transformation is as follows:

$$P \begin{cases} \xrightarrow{r_1} Q_1 \not\xrightarrow{\infty} \\ \xrightarrow{r_2} Q_2 \begin{cases} \xrightarrow{\infty(p)} Q_{21} \not\xrightarrow{\infty} \\ \xrightarrow{\infty(1-p)} Q_{22} \not\xrightarrow{\infty} \end{cases} \end{cases} \quad \text{becomes} \quad P \begin{cases} \overset{r_1}{\Longrightarrow} Q_1 \\ \overset{r_2 p}{\Longrightarrow} Q_{21} \\ \overset{r_2(1-p)}{\Longrightarrow} Q_{22} \end{cases}$$

In general, a sequence of reductions $P \xrightarrow{r} P_1 \xrightarrow{\infty(p_1)} \ldots P_n \xrightarrow{\infty(p_n)} Q \not\xrightarrow{\infty}$ reduces to $P \overset{rp_1\ldots p_n}{\Longrightarrow} Q$. However, we must beware of merging initially distinct states. Indeed, in the previous example, if $Q_{22} \equiv Q_1$ then the CTMC should have transitions $P \overset{r_1+r_2(1-p)}{\Longrightarrow} Q_1$ and $P \overset{r_2 p}{\Longrightarrow} Q_{21}$. In order to infer these transitions correctly, the elimination procedure defines labeled transitions $\underset{w}{\Longrightarrow}$ with *labels* $w \in (\mathbb{N} \cup \mathbb{N}^4)^{\star}$ representing *paths* in the labeled derivation trees of $\xrightarrow{r}$.

---

[1] We assume if X is exponentially distributed with parameter ∞ then $Pr(X = 0) = 1$.

For any P such that $P \xrightarrow{\infty}$, $(\mathscr{P}_{/\equiv}, (\xRightarrow{r})_{r \in \mathbb{R}^+}, P_{/\equiv})$ is a Markovian transition system[2] with sojourn time parameters and transition probabilities:

$$\Downarrow P = \sum_{P \xRightarrow{r} Q} r \qquad \text{and} \qquad Pr(P \Rightarrow Q) = \begin{cases} r / \sum_{P \xRightarrow{r'} Q'} r' & \text{if } P \xRightarrow{r} Q \\ 0 & \text{otherwise} \end{cases}$$

In order to show that this defines a Markovian model for the reduction semantics with immediate transitions, we show that their dynamics coincide, that is: the sojourn time parameters and the transition probabilities with respect to $\xrightarrow{r}$ are identical to those of $\Rightarrow$. However, transition probabilities can be compared only for processes performing timed transitions. We thus define a suitable transition probability $Pr(P \twoheadrightarrow Q)$ for $P \xrightarrow{\infty}$ and $Q \xrightarrow{\infty}$, that is the probability to reach Q from P by a sequence of transitions made of one timed transition and possibly several intermediate immediate transitions. Formally, $Pr(P \twoheadrightarrow Q)$ is the sum of the probabilities of all such sequences:

$$Pr(P \twoheadrightarrow Q) = \sum_{P \xrightarrow{r} Q_1 \xrightarrow{\infty (p_1)} \dots Q_n \xrightarrow{\infty (p_n)} Q \xrightarrow{\infty}} \left(Pr(P \to Q_1) \times \prod_{i=1}^n p_i \right)$$

Theorem 1. *If $P \xrightarrow{\infty}$ and if no infinite sequence of immediate transitions is reachable from P, then*

- *(Timed correctness)* $\Downarrow P = \Downarrow P$,
- *(Probabilistic correctness)* $Pr(P \twoheadrightarrow Q) = Pr(P \Rightarrow Q)$.

7   Encoding Input Patterns

We now encode SPiCO back into the stochastic π-calculus. The latter can be identified as the special case with a unique function name per arity (we assume arities bounded by some max): $\mathscr{F}' = \{\text{UNIT}_i \mid 0 \le i \le max\}$. In what follows, we write UNIT instead of UNIT_i.

We assume a total ordering $<$ on a finite set of function names $\mathscr{F}$. This means that $\mathscr{F}$ has a unique representation $\mathscr{F} = \{f_1, \dots, f_n\}$ with $f_1 < \dots < f_n$. Our encoding uses channel names from the set $\mathscr{N} \times \mathscr{F}$. We denote elements (x, f) of this set by x_f. For each channel x we define a sequence of n channels $x_{\mathscr{F}}$ as follows: $x_{\mathscr{F}} =_{\text{def}} x_{f_1}, \dots, x_{f_n}$. Channels in the target language are associated a rate (that may be infinite) by means of the encoding of ϱ defined as $[\![\varrho]\!](x_f) = \varrho(x)(f)$. We write $\tilde{x}, \tilde{y}$ for the concatenation of two sequences $\tilde{x}$ and $\tilde{y}$. If $\tilde{x} = x_1, \dots, x_n$ then we let $\tilde{x}_{\mathscr{F}} =_{\text{def}} x_{1\mathscr{F}}, \dots, x_{n\mathscr{F}}$. The encoding is given in Table 6.

The following theorem states the correctness of our encoding. It allows us to run simulations of models expressed in SPiCO, via an implementation of the original stochastic π-calculus, as implemented in the SPiM system [19].

Table 6. Encoding of input patterns

$$[\![\mathbf{new}\ x{:}\rho.\ P]\!] =_{\mathrm{def}} \mathbf{new}\ x_{f_1}{:}\rho(f_1). \cdots .\mathbf{new}\ x_{f_n}{:}\rho(f_n).\ [\![P]\!]$$
$$[\![P_1 \mid P_2]\!] =_{\mathrm{def}} [\![P_1]\!] \mid [\![P_2]\!] \qquad\qquad\qquad [\![A(\tilde{y})]\!] =_{\mathrm{def}} A(\tilde{y}_{\mathscr{F}})$$
$$[\![C_1 + \cdots + C_n]\!] =_{\mathrm{def}} [\![C_1]\!] + \cdots + [\![C_n]\!] \qquad [\![A(\tilde{x}) \triangleq P]\!] =_{\mathrm{def}} A(\tilde{x}_{\mathscr{F}}) \triangleq [\![P]\!]$$
$$[\![x?f(\tilde{y}).P]\!] =_{\mathrm{def}} x_f?(\tilde{y}_{\mathscr{F}}).[\![P]\!] \qquad\qquad [\![x!f(\tilde{y}).P]\!] =_{\mathrm{def}} x_f!(\tilde{y}_{\mathscr{F}}).[\![P]\!]$$

Theorem 2. *The encoding defines a stochastic bisimulation: for all processes P, Q and finite sets of definitions Δ, and all rates $s \in \mathbb{R}^+ \cup \{\infty(p) \mid p \in]0,1]\}$ it holds that $P \xrightarrow{s} Q$ relative to Δ if and only if $[\![P]\!] \xrightarrow{s} [\![Q]\!]$ relative to $[\![\Delta]\!]$.*

The statement $P \xrightarrow{s} Q$ relative to Δ means that there exists some function $\varrho : \mathscr{N} \to \mathscr{F} \to (\mathbb{R}^+ \cup \{\infty\})$ such that $P \xrightarrow{s} Q$ relative to Δ and ϱ. The values $\varrho(x)$ will be the rate ρ assigned to x in the declaration $\mathbf{new}\ x{:}\rho$. It holds for all ρ and x that $\varrho(x) = \rho$ iff $[\![\varrho]\!](x_f) = \rho(f)$ for all $f \in \mathscr{F}$.

The statement $[\![P]\!] \xrightarrow{s} [\![Q]\!]$ relative to $[\![\Delta]\!]$ means that there exists some function $\varrho' : \{x_f \mid f \in \mathscr{F}, x \in \mathscr{N}\} \to (\mathbb{R}^+ \cup \{\infty\})$ such that $[\![P]\!] \xrightarrow{s} [\![Q]\!]$ relative to $[\![\Delta]\!]$ and ϱ'. The situation differs in that there exists only a single function UNIT for all arities. We are a little sloppy in identifying a constant function with its constant value, i.e. $\varrho'(x_f) = \varrho'(x_f)(\text{UNIT})$.

8   Conclusion and Future Work

We presented SPiCO, a novel higher-level modeling language for systems biology. SPiCO provides multi-profile objects with static inheritance. It supports the paradigm of modeling "molecules as concurrent objects". The core of SPiCO is a novel stochastic π-calculus with input patterns. We presented its stochastic semantics in terms of CTMCs and showed how to compile it into the biochemical stochastic π-calculus, so that the semantics is preserved. In future work, we plan to finalize SPiCO's language specification and to provide an implementation.

References

1. Baldamus, M., Parrow, J., Victor, B.: A fully abstract encoding of the π-calculus with data terms. In: Caires, L., Italiano, G.F., Monteiro, L., Palamidessi, C., Yung, M. (eds.) ICALP 2005. LNCS, vol. 3580, pp. 1202–1213. Springer, Heidelberg (2005)
2. Bernardo, M., Donatiello, L., Gorrieri, R.: MPA: A stochastic process algebra. Technical Report UBLCS-94-10, University Bologna (1994)
3. Cardelli, L., calculi, B.: Interactions of biological membranes. In: Database Machine Performance: Modeling Methodologies and Evaluation Strategies. LNCS (LNBI), vol. 3082, pp. 257–278. Springer, Heidelberg (2005)
4. Chabrier-Rivier, N., Fages, F., Soliman, S.: The biochemical abstract machine BioCham. In: CMSB 2004. LNCS, vol. 3082, pp. 172–191. Springer, Heidelberg (2005)

5. Ciocchetta, F., Priami, C.: Biological transactions for quantitative models. In: MeCBIC. ENTCS (2006) (to appear)
6. Duchier, D., Kuttler, C.: Biomolecular agents as multi-behavioural concurrent objects. In: Proc. MTCoord, vol 150 of ENTCS, pp. 31–49 (2005)
7. Gillespie, D.T.: A general method for numerically simulating the stochastic time evolution of coupled chemical reactions. J Comp Phys 22, 403–434 (1976)
8. Hermanns, H. (ed.): Interactive Markov Chains. LNCS, vol. 2428. Springer, Heidelberg (2002)
9. Hillston, J.: A Compositional Approach to Performance Modelling. In: PhD thesis, University of Edinburgh, Cambridge University Press, Cambridge (1996)
10. Hucka, M., et al.: The systems biology markup language (SBML). Bioinformatics 19, 524–531 (2003)
11. Kuttler, C., Niehren, J.: Gene regulation in the pi calculus: Simulating cooperativity at the lambda switch. In: Priami, C., Ingólfsdóttir, A., Mishra, B., Nielson, H.R. (eds.) Transactions on Computational Systems Biology VII. LNCS (LNBI), vol. 4230, pp. 24–55. Springer, Heidelberg (2006)
12. Kuttler, C.: Bacterial transcription and translation in the pi calculus. In: Priami, C., Plotkin, G. (eds.) Transactions on Computational Systems Biology VI. LNCS (LNBI), vol. 4220, pp. 113–149. Springer, Heidelberg (2006)
13. Kuttler, C.: Modeling Bacterial Gene Expression in a Stochastic Pi Calculus with Concurrent Objects. PhD thesis, University Lille 1 (2006)
14. Kuttler, C., Lhoussaine, C., Niehren, J.: A stochastic pi-calculus for concurrent objects. INRIA technical report 6076 (2006)
15. Kwiatkowska, M., Norman, G., Parker, D., Tymchyshyn, O., Heath, J., Gaffney, E.: Simulation and verification for computational modelling of signalling pathways. Winter Simulation Conference (2006) (to appear)
16. Lecca, P., Priami, C., Quaglia, P., Rossi, B., Laudanna, C., Constantin, G.: A stochastic process algebra approach to simulation of autoreactive lymphocyte recruitment. SCS Simulation 80(6), 273–288 (2004)
17. Milner, R., Parrow, J., Walker, D.: A calculus of mobile processes (I and II). Information and Computation 100, 1–77 (1992)
18. Paulino, H., Marques, P., Lopes, L., Vasconcelos, V.T., Silva, F.: A multi-threaded asynchronous language. In: Malyshkin, V. (ed.) PaCT 2003. LNCS, vol. 2763, pp. 316–323. Springer, Heidelberg (2003)
19. Phillips, A., Cardelli, L.: A correct abstract machine for the stochastic pi-calculus. In: Proc. Workshop on Concurrent Models in Molecular Biology (2004)
20. Priami, C.: Stochastic π-calculus. Computer Journal 6, 578–589 (1995)
21. Priami, C., Quaglia, P.: Beta binders for biological interactions. In: Simple Program Schemes and Formal Languages. LNCS (LNBI), vol. 3082, pp. 20–33. Springer, Heidelberg (2005)
22. Priami, C., Regev, A., Shapiro, E., Silverman, W.: Application of a stochastic name-passing calculus to representation and simulation of molecular processes. Information Processing Letters 80, 25–31 (2001)
23. Ravara, A., Vasconcelos, V.T.: Typing non-uniform concurrent objects. In: Palamidessi, C. (ed.) CONCUR 2000. LNCS, vol. 1877, pp. 474–488. Springer, Heidelberg (2000)
24. Regev, A., Panina, E.M., Silverman, W., Cardelli, L., Shapiro, E.: BioAmbients. TCS 325(1), 141–167 (2004)
25. Vasconcelos, V.T., Tokoro, M.: A typing system for a calculus of objects. In: Nishio, S., Yonezawa, A. (eds.) Object Technologies for Advanced Software. LNCS, vol. 742, pp. 460–474. Springer, Heidelberg (1993)

Modeling Static Biological Compartments with Beta-binders

Maria Luisa Guerriero[1], Corrado Priami[1,2], and Alessandro Romanel[1,2]

[1] Dipartimento di Informatica e Telecomunicazioni, Università di Trento, Italy
[2] The Microsoft Research - University of Trento
Centre for Computational and Systems Biology, Italy
guerrier@dit.unitn.it, {priami,romanel}@cosbi.eu

Abstract. We investigate the modeling of biological systems with static compartments through Beta-binders, a recently developed process calculus. Biological entities are represented as bio-processes and the calculus is extended with the notion of compartment. Entities can either be internal to compartments or reside on compartment borders. Movement in and out of compartments is requested by internal objects and mediated by border objects. The extended calculus is equipped with the notion of locality, and various kinds of relations between actions are defined. Moreover, we compare our proposal with similar formalisms and we show how to use the proposed calculus for modeling and analyzing the cAMP-signaling pathway in OSNs.

1 Introduction

Compartments are present in all biological systems: a cell is a compartment, which in turn contains other compartments (the most important of which is the nucleus). Compartments are fundamental for the evolution of biological systems, because they provide a means for isolating their content from the external environment, still allowing some exchange of information, mainly through membrane proteins.

Several languages have been proposed to model biological compartments (e.g. *Brane calculi* [1], *BioAmbients* [2] and *Beta-binders* [3,4]). All of them have some differences in the considered notion of compartment and in the kinds of operations allowed (see Sect. 6 for a discussion). We focus here on Beta-binders, a process calculus with a two level syntax. The main objects, called bio-processes, are boxes with typed interfaces and whose behavior is driven by simplified π-calculus [5] like processes that they enclose.

In Beta-binders the nesting of boxes is not allowed, but typed interfaces ensure that a virtual form of nesting can be represented. Modeling complex hierarchies, however, is quite a difficult task that is simplified by our current proposal. We rely on a general interpretation of bio-processes as structured communicating objects and we propose an extension with the notion of *static compartments*, that permits an intuitive representation of hierarchical structures, still forbidding the explicit nesting of boxes.

H. Anai, K. Horimoto, and T. Kutsia (Eds.): AB 2007, LNCS 4545, pp. 247–261, 2007.
© Springer-Verlag Berlin Heidelberg 2007

The structure of compartments and boxes allows us to consider *spatial* relations between events, e.g. the location where a protein-protein interaction occurs. Therefore, we enrich the calculus with some locality relations, both on compartments and on boxes. We adopt the transition system-based technique used in [6,7] to define a locality relation for the π-calculus. Examples of use of locality relations in modeling biochemical systems are in [8].

In the next section we briefly describe a slight variant of Beta-binders. In Sect. 3 we introduce the notion of compartments and in Sect. 4 we present the labeled semantics of Beta-binders; in Sect. 5 some locality relations are defined. In Sect. 6 we compare our proposal with some existing works, while the application of our proposal to a model of the cAMP-signaling pathway in OSNs is shown in Sect. 7. Finally some concluding remarks are presented.

2   Beta-binders

Beta-binders [3,4] is a bio-inspired process calculus developed to better adhere to the structure and dynamics of biological systems. By introducing the concept of *affinity*, the calculus relaxes the *key-lock* model of interaction, commonly assumed in classical process calculi, and hence it permits us to model more correctly domains and interactions between enzymes and small molecules based on their types and affinities. In Beta-binders, *pi-processes* are encapsulated into *boxes* with interaction capabilities, also called *bio-processes*. Like the π-calculus, Beta-binders is based on the notion of *naming*. Thus, we assume the existence of a countably infinite set $\mathcal{N}$ of names (ranged over by lower-case letters).

The *pi-processes* wrapped into boxes are given by the following context free grammar:

$$P ::= \mathsf{nil} \mid \pi.P \mid P|P \mid !\pi.P$$

$$\pi ::= x(y) \mid \overline{x}\langle y \rangle \mid \mathsf{expose}(x, \Gamma) \mid \mathsf{hide}(x) \mid \mathsf{unhide}(x) \ .$$

The π-calculus syntax is enriched by the last three prefixes for π to manipulate the interaction *sites* of the boxes. The object y in the input prefix $x(y)$ as well as the object x in $\mathsf{expose}(x, \Gamma)$ prefix act as binding occurrences. Hence we can define free fn and bound bn names as usual. Bio-processes are defined as pi-processes prefixed by specialized binders that represent interaction capabilities. An *elementary beta binder* has the form $\beta(x : \Gamma)$ (active) or $\beta^h(x : \Gamma)$ (hidden) where the name x is the subject of the beta binder and Γ represents the type of x. With $\widehat{\beta}$ we denote either β or β^h. A *well-formed* beta binder (ranged over by $\mathbf{B}, \mathbf{B}_1, \mathbf{B}', \cdots$) is a non-empty string of elementary beta binders whose subjects and types are all distinct. The function $\mathsf{sub}(\mathbf{B})$ returns the set of all the beta binder subjects in $\mathbf{B}$. Moreover, $\mathbf{B}^*$ denotes either a well-formed beta binder or the empty string. The function $\alpha(\Delta, \Gamma) \rightarrow \mathbb{R}$ returns the affinity of the types Δ and Γ. Types are any algebraic structure for which there exists a decidable equality procedure. Hereafter, we also assume that substitution is not defined over the elements in a type. Note also that we do not have restriction in pi-processes and ! is guarded. This choice is done to adhere to the implementation of the language developed so far [9], but the general case works perfectly with the following development in this paper.

Bio-processes (ranged over by B, B_1, B', $\cdots$) are generated by the following context free grammar:

$$B ::= \text{ Nil } | \mathbf{B}[\,P\,] \mid B \parallel B.$$

The system is a parallel composition of bio-processes that can be either the deadlock bio-process Nil or the elementary bio-process $\mathbf{B}[\,P\,]$. The semantics of bio-processes is given in [3] in terms of a *reduction relation* ($\longrightarrow$), which uses a *structural congruence relation* ($\equiv$). We postpone the formal definitions of these relations to the next sections. For their standard definitions, see [3].

3 Compartments

The way in which a biological system is modeled here with Beta-binders is as a composition of boxes, where each box represents a biological entity. Although the nesting of boxes is forbidden, the typing for sites and the operational semantics ensures that a virtual form of nesting can be represented [4]. This model might be too abstract, but it has been chosen to keep the formalism as simple as possible.

Consider for example the system

$$S = B \parallel B' \parallel B''$$

where $B = \beta(s : \varDelta_2)[\,s(v)\,.R\,]$, $B' = \beta(x : \varDelta_0)[\,\overline{x}\langle z\rangle\,.\,\text{hide}(x)\,.P\,]$ and $B'' = \beta(y : \varDelta_1)[\,y(v)\,.Q\,]$ and where $\alpha(\varDelta_0, \varDelta_1) > Th \wedge \alpha(\varDelta_j, \varDelta_2) < Th$ with $j \in \{0, 1\}$.[1] The affinity between the types exposed by the boxes could give us an idea of how the boxes are grouped in compartments. In fact, we could imagine that the first and the second boxes are in the same compartment and that the third box is in another one. However, this kind of virtual nesting is ambiguous and for each defined system several different hierarchical structures can be deduced. In fact, all the following three compartmentalization would be valid.

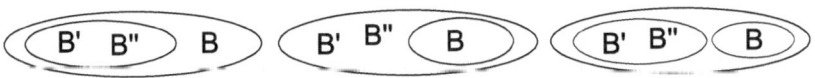

Moreover, consider the movement of objects across compartments. Since types encode compartments, moving an object from a compartment to another one means changing the types of the sites properly, using sequences of hide, unhide and expose operations. As the complexity of the model grows, the number of necessary actions makes this approach not practical and difficult to manage.

---

[1] The value Th represents a context dependent threshold over which two types are considered compatible.

For these reasons we decided to introduce a finer and more explicit notion of compartments. Since we do not want to diverge from the original language, we decided to maintain the representation of systems as parallel composition of bio-processes, enriching them statically with labels acting as unique names which specify their location.

3.1 The Abstraction

Our goal is to provide a simple framework for modeling systems with static compartments and movements of components across compartments. A component is a structured object that can interact with other components through an affinity interaction model. Moreover, the movement of components between compartments is mediated by other components lying on compartment borders. From a biological point of view this can be seen as a system where molecules and complexes can change compartment through interaction with transmembrane proteins. However, since compartmentalization and movement of components across compartments play a critical role in computational systems, our approach can be applied in different contexts and at different levels of abstraction.

3.2 Static Compartment Hierarchy

Consider the system represented in Fig. 1(a). There is an outside compartment (S) that represents the whole system. The system contains three sub-compartments (A, B, C). Moreover, the compartment B contains another compartment (D). The rectangles and the triangles represent the components of the system. In particular, rectangles are components internal to compartments, while triangles are components that reside on compartment borders. We call *i-components* the internal components and *b-components* the border ones. We introduce the distinction between i- and b-components to be closer to real biological systems, in which objects residing on membranes and objects residing inside compartments have different and specific functions.

(a)   (b)

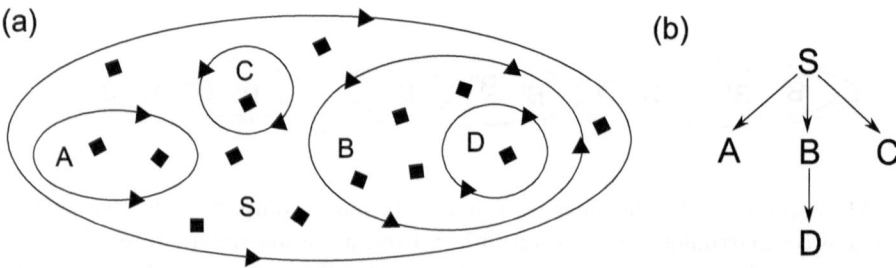

Fig. 1. (a) System with static compartments; (b) The tree representation of the hierarchical structure of the compartments

The static hierarchical structure of the system can be seen as a tree (Fig. 1(b)) where the nodes represent the compartments, and the numbers on the edges represent the numbering of the children.

Instead of using specific labels (e.g. S, A, ...) we can identify each compartment of the system with a sequence of natural numbers representing the position of the compartment inside the tree structure of the system, similarly to the Dewey's indexes. Thus, the compartments of the system in Fig. 1 can be identified with the following sequences:

$$S \to 0 \qquad A \to 0,0 \qquad B \to 0,1 \qquad C \to 0,2 \qquad D \to 0,1,0.$$

Since each component of the system is represented by a bio-process and resides in a particular compartment, we modify the syntax of Beta-binders by labeling bio-processes with the identifier of the compartment in which the bio-process resides. Moreover, to distinguish between components lying inside a compartment and components lying on compartment borders, we add to each bio-process a special marker representing the component type. Formally, the definition of bio-processes is modified as follows:

$$B ::= \mathsf{Nil} \mid \mathbf{B}[\, P\,]_s^\kappa \mid B \parallel B \qquad\qquad \kappa ::= n \mid \kappa, n$$

where $n \in \mathbb{N}$ specifies the position in the static structure and $s \in \{i, b\}$ denotes whether the component is an internal or a border one. As an example, a component lying on the border of the compartment D is represented with a bio-process $B = \mathbf{B}[\, P\,]_b^{0,1,0}$.

3.3   Movements Across Compartments

An i-component can move across a compartment border only through the interaction with a b-component residing on that border. This assumption mimics the role of transmembrane proteins in biological compartments. Since the calculus is based on a binary synchronous communication model, we still use affinity to mediate movement. Therefore, we modify the syntax of Beta-binders by adding the following new complementary prefixes:

$$\pi ::= \cdots \mid \mathsf{move}(x) \mid \mathsf{in}(x) \mid \mathsf{out}(x)$$

where $x \in \mathcal{N}$. The move action synchronizes with in or out actions, thus giving to i-components the ability to move across compartment borders, and b-components the ability to control the flow direction. As an example, consider the system in Fig. 2(a), described in Beta-binders by the bio-process

$$S \;=\; (B_1 = \mathbf{B}_1[\, P_1\,]_i^0) \parallel (B_2 = \mathbf{B}_2[\, P_2\,]_b^{0,0}) \;.$$

Intuitively, B_1 can move into the sub-compartment interacting through a complementary $\mathsf{move}(x)/\mathsf{in}(y)$ action with B_2, where x and y are subjects of binders with affine types. The new configuration of the system (Fig. 2(b)) after the movement of B_1, is described in Beta-binders by the bio-process

$$S' \;=\; (B_1' = \mathbf{B}_1[\, P_1'\,]_i^{0,0}) \parallel (B_2' = \mathbf{B}_2[\, P_2'\,]_b^{0,0}) \;.$$

The detailed semantics is presented in the next section.

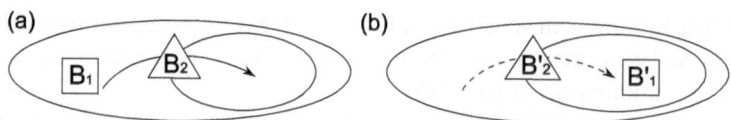

Fig. 2. Example of movement across compartment

4   Labeled Semantics

To introduce locality relations we first enrich the language and its semantics with labels that allow us to uniquely identify the location of bio-processes and compartments.

We define $\vartheta \in \{\|_0, \|_1\}^*$ and we use it to label bio-processes. We statically replace each bio-process $\mathbf{B}[P]_s^\kappa$ with a labeled process $\vartheta\mathbf{B}[P]_s^\kappa$ (where ϑ provides a linear encoding of the syntactical location of the sub-tree of $\mathbf{B}[P]_s^\kappa$ in the syntax tree of the whole system). We chose this approach to take advantage of the syntax of the calculus and ease the implementation of the naming structure. We could have used any unique name generator just to distinguish the locations of bio-processes.

For instance, the bio-process $\beta^h(y : \Sigma)[P_0|P_1]_{s_0}^{\kappa_0} \| \beta(z : \Sigma)[Q_0|Q_1]_{s_1}^{\kappa_1}$ is mapped to $\|_0 \beta^h(y : \Sigma)[P_0|P_1]_{s_0}^{\kappa_0} \| \|_1 \beta(z : \Sigma)[Q_0|Q_1]_{s_1}^{\kappa_1}$.

Each transition is labeled by a pair $\phi = \langle \theta; \kappa \rangle$, where κ is defined as in Sect. 3.2 and θ is defined by the following BNF-like grammar:

$$\theta ::= \vartheta\mu \mid \vartheta\rho \mid \vartheta\langle x(w), \overline{x}\langle z \rangle\rangle \mid \vartheta\langle \|_j \vartheta_0 'x(w)`, \|_{1-j}\vartheta_1 '\overline{y}\langle z \rangle`\rangle \mid$$
$$\vartheta \langle \|_0 \vartheta_0 \mathsf{join}\, P_0, \|_1\vartheta_1\mathsf{join}\, P_1\rangle \mid \vartheta\langle \|_j\vartheta_0\psi, \|_{1-j}\vartheta_1 \,\mathsf{move}(x)\rangle$$

where $\mu ::= a|c|d$ (with $a ::= \mathsf{expose}(x, \Gamma) \mid \mathsf{hide}(x) \mid \mathsf{unhide}(x)$, $c ::= x(w) \mid \overline{x}\langle y \rangle$, and $d ::= 'x(w)` \mid '\overline{x}\langle y \rangle`)$, $\rho ::= \mathsf{split}\langle P_0, P_1\rangle \mid \mathsf{join}\, P$ and $\psi ::= \mathsf{in}(x) \mid \mathsf{out}(x)$.

Table 1. Laws for structural congruence

(a) Pi-processes	(b) Boxes and bio-processes
	$\mathbf{B}[P_1] \equiv \mathbf{B}[P_2]$ provided $P_1 \equiv P_2$
$P_1 \equiv P_2$, provided P_1 α-converse of P_2	$\mathbf{B}^*\hat{\beta}(x : \Gamma)[P] \equiv \mathbf{B}^*\hat{\beta}(y : \Gamma)[P\{y/x\}]$
	provided y fresh in the system
$P_1|(P_2|P_3) \equiv (P_1|P_2)|P_3$	$\mathbf{B}_1\mathbf{B}_2[P] \equiv \mathbf{B}_2\mathbf{B}_1[P]$
	$B_1 \| B_2 \equiv_c B_2 \| B_1$
$P_1|P_2 \equiv P_2|P_1$	$B_1 \| (B_2 \| B_3) \equiv_c (B_1 \| B_2) \| B_3$
	$B \| \mathrm{Nil} \equiv_c B$
$P|\mathrm{nil} \equiv P$	$\vartheta\mathbf{B}_1[P_1]_s^\kappa \equiv_c \vartheta\mathbf{B}_2[P_2]_s^\kappa$
	provided $\mathbf{B}_1[P_1] \equiv \mathbf{B}_2[P_2]$

Table 2. Axioms and rules for the reduction relation

(intra)

$$P \equiv x(w) . P_0 \,\big|\, \overline{x}\langle z\rangle . P_1 \,\big|\, P_2$$

$$\vartheta\mathbf{B}\Big[\,P\,\Big]_s^{\kappa} \xrightarrow{\langle \vartheta\,\langle x(w),\overline{x}\langle z\rangle\rangle;\kappa\rangle} \vartheta\mathbf{B}\Big[\,P_0\{z/w\}\,\big|\,P_1\,\big|\,P_2\,\Big]_s^{\kappa}$$

(inter)

$$P \equiv x(w) . P_1 \,\big|\, P_2 \qquad\qquad Q \equiv \overline{y}\langle z\rangle . Q_1 \,\big|\, Q_2 \qquad , \text{ where}$$

$$X \xrightarrow{\langle \vartheta\,\langle\|_j\vartheta_0\,'x(w)^{\backslash},\|_{1-j}\vartheta_1\,'\overline{y}\langle z\rangle^{\backslash}\rangle;\kappa_l\rangle} Y$$

$$X = \vartheta\|_j\vartheta_0\,\beta(x:\Gamma)\,\mathbf{B}_0^*\Big[\,P\,\Big]_{s_0}^{\kappa_0} \;\big\|\; \vartheta\|_{1-j}\vartheta_1\,\beta(y:\Sigma)\,\mathbf{B}_1^*\Big[\,Q\,\Big]_{s_1}^{\kappa_1},$$

$$Y = \vartheta\|_j\vartheta_0\,\beta(x:\Gamma)\,\mathbf{B}_0^*\Big[\,P_1\{z/w\}\,\big|\,P_2\,\Big]_{s_0}^{\kappa_0} \;\big\|\; \vartheta\|_{1-j}\vartheta_1\,\beta(y:\Sigma)\,\mathbf{B}_1^*\Big[\,Q_1\,\big|\,Q_2\,\Big]_{s_1}^{\kappa_1}$$

provided $\alpha(\Gamma,\Sigma) \geq Th$ and $(\kappa_l = \kappa_{1-l} \vee (\kappa_l = \kappa_{1-l}, n \wedge s_l = \mathrm{b} \wedge s_{1-l} = \mathrm{i}))$

(expose)

$$P \equiv \mathsf{expose}(x,\Gamma) . P_1 \,\big|\, P_2$$

$$\vartheta\mathbf{B}\Big[\,P\,\Big]_s^{\kappa} \xrightarrow{\langle \vartheta\,\mathsf{expose}(x,\Gamma);\kappa\rangle} \vartheta\mathbf{B}\,\beta(y:\Gamma)\,\Big[\,P_1\{y/x\}\,\big|\,P_2\,\Big]_s^{\kappa} \quad , \; y \text{ fresh in the system}$$

(hide)

$$P \equiv \mathsf{hide}(x) . P_1 \,\big|\, P_2$$

$$\vartheta\mathbf{B}^*\,\beta(x:\Gamma)\Big[\,P\,\Big]_s^{\kappa} \xrightarrow{\langle \vartheta\,\mathsf{hide}(x);\kappa\rangle} \vartheta\mathbf{B}^*\,\beta^{\mathrm{h}}(x:\Gamma)\Big[\,P_1\,\big|\,P_2\,\Big]_s^{\kappa}$$

(unhide)

$$P \equiv \mathsf{unhide}(x) . P_1 \,\big|\, P_2$$

$$\vartheta\mathbf{B}^*\,\beta^{\mathrm{h}}(x:\Gamma)\Big[\,P\,\Big]_s^{\kappa} \xrightarrow{\langle \vartheta\,\mathsf{unhide}(x);\kappa\rangle} \vartheta\mathbf{B}^*\,\beta(x:\Gamma)\Big[\,P_1\,\big|\,P_2\,\Big]_s^{\kappa}$$

(join)

$$\vartheta\vartheta_0\mathbf{B}_0\Big[\,P_0\,\Big]_s^{\kappa} \;\big\|\; \vartheta\vartheta_1\mathbf{B}_1\Big[\,P_1\,\Big]_s^{\kappa} \xrightarrow{\langle \theta;\kappa\rangle} \vartheta\vartheta_0\mathbf{B}\Big[\,P_0\sigma_0\,\big|\,P_1\sigma_1\,\Big]_s^{\kappa} \;\big\|\; \vartheta\vartheta_1\,\mathsf{Nil}$$

where $\theta = \vartheta\,\langle \vartheta_0\mathsf{join}\,P_0,\vartheta_1\mathsf{join}\,P_1\rangle$, $\vartheta_0 = \|_0\vartheta_0'$, $\vartheta_1 = \|_1\vartheta_1'$

provided $f_{\mathsf{join}}(\mathbf{B}_0,\mathbf{B}_1,P_0,P_1) = (\mathbf{B},\sigma_0,\sigma_1)$

(split)

$$\vartheta\mathbf{B}\Big[\,P_0\,\big|\,P_1\,\Big]_s^{\kappa} \xrightarrow{\langle \vartheta\,\mathsf{split}\langle P_0,P_1\rangle;\kappa\rangle} \vartheta\|_0\mathbf{B}_0\Big[\,P_0\sigma_0\,\Big]_s^{\kappa} \;\big\|\; \vartheta\|_1\mathbf{B}_1\Big[\,P_1\sigma_1\,\Big]_s^{\kappa}$$

provided $f_{\mathsf{split}}(\mathbf{B},P_0,P_1) = (\mathbf{B}_0,\mathbf{B}_1,\sigma_0,\sigma_1)$

(in)

$$P \equiv \mathsf{in}(x) . P_1 \,\big|\, P_2 \qquad\qquad Q \equiv \mathsf{move}(x) . Q_1 \,\big|\, Q_2 \qquad , \text{ where}$$

$$X \xrightarrow{\langle \vartheta\,\langle\|_j\vartheta_0\,\mathsf{in}(x),\|_{1-j}\vartheta_1\,\mathsf{move}(x)\rangle;\kappa,n\rangle} Y$$

$$X = \vartheta\|_j\vartheta_0\,\beta(x:\Gamma)\,\mathbf{B}_0^*\Big[\,P\,\Big]_{\mathrm{b}}^{\kappa,n} \;\big\|\; \vartheta\|_{1-j}\vartheta_1\,\beta(y:\Sigma)\,\mathbf{B}_1^*\Big[\,Q\,\Big]_j^{\kappa},$$

$$Y = \vartheta\|_j\vartheta_0\,\beta(x:\Gamma)\,\mathbf{B}_0^*\Big[\,P_1\,\big|\,P_2\,\Big]_{\mathrm{b}}^{\kappa,n} \;\big\|\; \vartheta\|_{1-j}\vartheta_1\,\beta(y:\Sigma)\,\mathbf{B}_1^*\Big[\,Q_1\,\big|\,Q_2\,\Big]_{\mathrm{i}}^{\kappa,n}$$

provided $\alpha(\Gamma,\Sigma) \geq Th$

(out)

$$P \equiv \mathsf{out}(x) . P_1 \,\big|\, P_2 \qquad\qquad Q \equiv \mathsf{move}(x) . Q_1 \,\big|\, Q_2 \qquad , \text{ where}$$

$$X \xrightarrow{\langle \vartheta\,\langle\|_j\vartheta_0\,\mathsf{out}(x),\|_{1-j}\vartheta_1\,\mathsf{move}(x)\rangle;\kappa,n\rangle} Y$$

$$X = \vartheta\|_j\vartheta_0\,\beta(x:\Gamma)\,\mathbf{B}_0^*\Big[\,P\,\Big]_{\mathrm{b}}^{\kappa,n} \;\big\|\; \vartheta\|_{1-j}\vartheta_1\,\beta(y:\Sigma)\,\mathbf{B}_1^*\Big[\,Q\,\Big]_j^{\kappa,n},$$

$$Y = \vartheta\|_j\vartheta_0\,\beta(x:\Gamma)\,\mathbf{B}_0^*\Big[\,P_1\,\big|\,P_2\,\Big]_{\mathrm{b}}^{\kappa,n} \;\big\|\; \vartheta\|_{1-j}\vartheta_1\,\beta(y:\Sigma)\,\mathbf{B}_1^*\Big[\,Q_1\,\big|\,Q_2\,\Big]_{\mathrm{i}}^{\kappa}$$

provided $\alpha(\Gamma,\Sigma) \geq Th$

(redex)

$$\frac{B \xrightarrow{\phi} B'}{B \,\big\|\, B'' \xrightarrow{\phi} B' \,\big\|\, B''}$$

(struct)

$$\frac{B_1 \equiv_c R_1' \qquad R_1' \xrightarrow{\phi} B_2}{B_1 \xrightarrow{\phi} B_2}$$

The first pair of labels is used to denote intra-communications (communications within one bio-process), while the second one is used to denote inter-communications (communications between different bio-processes); the third and the fourth ones are used to denote join and movement operations, respectively. Note that the definition of d allows us to distinguish between the input/output actions used in intra-communications ($x(w)$ / $\overline{x}\langle y\rangle$) and the ones used in inter-communications ($'x(w)'$ / $'\overline{x}\langle y\rangle'$).

We introduce two new sets of labels, with metavariable γ and δ respectively, that will be useful in the following:

$$\gamma ::= a \mid \langle c_0, c_1\rangle \qquad\qquad \delta ::= d \mid \psi \mid \mathsf{move}(x) \ .$$

Definition 1. *The* structural congruence *over pi-processes, denoted by* $\equiv$, *is the smallest relation which satisfies the laws in Table 1(a). The* structural congruence *over bio-processes is identified by two relations, denoted respectively by* $\equiv$ *and* $\equiv_c$, *that are the smallest ones which satisfy the laws in Table 1(b).*

Definition 2. *The* reduction relation $\longrightarrow$ *is the smallest relation over bio-processes defined by the axioms and rules in Table 2.*

As in [3], f_{join} and f_{split} functions are user defined λ-calculus functions which describe the aggregation and disaggregation of boxes and depend on the structure of bio-processes.

5   Locality Relations

To define some locality relations on Beta-binders transitions we first need two auxiliary functions for each transition ϕ: $\mathsf{act}(\phi)$ specifies the action executed, and $\mathsf{comp}(\phi)$ specifies the compartment in which the action is executed.

$$\mathsf{act}(\langle\theta, \kappa\rangle) = \theta \qquad\qquad \mathsf{comp}(\langle\theta, \kappa\rangle) = \kappa \ .$$

We first define some relations concerning compartments (i.e. the compartments in which the actions triggering the transitions occur). Finally, we consider the level of bio-processes (i.e. the bio-processes involved in the transitions).

Based on the definition of localities described in [10], compartments are *static* localities: they do not change dynamically during execution, and hence they represent the sites at which events occur. Therefore, the relations introduced in the next section refer to the relative positions of the considered transitions. ϑ labels of bio-processes are, instead, *dynamical* localities: in fact they are built incrementally when actions are performed (actually, only split operations modify the labels by adding sublabels to the labels of the created bio-processes), and hence they represent, for each action, the ones that locally precede it.

5.1   Compartments Locality Relations

In this section a set of significant locality relations between transitions is defined, assuming a computation $B_0 \xrightarrow{\phi_0} B_1 \xrightarrow{\phi_1} \cdots \xrightarrow{\phi_n} B_{n+1}$, according to the relative position of the compartments in which the transitions occur. We end this section by discussing how our relations can be useful in biological systems.

Definition 3 (Same-compartment relation). *We say that ϕ_n has a same-compartment dependency on ϕ_h (denoted with $\phi_h \asymp \phi_n$) if $h < n$ and $\mathsf{comp}(\phi_h) = \mathsf{comp}(\phi_n)$.*

With this relation we underline the fact that the two actions ϕ_h and ϕ_n occur in the same compartment.

Definition 4 (Son-father relation). *We say that ϕ_n has a son-father dependency on ϕ_h (denoted with $\phi_h \curlywedge \phi_n$) if $h < n$ and $(\mathsf{comp}(\phi_n), m) = \mathsf{comp}(\phi_h)$ ($m \in \mathbb{N}$).*

This means that the compartment in which the action ϕ_h occurs is the son of the compartment in which the action ϕ_n occurs.

Definition 5 (Father-son relation). *We say that ϕ_n has a father-son dependency on ϕ_h (denoted with $\phi_h \curlyvee \phi_n$) if $h < n$ and $(\mathsf{comp}(\phi_h), m) = \mathsf{comp}(\phi_n)$ ($m \in \mathbb{N}$).*

In this case, the compartment in which the action ϕ_h occurs is the father of the compartment in which the action ϕ_n occurs.

Son-father and *father-son* relations can be easily generalized to *sub-compartment* and *super-compartment* relations respectively, by considering their transitive closures.

Definition 6 (Sub-compartment and super-compartment relations).
Let $\overline{\curlywedge} \triangleq (\curlywedge)^*$ be the transitive closure of $\curlywedge$. We say that ϕ_n has a sub-compartment dependency on ϕ_h if $\phi_h \overline{\curlywedge} \phi_n$.
Let $\underline{\curlyvee} \triangleq (\curlyvee)^*$ be the transitive closure of $\curlyvee$. We say that ϕ_n has a super-compartment dependency on ϕ_h if $\phi_h \underline{\curlyvee} \phi_n$.

The relation $\overline{\curlywedge}$ means that the compartment in which the action ϕ_h occurs is a sub-compartment of the compartment in which the action ϕ_n occurs, and $\underline{\curlyvee}$ means that the compartment in which the action ϕ_n occurs is a sub-compartment of the compartment in which the action ϕ_h occurs.

Note that $\phi_h \overline{\curlywedge} \phi_n \Rightarrow h < n$ and $(\mathsf{comp}(\phi_n), \kappa) = \mathsf{comp}(\phi_h)$, and that $\phi_h \underline{\curlyvee} \phi_n \Rightarrow h < n$ and $(\mathsf{comp}(\phi_h), \kappa) = \mathsf{comp}(\phi_n)$.

In the area of dynamical modeling of biological systems, locality relations can be useful for analyzing the spatial distribution of entities. For example, when observing transitions originated by an interesting event, we could be interested in investigating what happened in the same compartment previously. The relation $\asymp$ can be used for that. Similarly, we could use the other relations to study what happened in super- or sub-compartments.

5.2   Inter-box Locality Relation

In this section we define the inter-box locality relation between pairs of transitions in a computation: an inter-box locality relation exists between an activity

A and an activity B, if A and B are executed by pi-processes in the same bio-process. Our labels can be used as unique names for the transitions as they are linearizations encoding the position of the prefixes and processes originating the transitions in the syntax tree.

Definition 7 (Direct inter-box locality relation). *Given a computation* $B_0 \xrightarrow{\phi_0} B_1 \xrightarrow{\phi_1} \cdots \xrightarrow{\phi_n} B_{n+1}$, *we say that* ϕ_n *has a direct inter-box locality dependency on* ϕ_h *if* $h < n$ *and* $\mathsf{act}(\phi_h) \lessdot \mathsf{act}(\phi_n)$ *can be derived by repeated applications of the following rules, where* $j \in \{0, 1\}$.

1. $\|_j \theta \lessdot \|_j \theta'$ *if* $\theta \lessdot \theta'$
2. $\gamma \lessdot \gamma'$
3. $\langle \|_j \vartheta_0 \delta_0, \|_{1-j} \vartheta_1 \delta_1 \rangle \lessdot \langle \|_l \vartheta'_0 \delta'_0, \|_{1-l} \vartheta'_1 \delta'_1 \rangle$
   if $((\|_j \vartheta_0 = \|_l \vartheta'_0 \wedge \|_{1-j} \vartheta_1 = \|_{1-l} \vartheta'_1) \vee (\|_j \vartheta_0 = \|_{1-l} \vartheta'_1 \wedge \|_{1-j} \vartheta_1 = \|_l \vartheta'_0))$.

The rules listed above are applied recursively to a pair of actions θ_h, θ_n in order to verify whether there is an inter-box locality dependency between them. Since inter-box locality only concerns the bio-processes (and not their internal structure), the recursive step is implemented by removing the common prefixes of θ_h and θ_n through rule 1, as long as they are relative to the labels of bio-processes ($\|_0$ and $\|_1$). Then, at the end of the recursive steps (i.e. if θ_h and θ_n refer to the same bio-process), either rule 2 or rule 3 could be applied. Rule 2 states that actions on beta binders (i.e. expose, hide and unhide) and communications have an inter-box locality dependence on other actions on beta binders and other communications executed by pi-processes in the same bio-process. Rule 3 states that inter-communications and transport operations have an inter-box locality dependence on other inter-communications and other transport operations if both operations are between the same bio-processes (i.e. each partner of one operation is executed by the same bio-process of one of the partners of the other operation).

Note that our mechanism is not affected by the associativity and commutativity of $\|$, because the ϑ labels are attached statically to processes and updated in the operational semantics by the rules that affect the structure of the system.

The definition of the inter-box locality relation between two transitions of a computation is obtained by taking into account the transitive closure of the direct inter-box locality relation.

Definition 8 (Inter-box locality relation). *Let* $< \triangleq (\lessdot)^*$ *be the transitive closure of* $\lessdot$. *Then, given a computation* $B_0 \xrightarrow{\phi_0} B_1 \xrightarrow{\phi_1} \cdots \xrightarrow{\phi_n} B_{n+1}$, *we say that* ϕ_n *has an inter-box locality dependency on* ϕ_h *if* $\mathsf{act}(\phi_h) < \mathsf{act}(\phi_n)$.

From a biological point of view, when observing an action performed by a bio-process, we could be interested in investigating other actions performed previously by the same bio-process. The relation $<$ can be used for that. Inter-box relation, together with compartments locality relations, can be useful for analyzing the spatial distribution of the actions executed by a bio-process.

6   Related Works

Several languages have been proposed to model biological compartments; the most common ones are *Brane calculi* [1] and *BioAmbients calculus* [2].

Differently from these calculi, the main aim of our work is to represent static compartments and movements of objects across them; hence, we give merely an informal comparison on the usability of the languages with respect to different biological domains.

6.1   Brane Calculi

The main feature of *Brane Calculi* is that membranes are considered active elements, and hence the whole computation happens *on* membranes: they can move, merge, split, enter in and exit from other membranes. A system is represented as a set of nested membranes, and a membrane as a set of actions; actions carry out the mentioned membrane transformations. The main events that can be directly modeled are phagocytosis (the engulfment of a membrane by another one) and exocytosis (the expulsion of a membrane by another one). Moreover, operations such as mitosis (the splitting of one membrane in two membranes) and mating (the merging of two membranes) can also be described. On-membrane and cross-membrane communications can also be modeled.

Being Brane Calculi primarily concerned on membrane interaction, it permits to easily model membrane operations; on the other hand, it does not take the internal structure of membrane-bound compartments into account, therefore it is not easy to describe events such as protein activation, phosphorylation, etc. Beta-binders, instead, is primarily focused on interaction between internal processes, hence compartments are used to describe the relative positions of the interacting bio-processes and to forbid interactions between processes which are in different compartments; hence, compartments (i.e. compartmental membranes) are static containers (it is not possible to create, destroy, or merge them) and bio-processes (i.e. proteins) can move across their borders. Therefore, operations involving membrane fusion, such as phagocytosis, exocytosis, mitosis and mating, cannot be modeled in Beta-binders by operations on compartments. For example, if compartments represent cells, it is not possible to merge compartments to model cell mating; however, we point out that it is sufficient to change the level of abstraction, i.e. to represent cells with bio-processes and use f_{join} and f_{split} functions to model such operations. Events that are not directly related to cellular membranes (e.g. phosphorylation) can easily be modeled by standard Beta-binders communications and operations on bio-processes interfaces.

Finally, in Brane Calculi everything is interpreted as a membrane, which means that membrane-bound cellular compartments (e.g. cells and organelles) and molecular compartments (e.g. proteins) are modeled in the same way: this seems strange from a conceptual point of view. The proposed Beta-binders extension, instead, provides a double layer of compartmentalization (bio-processes and compartments), which permits a clear distinction between the two compartments types: when modeling cellular processes, cellular compartments are represented by compartments, while proteins are represented by bio-processes.

6.2   BioAmbients

BioAmbients calculus is an extension of the work described in [11], enriched with a concept of compartments similar to the one of *Ambient calculus* [12]. A system is represented as a set of nested ambients, and an ambient is a bounded compartment containing processes whose actions specify the evolution of the system. Ambients can enter in and exit from other ambients (phagocytosis and exocytosis) and they can merge together (mating). π-calculus-style communications can occur within an ambient, between sibling ambients, and between father-child ambients.

Similarly to membranes in Brane calculi, ambients are used to represent both membrane-bound cellular compartments and molecular compartments (proteins and protein complexes). Moreover, BioAmbients does not provide an explicit way to model membrane proteins (they are implicitly considered by the primitives through which an ambient can allow another one to enter, exit or merge with). Hence, it is not easy to model complex interactions between membrane proteins and internal proteins: the movement of an ambient in or out of another one is obtained by complementary actions executed by the two ambients. For example, there is no way to describe the expulsion of a molecule (an ambient "m") from a cell (an ambient "c" containing "m"), mediated by a membrane protein (an ambient "p" lying inside "c"). This, instead, can be easily done in Beta-binders. Finally, in BioAmbients it is not possible to move processes which are not lying in some ambient; hence, in order to describe the movement of small molecules across cellular membranes, they need to be enclosed within an ambient.

As previously stated, operations involving membrane fusion cannot be modeled in Beta-binders by operations on compartments (though they can be modeled by operations on bio-processes). In BioAmbients, instead, it is easy to model them (except mitosis, whose description in BioAmbients is not straightforward).

Entities in the same compartment can interact in BioAmbients though a *local* communication on a channel. In Beta-binders, interaction is also done through inter-communications if the interfaces of the two entities are compatible.

Finally, it is not easy to model in BioAmbients events such as protein activations (in particular multi-step chains of proteins activations), whereas this can easily be done in the proposed Beta-binders extension (as shown in the example described in the following section).

7   Example: The cAMP-Signaling Pathway in OSNs

In this section we present how the extended Beta-binders formalism can be used for modeling a biological system that includes membranes and membrane proteins. In particular, we model the cAMP-signaling pathway in olfactory sensory neurons (OSNs). The pathway describes how G protein-coupled receptors indirectly modulate the activity of ion channels via the action of second messengers (Fig. 3).

An odorant ligand O can bind with an odorant receptor OR through the reversible reaction $r1$, activating it. The active OR stimulates ($r2$) the G-protein GDP$\alpha\beta\gamma$ (denoted with GDPabg), causing the dissociation of the trimer in two

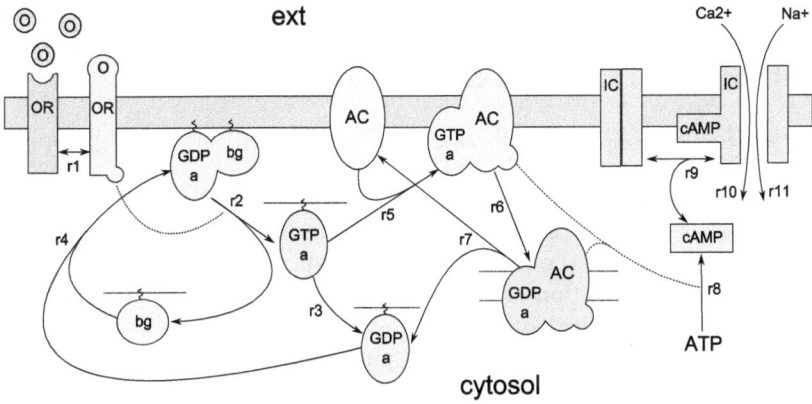

Fig. 3. The cAMP-signaling pathway in OSNs

active subunits GTPa and bg. At this point, GTPa can either hydrolyze ($r3$), returning GDPa, or activate the adenylyl cyclase (AC), his target protein ($r5$). If the reaction $r3$ takes place, the subunit GDPa reassociates with the subunit bg ($r4$). If, instead, the reaction $r5$ takes place, the activation of AC produces, through the synthesis of ATP ($r8$), an increase in the concentration of the second messenger cAMP. A cAMP molecule can open, through a reversible binding ($r9$), the ion-channel IC, allowing Na^+ and Ca^{2+} molecules to enter. However, the hydrolysis of GTP to GDP causes GTPa to dissociate from AC and reassociate with bg. For a more detailed description of the pathway we refer the reader to [13].

Table 3. Specification of the model

$O = \overline{x}\langle z\rangle . x(z) . O$

$OR = x(z) . \text{unhide}(y) . \text{hide}(y) . \overline{x}\langle z\rangle . OR$

$A = \overline{y}\langle z\rangle . A$

$GDP = x(z) . GTP$

$GTP = y(z) . GDP$

$P_\alpha = \overline{y}\langle z\rangle . P_\alpha$

$AC = y(z) . AC$

$cAMP = \overline{x}\langle z\rangle . x(z) . cAMP$

$IC = x(z) . \text{unhide}(y) . \text{hide}(y) . \overline{x}\langle z\rangle . IC$

$M = \text{in}(y) . M$

$Na^+ = \text{move}(x) . Na^+$

$Ca^{2+} = \text{move}(x) . Ca^{2+}$

$B_O = \|_0\, \beta(x : \Delta_0)[O]_i^0$

$B_{OR} = \|_1\|_0\, \beta(x : \Delta_0)\, \beta^h(y : \Delta_1)[OR|A]_b^{0,0}$

$B_G = \|_1^2\|_0\, \beta(x : \Delta_1)[GDP|P_\alpha|P_{\beta\gamma}]_h^{0,0}$

$B_{AC} = \|_1^3\|_0\, \beta^h(y : \Delta_2)[AC]_b^{0,0}$

$B_{ATP} = \|_1^4\|_0\, \beta(x : \Delta_3)\, \beta(y : \Delta_2)[\overline{y}\langle z\rangle . cAMP]_i^{0,0}$

$B_{IC} = \|_1^5\|_0\, \beta(x : \Delta_3)\, \beta^h(y : \Delta_4)(IC|M]_b^{0,0}$

$B_{Na^+} = \|_1^6\|_0\, \beta(x : \Delta_5)[Na^+]_i^0$

$B_{Ca^{2+}} = \|_1^i\, \beta(x : \Delta_6)[Ca^{2+}]_i^0$

$f_{\text{split}_G}(B, P_0, P_1) =$
if$(B[P_0|P_1] \equiv \beta(x : \Delta_1)[(GTP|P_\alpha)|P_{\beta\gamma}])$
  then$(\beta^h(x : \Delta_1), \beta^h(x : \Delta_1), id, id)$
  else $\bot$

$f_{\text{split}_{AC}}(B, P_0, P_1) =$
if$(B[P_0|P_1] \equiv \beta^h(x : \Delta_1)\, \beta(y : \Delta_2)[(GDP|P_\alpha)|AC]$
  then$(\beta^h(x : \Delta_1), \beta^h(y : \Delta_2), id, id)$
  else $\bot$

$f_{\text{join}_G}(B_0, B_1, P_0, P_1) =$
if$(B_0[P_0] \equiv \beta^h(x : \Delta_1)[GDP|P_\alpha] \wedge$
  $B_1[P_1] \equiv \beta^h(x : \Delta_1)[P_{\beta\gamma}])$
  then$(\beta(x : \Delta_1), id, id)$
  else $\bot$

$f_{\text{join}_{AC}}(B_0, B_1, P_0, P_1) =$
if$(B_0[P_0] \equiv \beta^h(x : \Delta_1)[GTP|P_\alpha] \wedge$
  $B_1[P_1] \equiv \beta^h(y : \Delta_2)[AC])$
  then$(\beta^h(x : \Delta_1)\, \beta(y : \Delta_2), id, id)$
  else $\bot$

Table 3 shows the specification of the Beta-binders model of the presented pathway.[2] Moreover, $\alpha(\Delta_k, \Delta_j) > Th$ iff $(k = j) \vee (k = 4 \wedge j \in \{5, 6\})$. The parallel composition of all the defined bio-processes

$$S = B_O \parallel B_{OR} \parallel B_G \parallel B_{AC} \parallel B_{ATP} \parallel B_{IC} \parallel B_{Na^+} \parallel B_{Ca^{2+}}$$

represents the initial configuration of the system, denoted with S. All the communications enabled by the bio-processes represent the reactions $r1, \cdots, r11$ shown in Fig. 3, and all the intermediate configurations that the system S can reach through the execution of communications represent all the possible configurations of the biological system.

Now consider one of the possible computations, in which, starting from the initial configuration S, the ion-channel IC is activated, causing the entrance of a Ca^{2+} molecule:

$\phi_1 :$ $\langle\langle(\|_1\|_0 \, 'x(z)`, \|_0 \, '\overline{x}(z)`); 0\rangle$ $\phi_6 :$ $\langle\|_1^2\langle\|_0^2 \, 'y(z)`, \|_1\|_0 \, '\overline{y}(z)`\rangle; 0, 0\rangle$

$\phi_2 :$ $\langle\|_1\|_0 \, \text{unhide}(y); 0, 0\rangle$ $\phi_7 :$ $\langle\|_1^4\langle\|_0 \, '\overline{x}(z)`, \|_1\|_0 \, 'x(z)`\rangle; 0, 0\rangle$

$\phi_3 :$ $\langle\|_1\langle\|_0 \, '\overline{y}(z)`, \|_1\|_0 \, 'x(z)`\rangle; 0, 0\rangle$ $\phi_8 :$ $\langle\|_1^5\|_0 \, \text{unhide}(y); 0, 0\rangle$

$\phi_4 :$ $\langle\|_1^2\|_0 \, \text{split}\langle \text{GTP} \mid P_\alpha, P_{\beta\gamma}\rangle; 0, 0\rangle$ $\phi_9 :$ $\langle\|_1^5\langle\|_0 \, \text{in}(y), \|_1^2 \, \text{move}(x)\rangle; 0, 0\rangle$.

$\phi_5 :$ $\langle\|_1^2\langle\|_0^2\text{join GTP} \mid P_\alpha, \|_1\|_0\text{join AC}\rangle; 0, 0\rangle$

By analyzing the computation with the locality relations previously defined, we can observe, for example, that ϕ_4 has a *father-son* dependency on ϕ_1 (denoted with $\phi_1 \curlyvee \phi_4$), while ϕ_6 has a *same-compartment* dependency on ϕ_9 (denoted with $\phi_6 \asymp \phi_9$).

From a biological point of view, these relations state that a spatial relation $\curlyvee$ and a spatial relation $\asymp$ exist, respectively, between the dissociation of the G-protein $GTP\alpha\beta\gamma$ and the activation of the receptor OR, and between the entrance of a Ca^{2+} molecule and the activation of the target protein AC.

8   Conclusions and Further Work

We extend Beta-binders with compartments and localities. Since the nesting of boxes is forbidden, modeling hierarchies of compartments in the standard version of the calculus is not primitive. We overcome this limitation by introducing the concept of static compartments. Compartments allow the representation of operations such as the movement of objects across compartment borders and the communication between internal and border objects.

Finally, some locality definitions have been introduced (both on compartments and on bio-processes/boxes) which can be useful when studying the spatial distributions of objects (and events) in complex systems.

---

[2] With $\|_k^n$ we indicate the sequence $\underbrace{\|_k \cdots \|_k}_{n}$.

Many further extensions are possible. One is to differentiate the various types of border objects: transmembrane proteins and those lying on the internal/external side of the membrane. The definition of the calculus should be slightly modified in order to take those differences into account. In addition to this, most proteins cannot move freely on the membrane, hence interactions between proteins on the same membrane are not always possible: the position of the proteins is important. This could be considered by introducing additional constraints, based on the position of bio-processes, which permits interactions only between "near" (according to some definition of distance) bio-processes.

A simulator for the extended calculus is under development. This will allow us to test our framework on large scale biological problems.

References

1. Cardelli, L.: Brane Calculi - Interactions of Biological Membranes. In: Danos, V., Schachter, V. (eds.) CMSB 2004. LNCS (LNBI), vol. 3082, pp. 257–278. Springer, Heidelberg (2005)
2. Regev, A., Panina, E.M., Silverman, W., Cardelli, L., Shapiro, E.Y.: BioAmbients: an Abstraction for Biological Compartments. Theoretical Computer Science 325(1), 141–167 (2004)
3. Priami, C., Quaglia, P.: Beta binders for biological interactions. In: Danos, V., Schachter, V. (eds.) CMSB 2004. LNCS (LNBI), vol. 3082, pp. 20–33. Springer, Heidelberg (2005)
4. Priami, C., Quaglia, P.: Operational patterns in Beta-binders. Transactions on Computational Systems Biology 1, 50–65 (2005)
5. Milner, R.: Communicating and mobile systems: the π-calculus. Cambridge Universtity Press, Cambridge (1999)
6. Degano, P., Priami, C.: Non-interleaving semantics for mobile processes. Theoretical Computer Science 216(1–2), 237–270 (1999)
7. Degano, P., Gadducci, F., Priami, C.: Causality and Replication in Concurrent Processes. In: Broy, M., Zamulin, A.V. (eds.) PSI 2003. LNCS, vol. 2890, pp. 307–318. Springer, Heidelberg (2004)
8. Curti, M., Degano, P., Priami, C., Baldari, C.T.: Modelling biochemical pathways through enhanced π-calculus. Theoretical Computer Science 325(1), 111–140 (2004)
9. Romanel, A., Dematté, L., Priami, C.: The Beta Workbench. Technical report TR-03-2007, The Microsoft Research - University of Trento Centre for Computational and Systems Biology (2007)
10. Bergstra, J.A., Ponse, A., Smolka, S.A. (eds.): Handbook of Process Algebra. Elsevier, Amsterdam (2001)
11. Regev, A., Silverman, W., Shapiro, E.: Representation and simulation of biochemical processes using the π-calculus process algebra. In: Proceedings of Pacific Symposium on Biocomputing (PSB'01), vol. 6, pp. 459–470 (2001)
12. Cardelli, L., Gordon, A.D.: Mobile Ambients. In: Nivat, M. (ed.) ETAPS 1998 and FOSSACS 1998. LNCS, vol. 1378, pp. 140–155. Springer, Heidelberg (1998)
13. Alberts, B., Johnson, A., Lewis, J., Raff, M., Roberts, K., Walter, P.: Molecular biology of the cell. Garland Science (2002)

Deducing Interactions in Partially Unspecified Biological Systems*

Paolo Baldan[1], Andrea Bracciali[2], Linda Brodo[3], and Roberto Bruni[2]

[1] Dipartimento di Matematica Pura e Applicata, Università di Padova, Italia
[2] Dipartimento di Informatica, Università di Pisa, Italia
[3] Dipartimento di Scienze dei Linguaggi, Università di Sassari, Italia
baldan@math.unipd.it, {braccia,bruni}@di.unipi.it, brodo@uniss.it

Abstract. We show how a symbolic approach to the semantics of process algebras can be fruitfully applied to the modeling and analysis of partially unspecified
biological systems, i.e., systems whose components are not fully known, cannot
be described entirely, or whose functioning is not completely understood. This
adds a novel deductive perspective to the use of process algebras within systems
biology: the investigation of the behavioural or structural properties that unspecified components must satisfy to interact within the system. These can be computationally inferred, extending the effectiveness of the in silico experiments. The
use of the approach is illustrated by means of case studies.

1 Introduction and Motivations

The convergence of mathematical, technical and natural sciences yields multidisci-
plinary approaches that can help in better understanding biological phenomena. The
formal modeling of such phenomena has recently gained a lot of attention, see e.g.
[31,34,20,21,17]. Among these approaches, *process algebras* provide expressive descriptions, enjoy friendly syntax, compositionality and generally support software simulation. To some extent, they appear as easily accessible formalisms, particularly suited
for such interdisciplinary research, that favor cross-fertilisation between the two fields:
existing calculi have been sometimes applied rather directly, like in the case of stochastic semantics for the Pi-calculus [27,29], while in other cases new language primitives
have been specifically designed to capture molecular and biological interaction, like the
explicit treatment of membrane nesting [30], membrane activity [7], probability-based
reactions [25], active sites in a protein [11] and structure-determined reactions [28,12].
These linguistic abstractions are generally complemented with suitable formal semantics that may describe system behaviour both qualitatively and quantitatively, e.g., in
terms of happening reactions and their dynamic constants (i.e., stochastic semantics
[26,32] based on Gillespie's algorithm [16]). Often, executable counterparts are provided so that system properties can be both assessed theoretically and verified by means
of *in silico* simulations. Encouraging results, e.g. in terms of the coherence between *in
silico* and *in vitro* experiments, have been obtained [22,4,10].

---

* This work has been partially supported by the MIUR project *Bisca*.

H. Anai, K. Horimoto, and T. Kutsia (Eds.): AB 2007, LNCS 4545, pp. 262–276, 2007.

Within this line of research, that exploits the analogy between biological and software systems, we address the problem of dealing with qualitative analysis of *open systems*. In the context of computer science open systems account for components that are not fully specified or may dynamically join the system at a later stage, such as applications that access services on the network, or proprietary software components. In the biological setting open systems may play the part of not fully understood cellular and chemical compounds.

Here we apply *symbolic transition systems* [2,3], originally developed for software open systems, to the modeling and analysis of biological systems. We aim to show that: *(i)* the symbolic model adds a deductive dimension to the *in silico* experiments, allowing one *to derive the (most general) features that unknown components exhibit when interacting within a system*, and *(ii)* the framework is *language-independent*, in the sense that it applies to a variety of different modeling problems at different levels of abstraction.

The main ingredients of the approach are an algebraic syntax, an operational semantics in terms of suitable labelled transitions and the possibility to deal uniformly with open and closed systems. The approach will be exemplified on a few case studies, related to different levels of abstractions, granularity, and aspects of interest.

Next, we give an informal account of the modeling of a small scenario comprising a virus v and a cell c. Take the (closed) system

$$E = v[in\ c.\ rna] \mid c[open\ v.\ (prot \mid rna^\perp)] \tag{1}$$

On the biological side, terms like $v[\ldots]$ can be understood as membraned components, while action prefixes like *in c* model reaction capabilities. The name of the membrane identifies the kind of the component. For instance, the virus v is ready to sequentially execute the actions *in c* and *rna*, modeling, respectively, the capability to enter a cell of kind c and then to communicate some RNA information. The cell c "reacts" and opens the membrane of v (action *open v*, i.e., the system comprises some sort of location-awareness by means of membrane names).

The evolution of the system is modeled via labeled transitions from one configuration to the next. The labels record events that are visible to an external observer. The special label τ is used when the corresponding event is an internal reaction, transparent to the outside. In our example, the virus can enter the cell, the membrane v is opened and the RNA interaction takes place: the compound $prot \mid rna^\perp$ interacts, without further consequences in this example, with the virus RNA by means of the complementary action $rna^\perp$ (*prot* information is disregarded by the virus).

$$E \ \to_\tau c[v[rna] \mid open\ v.(prot \mid rna^\perp)] \ \to_\tau c[rna \mid prot \mid rna^\perp] \ \to_\tau c[prot] \tag{2}$$

The infinite set of transitions relative to all the terms of the calculus can be finitely specified by a set of structured operational semantics (sos) rules. For instance, all the transitions about a membraned component $m[in\ n.Q \mid R]$ entering the membrane $n[P]$, or a component $open\ n.Q$ destroying the membrane of $n[P]$ are respectively modelled by rules *(in)* and *(open)* in Fig. 1, valid for all m, n, P, Q, R. Analogously, if any two components P_1 and P_2 can exhibit complementary actions α and $\alpha^\perp$, then by rule *(comm)* their reaction generates a τ transition.

Imagine that the content of the virus cannot be fully characterised, e.g., because not fully understood. In this case we regard E as an *environment*: an open biological system

$$m[in\ n.Q \mid R] \mid n[P] \to_\tau n[m[Q \mid R] \mid P] \quad (in) \qquad \overline{n[P] \mid open\ n.Q \to_\tau P \mid Q} \quad (open)$$

$$\frac{P_1 \to_\alpha Q_1 \quad P_2 \to_{\alpha^\perp} Q_2}{P_1 \mid P_2 \to_\tau Q_1 \mid Q_2} \quad (comm)$$

Fig. 1. Rules for membraned components

modeled as a term with place-holders X, whose unknown components could be disclosed only dynamically (e.g. when they react to certain stimuli) or where *components* (i.e., closed systems) or other sub-environments can be dynamically plugged in. That is

$$E[X] = v[\ X\ ] \mid c[open\ v.(prot \mid rna^\perp)] \qquad (3)$$

One possibility is to study the closures $E[p]$ of $E[X]$ w.r.t. all the possible closed components p. When simulation is attempted *in silico*, then infinitely many p must be considered. Moreover, conceptually, this approach prevents the dynamic disclosure of environments to be considered, since they are fully exposed at the beginning.

Symbolic transition systems (STS) allow environments as states and logic formulae as transition labels. They exploit the idea that the behaviour of $E[X]$ depends on the applicable semantic rules, which can be partly determined by means of the known structure of $E[X]$ itself, and may, in turn, impose a requirement over X in order make the rule applicable. The formulae of STS transitions, which annotate unknown components with their relevant behavioural or structural requirements, can be composed throughout an execution trace of the environment and represent the "inferred" constraints that an unknown component must fulfill to drive the system to a given state. This allows us to attack problems like predicting the environmental conditions that let a virus reproduce. For instance, the open system $E[X]$ can evolve via suitable "abstractions" of the transitions in (2) for the closed E:

$$E[X] \overset{in\ c.Y\mid Z}{\longrightarrow}_\tau c[v[Y \mid Z] \mid open\ v.(prot \mid rna^\perp)] \overset{Y,Z}{\longrightarrow}_\tau c[Y \mid Z \mid prot \mid rna^\perp] \overset{\diamond rna W,Z}{\longrightarrow}_c [W \mid Z \mid prot]$$

$$(4)$$

The first one exhibits the formula *in* $c.Y \mid Z$: the unspecified component X should "at least" be able to perform *in c* (hence X must "know" c) and then behave as $Y \mid Z$, as required by rule (*in*). The second one imposes no constraints since the environment evolves autonomously, the third one requires on Y the capability to interact by means of *rna*.

Composition of formulae is relevant for the analysis of the evolution of partially specified bio-environments. Indeed, the formula *in* $c.(\diamond rna.W \mid Z)$, obtained by composing the formulae along the execution trace, generalises the capabilities required to X in order to carry on the overall interaction within the environment (and it is satisfied by the component $p = in\ c.rna \equiv in\ c.(rna.0 \mid 0)$ that instantiates $E[X]$ to $E[p] = E$).

Synopsis. In § 2 we recall the basics of STS. The framework for the analysis of bio-processes is illustrated in § 3 by discussing, in two examples, how it can be used to reason with incomplete information. The first example is based on an original formalisation of the life cycle of the λ-phage virus in BioAmbients. The second example deals with a model of viral cell infection, originally from [1] and used in [7] to introduce Brane Calculi. Concluding remarks and future perspectives are in § 4.

2   Symbolic Operational Semantics

This section recalls the key definitions about symbolic transition systems (see [2,3] for a more comprehensive formal presentation).

Definition 1 (Symbolic Transition System). *A symbolic transition system (STS) S is a set of transitions*

$$C[X_1,\ldots,X_n] \xrightarrow{(\varphi_1,\ldots,\varphi_n)}_a D[Y_1,\ldots,Y_m]$$

where $C[X]$ and $D[Y]$ are environments, a an action label and φ_i are formulae over variables $\{Y_1,\ldots,Y_m\}$ (in a suitable logic, as defined below).

Informally, a symbolic transition represents the fact that the environment $C[X]$ can exhibit an action a and evolve to $D[Y]$ whenever the holes X are filled with any components satisfying φ. The label φ should encode the "least necessary" conditions that components should fulfill for properly taking part to the transition. For the sake of this presentation, following [2], we exploit the STS logic SL, defined below, with action and structural modalities in the style of the ambient logic [9]. However, different choices are conceivable, depending on the calculus of interest.

Spatial modalities emerge when, in order to perform a transition, an environment $E[X]$ must match the left-hand side of the conclusion of a rule. This may require a certain structure to the components that may possibly be plugged in, hence requiring the constructors of the calculus, like $\_ \mid \_$ or $n[\_]$, to appear as terms of the logic, which we call *spatial operators*. Furthermore, the premises of the matched rule must be satisfiable. Such premises may typically require each plugged component to be able to exhibit some behaviour, as in rule (*comm*). Hence, the logic also includes modal operators $\diamond a \_$ expressing the capability to perform an action a. A formula which does not impose any constraint on the component is represented as a logical variable X, called the *residual placeholder*.

Definition 2 (SL). *The formulae of the STS logic SL are*

$$\varphi ::= X \mid \diamond a \varphi \mid f(\varphi_1,\ldots,\varphi_n)$$

where X is a residual pleaceholder, a is an action and f is a spatial operator. A component p satisfies the formula φ, if $p \models \varphi$ holds according to the following rules:

$$p \models X$$
$$p \models f(\varphi_1,\ldots,\varphi_n) \quad \text{if } \exists p_1,\ldots,p_n.\ p \equiv f(p_1,\ldots,p_n) \wedge \forall\, i.\ p_i \models \varphi_i$$
$$p \models \diamond a \varphi \quad\quad\quad \text{if } \exists q.\ p \to_a q \wedge q \models \varphi$$

For example, the component $p = c^\perp.0 \mid a.b.0$ satisfies the formula $\diamond a X$, namely $p \models \diamond a X$, because $p \to_a c^\perp.0 \mid b.0$ (since $a.b.0 \to_a b.0$) and $c^\perp.0 \mid b.0 \models X$.

Given a formula φ and n components $q_1,\ldots,q_n$, we write $\varphi[q_1/X_1,\ldots,q_n/X_n]$ for the formula obtained from φ by replacing variables X_i with components q_i. Analogously, we denote by $\varphi[\varphi_1/X_1,\ldots,\varphi_n/X_n]$ the formula obtained from φ by replacing variables X_i with formulae φ_i. Variables in formulae stand for the residual q of p, after that p has exhibited the capabilities and/or the structure imposed by the formula.

Definition 3 (Satisfaction with residuals). *Let p be a component, $q = q_1, \ldots, q_n$ a tuple of components and $\varphi \in$ SL a formula whose variables are contained in $\{X_1, \ldots, X_n\}$. Then, we say that p satisfies φ with residuals $q_1, \ldots, q_n$, written $p \models \varphi; q$, whenever $p \models \varphi[q_1/X_1, \ldots, q_n/X_n]$.*

For example, if $p = n[a^\perp.0 \mid a.b.0]$ and $\varphi = n[\diamond a^\perp X_1 \mid a.X_2]$, then trivially $p \models \varphi$, and if $q_1 = 0$ and $q_2 = b.0$ then $p \models \varphi; (q_1, q_2)$. We shall write $\boldsymbol{p} \models \boldsymbol{\varphi}; \boldsymbol{q}$ where $\boldsymbol{\varphi} = \varphi_1, \ldots, \varphi_k$ and $\boldsymbol{p} = p_1, \ldots, p_k$ are tuples of formulae and components, respectively, obviously meaning that $p_i \models \varphi_i[q_1/X_1, \ldots, q_n/X_n]$, $\forall i \in \{1, \ldots, k\}$.

The proper correspondence between the transitions of environments and those of their closed instances, i.e. components, is established by suitable *soundness* and *completeness* properties (see [2,3] for their formal definition). Notably, some sort of "standard" sound and complete STS can be derived for a large class of process calculi (whose semantics is given by SOS rules in suitable formats [3]). Such STS can be constructed by means of a unification-based procedure. All the symbolic transitions spawn from $E[X]$ in (3) are sound.

3   Reasoning with Incomplete Information

We apply now the STS framework to two case studies (the Appendix reports the transitions of their closed specifications). The first one is an original formalization of a pattern of protein interaction relative to the λ–phago virus. Starting from an incomplete BioAmbients specification of the system, the behaviour of one of the proteins can be inferred by reasoning symbolically on the dynamics of the system. The second example, split in two parts, consists of the symbolic reading of biological interaction, also used to introduce the Brane Calculi in [7]. Here, pretending that cell reactions to viruses are not fully understood, we infer the same behaviour described in the original example. Moreover, without changing the experiment, we additionally deduce the (known) mechanisms allowing a given protein to block the virus. These examples are aimed at illustrating the applicability of our approach to different levels of abstraction, and its versatility in supporting the right representation language according to the problem at hand.

3.1   Protein Interaction: λ-Phage Life-Cycle

λ-phage simplified life cycle. We consider a simplified representation of the λ-*phage* virus. This virus replicates by binding with the *E.coli* bacterium and injecting its DNA into the bacterium cell. Then, either the virus replicates in several copies until the bacterium membrane is destroyed and the copies released (*lytic pathway*), or the virus DNA merges into the bacterium DNA, the infected bacterium cell multiplies, and its offspring may themselves eventually end up in a lytic pathway (*lysogenic pathway*). The pathway selection is determined by the interaction of the CRO, CI, CII, CIII and HFL proteins in the bacterium cell.

We study the system assuming the following knowledge (see Fig. 2). A high concentration of CI determines the lysogenic cycle, its absence the lytic one. The production of CI is promoted by CII, if it is not inhibited. The role of the bacterial protein HFL is

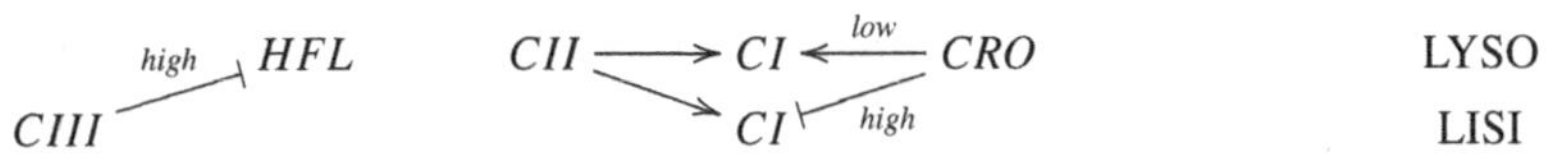

$high$ ↗ HFL $CII \rightleftarrows CI \xleftarrow{low} CRO$ LYSO

$CIII$ CI ↘ $high$ LISI

Fig. 2. Hypothesis on inhibition and activation roles of CRO and $CIII$ proteins

[VIRUS] = [merge$^+$ virus.([C3] | [C2] | [C1] | [CRO]) | [DNAλ]]

C3 = l_c3!.0 + accept h_c3. pro_c2!.0   C2 = pro_c2?. pro_c1!.0 + enter c2.0

C1 = pro_c1?.(h_cro?. lysi!.0 + l_cro?. lyso!.0)   CRO = l_cro!.0 + h_cro.0

DNAλ = (lyso?.enter dnae.0) + lysi?.($_\lambda$[exit newph.VIRUS] | expel newph)

[ECOLI] = [merge$^-$virus | $_{Dna_e}$[accept dnae] | [HFL]]   **HFL = enter h_c3.0 + X**

Fig. 3. Partial specification of [VIRUS] and [ECOLI]

not fully understood, but we know that it can be inhibited by a high concentration of
CIII. Moreover, a low concentration of CRO directly stimulates the production of CI,
while a high concentration of it destroys CI. Hence, the lysogenic cycle (top row) can
be characterised as low CRO and high CIII concentrations, while the lytic one (bottom
row) seems to depend on high CRO, exclusively.

BioAmbients uncomplete specification. Under the above hypotheses, the virus and
the bacterium can be naturally represented in the BioAmbient calculus as two membraned systems, as shown in Fig. 3 (for a formal description of the BioAmbient calculus we refer the reader to [30]). Proteins are represented as membranes (written
[. . .]) "delimiting" the behaviour they can express. They interact at the same level
of nesting: activation is modeled as communication (input/output pairs of actions
[..*pro_c2*?..]||[..*pro_c2*!..]) and inhibition as encapsulation ($_1$[..*enter a*..]| $_2$[..*accept a*..]
that evolves in $_2$[..$_1$[..]..]..]), since this technically blocks the capability of the enclosed
protein to communicate in its original environment. The virus consists of the capability
to penetrate a suitable membraned environment ([*merge$^+$ virus*..]), i.e., the bacterium
cell ([*merge$^-$ virus*..]), and then expose its DNA and express proteins. CIII can either
signal a low concentration or enclose HFL (or any compatible protein) then activating
CII by means of a suitable communication. Once activated, CII promotes CI. Moreover,
the possibility of CII being itself inhibited has also been modeled (*enter c2*). Sensitivity
to high or low concentrations of CRO, modeled by means of suitable communications,
causes CI to emit either the lysogenic or the lytic activation signal. This is received by
the virus DNA which, accordingly, either enters the bacterium DNA, or expels into the
bacterium cell a copy of the virus. The bacterium is modeled as a membrane that can
be injected by a virus and contains membraned DNA, which can be accessed by other
suitable DNA, and the HFL protein. Importantly, this is represented as a partially specified component, which, as we know, can be inhibited by CIII (*enter h_c3*) but also could
alternatively exhibit a behaviour we are not able to specify at the present, represented
as variable X.

$$_{Ecoli}[([\textbf{C3}] \mid [C2] \mid [C1] \mid [CRO])|[DNA\lambda] \mid_{Dna_e} [accept\ dnae] \mid [\textbf{enterh\_c3.0} + \textbf{X}]]$$

$$\xrightarrow{(l\_c3?.Y_1+Y_2)|Y_3}{}_{Ecoli}[_{CIII}[0] \mid [C2] \mid [C1] \mid [CRO])|[DNA\lambda] \mid_{Dna_e}[accept\ dnae] \mid_{hfl}[\textbf{Y}_1 \mid \textbf{Y}_3]]$$

$$\xrightarrow{(Y_4+accept\ c2.Y_5|Y_6),Y_3}{}_{Ecoli}[(_{CIII}[0] \mid [C1] \mid [CRO])|[\textbf{DNA}\lambda] \mid_{Dna_e} [accept\ dnae] \mid_{hfl}[_{CII}[0]|(\textbf{Y}_5 \mid \textbf{Y}_6) \mid \textbf{Y}_3]]$$

$$\xrightarrow{(Y_7+lysi!.Y_8),Y_6,Y_3}{}_{Ecoli}[(_{CIII}[0] \mid [C1] \mid [CRO])|[_\lambda[\textbf{exit newph.VIRUS}] \mid \textbf{expel newph}] \mid$$
$$_{Dna_e}[accept\ dnae] \mid_{hfl}[_{CII}[0]|(Y_8 \mid Y_6) \mid Y_3]]$$

$$\xrightarrow{Y_8,Y_6,Y_3}{}_{Ecoli}[(_{CIII}[0] \mid [C1] \mid [CRO])|_\lambda[\textbf{VIRUS}] \mid_{Dna_e}[accept\ dnae] \mid_{hfl}[_{CII}[0]|(Y_8 \mid Y_6) \mid Y_3]]$$

Fig. 4. A symbolic trace for $_\lambda[VIRUS] \mid {}_{Ecoli}[ECOLI]$

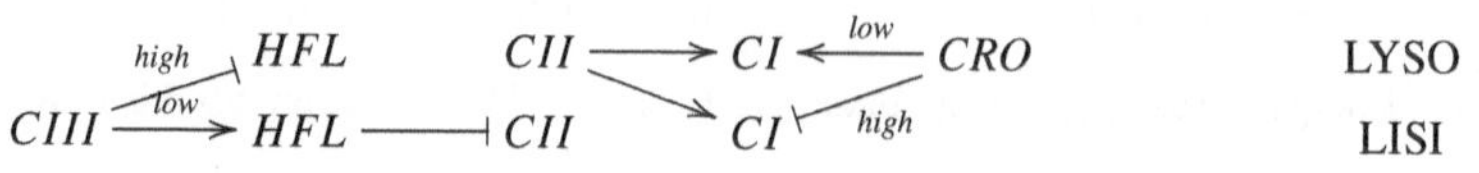

Fig. 5. Full specification of the inhibition and activation schema in Fig. 2

Symbolic transition system. We study the possible evolutions of the open system $_\lambda[VIRUS] \mid {}_{Ecoli}[ECOLI]$ (bio-ambients are sometimes labeled for clarity) in order to understand the possible interactions of HFL within it. In the corresponding STS we can find the trace reported in Fig. 4. (As mentioned in § 2, the labels can be automatically constructed on the basis of the BioAmbient proof rules, while the logic simply consists of the modal operator ◇ _, which stands for the possibility of performing an unlabelled transition, and of the spatial operators deriving from the syntax of the calculus). The composition of the formulae over the trace yields an interesting characterisation of the behaviour of HFL:

$$(l\_c3?.(Y_4 + accept\ c2(Y_7 + lysi!.Y_8) \mid Y_6) + Y_2) \mid Y_3$$

It is possible to see that this is a correct abstraction of the actually known HFL (see (10) in Appendix A) and the symbolic trace in Fig. 4 is an abstraction of the corresponding ground trace (see (12)-(16) in Appendix A). Finally, the picture of protein interactions of Fig. 2 can be completed with the relation between low $CIII$, HFL and CII, as shown in Fig. 5, although the partial specification adopted did not have any information about this specific point.

3.2 Cellular Interaction: Membrane Trepassing

We model an abstraction of a virus replicating its RNA by exploiting a host cell (this has been more exhaustively treated in [7]). The virus membrane complex contains the *capsid*, another membrane complex, which encloses the *nucleocapsid*, i.e. the cytoplasme containing the viral RNA. Here we model the endocytic pathway: the virus penetrates the cell membrane.

We assume that the behaviour of the virus membrane is not known and we deduce it from the operational rules describing the behaviour of the cell. We use fBC, a simplified version of the Brane Calculi, which more suitably models this example. Indeed, fBC focuses on membrane interactions, within the scope of this section, and molecular interactions in the next one. Brane Calculi are intended to model biological interactions inspired by endocytosis/exocytosis, indicated in [7] as *bitonal interactions*, since,

$$P, Q ::= \Diamond \mid \sigma[P] \mid P \circ Q \mid r \qquad \sigma, \rho, \tau ::= 0 \mid a.\tau \mid \sigma|\tau \mid \dots$$

$$r, s ::= \Diamond \mid m \circ r \qquad a ::= p_n, \ p_n^\perp(\sigma), \ e_n, \ e_n^\perp, m_n, \ m_n^\perp \dots$$

$$p_n.\sigma \mid \sigma_0 \ [P] \circ p_n^\perp(\rho).\tau \mid \tau_0 \ [Q] \quad \rightarrow \quad \tau \mid \tau_0 \ [\rho \ [\sigma \mid \sigma_0 \ [P]] \circ Q] \qquad (phago)$$

$$e_n^\perp.\tau \mid \tau_0 \ [e_n.\sigma \mid \sigma_0 \ [P] \circ Q] \quad \rightarrow \quad P \circ \sigma \mid \sigma_0 \mid \tau \mid \tau_0 \ [Q] \qquad (exo)$$

$$m_n.\sigma|\sigma_0[P] \circ m_n^\perp.\tau|\tau_0[Q] \rightarrow \sigma|\sigma_0|\tau|\tau_0[P \circ Q] \quad (mate) \qquad \frac{P \rightarrow Q}{P \circ R \rightarrow Q \circ R} \ (par)$$

$$r_1 \circ r_1(r_2) \rightrightarrows s_1(s_2).\sigma \mid \sigma_0[r_2 \circ P] \ \rightarrow \ s_1 \circ \sigma \mid \sigma_0[s_2 \circ P] \quad (b\&r) \qquad \frac{P \rightarrow Q}{\sigma \ [P] \rightarrow \sigma \ [Q]} \ (mem)$$

Fig. 6. Syntax and operational semantics of fBC

informally speaking, they preserve a periodicity between inner and outer areas of membranes.

The calculus fBC (see Fig. 6) can be understood as an extension of BioAmbients, where membranes exhibit themselves a behaviour. The basic membrane complex $\sigma[P]$ consists of an active external membrane layer σ and of complex P inside the membrane ($\Diamond$ is the null membrane complex). Other complexes can be obtained by the composition of P and Q, written $P \circ Q$, or as a multiset of molecules, $m_1 \circ \dots \circ m_k$. Interaction between membrane complexes happens through the active membrane layer σ, which can be halted 0, an action prefixed to an active layer $a.\sigma$ and the parallel composition of active layers $\sigma|\tau$. Membranes behave as follows. $\sigma[P]$ can enter $\tau[Q]$, if σ can execute a p_n action and τ the corresponding coaction $p_n^\perp(\rho)$ (with the same n) and $\sigma[P]$ is enclosed within the active membrane ρ, according to the spirit of bitonal reactions (*phago*). In $\tau[\sigma[P] \circ Q]$ the subsystem P can leave the $\tau[\dots]$ membrane complex if σ and τ are ready to execute, respectively, e_n and $e_n^\perp$ (*exo*). Finally, $\sigma[P]$ and $\tau[Q]$ merge in $\sigma|\tau[P \circ Q]$ if the membranes can execute m_n and $m_n^\perp$ respectively (*mate*).

fBC formalisation. Via a phagocytosis the virus enters the cell wrapped by a membrane. Then, the external membrane of the virus merges with a component of the cell, the *endosome*. Finally, through an exocytosis, the viral *nucleocapsid*, and the viral RNA it contains, is released directly in the *cytosol* of the cell (a possible formalisation in fBC is reported in Appendix B).

Let us suppose now that the mechanisms in the virus membrane are not very well understood. We represent this with the following partial specification, having variable Y in place of the virus membrane.

virus – $Y[$ nucap $]$ $nucap = capsid \ [\ vRNA \ ]$ $capsid = p_b \mid bud \mid disasm$

$cell = p_a^\perp(m_a) \mid e_b^\perp \ [\ cytosol \ ]$ $cytosol = endosome \circ CC$ $endosome = m_a^\perp \mid e_a^\perp \ [\ \Diamond \ ]$

The virus content *nucap* is known and it will take part to later stages. It consists of a membrane complex, which contains the RNA and whose active part, *capsid*, is ready to

$$\varphi ::= X \mid \psi[\varphi] \mid \varphi \circ \varphi \mid \diamond\!\!-\varphi \mid \theta \qquad \text{(complexes)} \qquad \psi ::= Y \mid \psi|\psi \mid \alpha.\psi \text{ (membranes)}$$
$$\alpha ::= p_n \mid p_n^{\perp}(\psi) \mid e_n \mid e_n^{\perp} \mid ; \theta(\theta) \rightrightarrows \theta(\theta) \text{ (actions)} \qquad \theta ::= Z \mid m \mid \theta|\theta \text{ (molecules)}$$

Fig. 7. The logic associated to the calculus fBC

execute a phago action p_b, a *disasm* set of actions that will be defined later, and a *bud* action that is not relevant here. The cell membrane is ready for a phago $p_a^{\perp}(m_a)$ and an exo $e_b^{\perp}$ action (for the reproduced virus eventually leaving the cell). Its content, *cytosol*, consists of a part, denoted CC, here not relevant, and the *endosome*, i.e. a membrane complex that can merge $m_a^{\perp}$ with what has been phago-ed and it can uncoat its content $e_a^{\perp}$, in case a suitable coaction can be provided, possibly by the virus.

Symbolic transition system. Also in this example, the associated logic is straight-forwardly induced by the syntax of fBC, as shown in Fig. 7. Note that, being fBC se-mantics unlabelled, the modality $\diamond\!\!-\psi$ simply stands for the capability of executing any action (e.g., as it may be required by the (*mem*) rule, see Fig. 6). Then, we study the environment:

$$F[Y] = Y[\,nucap\,] \circ p_a^{\perp}(m_a)|e_b^{\perp}[\,cytosol\,]$$

where Y stands for the unknown virus membrane. A possible symbolic trace of the sTs of $F[X]$ is:

$$F[Y] \xrightarrow{p_a.Y_1|Y_2} e_b^{\perp}[\,m_a[Y_1|Y_2[nucap]] \circ m_a^{\perp} \mid e_a^{\perp}\,[\,\diamond\,] \circ CC\,]] \qquad (5)$$
$$\xrightarrow{Y_1,Y_2} e_b^{\perp}\,[\,e_a^{\perp}\,[\,Y_1|Y_2[\,nucap\,]] \circ CC] \qquad (6)$$
$$\xrightarrow{e_a.Y_3|Y_4,Y_2} e_b^{\perp}[\,Y_3|Y_4|Y_2[\diamond] \circ nucap \circ CC] \qquad (7)$$

The first symbolic transition (5) constrains the virus membrane to be able to perform a phago p_a action in order to enter the cell via endocytosis, $Y = p_a.Y_1 \mid Y_2$. The re-quirement is specific for the action offered by the cell membrane. The second symbolic transition (6) does not involve the virus membrane, since the *nucap* of the virus can merge with the *cytosol* of the cell without imposing any further condition on the viral membrane. The formula hence reverts to identity Y_1, Y_2: no requirements over the un-specified components, since the rest of the system is able to evolve autonomously. The last transition (7) requires $e_a.Y_3|Y_4, Y_2$, i.e. the (current state of the) virus membrane should be able to exhibit an action e_a in order to uncoat its content *nucap* via exocy-tosis. The constraint $Y_1 = e_a.Y_3|Y_4$ is the most general, coherently with the semantic rules.

Inferred information about the unknown components, when they contribute to the overall system behaviour, can be gathered by composing the logical formulae used as labels: any virus whose membrane satisfies $\psi_{\text{STS}} = p_a.(e_a.Y_3|Y_4)|\,Y_2$ will be able to enter in the cell and release its *nucap*. Note that ψ_{STS} characterises a general class of compo-nents which allow for the interaction of interest. For instance, not only the membrane of the virus $p_a.e_a$ [*nucap*] but also the membranes $p_a.(e_a \mid p_a.e_a)$ or $m_n \mid p_a.e_a$ satisfy ψ_{STS} and, once plugged in Y, are sufficient to drive the system through the same kind of behaviour (and maybe others).

3.3 Biochemical Interaction: Viral RNA Replication

Once inside the cell, the virus capsid is removed (uncoating process) and the virus RNA replicates. Besides this process, also discussed in [7], we address virus neutralisation by the cell. Here, we assume that the mechanisms in the cell content are not fully understood and we show how information about the cell content, relevant for virus replication or neutralisation, can be inferred.

The fBC calculus, in order to express biochemical phenomena, includes a *bind and release* action ($b\&r$) to let membranes interact with molecules. The action, denoted $r_1(r_2) \rightrightarrows s_1(s_2)$, can be executed by a membrane if the molecules r_2 are contained in the membrane complex and the molecules r_1 are present outside it. Its effect is to substitute the molecules r_1 with the molecules s_1 outside, and r_2 with the s_2 inside.

We study again a partially specified cell, and then we show how the inferred constraints are coherent with actual components that can reasonably play the part of the unspecified ones.

Virus replication. We assume that the virus $p_a.e_a$ [*nucap*], which fulfills the characterisation inferred in § 3.2, has entered the cell, reaching the state $e_b^{\perp}[\ p_b|bud|disasm[vRNA]\ \circ CC]$ (a coherent instance of the one in (7)). The action *disasm*, responsible of uncoating the *vRNA*, is specified as a $b\&r$ action activated by the presence of an outer trigger. It moves the inner *vRNA* outside: $disasm = disTrg(vRNA) \rightrightarrows vRNA(\diamond)$. Moreover, we suppose that the remaining cell content CC is not fully understood, i.e. we focus on the environment

$$G[X] = e_b^{\perp}[p_b \mid bud \mid disTrg(vRNA) \rightrightarrows vRNA(\diamond)[vRNA] \circ X]$$

A possible symbolic trace of $G[X]$ is the following:

$$G[X] \xrightarrow{disTrg \circ X_0} e_b^{\perp}[p_b \mid bud[\diamond] \circ vRNA \circ X_0] \xrightarrow{\xi} e_b^{\perp}[p_b|bud[\diamond] \circ Z_2 \circ Y_4|Y_5[Z_3 \circ X_6] \circ X_7] \quad (8)$$

The applicable ($b\&r$) rule justifies the first symbolic transition with a spatial constraint requiring that X contains at least a *disTrg* molecule in order to trigger the removal of the viral *capsid*. The second symbolic transition is justified by the *vRNA* molecule, now free within the cell, used as a trigger for another application of ($b\&r$), where $\xi = vRNA(Z_1) \rightrightarrows Z_2(Z_3).Y_4|Y_5[Z_1 \circ X_6] \circ X_7$. The formula ξ implies that, triggered by the outer presence of *vRNA*, a set of molecules Z_2 can be released within the cell, so that *vRNA* replication can be supported by the cell. The composition of the formulae in (8) yields

$$\Psi_{\text{STS}} = disTrg \circ vRNA(Z_1) \rightrightarrows Z_2(Z_3).Y_4|Y_5[Z_1 \circ X_6] \circ X_7$$

As expected, Ψ_{STS} characterises the mechanisms of virus replication as modeled in [7], which we are following. There, CC is read as providing the suitable triggering and replication (two *vRNA* released) capability:

$$CC = disTrg \circ vRNArepl \circ CC' \quad vRNArepl = vRNA(\diamond) \rightrightarrows vRNA \circ vRNA(\diamond)\ [\diamond]$$

Importantly, the above definition satisfies the characterisation Ψ_{STS} obtained by reasoning symbolically, i.e., $CC \models \Psi_{\text{STS}}$ (where Z_2 stands for $vRNA \circ vRNA$, Z_1, Z_3, X_6, X_7 for $\diamond$, and Y_4, Y_6 for 0). Moreover, the behaviour of $G[CC]$ comprises a trace that is an instance of the symbolic (8), leading, as expected, to a state where *vRNA* has been replicated: $e_b^{\perp}[\ p_b \mid bud[\diamond] \circ vRNA \circ vRNA \circ CC'\ ]$.

Virus neutralisation. It is interesting to observe another possible evolution of $G[X]$, justified in the corresponding STS by a (*phago*) rule

$$G[X] \xrightarrow{\psi} e_b^{\perp}[Y_2|Y_3[\ Y_1[\quad bud \mid disasm[vRNA]\quad] \circ X_4]] \tag{9}$$

with $\psi = p_b^{\perp}(Y_1).Y_2|Y_3[X_4]$. In this case the p_b action of the virus membrane has been exploited to trap the virus *nucap* within a membraned complex ($Y_1[\ ]$).

This evolution, not considered in [7], mimics the presence of a *Mx*-like protein in the cell that inhibits the replication of the virus. This type of proteins seems to play an antiviral activity by trapping the viral capsid and moving it in a location of the cell where the mechanism for the generation of new virus particles becomes unavailable [18]. The simplest representation of the *Mx* protein can be drawn from ψ as $Mx = p_b^{\perp}(0)[\diamond]$: by means of $p_b^{\perp}(0)$ the viral *nucap* is trapped within an empty membrane. *Mx* seems to be a coherent simplification of the actual known behaviour of the protein, while $\psi = p_b^{\perp}(Y_1).Y_2|Y_3[X_4]$ characterises the cell components (including the protein *Mx*, but possibly others) capable of trapping the virus within a membrane. Adding the *Mx* protein to a cell with the RNA replication mechanism determines a trace reaching a state where the virus has been phago-ed in a membrane where it can not reproduce (the *Mx* membrane has been annotated for readability):

$$e_b^{\perp}[\ 0\overline{[}\ \overline{0}[\ bud \mid disasm[vRNA]\ ]\ \overline{]} \circ RC\ ].$$

While the behaviour of *Mx* proteins has been studied elsewhere, it is worth noting that here it has been inferred from the general rules defining the calculus and an incomplete initial specification. The same specification has led to the inference of the virus replication mechanism. This experiment shows how symbolically reasoning appears as a deductive mechanism, suitable to infer unknown information.

3.4   Discussion

We have presented two proof of concept examples of the application of STS to biological problems. The former is an original formulation of biological interaction that highlights the problem of understanding the interplay of a protein network. The latter is a paradigmatic example of the interaction between a virus and its host cell. These examples have illustrated how, reasoning in presence of incomplete specifications, it has been possible to deduce new knowledge about the studied systems, like

- the emergence of unspecified interactions between bio-components, e.g. the possibility of a *HFL*-like protein to determine the lytic or lysogenic cycle in the λ−phage virus life-cycle;
- the constraints over the behaviour and the structure of bio-components needed to participate to the evolution of a system, e.g. the need of the cell content to provide a trigger for *vRNA* replication;
- the discovery of possible components or behaviour not explicitly foreseen in the initial specification of the system under analysis, e.g. the existence and behaviour of an *Mx*-like protein blocking virus replication.

This poses the problem of characterising the (reachability of) relevant states and evolution traces of the partially specified bio-environments, and, analogously, of proving bio-system properties. For instance one might want to exploit the synthesized STS for the automated state-space exploration in order to characterise "unknown" dangerous bioagents that can compromise the regular activity of a cellular system. An interesting approach in this sense is the definition of a modal logic and a model checking algorithm for Brane Calculi [24] (along the line of Ambient Logic [9]), which defines spatial and temporal properties on membrane systems. Similarly, [13,14,35] exploit Pathway logic and matching algorithms for model checking the evolutions of biological systems. However, beyond similarities, e.g. formulas as labels and unification/matching algorithms, our theoretical framework poses the problem of model checking *open-ended* systems. This is an interesting problem under investigation, whose scope is beyond this paper.

Formulae relative to traces play the part of (minimal) necessary conditions that unspecified components must fulfill to *possibly* drive the system through the trace. Trivially, if $p \models \phi$ and $q \models \psi$ then $p + q$ satisfies both of them and can lead the system through possibly completely unrelated evolutions. This suggests, in general, the difficulties in associating processes to desired behaviours. Besides, while $p + q$ is definitely a process in abstract terms, it might not be feasible in biological terms, so that process characterisation could also require domain specific solutions. This issue is under investigation.

4   Concluding Remarks and Future Work

We have proposed the application of a symbolic approach to the modeling of open biological processes, where some components or features are unknown. An open biological system is seen as a partially specified process of a given calculus tailored to biological processes. Its semantics is given in terms of symbolic transition systems (STS), whose transitions are labeled with logical formulae that express the structural and/or behavioural requirements over the unknown components that let the system evolve. STS can be effectively generated from the SOS rules of the process calculus by using a unification-based approach, supporting *in silico* analysis of complex biological systems. Overall, this provides a formal and computational framework capable to infer information about components that are not fully understood beforehand, and that permits the choice of the more appropriate representation language.

To the best of our knowledge, our open-ended and inferential modeling is original in the context of bio process algebras. Indeed, in the literature, simulations and analysis have been carried out starting from completely specified models. Hence, no further information about the behaviour of the system can be inferred, in the sense we do it. A related approach, but in a different perspective, can be [6], where temporal logic is taken as a specification language and machine learning techniques are used to revise the reaction rules initially (fully) available. Another way to deal with incomplete information are discrete approximations, as claimed in [5], but in the significantly different context of the numerical approaches. An analogous unification-based semantics construction is given in [36], in the different context of model-checking for nominal calculi, where unknown can be the communication network.

Beyond the discussed ongoing extensions, a challenging major direction to extend our approach is the use of quantitative and stochastic information, e.g. probabilities and rate constants of reactions, as in [27,29].

Acknowledgments. We are grateful to Pierpaolo Degano and Corrado Priami whose comments have helped us in improving the content and presentation of this paper.

References

1. Alberts, B., Bray, D., Lewis, J., Raff, M., Roberts, K., Watson, J.D.: Molecular Biology of the Cell. Garland (1994)
2. Baldan, P., Bracciali, A., Bruni, R.: Bisimulation by unification. In: Kirchner, H., Ringeissen, C. (eds.) AMAST 2002. LNCS, vol. 2422, pp. 254–270. Springer, Heidelberg (2002)
3. Baldan, P., Bracciali, A., Bruni, R.: Symbolic equivalences for open systems. In: Priami, C., Quaglia, P. (eds.) GC 2004. LNCS, vol. 3267, pp. 1–17. Springer, Heidelberg (2005)
4. Calder, M., Hilston, J., Gilmore, S.: Modelling the influence of RKIP on the ERK signalling pathway using the stochastic process algebra PEPA. In: Priami, C., Ingólfsdóttir, A., Mishra, B., Nielson, H.R. (eds.) Transactions on Computational Systems Biology VII. LNCS (LNBI), vol. 4230, pp. 1–23. Springer, Heidelberg (2006)
5. Calder, M., Vyshemirsky, V., Gilbert, D., Orton, R.: Analysis of signalling pathways using the PRISM model checker. In: Priami, C., Plotkin, G. (eds.) Transactions on Computational Systems Biology VI. LNCS (LNBI), vol. 4220, pp. 179–190. Springer, Heidelberg (2006)
6. Calzone, L., Chabrier-Rivier, N., Fages, F., Soliman, S.: Machine learning biochemical networks from temporal logic properties. In: Priami, C., Plotkin, G. (eds.) Transactions on Computational Systems Biology VI. LNCS (LNBI), vol. 4220, pp. 68–94. Springer, Heidelberg (2006)
7. Cardelli, L.: Brane calculi-interactions of biological membranes. In: Danos, V., Schachter, V. (eds.) CMSB 2004. LNCS (LNBI), vol. 3082, pp. 257–280. Springer, Heidelberg (2005)
8. Cardelli, L., Gordon, A.: Mobile ambients. Th. Comp. Sci. 240(1), 177–213 (2000)
9. Cardelli, L., Gordon, A.: Anytime, anywhere. Modal logics for mobile ambients. In: Proc. POPL'00, pp. 365–377. ACM, New York (2000)
10. Chiarugi, D., Curti, M., Degano, P., Marangoni, R.: VICE: A VIrtual CEll. In: Danos, V., Schachter, V. (eds.) CMSB 2004. LNCS (LNBI), vol. 3082, pp. 207–220. Springer, Heidelberg (2005)
11. Danos, V., Laneve, C.: Formal molecular biology. Th. Comp. Sci. 325(1), 69–110 (2004)
12. Degano, P., Prandi, D., Priami, C., Quaglia, P.: Beta-binders for biological quantitative experiments. In: Proc. QAPL'06 ENTCS, vol. 164(3), pp. 101–117. Elsevier, North-Holland (2006)
13. Eker, S., Knapp, M., Laderoute, K., Lincoln, P., Meseguer, J., Sonmez, K.: Pathway Logic: Symbolic analysis of biological signaling. In: Proc. Pacific Symp. on Biocomputing, pp. 400–412 (2002)
14. Eker, S., Knapp, M., Laderoute, K., Lincoln, P., Talcott, C.: Pathway Logic: Executable models of biological networks. In: Proc. Rewriting Logic and its Applications, ENCS, 71 (2002)
15. Gadducci, F., Montanari, U.: The Tile Model. In: Proof, Language and Interaction: Essays in Honour of Robin Milner, pp. 133–166. MIT Press, Cambridge (1998)
16. Gillespie, D.T.: Exact stochastic simulation of coupled chemical reactions. J. of Physical Chemistry 81(25), 2340–2361 (1977)
17. Gutnik, B., Pinto, D., Ermentrout, B.: Mathematical neuroscience: from neurons to circuits to systems. J. of Physiology 97, 209–219 (2003)

18. Haller, O., Kochs, G.: Interferon-induced Mx proteins: Dynamin-like GTPases with antiviral activity. Traffic 3, 710–717 (2002)
19. Hennessy, M., Lin, H.: Symbolic bisimulations. Th. Comp. Sci. 138, 353–389 (1995)
20. de Jong, H.: Modeling and simulation of genetic regulatory systems: a literature review. J. of Comp. Biology 9(1), 67–103 (2002)
21. Kitano, H.: Systems Biology: a brief overview. Science 295(5560), 1662–1664 (2002)
22. Lecca, P., Priami, C., Quaglia, P., Rossi, B., Laudanna, C., Costantin, G.: A stochastic process algebra approach to simulation of autoreactive lymphocyte recruitment. SIMULATION 80(4), 273–288 (2004)
23. Leifer, J., Milner, R.: Deriving bisimulation congruences for reactive systems. In: Palamidessi, C. (ed.) CONCUR 2000. LNCS, vol. 1877, pp. 243–258. Springer, Heidelberg (2000)
24. Miculan, M., Bacci, G.: Modal logics for Brane Calculus. In: Priami, C. (ed.) CMSB 2006. LNCS (LNBI), vol. 4210, pp. 1–16. Springer, Heidelberg (2006)
25. Nagasaki, M., Onami, S., Miyano, S., Kitano, H.: Bio-calculus: its concept and molecular interaction. Genome Informatics 10, 133–143 (1999)
26. Phillips, A., Cardelli, L.: A correct abstract machine for the stochastic pi-calculus. In: Proc. Bioconcur'04 (2004)
27. Priami, C.: Stochastic π-calculus. The Computer Journal 38(6), 578–589 (1995)
28. Priami, C., Quaglia, P.: Beta-binders for biological interactions. In: Danos, V., Schachter, V. (eds.) CMSB 2004. LNCS (LNBI), vol. 3082, pp. 21–34. Springer, Heidelberg (2005)
29. Priami, C., Regev, A., Shapiro, E., Silvermann, W.: Application of a stochastic name-passing calculus to representation and simulation of molecular processes. Information Processing Letters 80, 25–31 (2001)
30. Regev, A., Panina, E., Silverman, W., Cardelli, L., Shapiro, E.: Bioambients: An abstraction for biological compartements. Th. Comp. Sci. 325(1), 141–167 (2004)
31. Regev, A., Shapiro, E.: Cellular Abstractions: Cells as Computation. Nature 419, 343 (2002)
32. Regev, A., Silverman, W., Shapiro, E.: Representation and simulation of biochemical processes using the π-calculus process algebra. In: Proc. Pacific Symp. on Biocomputing, pp. 459–470 (2001)
33. Rensink, A.: Bisimilarity of open terms. Inform. and Comput. 156(1-2), 345–385 (2000)
34. Surridge, C.: Computational biology. Nature Insight 420(6912), 206–246 (2002)
35. Talcott, C., Eker, S., Knapp, M., Lincoln, P., Laderoute, K.: Pathway logic modeling of protein functional domains in signal transduction. In: Proc. Pacific Symp. on Biocomputing (2004)
36. Yang, P., Ramakrishnan, C., Smolka, S.: A logical encoding of the pi-calculus: Model checking mobile processes using tabled resolution. In: Zuck, L.D., Attie, P.C., Cortesi, A., Mukhopadhyay, S. (eds.) VMCAI 2003. LNCS, vol. 2575, pp. 86–101. Springer, Heidelberg (2002)
37. Winnacker, E.L.: From Genes to Clones VCH Publishers (1987)

A λ-Phage Life Cycle

We report three traces, starting from the completely specified model of the λ-phage life cycle. They express three pathways illustrated in [37], which lead to the lysogenic and lytic cycles. We assume the behaviour of HFL to be:

$$HFL = enterh\_c3.0 + l\_c3?.acceptc2.lysi!.0 \tag{10}$$

All executions are preceded by the initial phase of virus injection: the two reductions in (11) concerning the merging of the virus and the bacterium membranes. For readability we use the abbreviation: $DNA_e = accept\ dnae.0$.

$$_\lambda[VIRUS] \mid_{Ecoli} [ECOLI] \rightarrow_{Ecoli} [([C3] \mid [C2] \mid [C1] \mid [CRO])\mid[DNA\lambda] \mid [DNA_e] \mid [HFL]] \quad (11)$$

In case of a low concentration of CIII the simulation starts at (12):

$$_{Ecoli}[([\mathbf{C3}] \mid [C2] \mid [C1] \mid [CRO])\mid[DNA\lambda] \mid [DNA_e] \mid [\mathbf{HFL}]] \quad (12)$$
$$\rightarrow_{Ecoli} [_{CIII}[0] \mid [\mathbf{C2}] \mid [C1] \mid [CRO])\mid[DNA\lambda] \mid [DNA_e] \mid {}_{hfl}[\mathbf{accept\ c2.lysi!.0}]] \quad (13)$$
$$\rightarrow_{Ecoli} [([0] \mid [C1] \mid [CRO]) \mid [\mathbf{DNA}\lambda] \mid [DNA_e] \mid {}_{hfl}[_{CII}[0]|\mathbf{lysi!.0}]] \quad (14)$$
$$\rightarrow_{Ecoli} [([0] \mid [C1] \mid [CRO])\mid_\lambda[\mathbf{exit\ newph.VIRUS}] \mid \mathbf{expel\ newph}] \mid [DNA_e] \mid_{hfl} {}_{CII}[0]|0]] \quad (15)$$
$$\rightarrow_{Ecoli} [([0] \mid [C1] \mid [CRO])\mid_\lambda[\mathbf{VIRUS}] \mid [DNA_e] \mid_{hfl} [{}_{CII}[0]|0]] \quad (16)$$

A high concentration of CIII and a low of CRO leads the system to lysogeny, the simulation starts at (17):

$$_{Ecoli}[([\mathbf{C3}] \mid [C2] \mid [C1] \mid [CRO])\mid[DNA\lambda] \mid [DNA_e] \mid [\mathbf{HFL}]] \quad (17)$$
$$\rightarrow_{Ecoli} [_{CIII}[\mathbf{pro\_c2!.0}| {}_{HFL}[]] \mid [\mathbf{C2}] \mid [C1] \mid [CRO])\mid[DNA\lambda] \mid [DNA_e]] \quad (18)$$
$$\rightarrow_{Ecoli} [([0| {}_{HFL}[0]] \mid [\mathbf{pro\_c1.0}] \mid [C1] \mid [CRO]) \mid [DNA\lambda] \mid [DNA_e]] \quad (19)$$
$$\rightarrow_{Ecoli} [([0| {}_{HFL}[0]] \mid [0] \mid [\mathbf{high\_cro?.lysi!.0 + l\_cro?.lyso!.0}] \mid [CRO]) \mid [DNA\lambda] \mid [DNA_e]] \quad (20)$$
$$\rightarrow_{Ecoli} [([0| {}_{HFL}[0]] \mid [0] \mid [\mathbf{lyso!.0}] \mid [0]) \mid [\mathbf{DNA}\lambda] \mid [DNA_e]] \quad (21)$$
$$\rightarrow_{Ecoli} [([0| {}_{HFL}[0]] \mid [0] \mid [0] \mid [0]) \mid [\mathbf{enter\ dnae.0}] \mid [\mathbf{DNA_e}]] \quad (22)$$
$$\rightarrow_{Ecoli} [({}_{CIII}[0| {}_{HFL}[0]] \mid {}_{CII}[0] \mid {}_{CI}[0] \mid {}_{CRO}[0]) \parallel {}_{Dnae}[{}_{Dna\lambda}[0]]] \quad (23)$$

Finally, a high concentration of CRO leads the system to lysis, even with a high concentration of CIII, simulation starts at (24), continuing from (20):

$$_{Ecoli}[([0| {}_{HFL}[0]] \mid [0] \mid [\mathbf{high\_cro?.lysi!.0 + l\_cro?.lyso!.0}] \mid [CRO]) \mid [DNA\lambda] \mid [DNA_e]] \quad (24)$$
$$\rightarrow_{Ecoli} [([0| {}_{HFL}[0]] \mid [0] \mid [\mathbf{lysi!.0}] \mid [0]) \mid [DNA\lambda] \mid [DNA_e]] \quad (25)$$
$$\rightarrow_{Ecoli} [([0| {}_{HFL}[0]] \mid [0] \mid [0] \mid [0]) \mid [{}_\lambda[\mathbf{exit\ newph.VIRUS}] \mid \mathbf{expel\ newph}] \mid [DNA_e]] \quad (26)$$
$$\rightarrow_{Ecoli} [([0| {}_{HFL}[0]] \mid [0] \mid [0] \mid [0]) \mid {}_\lambda[VIRUS] \mid [DNA_e]] \quad (27)$$

B   Virus Entering a Cell

We report the fully specified model of the virus entering a cell.

$$virus = p_a.e_a [nucap] \qquad nucap = capsid [vRNA] \qquad capsid = p_b \mid bud \mid disasm$$

$$cell = p_a^\perp(m_a) \mid e_b^\perp [cytosol] \quad cytosol = endosome \circ CC \quad endosome = m_a^\perp \mid e_a^\perp [\diamond]$$

The initial configuration of the system gives origin to the following simulation that exhibits the expected behaviour:

$$p_a.e_a[nucap] \circ p_a^\perp(m_a)|e_b^\perp[m_a^\perp \mid e_a^\perp [\diamond] \circ CC] \quad (28)$$
$$\rightarrow e_b^\perp[m_a[e_a[nucap]] \circ m_a^\perp \mid e_a^\perp [\diamond] \circ CC] \quad (29)$$
$$\rightarrow e_b^\perp[e_a^\perp[e_a[nucap]] \circ CC] \rightarrow e_b^\perp[0[\diamond] \circ nucap \circ CC] \quad (30)$$

Reduction of Algebraic Parametric Systems by Rectification of Their Affine Expanded Lie Symmetries

Alexandre Sedoglavic

ALIEN Project, INRIA Futurs & LIFL (CNRS, UMR 8022),
Université des Sciences et Technologies de Lille, 59655 Villeneuve d'Ascq, France

Abstract. Lie group theory states that knowledge of a m-parameters solvable group of symmetries of a system of ordinary differential equations allows to reduce by m the number of equations. We apply this principle by finding some *affine derivations* that induces *expanded* Lie point symmetries of considered system. By rewriting original problem in an invariant coordinates set for these symmetries, we *reduce* the number of involved parameters. We present an algorithm based on this standpoint whose arithmetic complexity is *quasi-polynomial* in input's size.

1 Introduction

Before analysing a biological model described by an algebraic system, it is useful to reduce the number of relevant parameters that determine the dynamics.

Example 1. In order to give an example of such a reduction, let us consider the following Verhulst's logistic growth model with linear predation (see § 1.1 in [1]):

$$\dot{x} = (a - bx)x - cx, \quad \dot{a} = \dot{b} = \dot{c} = 0, \quad \dot{t} = 1. \tag{1}$$

Assuming that $a \neq c$ and $b \neq 0$, one can represent the flow (t, x) of (1) using parameterization:

$$t = \mathrm{t}/(a - c), \qquad x = (a - c)\mathrm{x}/b, \tag{2}$$

where (t, x) is the flow of the following *simpler* differential equation:

$$\dot{\mathrm{x}} = (1 - \mathrm{x})\mathrm{x}. \tag{3}$$

In this formulation of (1), parameters a and c were lumped together into $a - c$ and its state variables x and t were *nondimensionalise*.

Usually, presentation of this kind of simplification relies on rules of thumbs (for example, the knowledge of units in which is expressed the problem when dimensional analysis is used) and thus, there is—up to our knowledge—no complexity results on these kind of reduction methods (see [2] and references therein).

However, these reductions are generally based on the existence of Lie point symmetries of the considered problem (for reduction based on dimensional analysis, see § 1.2 in [3] and Theorem 3.22 in [4]).

H. Anai, K. Horimoto, and T. Kutsia (Eds.): AB 2007, LNCS 4545, pp. 277–291, 2007.

Example 1 (continued). The following continuous groups of transformations:

$$\mathcal{T}_\lambda : \begin{matrix} t \to t, \\ x \to x, \end{matrix} \begin{matrix} a \to a - \lambda, \\ b \to b, \\ c \to c - \lambda, \end{matrix} \qquad \mathcal{S}_{(\mu,\nu)} : \begin{matrix} t \to t/\nu, \\ x \to \mu x, \end{matrix} \begin{matrix} a \to \nu a, \\ b \to \nu b/\mu, \\ c \to \nu c, \end{matrix} \qquad (4)$$

leave invariant system (1) and its solutions. These symmetries are called *expanded* because they act on the expanded space of variables that includes the system parameters in addition to independent and dependent variables.

The system (3) is obtained by *factoring out* the actions of the symmetries (4) of the original system (1) and thus, it is invariant under these actions. The relations (2) parameterize the solutions of the original system (1) in function of the solution (t, x) of invariant system (3) and of the free parameters a, b, c; they are defined by the composition $(\mathcal{S}_{(a/b,a)} \circ \mathcal{T}_c)(t, x, a, b, c)$.

The aim of this note is to show how Lie theory unifies and extends the classical methods (exact lumping, dimensional analysis, etc.) used to simplify parametric algebraic (differential) systems. We adopt a presentation based on algebraic tools closer to the actual computations (mainly Jordan normal form and linear algebra) on which are based our reduction process.

1.1   Related Works

The literature on investigation of the invariants of Lie group action and their applications to algebraic systems is far too vast to be reviewed properly here. Nevertheless, let us notice that some simplification process (dimensional analysis for example) are known to be based on classical Lie theory (see § 1.2 in [3] and Theorem 3.22 in [4]) but, to the best of our knowledge, the use of more general transformations than scaling (and their application to simplification of algebraic parametric systems) are not described elsewhere in the literature.

The book [5] shows various applications of invariants in the study of dynamical systems under a computer algebra viewpoint. Section 4.1 for example, shows how the knowledge of invariants of a given dynamical system could simplify further computation on it by reducing the degree of involved polynomial expressions. In [6] authors show, given a rational group action, how to compute a complete set of its invariants.

We adopt the same general philosophy—determine some system's symmetries and use their invariants—but our purposes are more to reduce the number of variables involved in these expressions than their degrees (that is for us only a byproduct). Furthermore, while the symmetries considered in [5,6] are quite general—and thus, their computation of invariants are exclusively done using Gröbner bases computations—we restrict ourselves to the use of *affine* Lie symmetries and thus, the required operations are restricted to linear algebra over a number field and univariate polynomial factorization.

1.2   Main Steps and Tools of the Reduction Process

Let us present now our simplification process through our introducing example:

Example 1 (continued). Reduction of Example 1 is classically done as follow

Step 1. Determine affine infinitesimal generators that induce expanded Lie symmetries of the considered system. If there is no such derivations, our reduction process stops. In our example 1, they are:

$$\delta_1 := \tfrac{\partial}{\partial a} + \tfrac{\partial}{\partial c}, \quad \delta_2 := x\tfrac{\partial}{\partial x} - b\tfrac{\partial}{\partial b}, \quad \delta_3 := a\tfrac{\partial}{\partial a} + b\tfrac{\partial}{\partial b} + c\tfrac{\partial}{\partial c} - t\tfrac{\partial}{\partial t}. \qquad (5)$$

The section 3.1 describes the required computations and shows that their complexity is quasi-polynomial in input size.

Step 2. Choose a generator that is a symmetry of the others (above infinitesimal generators form a solvable Lie algebra and our reduction process have to take this property into account by computing associated *structure constants*). As $[\delta_1, \delta_2] = [\delta_2, \delta_3] = 0$ and $[\delta_1, \delta_3] = \delta_1$, we could choose δ_1 or δ_2. Determine a *preprincipal element* ($\varrho := b$) associated with a *principal element* ($\rho := \ln b$) of this last generator i.e. an element defining a coordinates set in which the derivation δ_2 is rectified (equal to the translation $\partial/\partial\rho$).

Step 3. The chosen principal element induces an *invariantization* of the considered system i.e. the system $\dot{x} = (a - x)x - cx$ which is the intersection of the original system with the algebraic hyperplane defined by the relation $\varrho - 1 = 0$; the resulting system is invariant under the action $S_\mu : x \to \mu x$, $b \to b/\mu$ of the one-parameter group of symmetry induced by δ_2. The solution x of the original system is then by $S_b(\mathrm{x})$ where x is a solution of the invariantized system and b is a free parameter.

Step 4. Repeat Step 1 (supplementary affine symmetries could appear after Step 3).

Let us stress that it is generally hard to find a general infinitesimal generator of a system's symmetry (Step 1) and to give an explicit representation of an invariant coordinates set (Step 2) for it. Thus, we restrict ourself to Lie symmetries associated to affine infinitesimal generators (whose coefficients are linear functions) for which invariant coordinates computation is easy (for general case see [6] and references therein). Hence, we do not follow methods developed for general cases because their complexity are likely exponential in input's size while we focus our attention to method of quasi-polynomial complexity.

To conclude, remark that this reduction process works also for purely algebraic system (describing fixed point of a dynamical system for example).

Outline. In the next section, we recall some basic definitions concerning considered systems and related derivations. Then, we present the notion of principal element and show how it could be used in order to define a rectifying coordinates set for general derivations. In the second part of this note, we focus our attention on affine Lie point symmetries in order to propose a probabilistic strategy to compute them and their associated principal elements. We show how previously introduced notions are used in the reduction process by considering invariantization of purely algebraic (resp. differential) system and their parameterization. Finally, in conclusion we make some remarks and suggest possible further works.

2   Considered Systems and Associated Derivations

2.1   Some Algebraic Systems Used in Analysis of Biological Model

Note 2. Notations — Hereafter, we consider an explicit algebraic ordinary differential system Σ bearing on n state variables $X := (x_1, \ldots, x_n)$ and depending on ℓ parameters $\Theta := (\theta_1, \ldots, \theta_\ell)$:

$$\Sigma \qquad \begin{cases} \dot{X} = F(t, X, \Theta), \\ \dot{t} = 1, \quad \dot{\Theta} = 0. \end{cases} \tag{6}$$

Denoting the set $\{1, \ldots, n\}$ by N, the letter $\dot{X}$ stands for first order derivatives of state variables $(\dot{x}_\jmath \,|\, \jmath \in N)$ w.r.t. time t and $F := (f_\jmath \,|\, \jmath \in N)$ is a finite subset of $\mathbb{K}(t, X, \Theta)$ where $\mathbb{K}$ is a subfield ($\mathbb{Q}$ for example) of $\mathbb{C}$. In order to determine the qualitative properties of the dynamical system (6), it is usual to consider the following systems for various subset J of N:

$$\Sigma_J \qquad \begin{cases} \dot{x}_\jmath = f_\jmath(t, X, \Theta), & \forall \jmath \in J, \\ \dot{x}_\imath = f_\imath(t, X, \Theta) = 0, & \forall \imath \in N \setminus J, \\ \dot{t} = 1, \quad \dot{\Theta} = 0, \end{cases} \tag{7}$$

in which some state variables are considered as parameters. In fact, for $\imath$ in N, the system $\Sigma_{\{\imath\}}$ defines the so-called $x_\imath$-nullcline of Σ and the purely algebraic system $\Sigma_\emptyset$ defines its fixed points (see examples of applications in [1]).

Remark 3. In the sequel, we are going to avoid—as much as possible—any distinction between time, state variables and parameters i.e. we work in an expanded state space (see [7] for another application of this standpoint); hence, let us denote the set (t, X, Θ) by $Z := (z_\imath \,|\, 1 \le \imath \le 1 + \ell + n)$ and its cardinal by m.

2.2   Infinitesimal Generators, Associated Flows and Their Rectification

First let us recall some basic facts about derivations.

Definition 4. *Given a polynomial algebra $\mathbb{K}[Z]$, a derivation of $\mathbb{K}[Z]$ with constant field $\mathbb{K}$ is an additive mapping $\delta : \mathbb{K}[Z] \to \mathbb{K}[Z]$ that satisfies Leibniz rules:*

$$\forall (f_1, f_2) \in \mathbb{K}[Z]^2, \quad \delta(f_1 f_2) = f_1 \delta f_2 + f_2 \delta f_1, \tag{8}$$

and have $\mathbb{K}$ in its kernel. We denote by $\mathrm{Der}_\mathbb{K} \mathbb{K}[Z]$ the set of all such derivations. The Lie bracket is defined by the $\mathbb{K}$-bilinear map:

$$\begin{aligned} [\,,\,] : \mathrm{Der}_\mathbb{K} \mathbb{K}[Z] \times \mathrm{Der}_\mathbb{K} \mathbb{K}[Z] &\to \mathrm{Der}_\mathbb{K} \mathbb{K}[Z], \\ (\delta_1, \delta_2) \qquad &\to \delta_1 \delta_2 - \delta_2 \delta_1. \end{aligned} \tag{9}$$

This map is skew-symmetric and satisfies the following Jacobi identity:

$$\forall (\delta_1, \delta_2, \delta_3) \subset \mathrm{Der}_\mathbb{K} \mathbb{K}[Z], \quad [\delta_1, [\delta_2, \delta_3]] + [\delta_2, [\delta_3, \delta_1]] + [\delta_3, [\delta_1, \delta_2]] = 0. \tag{10}$$

The set $\mathrm{Der}_\mathbb{K} \mathbb{K}[Z]$ is a $\mathbb{K}$ vector-space spanned by the set of canonical derivations $\{\partial/\partial z_1, \ldots, \partial/\partial z_m\}$. It is also a Lie algebra with Lie bracket as product.

Remark 5. An algebraic system Σ_J defined by (7) could be seen as a derivation:

$$D_J := \tfrac{\partial}{\partial t} + \sum_{j \in J} f_j \tfrac{\partial}{\partial x_j}, \tag{11}$$

associated to the algebraic relations $\{f_i = 0, \forall i \in N \setminus J\}$.

Derivations considered as infinitesimal generators. The *exponentiation* of a derivation δ induces several morphisms as shown by following definitions:

Definition 6. *Given a derivation δ and τ one of its constant ($\delta\tau = 0$), one can define the exponential map $e^{\tau\delta} := \sum_{i \in \mathbb{N}} \tau^i \delta^i / i!$ from $\mathbb{K}[Z]$ into the algebra $\mathbb{K}[[\tau, Z]]$ of power series in the indeterminates (τ, Z).*

1. *This map is a morphism that associates to any f in $\mathbb{K}[Z]$ its Lie series defined by the formal power series $\sum_{i \in \mathbb{N}} \tau^i \delta^i f / i!$.*
2. *The derivation δ is called the* infinitesimal generator *of $e^{\tau\delta}$.*
3. *The formal power series $e^{\tau\delta} Z$ are solutions of the vector field associated to δ. These series form the* formal flow *of δ; this derivation induces an infinitesimal transformation from $\mathbb{K} \times \mathbb{K}^m$ into $\mathbb{K}^m$ that associates, under suitable condition of convergence, the evaluation $(e^{\tau\delta} Z)(\mathcal{V})$ to any parameter τ in $\mathbb{K}$ and any initial point $\mathcal{V}$ in $\mathbb{K}^m$; this map is the action of the flow $e^{\tau\delta}$ on $\mathbb{K}^m$.*

Example 7. Hence, the $\mathbb{C}(x)$-morphism $\sigma_\tau : x \to e^\tau x$ could be defined by the exponential map $\sigma_\tau := e^{\tau\delta}$ where δ denotes the derivation $x\partial/\partial x$ acting on the field $\mathbb{C}(x)$. The set $\{\sigma_\tau \,|\, \tau \in \mathbb{C}\}$ is a one-parameter group of automorphisms.

Lemma 8. *Given two derivations ∂ and δ, the Baker Campbell Hausdorff formula states that the relation $e^\partial e^\delta = e^\delta e^\partial e^{[\partial,\delta]}$ holds.*

The next section presents how some derivations could be expressed as translation in a suitable coordinates set.

Some Algebraic Tools for Rectification of an Infinitesimal Generator

Principal element. Forthcoming manipulations are based on the existence of a special element that behaves as a *time* variable for considered derivation as shown by the following definition:

Definition 9. *An ρ element in an algebra A is principal for a derivation δ acting on A if the relation $\delta\rho = 1$ holds.*

To determine a principal element ρ of a derivation δ, one could solve the following partial differential equation $\delta\rho = 1$. As this is not a trivial task and as not every derivation has such an element, we are going in the sequel to restrict our manipulation to the following kind of principal elements:

Lemma 10. *Given a derivation δ of $\mathbb{K}[Z]$, if there exists an element ϱ in $\mathbb{K}[Z]$,*

1. *such that the relations $\delta\varrho \neq 0$ and $\delta^2\varrho = 0$ hold, then the fraction $\rho := \varrho/\delta\varrho$*
2. *and a constant λ of δ such that the relation $\delta\varrho = \lambda\varrho$ holds, then for any constant c of this derivation, the transcendental element $\rho := \log(c\varrho)/\lambda$*

is a principal element ρ of δ. The element ϱ is called the preprincipal element *of δ associated to ρ.*

Remark 11. As we adopt an algebraic standpoint in this note, this lemma requires to consider in the sequel the localization $\mathbb{K}[\rho, Z]_{(\wp)}$ of $\mathbb{K}[\rho, Z]$ at the multiplicative closed set $\wp := \{(\delta\varrho)^{\imath} \mid \imath \in \mathbb{N}\}$ (resp. $\wp := \{\varrho^{\imath} \mid \imath \in \mathbb{N}\}$). In fact, given any canonical derivation $\partial/\partial z$ of $\mathbb{K}[Z]$, there exists one, and only one, canonical derivation of $\mathbb{K}[\rho, Z]_{(\wp)}$ extending $\partial/\partial z$ and such that the following usual relations $\partial\rho/\partial z = \partial\varrho/\partial z - (\rho/(\delta\varrho)^2)\partial\delta\varrho/\partial z$ (resp. $\partial\rho/\partial z = (1/\varrho)\partial\varrho/\partial z$) are well defined in $\mathbb{K}[\rho, Z]_{(\wp)}$ (for the sake of simplicity, we use the same notation for derivations acting on $\mathbb{K}[Z]$ and their extension to derivation acting on $\mathbb{K}[\rho, Z]_{(\wp)}$). This shows that the Lie algebra $\mathrm{Der}_{\mathbb{K}}\mathbb{K}[\rho, Z]_{(\wp)}$ is well defined.

Example 12. For any element h in $\mathbb{K}(Z)$, we consider the logarithm $\log h$ i.e a transcendental field extension $\mathbb{K}(Z, \log h)$ and the associated derivation extension such that $\delta \log h = \delta h/h$. Hence, the derivation $\delta := x\partial/\partial x$ acting on $\mathbb{C}(x)$ has a unique extension to a derivation $\overline{\delta}$ acting on $\mathbb{C}(x, \log(x))$ such that the relation $\overline{\delta} \log(x) = 1$ holds.

Construction of a rectifying coordinate ring. Principal elements of a derivation δ allow to construct a rectifying field in which δ acts as a simple translation.

Lemma 13. *Given a derivation δ and one of its principal element ρ, let us define the following formal operator:*

$$\pi_{\delta,\rho} := \sum_{\imath \in \mathbb{N}} (-\rho)^{\imath} \frac{\delta^{\imath}}{\imath!}. \tag{12}$$

As δ is a derivation, this operator induces a homomorphism and the following exact sequence:

$$0 \to \ker \pi_{\delta,\rho} \to \mathbb{K}[\rho, Z]_{(\wp)} \xrightarrow{\pi_{\delta,\rho}} \mathbb{K}[[\rho, Z]]_{(\wp)} \to \mathbb{K}[\zeta] \to 0 \tag{13}$$

where the variables set ζ denotes the set $\pi_{\delta,\rho}Z$ of formal power series.

Remark 14. To prove that the map $\pi_{\delta,\rho}$ is a homomorphism, one can use the same argument then whose used in the proof stating the same property for the exponential map e^{δ}. By construction $\pi_{\delta,\rho}\rho$ is equal to 0. Thus, the kernel $\ker \pi_{\delta,\rho}$ contains the ideal $\rho\mathbb{K}[\rho, Z]_{(\wp)}$ and is not trivial (see also Proposition 30).

Using exact sequence (13), we could define a coordinate ring $\mathbb{K}[\zeta]$ that is isomorphic to the quotient algebra $\mathbb{K}[\rho, Z]_{(\wp)}/(\ker \pi_{\delta,\rho})\mathbb{K}[\rho, Z]_{(\wp)}$ and a *rectifying* ring $\mathbb{K}[\zeta, \rho]$ that is its finitely generated extension. To explain this terminology, first remark that the derivation δ acting on $\mathbb{K}[\rho, Z]_{(\wp)}$ could be easily extended to a derivation acting on $\mathbb{K}[[\rho, Z]]_{(\wp)}$ and thus to $\mathbb{K}[\rho, \zeta]$. The following lemma states that the derivation δ is *rectified* when we consider its action on $\mathbb{K}[\rho, \zeta]$:

Lemma 15. *With previously introduced notations, the following relations hold:*

$$\delta\rho = 1, \quad \delta\zeta = 0. \tag{14}$$

Sketch of proof. The first relation is the definition of a principal element. Elements ζ are defined by the series $\sum_{\imath \in \mathbb{N}} (-\rho)^\imath \delta^\imath Z / \imath!$. By Leibniz' rule, we have:

$$\forall \imath \in \mathbb{N}, \quad \delta\left((-\rho)^\imath \tfrac{\delta^\imath}{\imath!}\right) = (-1)^\imath \left(\rho^\imath \tfrac{\delta^{\imath+1}}{\imath!} + \rho^{\imath-1} \tfrac{\delta^\imath}{(\imath-1)!}\right), \tag{15}$$

and thus, derivation's linearity proves the last relations.

The morphism $\pi_{\delta,\rho}$ induces a coordinates change allowing to express the derivation δ as a simple translation $\partial/\partial\rho$ in this new coordinates set. We do not describe further this coordinates change, because we are just going to use some of its properties and not its exact formulation. Forthcoming considerations are based on the fact that the relations $\delta\zeta = 0$ imply that the relations $e^{\tau\delta}\zeta = \zeta$ hold. Thus, the morphism $\pi_{\delta,\rho}$ maps the coordinate ring $\mathbb{K}[Z]$ of the ambient space $\mathbb{K}^m$ onto a coordinate ring invariant under the action of the flow $e^{\tau\delta}$.

2.3   Expanded Lie Point Symmetries and Their Determining System

Let us define now the derivations used in the sequel.

Definition 16. *Given a derivation δ, an algebraic system Σ_J and the associated derivation D_J, δ is an infinitesimal generator of an expanded Lie point symmetry of Σ_J if there exists a constant λ in $\mathbb{K}$ such that the following relations hold:*

$$D_J\delta(t) = \frac{\partial\delta(t)}{\partial t} + \sum_{j \in J} f_j \frac{\partial\delta(t)}{\partial x_j} = -\lambda, \tag{16}$$

$$\sum_{z \in Z} \delta(z)\frac{\partial f_\imath}{\partial z} - \frac{\partial\delta(x_\imath)}{\partial t} - \sum_{j \in J} f_j \frac{\partial\delta(x_\imath)}{\partial x_j} = \lambda f_\imath, \quad \forall \imath \in J, \tag{17}$$

$$\delta f_\imath = \sum_{z \in Z} \delta(z)\frac{\partial f_\imath}{\partial z} = \lambda f_\imath, \quad \forall \imath \in N \setminus J, \tag{18}$$

$$D_J\delta(\theta) = \frac{\partial\delta(\theta)}{\partial t} + \sum_{j \in J} f_j \frac{\partial\delta(\theta)}{\partial x_j} = 0, \quad \forall\theta \in \Theta. \tag{19}$$

These relations form the determining system *of Σ_J expanded Lie point symmetries.*

Remark 17. Solution space structure — Derivations δ satisfying (16)–(19) form a Lie sub-algebra of $\mathrm{Der}_{\mathbb{K}}\mathbb{K}[Z]$ denoted by $\mathrm{LieSym}(\Sigma)$. Furthermore, if J_1 is a subset of J_2, the Lie algebra $\mathrm{LieSym}(\Sigma_{J_1})$ is a sub-algebra of $\mathrm{LieSym}(\Sigma_{J_2})$.

Remark 18. Considered Lie symmetries vs general Lie symmetries — The definition 16 is designed for our algebraic purposes but is only a restriction of the general definition of Lie point symmetries (see [4]). In fact, remark that if the considered algebraic system Σ_J is

- a vector field ($J = N$), this definition reduces to the classical one of Lie point symmetries based on the Lie bracket i.e. $[D, \delta] = \lambda D$ with the restriction that λ is not a general constant of the derivation D but a constant in $\mathbb{K}$;

– a purely algebraic system $(J = \emptyset)$, this definition is much more restrictive than the classical definition presented in Section 2.1 of [4]. In fact, let us consider the system $f_1 := x_1^2 + y_1^2 - 1$, $f_2 := x_2^2 + y_2^2 - 1$, $f_3 := x_2 y_1 - y_2 x_1$ and the derivation $\delta := x_2 \partial/\partial y_1 - y_2 \partial/\partial x_1 + x_1 \partial/\partial y_2 - y_1 \partial/\partial x_2$. The relations $\delta f_1 = 2f_3$, $\delta f_2 = -2f_3$, $\delta f_3 = f_1 - f_2$ show that the derivation δ leaves invariant the ideal spanned by $\{f_1, f_2, f_3\}$ and thus according to [4], δ is the infinitesimal generator of a one-parameter group of Lie symmetry (a family of morphism $e^{\tau \delta}$ parameterized by a constant τ) that leaves the variety associated to $\{f_1, f_2, f_3\}$ invariant ($e^{\tau \delta} f_1 = f_1 \cos^2 \tau + f_2 \sin^2 \tau + 2f_3 \cos \tau \sin \tau$). But δ does not satisfy Definition 16. This kind of derivations are not taken into account in this note because the general definition of a Lie point symmetry for an algebraic system $F = 0$ implies that we must perform our computations in the quotient algebra $\mathbb{K}[Z]/F\mathbb{K}[Z]$; as this task is not in the complexity class considered here, we made a first restriction to the set of Lie symmetry used in our work by only considering solution of system $(16)-(19)$.

Furthermore, there is little hope to solve the general partial differential problem $(16)-(19)$; thus, we restrict our solution space to *affine* infinitesimal generators for which the associated determining equations form a linear system.

3 Affine Derivation and Associated Invariantization

Definition 19. *Let us denote by* $\mathrm{AffDer}_{\mathbb{K}}\mathbb{K}[Z]$ *the following set of derivations:*

$$\left\{ \delta = \sum_{z \in Z} \delta(z) \frac{\partial}{\partial z} \,\middle|\, \delta(z_1) := b_{z_1} + \sum_{z_2 \in Z} a_{z_1 z_2} z_2, \; \left(b_{z_1}, a_{z_1 z_2} \mid z_2 \in Z \right) \in \mathbb{K}^{m+1} \right\}.$$

(20)

Note 20. Notation — Given a derivation δ in $\mathrm{AffDer}_{\mathbb{K}}\mathbb{K}[Z]$, we are going in the sequel to consider Z as a vector and use the following matricial notations:

$$\mathcal{A}_\delta = (a_{z_1 z_2})_{(z_1, z_2) \in Z^2}, \qquad \mathcal{B}_\delta = (b_z)_{z \in Z}, \qquad \delta Z = \mathcal{A}_\delta Z + \mathcal{B}_\delta. \qquad (21)$$

3.1 Determining System Defining Affine Infinitesimal Generators

Lemma 21. *For an affine infinitesimal generator* δ *in* $\mathrm{AffDer}_{\mathbb{K}}\mathbb{K}[Z]$, *the associated determining system* $(16)-(19)$ *reduces to the following linear system:*

$$\begin{pmatrix} 0 & \cdots & 0 \\ \frac{\partial f_1}{\partial z_1} & \cdots & \frac{\partial f_1}{\partial z_m} \\ \vdots & & \vdots \\ \frac{\partial f_n}{\partial z_1} & \cdots & \frac{\partial f_n}{\partial z_m} \\ 0 & \cdots & 0 \\ \vdots & & \vdots \\ 0 & \cdots & 0 \end{pmatrix} (\mathcal{A}_\delta Z + \mathcal{B}_\delta) - \mathcal{A}_\delta \begin{pmatrix} 1 \\ f_1 \mathbb{1}_{1 \in J} \\ \vdots \\ f_n \mathbb{1}_{n \in J} \\ 0 \\ \vdots \\ 0 \end{pmatrix} = \lambda \begin{pmatrix} 1 \\ f_1 \\ \vdots \\ f_n \\ 0 \\ \vdots \\ 0 \end{pmatrix}, \qquad (22)$$

where $\mathbb{1}_{\iota \in J}$ *is equal to* 1 *if the index* ι *is in* J *and* 0 *otherwise.*

*Remark 22. Probabilistic resolution of determining system defining affine deriva-
tion* — The system (22) could be rewritten in a the more convenient matricial
notation $M(Z)K = 0$ where $M(Z)$ is a $m \times (m+1)m$ matrix with coefficients
in $\mathbb{K}[Z]$ and K is a vector whose $(m+1)m$ coefficients are the coefficients of $\mathcal{A}_\delta$
and $\mathcal{B}_\delta$. Affine derivations that are solution of this determining system, are given
by the kernel of $M(Z)$ in a field $\mathbb{K}$. Kernel computation could be done by the following probabilistic method. Indeterminates Z are specialized in matrix $M(Z)$
to some random value in $\mathbb{K}^m$ in order to obtain a matrix M_1 over the field $\mathbb{K}$; the
resulting linear system M_1K could be underdetermined and thus, several spe-
cializations should be considered in order to obtain a linear system L_i defined
by $M_1K = \cdots = M_iK = 0$. The rank r_i of L_i increases with i and the specialization process could be stopped when $r_j = r_{j+1}$; the considered system L_j could
then be solved using a numerical method. The specialization set for which this
process fails to find a correct solution is a zero-dimensional algebraic variety and
thus, its probability of failure is low.

However, there is an infinite way to choose a basis of the kernel computed
above. But, one can use Lenstra, Lenstra and Lovász' basis reduction algorithm
in order to obtain a reduced basis in the sense that less variables are involved in
each infinitesimal generators definition.

To conclude, remark that some solutions of system (22) are spurious for our
purposes since they describe the same flow and should be discarded. In fact, consider the problem Σ defined by $\dot{x} = \theta x$: the base field of LieSym(Σ) is the con-
stant field of the derivation $D := \partial/\partial t + \theta x \partial/\partial x$; thus, as $\partial/\partial t$ is in LieSym(Σ)
and θ is a constant of D, $\theta\partial/\partial t$ is another infinitesimal generators representing
the same Lie symmetry then $\partial/\partial t$. These two derivations define the same Lie
symmetry but are given by two different solutions of system (22).

3.2   Principal Element Computation for Affine Derivation

Lemma 23. *Given a derivation δ in* AffDer$_\mathbb{K}\mathbb{K}[Z]$. *If there exits a vector of δ's
constants denoted by $C := (c_1, \ldots, c_m)$ such that the relations*

1. ${}^tC\mathcal{A}_\delta = 0$ *and* ${}^tC\mathcal{B}_\delta \neq 0$ *hold, then the fraction* ${}^tCZ/({}^tC \cdot (\mathcal{A}_\delta Z + \mathcal{B}_\delta))$
2. ${}^t\mathcal{A}_\delta C = \lambda C$ *and* $\delta\lambda = 0$ *hold, then the element* $\left(\log({}^tC \cdot (Z + \mathcal{B}_\delta/\lambda))\right)/\lambda$

is a principal element of δ.

Sketch of proof. 1) Consider the polynomial ${}^tCZ$ denoted by ϱ. Using nota-
tion (21), remark that $\delta\varrho$ is equal to the linear combination ${}^tC(\mathcal{A}_\delta Z + \mathcal{B}_\delta)$.
Thus, the conditions on C given in the first item show that $\delta\varrho$ is a constant
different of 0 and thus $\delta^2\varrho$ is equal to 0. The first assertion of Lemma 10 is
sufficient to conclude in that case. 2) Consider the polynomial ${}^tC(Z + \mathcal{B}_\delta/\lambda)$ denoted by ϱ. With the hypothesis of the second item, the element $\delta\varrho$ is equal to $\lambda\varrho$
and thus, the second assertion of lemma 10 is satisfied and the transcendental
element $\log(\varrho)/\lambda$ is a principal element of δ.

Remark 24. Computational strategy — This lemma shows that in order to find a principal element for a derivation δ in $\text{AffDer}_{\mathbb{K}}\mathbb{K}[Z]$, one have first to check condition 1) and if it is not satisfied, one have to find an eigenvector of $\mathcal{A}_\delta$.

3.3   Flow of a Affine Derivation and Resulting Quotient Space

Finding a coordinates change required to place a given derivation in rectified form is essentially the same problem as solving it in the first place. This could be easily done for affine derivation using the Jordan normal form.

Remark 25. Jordan normal form — Given a $m \times m$-matrix $\mathcal{A}_\delta$ associated to a derivation δ, if its minimal polynomial $p(\xi)$ is $\prod_{i=1}^{w} p_i$ with $p_i = (\xi - \lambda_i)^{\alpha_i}$ and $\sum_{i=1}^{w} \alpha_i = m$, then there exists a change of coordinates $\mathcal{P}$ such that:

$$\mathcal{A}_\delta = \mathcal{P} \begin{pmatrix} J_1 & 0 & \cdots & 0 \\ 0 & J_2 & \ddots & \vdots \\ \vdots & \ddots & \ddots & 0 \\ 0 & \cdots & 0 & J_w \end{pmatrix} \mathcal{P}^{-1}, \text{ with } J_i := \lambda_i \text{Id}_{\alpha_i \times \alpha_i} + \begin{pmatrix} 0 & 1 & 0 & \cdots & 0 \\ 0 & 0 & 1 & \ddots & \vdots \\ \vdots & & \ddots & \ddots & 0 \\ 0 & \cdots & & 0 & 1 \\ 0 & \cdots & & 0 & 0 \end{pmatrix} \quad (23)$$

if λ_i is different from 0 and $J_i := \lambda_i \text{Id}_{\alpha_i \times \alpha_i}$ otherwise; the symbol $\text{Id}_{\alpha_i \times \alpha_i}$ denotes the identity $\alpha_i \times \alpha_i$-matrix. This canonical form and Lemma 23 allow to compute preprincipal and associated principal element for any affine derivation δ.

Hypotheses 1. From now, we suppose that the base field $\mathbb{K}$ is $\mathbb{C}$ in order to contain all eigenvalues of the matrix $\mathcal{A}_\delta$ and to define the quantities related to principal elements (exponentials and logarithms).

Given an affine derivation δ, one of its preprincipal element ϱ and the associated principal element ρ, let us interpret geometrically the manipulation done in section 2.2.

Flow associated to a derivation and induced equivalence classes. To do so and following Definition 6-3, we consider the application defined by the linear system of ordinary differential equations associated to our affine derivation δ:

$$\begin{aligned} \Psi : \mathbb{K} \times \mathbb{K}^m &\to \quad \mathbb{K}^m, \\ (\tau, W) &\to \left(e^{\tau \delta} Z\right)(W) = \exp(\tau \mathcal{A}_\delta) W + \int_0^\tau \exp((\tau - s)\mathcal{A}_\delta) B_\delta ds. \end{aligned} \quad (24)$$

this application could be computed numerically or using the following relations:

$$\exp(\tau J_i) = \exp(\tau \lambda_i) \begin{pmatrix} 1 & \tau & \frac{\tau^2}{2} & \cdots & \frac{\tau^{\alpha_i-1}}{(\alpha_i-1)!} \\ 0 & 1 & \tau & \ddots & \vdots \\ \vdots & & \ddots & \ddots & \frac{\tau^2}{2} \\ \vdots & & & \ddots & \tau \\ 0 & \cdots & & 0 & 1 \end{pmatrix}. \quad (25)$$

Note 26. Orbits of the flow — The image of $\mathbb{K} \times \mathcal{W}$ by Ψ constitutes an orbit of the flow $e^{\tau\delta}$. This standpoint induces an equivalence relation $\sim$ among the point of $\mathbb{K}^m$, with $\mathcal{V}_1$ being equivalent to $\mathcal{V}_2$ if these points lie in the same orbit of Ψ. Let us denotes by $\mathcal{H}_\delta$ the set of equivalence classes a.k.a. the set of orbits. In the sequel, we suppose for the sake of conciseness that all the orbits have the same dimension i.e. we implicitly exclude from our statements the lower-dimensional orbits associated to the variety in $\mathbb{K}^m$ defined by the ideal $\{\delta(z) = 0 \,|\, z \in Z\}$.

All forthcoming manipulations rely on the following remark:

Remark 27. Invariantization — Any object (algebraic relations, derivations, etc.) that is invariant under the action of the flow $e^{\tau\delta}$ will have a counterpart on the lower-dimensional variety $\mathcal{H}_\delta$ whose representation—the *invariantization* of the considered object—completely characterize the original object.

Note 28. As a first illustration, let us remark that any function $f : \mathbb{K}^m \to \mathbb{K}$ invariant for the flow $e^{\tau\delta}$ is invariant along its orbits and therefore there is a well-defined induced function $\tilde{f} : \mathcal{H}_\delta \to \mathbb{K}$; conversely, given a function $\tilde{f} : \mathcal{H}_\delta \to \mathbb{K}$, there is an invariant function $f : \mathbb{K}^m \to \mathbb{K}$ defined by the relation $f(\mathcal{V}) := \tilde{f}(h)$ if $\mathcal{V}$ is in the orbit h. Hence, we obtain the following result:

Lemma 29. *There is a one-to-one correspondence between (polynomial) functions on $\mathbb{K}^m$ invariant under the action of the flow $e^{\tau\delta}$ and arbitrary (polynomial) functions on $\mathcal{H}_\delta$.*

In order to represent $\mathcal{H}_\delta$, one can first find an algebraic representation of the orbits of the flow and then use a cross section of these orbits (see [6] and references therein for more details). This general approach is based on Gröbner basis computation and treats general problems that exceed the scope of this note. Instead, we are going to use Lemmas 13 and 29 in order to give an algebraic description of $\mathcal{H}_\delta$ whose computation is based mainly on Jordan decomposition.

Algebraic representation of $\mathcal{H}_\delta$. Consider the formal operator $\pi_{\delta,\rho}$ introduced in Lemma 13. Lemma 15 implies that the image of $\pi_{\delta,\rho}$ is invariant under the action the flow $e^{\tau\delta}$. Thus, by describing the kernel of $\pi_{\delta,\rho}$, we obtain an algebraic description of functions on $\mathbb{K}^m$ invariant under the action of the flow $e^{\tau\delta}$. Lemma 29 shows that this description induces an algebraic representation of $\mathcal{H}_\delta$. The following lemma recapitulates these points when δ is an affine derivation:

Proposition 30. *Given an affine derivation δ, one of its preprincipal element ϱ and the associated principal element ρ, one can define a homomorphism $\pi_{\delta,\rho}$ on an algebra of constants $\mathbb{K}[\zeta]$ of δ using the formal operator (12) as follow:*

$$0 \to q_\rho \mathbb{K}[Z]_{(\wp)} \to \mathbb{K}[Z]_{(\wp)} \xrightarrow{\pi_{\delta,\rho}} \mathbb{K}[\zeta] \to 0 \tag{26}$$
$$\mathrm{p}(Z) \longrightarrow \mathrm{p}(\zeta)$$

where ζ is equal to $\pi_{\delta,\rho}Z$ and q_ρ is equal to ${}^tCZ$ if the preprincipal element ϱ is defined by the case 1) in Lemma 23 and to ${}^tC(Z + \mathcal{B}_\delta/\lambda) - 1$ otherwise. The set of equivalence classes $\mathcal{H}_\delta$ could be identify with the hyperplane $V(q_\rho)$ of dimension $m - 1$ defined in $\mathbb{K}^m$ by the linear form q_ρ. Furthermore,

1. *the map $\pi_{\delta,\rho}$ induces a projection that associates to any point $\mathcal{V}$ in $\mathbb{K}^m$ s.t. $\wp(\mathcal{V}) \neq 0$, the point $(\pi_{\delta,\rho}Z)(\mathcal{V}) := \sum_{i \in \mathbb{N}} ((-\rho(Z))^i (\delta^i Z)/i!)(\mathcal{V})$ of $V(q_\rho)$.*
2. *the points composing an orbit of $e^{\tau\delta}$ are projected to a single point in $V(q_\rho)$.*
3. *the orbit of $e^{\tau\delta}$ passing through a point in $V(q_\rho)$ is projected on this point.*

Sketch of proof. Consider a principal element ρ of δ and its defining preprincipal element ϱ such that $\delta\varrho = \mu\varrho$ (resp. $\delta\varrho \neq 0$ and $\delta^2\varrho = 0$). Then, the relation $\pi_{\delta,\rho}\varrho = \varrho e^{-\log(\varrho)}$ (resp. $\pi_{\delta,\rho}\varrho = 0$) holds and thus $\pi_{\delta,\rho}\varrho$ is equal to 1 (resp. 0) (to be more precise $\partial\pi_{\delta,\rho}\varrho/\partial z$ is equal to 0 for all z in Z and thus $\pi_{\delta,\rho}\varrho$ is in $\mathbb{K}$). If we denotes $\varrho - 1$ (resp. ϱ) by q_ρ, the ideal $q_\rho\mathbb{K}[Z]_{(\wp)}$ is include in ker $\pi_{\delta,\rho}$. Furthermore, as we suppose that the flow $e^{\tau\delta}$ acts regularly, its orbits have the same dimension 1 and thus, the associated invariant coordinates ring $\mathbb{K}[\zeta]$ is of dimension $m - 1$. Hence, the quotient algebra $\mathbb{K}[Z]_{(\wp)}/q_\rho\mathbb{K}[Z]_{(\wp)}$ is an algebraic description of $\mathbb{K}[\zeta]$. The assertion 1) is the definition of a projection on $V(q_\rho)$. 2) As the flow $e^{\tau\delta}$ is an homomorphism and relation $e^{\tau\delta}\delta = \delta e^{\tau\delta}$ holds, we have $\rho(e^{\tau\delta}Z)^i (\delta^i e^{\tau\delta}Z) = \rho(e^{\tau\delta}Z)^i (e^{\tau\delta}\delta^i Z) = e^{\tau\delta}(\rho(Z)^i \delta^i Z)$ for all integer i. In order to show that the relation $\rho(e^{\tau\delta}Z) = e^{\tau\delta}\rho(Z)$ holds, remark that as $e^{\tau\delta}$ is a homomorphism, $\varrho(e^{\tau\delta}Z)$ is equal to $e^{\tau\delta}\varrho(Z)$. If the principal element ρ is equal to $\log(\varrho)/\lambda$ (resp. $\varrho/\delta\varrho$) with $\delta\varrho = \lambda\varrho$ (resp. $\delta\varrho \neq 0$ and $\delta^2\varrho = 0$) then the transcendental element $\log(e^{\tau\delta}\varrho)/\lambda$ (resp. $e^{\tau\delta}(\varrho/\delta\varrho)$) is equal to $\log(\varrho e^{\tau\lambda})/\lambda$ (resp. $(\varrho + \tau\delta\varrho)/\delta(\varrho + \tau)$) and thus, $\rho(e^{\tau\delta}Z)$ is equal to $\rho + \tau$ which is also equal to $e^{\tau\delta}\rho$. Above relations show that $(\pi_{\delta,\rho}e^{\tau\delta}Z)(\mathcal{V})$ is equal to $(e^{\tau\delta}\pi_{\delta,\rho}Z)(\mathcal{V})$. As the flow leaves the image of $\pi_{\delta,\rho}$ invariant (i.e. $e^{\tau\delta}\pi_{\delta,\rho}Z = \pi_{\delta,\rho}Z$), this quantity is equal to $(\pi_{\delta,\rho}Z)(\mathcal{V})$. Hence, two points $\mathcal{V}$ and $(e^{\tau\delta}Z)(\mathcal{V})$ in the same orbit of $e^{\tau\delta}$ are projected onto the same point of $V(q_\rho)$. 3) If $\mathcal{W}$ is in $V(q_\rho)$ then $q_\rho(\mathcal{W})$ and $\rho(\mathcal{W})$ are equal to 0. Thus, by construction $(\pi_{\delta,\rho}Z)(\mathcal{W})$ is equal to $\mathcal{W}$.

Remark 31. The orbits of $e^{\tau\delta}$ cross the hyperplane $V(q_\rho)$ transversally ($\delta q_\rho \neq 0$). But, this hyperplane is not a generic cross-section of these orbits. The derivation $y\partial/\partial x - x\partial/\partial y$ is an infinitesimal generators of a Lie symmetry of the algebraic variety $V(f)$ defined by the polynomial $f : x^2 + y^2 - 1 = 0$. The preprincipal element described in Proposition 30 is $x - Iy$ (see Hypothesis 1) and thus the intersection of the hyperplane $V(x - Iy - 1)$ with $V(f)$ is the point $x = 1, y = 0$ while a generic linear cross section intersects $V(f)$ at 2 points.

In the sequel, we denote the hyperplane $V(q_\rho)$ by $\mathcal{H}_\delta$ (it is a covering algebraic variety of the invariants). Let us show now how works our reduction process.

3.4   Reduction Process: Invariantization and Parameterization

Algebraic systems. Consider the variety $V(F)$ in $\mathbb{K}^m$ defined by the ideal spanned by F in $\mathbb{K}[Z]_{(\wp)}$ (for the sake of simplicity, we suppose that F is prime). *Parameterization of an algebraic variety invariant under the action of $e^{\tau\delta}$.* If δ is an affine derivation such that the relation $\delta F = \lambda F$ holds, it is an infinitesimal generator of a Lie symmetry $e^{\tau\delta}$ that leaves the variety $V(F)$ invariant as shown by the following relations holding for all $\mathcal{W}$ in $\mathbb{K}^m$,

$$\mathcal{V} := (e^{\tau\delta}Z)(\mathcal{W}), \quad F(\mathcal{V}) := (F(e^{\tau\delta}Z))(\mathcal{W}) = (e^{\tau\delta}F(Z))(\mathcal{W}) = e^{\tau\lambda}F(\mathcal{W}). \quad (27)$$

As shown by Proposition 30, the hyperplane $\mathcal{H}_\delta$ is a linear cross-section of the orbit of $e^{\tau\delta}$ i.e. a variety that intersects these orbits in a single point. Furthermore, there is a variety W induced by $V(F)$ in $\mathcal{H}_\delta$ such that the variety $V(F)$ is the image of $\mathbb{K} \times W$ by the action of the flow $e^{\tau\delta}$. Let us described now W.

Invariantization of purely algebraic systems. The variety W is defined by the intersection $V(F \bmod q_\rho) \cap \mathcal{H}_\delta$ (if $q_\rho(W)$ and $(F \bmod q_\rho)(W)$ are equal to zero, then the relations $F(W) = 0$ hold). As q_ρ is linear, a description of W is obtained by a simple substitution in the equations describing $V(F)$ (compare with the replacement invariant studied in [6]) as shown by the following example:

Example 32. Let us consider the following purely algebraic system:

$$\Sigma: \quad (y - b)^2 + a^2 = l^2/4, \quad (x - a)^2 + b^2 = l^2/4, \quad x^2 + y^2 = l^2. \quad (28)$$

Using results of Section 3.2, we determine its expanded affine Lie symmetries:

$$\delta_1 := x\tfrac{\partial}{\partial x} + y\tfrac{\partial}{\partial y} + a\tfrac{\partial}{\partial a} + b\tfrac{\partial}{\partial b} + l\tfrac{\partial}{\partial l}, \; \delta_2 := -y\tfrac{\partial}{\partial x} + x\tfrac{\partial}{\partial y} + (b-y)\tfrac{\partial}{\partial a} + (x-a)\tfrac{\partial}{\partial b}. \quad (29)$$

As l is a preprincipal element of δ_1, the solutions Z of system (28) are represented by the parameterization $Z = e^{\tau\delta_1}Z_1$ where Z_1 are the solutions of an invariant system obtained by the intersection of (28) with the hyperplane $l = 1$:

$$\forall z \in \{x, y, a, b, l\}, \quad z = e^\tau z_1, \quad \begin{cases} (y_1 - b_1)^2 + a_1^2 = 1/4, \\ (x_1 - a_1)^2 + b_1^2 = 1/4, \\ x_1^2 + y_1^2 = 1. \end{cases} \quad l_1 = 1. \quad (30)$$

As δ_1 and δ_2 form an abelian Lie algebra, this last derivation is an infinitesimal generators of a Lie symmetry of δ_1; it could be used to reduce further the system (30) (the Lie algebra spanned by δ_1 and δ_2 is abelian and thus solvable). In fact, the linear form $a - Ib - x$ is a preprincipal element of δ_2 associated to the principal element $\log(a - Ib - x)/I$. Using symmetry δ_2, solutions of system (30) could be represented as follow:

$$\forall z_1 \in \{x_1, y_1, a_1, b_1\}, \quad z_1 = e^{\tau\delta_2}z_2, \quad \begin{cases} (y_2 - b_2)^2 + a_2^2 = 1/4, \\ (1 + Ib_2)^2 + b_2^2 = 1/4, \\ (a_2 - Ib_2 - 1)^2 + y_2^2 = 1. \end{cases} \quad x_2 = a_2 - Ib_2 - 1, \quad (31)$$

As noticed in remark 31, the cross-section $V(a_1 - Ib_1 - x_1 - 1)$ is not generic; while a linear generic cross section defines 4 points, its intersection with the variety defined by equations (30) reduces to the point $b_2 = 3I/8, a_2 = -5/8, y_2 = 3I/4$ of multiplicity 2. In this example, the variety associated to (28) is represented

- by a zero-dimensional algebraic system (31) that furnishes initial values Z_2
- to an explicit linear differential system whose solutions are $Z_1 = e^{\tau\delta_2}Z_2$ (this system associated to the derivation δ_2 is simple enough to be explicitly solved in closed form but in more complicated cases it could also be considered as a black box representation solved by purely numerical methods);
- these values Z_1 constitute an initial condition set of the linear differential system—induced by the derivation δ_1—such that resulting solutions Z parameterize the variety defined by system (31).

We show now that the same type of results exists for differential systems.

Differential Systems

Hypotheses 2. Restriction on symmetries specific to differential case — Lie symmetries of a given vector field D acting only on its state variables $(\dot z \neq 0)$ could be used for Lie based integration but not for the previous reduction process. Thus, we suppose that D have an expanded Lie symmetry that acts at least on one of its parameter $(Dz = 0)$ and that there is an associated principal elements such that the associated linear form q_ρ satisfies the relation $Dq_\rho = 0$.

Invariantization of an infinitesimal generator. Given any derivation D acting on $\mathbb{K}[Z]$, the sequence (26) induces a derivation $\overline D$ acting on $\mathbb{K}[\zeta]$ such that the relation $\pi_{\delta,\rho} \circ D = \overline D \circ \pi_{\delta,\rho}$ holds. The exponentiation of $\overline D$ (see Definitions 6) induces a flow $e^{\tau \overline D}$ on the hyperplane $\mathcal{H}_\delta$ that is the invariantization of the flow $e^{\tau D}$ acting on $\mathbb{K}^m$. Under above hypotheses, the flow $e^{\tau \overline D}$ is just the restriction of $e^{\tau D}$ on $\mathcal{H}_\delta$; in fact, as $Dq_\rho = 0$ the relation $e^{\tau D} q_\rho = q_\rho$ holds and thus, the flow $e^{\tau D}$ maps any point of $\mathcal{H}_\delta$ to another point of this hyperplane. The set of orbits of $e^{\tau D}$ in $\mathbb{K}^m$ is projected onto the set of orbits of $e^{\tau D}$ in $\mathcal{H}_\delta$. Let us see now the condition on δ and D that allows to parameterize the set of orbits of $e^{\tau D}$ in $\mathbb{K}^m$ by the set of orbits of $e^{\tau D}$ in $\mathcal{H}_\delta$ and the map $e^{\tau \delta}$.

Parameterization of vector field D invariant under the action of the flow $e^{\tau \delta}$. If δ is the infinitesimal generator of a symmetry of derivation D, according to Definition 16, the relation $[D, \delta] = \lambda D$ holds. The Baker Campbell Hausdorff formula (Lemma 8) shows that the relation $e^{\tau_1 \delta} e^{(1+\tau_1 \lambda)\tau_2 D} e^{-\tau_1 \delta} = e^{\tau_2 D}$ holds. This implies that any orbit of $e^{\tau D}$ in $\mathbb{K}^m$ is the image of an orbit of $e^{\tau D}$ in $\mathcal{H}_\delta$ by the flow $e^{\tau \delta}$. Let us explicit all the process described above through an example.

Example 33. Consider a FitzHugh Nagumo model (see § 7 in [1]):

$$\dot a = \dot b = \dot c = \dot d = 0, \quad \dot x = (x - x^3/3 - y + d)c, \quad \dot y = (x + a - by)/c. \quad (32)$$

The derivation $\delta := \partial/\partial y + b\partial/\partial a + \partial/\partial d$ is an infinitesimal generator of the following one-parameter group $y \to y + \lambda$, $a \to a + b\lambda$, $d \to d + \lambda$ that is composed of symmetries of the system (32). As the relation $\delta d = 1$ holds, d is a (pre)principal element of δ and the solutions $Z := \{x, y, a, b, c, d\}$ of system (32) are described by the parameterizations $Z = e^{d\delta} \mathsf{Z}$ where Z are solutions of a differential system on the hyperplane $V(d)$; hence, Z are given by the equations:

$$y = \mathsf{y} + d, \ a = \mathsf{a} - bd, \ \dot x = (x - x^3/3 - \mathsf{y})c, \ \dot{\mathsf{y}} = (x + \mathsf{a} - b\mathsf{y})/c. \quad (33)$$

4   Conclusion

In this note, we consider the computation of affine expanded Lie symmetries of a given algebraic system and show how this system could be rewrite in an invariant coordinates set for these symmetries in order to reduce the number of involved parameters. As this process is based on the computation of Jordan normal form and numerical linear algebra, its complexity is quasi-polynomial in input's size and likely polynomial for the great majority of practical cases.

Extension of the reduction process to more general types of derivations. The manipulation presented in previous sections for affine derivations could be used for non-affine symmetries that occurs in practice as shown below.

Example 34. Let us consider the following algebraic system:

$$f_1 : 6(z-y) - yz = 0, \quad f_2 : 3(z-x) - 2xz = 0, \quad f_3 : 2(y-x) - yx = 0, \quad (34)$$

in order to illustrate the limitation of our approach. The infinitesimal generator

$$\delta := x^2 \partial/\partial x + y^2 \partial/\partial y + z^2 \partial/\partial z, \\ (\delta f_1 = (z+y)f_1, \ \delta f_2 = (z+x)f_2, \ \delta f_3 = (x+y)f_3), \quad (35)$$

is associated to the following one-parameter group of automorphisms:

$$e^{\tau\delta} : x \to \tfrac{x}{1-x\tau}, \quad y \to \tfrac{y}{1-y\tau}, \quad z \to \tfrac{z}{1-z\tau}, \quad (36)$$

that is a one-parameter group of Lie point symmetries of the system (34). Remark that $\rho := -3xyz/(xz + xy + yz)$ is a principal element of (35) ($\delta\rho = 1$) and thus, all that we have done previously could be repeated. The invariant coordinate set:

$$\pi_{\delta,\rho} x = \tfrac{3xyz}{(xy - 2yz + zx)}, \quad \pi_{\delta,\rho} y = \tfrac{3xyz}{(xy + yz - 2zx)}, \quad \pi_{\delta,\rho} z = \tfrac{3xyz}{(2xy - yz - zx)}, \quad (37)$$

satisfies the relation $\pi_{\delta,\rho} x \, \pi_{\delta,\rho} y + \pi_{\delta,\rho} y \, \pi_{\delta,\rho} z + \pi_{\delta,\rho} z \, \pi_{\delta,\rho} x = 0$ and allows to represent the solutions set (x, y, z) of (34) by the 0 dimensional variety defined by the relations $f_1(x, y, z) = f_2(x, y, z) = f_3(x, y, z) = xy + yz + zx = 0$ and the parameterizations $x = e^{\tau\delta}x$, $y = e^{\tau\delta}y$, $z = e^{\tau\delta}z$ defined by the flow $e^{\tau\delta}$ of δ.

The results presented here could likely be extended for more general types of derivations but we do not know if the associated computations are feasible.

Acknowledgments. The author is grateful to G. Renault, É. Schost, and M. Safey El Din for many pleasant and useful discussions related to this note.

References

1. Murray, J.D.: Mathematical Biology. Interdisciplinary Applied Mathematics, vol. 17. Springer, Heidelberg (2002)
2. Khanin, R.: Dimensional Analysis in Computer Algebra. In: Mourrain, B. (ed.) Proceedings of the 2001 International Symposium on Symbolic and Algebraic Computation, London, Ontario, Canada, ACM, pp. 201–208. ACM press, New York (2001)
3. Bluman, G., Anco, S.: Symmetry and Integration Methods for Differential Equations. Applied Mathematical Sciences, vol. 154. Springer, Heidelberg (2002)
4. Olver, P.J.: Applications of Lie groups to differential equations, 2nd edn. Graduate Texts in Mathematics, vol. 107. Springer, Heidelberg (1993)
5. Gatermann, K.: Computer algebra methods for equivariant dynamical systems. Lecture Notes in Mathematics, vol. 1728. Springer, New York (2000)
6. Hubert, É., Kogan, I.: Rational invariants of an algebraic group action. Construction and rewriting. Journal of Symbolic Computation 42(1-2), 203 217 (2007)
7. Burde, G.I.: Expanded Lie group transformations and similarity reductions of differential equations. In: Nikitin, A.G., Boyko, V.M., Popovych, R.O. (eds.) Symmetry in nonlinear mathematical physics Part I. In: Proceedings of Institute of Mathematics of NAS of Ukraine, Kiev, Ukraine, vol 43, pp. 93–101 (2002)

Prefix Reversals on Binary and Ternary Strings[*]

Cor Hurkens[1], Leo van Iersel[1], Judith Keijsper[1], Steven Kelk[2], Leen Stougie[1,2],
and John Tromp[2]

[1] Technische Universiteit Eindhoven (TU/e), Den Dolech 2,
5612 AX Eindhoven, Netherlands
wscor@win.tue.nl, l.j.j.v.iersel@tue.nl, j.c.m.keijsper@tue.nl
[2] Centrum voor Wiskunde en Informatica (CWI), Kruislaan 413, 1098 SJ
Amsterdam, Netherlands
S.M.Kelk@cwi.nl, Leen.Stougie@cwi.nl, John.Tromp@cwi.nl

Abstract. Given a permutation π, the application of prefix reversal $f^{(i)}$ to π reverses the order of the first i elements of π. The problem of Sorting By Prefix Reversals (also known as *pancake flipping*), made famous by Gates and Papadimitriou (Bounds for sorting by prefix reversal, *Discrete Mathematics* 27, pp. 47-57), asks for the minimum number of prefix reversals required to sort the elements of a given permutation. In this paper we study a variant of this problem where the prefix reversals act not on permutations but on strings over a fixed size alphabet. We determine the minimum number of prefix reversals required to sort binary and ternary strings, with polynomial-time algorithms for these sorting problems as a result; demonstrate that computing the minimum prefix reversal distance between two binary strings is NP-hard; give an exact expression for the prefix reversal diameter of binary strings, and give bounds on the prefix reversal diameter of ternary strings. We also consider a weaker form of sorting called *grouping* (of identical symbols) and give polynomial-time algorithms for optimally grouping binary and ternary strings. A number of intriguing open problems are also discussed.

1 Introduction

For a permutation $\pi = \pi(0)\pi(1)\ldots\pi(n-1)$ the application of *prefix reversal* $f^{(i)}$, which we call *flip* for short, to π reverses the order of the first i elements: $f^{(i)}(\pi) = \pi(i-1)\ldots\pi(0)\pi(i)\ldots\pi(n-1)$. The problem of *Sorting By Prefix Reversals* (MIN-SBPR), popularised by Gates and Papadimitriou [11] and often referred to as the *pancake flipping problem*, is defined as follows: given a permutation π of $\{0, 1, \ldots, n-1\}$, determine its sorting distance i.e. the smallest number of flips required to transform π into the identity permutation $01\ldots(n-1)$.[1]

MIN-SBPR arises in the context of computational biology when seeking to explain the genetic difference between two given species by the most parsimonious (i.e. shortest) sequence of gene rearrangements. It is one of a family of

---

[*] This research has been funded by the Dutch BSIK/BRICKS project.
[1] We adopt the convention of numbering from 0 rather than 1.

H. Anai, K. Horimoto, and T. Kutsia (Eds.): AB 2007, LNCS 4545, pp. 292–306, 2007.
© Springer-Verlag Berlin Heidelberg 2007

genome rearrangement operations that also includes arbitrary (substring) reversals [13], *transpositions* (where two adjacent substrings are swapped) [6], and *translocations* (whereby, in the context of multiple chromosomes, the exchange of chromosome-ends is simulated) [2]. To extend biological relevance it is now commonplace that these operations are studied not only in isolation but also in (weighted) combination with each other [1]; MIN-SBPR is arguably most biologically applicable in the context of such weighted combinations. MIN-SBPR also has relevance in the area of efficient network design [14,16].

The computational complexity of MIN-SBPR, however, remains open. A recent 2-approximation algorithm [8] is currently the best-known approximation result[2]. Indeed, most studies to date have focused not on the computational complexity of MIN-SBPR but rather on determining the worst-case sorting distance $wc(n)$ over all length-n permutations i.e. the "worst case scenario" for length-n permutations. From [11] and [14] we know that $(15/14)n \leq wc(n) \leq (5n+5)/3$.

A natural variant of MIN-SBPR is to consider the action of flips not on permutations but on strings over fixed size alphabets. This shift is inspired by the biological observation that multiple "copies" of the same gene can appear at various places along the genome, although this does not lead to the bounded size alphabets that we will study here. The shift from permutations to strings alters the problem universe somewhat. With permutations, for example, the *distance problem*, i.e. given two permutations π_1 and π_2, determine the smallest number of flips required to transform π_1 into π_2, is equivalent to sorting, because the symbols can simply be relabelled to make either permutation equal to the identity permutation. For strings like 101, such a relabelling is not possible. Thus, the distance problem on string pairs appears to be strictly more general than the sorting problem on strings, naturally defined as putting all elements in non-descending order.

Indeed, papers by Christie and Irving [4] and Radcliffe, Scott and Wilmer [17] explore the consequences of switching from permutations to strings; they both consider arbitrary (substring) reversals and transpositions. It has been noted that, viewed as a whole, such rearrangement operations on strings have bearing on the study of orthologous gene assignment [3], especially where the level of symbol repetition in the strings is low. There is also a somewhat surprising link with the relatively unexplored family of *string partitioning* problems [12]. To put our work in context, we briefly describe the most relevant (for this paper) results from [4] and [17].

The earlier paper [4], gives, in both the case of reversals and transpositions, polynomial-time algorithms for computing the minimum number of operations to sort a given binary string, as well as exact, constructive diameter results on binary strings. Additionally, their proof that computing the reversal distance between strings is NP-hard, supports the intuition that distance problems are harder than sorting problems on strings. They present upper and lower bounds for computing reversal and transposition distance on binary strings.

---

[2] Although not explicitly described as such, the algorithm provided ten years earlier in [5] is a 2-approximation algorithm for the *signed* version of the problem.

The more recent paper [17] gives refined and generalised reversal diameter results for non-fixed size alphabets. It also gives a polynomial-time algorithm for optimally sorting a ternary (3 letter alphabet) string with reversals. The authors refer to the prefix reversal counterparts of these (and other) results as interesting open problems. They further provide an alternative proof of Christie and Irving's NP-hardness result for reversals, and sketch a proof that computing the *transposition distance* between binary strings is NP-hard. As we later note, this proof can also be used to obtain a specific reducibility result for prefix reversals. They also have some first results on approximation (giving a PTAS - a *Polynomial-Time Approximation Scheme* - for computing the distance between *dense instances*) and on the distance between random strings, both of which apply to prefix reversals as well.

In this paper we supplement results of [4] and [17] by their counterparts on prefix reversals. In Section 3 (*Grouping*) we introduce a weaker form of sorting where identical symbols need only be grouped together, while the groups can be in any order. For grouping on binary and ternary strings we give a complete characterisation of the minimum number of flips required to group a string, and provide polynomial-time algorithms for computing such an optimal sequence of flips. (The complexity of grouping over larger fixed size alphabets remains open but as an intermediate result we describe how a PTAS can be constructed for each such problem.) Grouping aids in developing a deeper understanding of sorting which is why we tackle it first. It was also mentioned as a problem of interest in its own right by Eriksson et al. [7]. Then, in Section 4 (*Sorting*), we give polynomial-time algorithms (again based on a complete characterisation) for optimally sorting binary and ternary strings with flips. (The complexity of sorting also remains open for larger fixed size alphabets. As with grouping we thus provide, as an intermediate result, a PTAS for each such problem.) In Section 5 we show that the flip diameter (i.e. the maximum distance between any two strings) on binary strings is $n-1$, and on ternary strings (for $n > 3$) lies somewhere between $n-1$ and $(4/3)n$, with empirical support for the former. In Section 6 we show that the flip distance problem on binary strings is NP-hard, and point out that a reduction in [17] also applies to prefix reversals, showing that the flip distance problem on *arbitrary* strings is polynomial-time reducible (in an approximation-preserving sense) to the binary problem. We conclude in Section 7 with a discussion of some of the intriguing open problems that have emerged during this work. Indeed, our initial exploration has identified many basic (yet surprisingly difficult) combinatorial problems that deserve further analysis.

2   Preliminaries

Let $[k]$ denote the first k non-negative integers $\{0, 1, ..., k-1\}$. A k-ary string is a string over the alphabet $[k]$, while a string s is said to be *fully k-ary*, or to *have arity k*, if the set of symbols occuring in it is $[k]$.

We index the symbols in a string s of length n from 1 through n: $s = s_1 s_2 \ldots s_n$. Two strings are *compatible* if they have the same symbol frequencies (and hence the same length), e.g. 0012 and 1002 are compatible but 0012 and 0112 are not. For a given string s, let $I(s)$ be the string obtained by sorting the symbols of s in non-descending order e.g. $I(1022011) = 0011122$. The prefix reversal (flip for short) $f^{(i)}(s)$ reverses the length i prefix of its argument, which should have length at least i. Alternatively, we denote application of $f^{(i)}(s)$ by underlining the length i prefix. Thus, $f^{(2)}(2012) = \underline{20}12 = 0212$ and $f^{(3)}(2012) = \underline{201}2 = 1022$. The *flip distance* $d(s, s')$ between two strings s and s' is defined as the smallest number of flips required to transform s into s', if they are compatible and ∞ otherwise. Since a flip is its own inverse, flip distance is symmetric.

The *flip sorting distance* $d_s(s) = d(s, I(s))$ of a string s is defined as the number of flips of an *optimal sorting sequence* to transform s into $I(s)$. An algorithm sorts s optimally if it computes an optimal sorting sequence for s.

In the next two sections we consider strings to be equivalent if one can be transformed into the other by repeatedly duplicating symbols and eliminating one of two adjacent identical symbols. As representatives of the equivalence classes we take the shortest string in each class. These are exactly the strings in which adjacent symbols always differ. We express all flip operations in terms of these *normalized* strings. E.g. we write $f^{(3)}(2012) = \underline{201}2 = 102$. A flip that brings two identical symbols together, thereby shortening the string by 1, is called a *1-flip*, while all others, that leave the string length invariant, are called *0-flips*.

We follow the standard notation for regular expressions: Superindex ^{i} on a substring denotes the number of repetitions of the substring, with * and $^+$ denoting 0-or-more and 1-or-more repetitions, respectively, ϵ denotes the empty string, brackets of the form {} are used to denote that a symbol can be exactly one of the elements within the brackets, and the product sign $\prod$ denotes concatenation of an indexed series. For example $\prod_{i=1}^{3}(10^i 2) = 102100210002$, and $\{1, 01\}^*\{\epsilon, 0\}$ denotes the set of binary strings with no 00 substring.

3   Grouping

The task of sorting a string can be broken down into two subproblems: *grouping* identical symbols together and putting the groups of identical symbols in the right *order*. Notice that first grouping and then ordering may not be the most efficient way to sort strings. Although grouping appears to be slightly easier than the sorting problem, essentially the same questions remain open as in sorting. Grouping binary strings is trivial and in Section 3.1 we give the grouping distances of all ternary strings. As a result we give polynomial time algorithms for binary and ternary grouping. For larger alphabets the grouping problem remains open; as an intermediate result we describe in Section 3.2 a PTAS for each such problem. While the problems of grouping and sorting are closely related for strings on small alphabets, the problems diverge when alphabet size approaches the string length, with permutations being the limit.

Recall that we consider only normalized strings, as representatives of equivalence classes. The *flip grouping distance* $d_g(s)$ of a fully k-ary string s is defined as the minimum number of flips required to reduce the string to one of length k.

3.1   Grouping Binary and Ternary Strings

Lemma 1. $d_g(s) \geq n - k$ *for any fully k-ary string s of length n.*

Proof. The proof follows from the observations that, after grouping, fully k-ary string s has length k and that each flip can shorten s by at most 1.   □

Lemma 2. $d_g(s) \leq n - 2$ *for any fully k-ary string s of length n.*

Proof. Consider the following simple algorithm. If the leading symbol occurs elsewhere then a 1-flip bringing them together exists, so perform this 1-flip. If not, then we use a 0-flip to put this symbol in front of a suffix in which we accumulate uniquely appearing symbols. Repeat until the string is grouped.

Clearly no more than $n - k$ 1-flips will be necessary. Also, no more than $k - 2$ 0-flips will ever be necessary, because after $k - 2$ 0-flips the prefix of the string will consist of only two types of symbol, and the algorithm will never perform a 0-move on such a string. Thus at most $(n - k) + (k - 2) = n - 2$ flips in total will be needed.   □

As a corollary we obtain the grouping distance of binary strings.

Theorem 1. $d_g(s) = n - 2$ *for any fully binary string s of length n.*   □

We will now define a class of bad ternary strings and prove that these are the only ternary strings that need $n - 2$ rather than $n - 3$ flips to be grouped.

Definition 1. *We define* bad *strings as all fully ternary strings of one of the following types, up to relabeling:*

I. *strings of length greater than 3, in which the leading symbol appears only once: $0(12)^{\geq 2}$ and $02(12)^{+}$*

II. *strings having identical symbols at every other position, starting from the last: $(\{0, 1\}2)^{+}$ and $(2\{0, 1\})^{+}2$*

III. *odd length strings whose leading symbol appears exactly once more, at an even position, and both occurrences are followed by the same symbol: $0(21)^{+}02(12)^{*}$*

IV. *the following strings:*
    $X_1 = 210212$, $X_2 = 021012$, $X_3 = 0120212$, $X_4 = 1201212$, $X_5 = 02101212$, $X_6 = 20210212$, $X_7 = 020210212$, $X_8 = 120120212$.

All other fully ternary strings are good. *Strings of type I, II and III, shortly I-, II-, and III-strings, respectively, are called generically bad, or g-bad for short.*

Lemma 3. $d_g(s) = n - 2$ *if ternary string s of length n is bad.*

Proof. Because of Lemmas 1 and 2, it suffices to show that in each case a 0-flip is necessary: I-strings admit only 0-flips. A 1-flip on a II-string leads to a II-string and eventually to a I-string. Any III-string admits only one 1-flip leading to a II-string. For IV-strings it is easy to check that each possible 1-flip leads to either a shorter IV-string, or to a I-,II-, or III-string. A full proof can be found in [15]. □

Lemma 4. $d_g(s) = n - 3$ *if ternary string s of length n is good.*

Proof. The proof is by induction on n. The induction basis for $n = 3$ is trivial. We show the statement for strings of length $n + 1$ by showing that if a bad string s' of length n can be obtained through a 1-flip from a good (*parent*) string s of length $n + 1$, then s admits another 1-flip which leads to a good string. Note that a 1-flip $f^{(i)}(s) = s'$ brings symbols s_1 and s_{i+1} together, hence $s_1 = s_{i+1} \neq s_i = s'_1$ which shows that the symbol deleted from *parent* s differs from the leading symbol of *child* s'. We enumerate all possible bad child strings s' and distinguish cases based on the leading symbol of good parent s.

For IV-strings, Table 1 lists all parents with, for each good parent, a 1-flip to a good string. It remains to prove that for each g-bad string all parents are either bad or have a g-1-flip, defined as a 1-flip resulting in a string that is not g-bad (i.e. either good or of type IV).

Type I, odd: $0(12)^{\geq 2}$ has possible parents starting with:
  1: $1(21)^i012(12)^j$ with $i + j > 0$:
   If $i > 0$ there is a g-1-flip $\underline{121}(21)^{i-1}012(12)^j = (21)^i012(12)^j$;
   If $i = 0$ and $j > 0$ there is a g-1-flip $\underline{1012}(12)^j = 210(12)^j$;
  2: $21(21)^i02(12)^j$ with $i + j > 0$.
   If $i > 0$ there is a g-1-flip $\underline{21}(21)^i02(12)^j = 1(21)^i02(12)^j$;
   If $i = 0$ and $j > 1$ there is a g-1-flip $\underline{210212}(12)^{j-1} = 120(12)^j$;
   If $i = 0$ and $j = 1$ the parent is $210212 = X_1$.
Type I, even: these strings are also of type II, see below.
Type II, odd: $(2\{0,1\})^+2$ has only parents of type II.
Type II, even: $02(\{0,1\}2)^*$ has possible parents starting with:
  2: $2(\{0,1\}2)^*$ is of type II;
  1: $12(\{0,1\}2)^*012(\{0,1\}2)^*$ with three cases for a possible third 1:
   None: parent is $12(02)^*012(02)^*$, which is of type III;
   Before 01: then there is a g-1-flip
     $\underline{12(\{0,1\}2)^*12}(\{0,1\}2)^*012(\{0,1\}2)^*$
     $= 2(\{0,1\}2)^*12(\{0,1\}2)^*012(\{0,1\}2)^*$;
   After 01: then there is a g-1-flip
     $\underline{12(\{0,1\}2)^*012}(\{0,1\}2)^*12(\{0,1\}2)^*$
     $= 2(\{0,1\}2)^*102(\{0,1\}2)^*12(\{0,1\}2)^*$.
Type III: $0(21)^+02(12)^*$ has possible parents starting with:
  1: $(12)^i01(21)^j02(12)^k$ with $i > 0$:
   If $i > 1$ there is a g-1-flip
     $\underline{12}(12)^{i-1}01(21)^j02(12)^k = 2(12)^{i-1}01(21)^j02(12)^k$;
   If $i = 1, j > 0$ there is a g-1-flip
     $\underline{120121}(21)^{j-1}02(12)^k = 21021(21)^{j-1}02(12)^k$;

If $i = 1, j = 0, k > 0$ there is a g-1-flip $\underline{120102}(12)^k = 20102(12)^k$;

If $i = 1, j = k = 0$ then the parent is $120102 = X_2$ (relabelled);

1: $(12)^+0(12)^+0(12)^+$: there is a g-1-flip
$\underline{(12)^+0(12)^+0(12)^+} = 0(21)^+20(12)^+$;

2: $\overline{2(12)^*0(21)^+02(12)^*}$: there is a g-1-flip
$2(12)^*0(21)^+02(12)^* = 0(12)^+0(21)^*2$;

2: $\overline{(21)^i20(12)^j02(12)^k}$ with $j > 0$:

If $i = 0, j = 1$ then the parent is $210212 = X_1$;

If $i + j > 1$ then $\underline{(21)^i2012}(12)^{j-1}02(12)^k = 102(12)^{i+j-1}02(12)^k$ is a
g-1-flip. ▢

Table 1. Type IV strings, their parents, and for each good parent, a 1-flip to a good string

IV-String	Parents
$X_1 = 210212$	$\underline{1210212}$, $0120212 = X_3$, $1201212 = X_4$
$X_2 = 021012$	$\underline{2021012}$, $\underline{1201012}$, $\underline{1012012}$, $\underline{2010212}$
$X_3 = 0120212$	$\underline{10120212}$, $\underline{21020212}$, $20210212 = X_6$, $\underline{12021012}$, $\underline{20212012}$
$X_4 = 1201212$	$\underline{21201212}$, $02101212 = X_5$, $\underline{21021212}$, $\underline{21210212}$
$X_5 = 02101212$	$\underline{202101212}$, $\underline{120101212}$, $\underline{101201212}$, $\underline{210120212}$, $\underline{121012012}$, 202010212
$X_6 = 20210212$	$020210212 = X_7$, $\underline{120210212}$, $\underline{012020212}$, $120120212 = X_8$
$X_7 = 020210212$	$\underline{2020210212}$, $\underline{2020210212}$, $\underline{1202010212}$, $\underline{2012020212}$, $\underline{1201202012}$, $\underline{2021021212}$
$X_8 = 120120212$	$\underline{2120120212}$, $\underline{0210120212}$, $\underline{2102120212}$, $\underline{0210210212}$, $\underline{2021021212}$, $\underline{2120210212}$

The following theorem results directly from the above lemmas.

Theorem 2. $d_g(s) = n - 2$ *if and only if fully ternary string s of length n is bad and $d_g(s) = n - 3$ otherwise. Moreover, there exists a polynomial time algorithm for grouping ternary strings with a minimum number of flips.*

Proof. The first statement is direct from Lemmas 3 and 4. In case string s is bad, which by Definition 1 can be decided in polynomial time, the algorithm implicit in the proof of Lemma 2 shows how to group s optimally in polynomial time. Otherwise, we repeatedly find a 1-flip to a good string as guaranteed by Lemma 4. The time complexity is $O(n^3)$, since grouping distance, number of choices for a 1-flip, and time to perform a flip and test whether its result is good are all $O(n)$. ▢

3.2   Grouping Strings over Larger Alphabets

Lemmas 1 and 2 say that $n - k \leq d_g(s) \leq n - 2$ for any fully k-ary string s. For any k there are fully k-ary strings that have flip grouping distance equal to

$n - 2$. For example the length $n = 2(k-1)$ string $1020\ldots(k-1)0$ requires for every 1-flip to bring a 0 to the front first and hence we need as many 0-flips as 1-flips, and $d_g(1020\ldots(k-1)0) \geq 2(k-2) = 2k-4 = n-2$. Computer calculations suggest that for $k = 4$ and $k = 5$, for n large enough, the strings with grouping distance $n-2$ are precisely those having identical symbols at every other position, starting from the last (i.e. type II of Definition 1). Proving (or disproving) this statement remains open, as well as finding a polynomial time algorithm for grouping k-ary strings for any fixed $k > 3$. We do, however, have the following intermediate result:

Theorem 3. *For every fixed k there is a PTAS for grouping k-ary strings.*

Proof. Follows from the algorithm in the proof of Lemma 2. We defer the details to a full version of the paper [15]. □

Clearly, there is a strong relationship between grouping and sorting. Understanding grouping may help us to understand sorting, and lead to improved bounds (especially as the length of strings becomes large relative to their arity), because for a k-ary string s, we have $d_g(s) \leq d_s(s) \leq d_g(s) + wc(k)$, with $wc(k)$ the flip diameter on permutations with k elements, as defined before.

Also $d_g(s) = \min\{d_s(t) : t \text{ a relabeling of } s\}$, which gives (for fixed k) a polynomial time reduction from grouping to sorting. Thus every polynomial time algorithm for sorting by prefix reversals directly gives a polynomial time algorithm for the grouping problem (for fixed k).

4   Sorting

In this section we present results on sorting similar to those on grouping in the previous section. Also flip sorting distance remains open for strings over alphabets of size larger than 3. As an intermediate result we thus provide at the end of this section a PTAS for each such problem.

Again a *1-flip* brings identical symbols together and thus shortens the representative of the equivalence class under symbol duplication. But since symbol order matters for sorting, relabelled strings are no longer equivalent. As in grouping, sorting of binary strings is straightforward:

Theorem 4. $d_s(s) = n-2$ *for every fully binary string s of length n with $s_n = 1$, and $d_s(s) = n-1$ otherwise.*

Proof. Exactly $n - 2$ 1-flips suffice and are necessary to arrive at length 2 string 01 or 10. If the last symbol is 0 an additional 0-flip is necessary putting a 1 at the end. All these flips can be $f^{(2)}$. □

From Lemma 1 we know that $d_g(s) \geq n - 3$ and hence $d_s(s) \geq n - 3$ for every ternary string s of length n. In the upper bound on $d_s(s)$ we derive below we focus on strings s ending in a 2 ($s_n = 2$), since sorting distance is invariant under appending a 2 to a string. It turns out that, when sorting a ternary string ending in a 2, one needs at most one 0-flip, except for the string 0212.

Lemma 5. $d_s(s) \leq n-2$ *for every fully ternary string s of length n with $s_n = 2$, except 0212.*

Proof. It is easy to check that 0212 requires 3 flips to be sorted. By induction on n we prove the rest of the lemma. The basis case of $n = 3$ is trivial. For a string s of length $n > 3$ we distinguish three cases:

- $s_{n-1} = 0$: If $s = 20102$ it is sorted in 3 flips: $\underline{20102} \to \underline{0102} \to \underline{102} \to 012$. Otherwise, by induction and relabeling $0 \leftrightarrow 2$, the string $s_1 \ldots s_{n-1}$ can be reduced to 210 in $n-3$ flips (to 20 or 10 by Theorem 4 if $s_1 \ldots s_{n-1}$ has only two symbols), and one more flip sorts s to 012.
- $s_{n-1} = 1$, $s_1 = 0$ and appears only once: Thus $s = 0(12)^{\geq 2}$ or $s = 02(12)^{\geq 2}$. Then s can be sorted with only one 0-flip: $0(12)^+12 \to 1(21)^+02 \to \ldots \to 2102 \to 012$ or, respectively, $\underline{02}(12)^{\geq 2} \to \underline{20(12)^+12} \to \overline{(12)^+102} \to \ldots \to 2102 \to 012$.
- $s_{n-1} = 1$, s_1 not unique:
  If $s = 12012$ then 3 flips suffice: $\underline{12012} \to \underline{21012} \to \underline{1012} \to 012$.
  Otherwise, since the other 2 parents of 0212 can flip to 1202, there is a 1-flip to a string $\neq 0212$ to which we can apply the induction hypothesis. $\square$

As in Section 3, we characterise the strings ending in a 2 that need $n-2$ rather than $n-3$ flips to sort.

Definition 2. *We define* bad *strings as all fully ternary strings ending in a 2 of the types:*

- I. $0(12)^{\geq 2}$
- II. $(\{0,1\}2)^+$ *and* $2(\{0,1\}2)^+$
- III. $(\{1,2\}0)^+2$ *and* $0(\{1,2\}0)^+$
- IV. $(\{1,2\}0)^+12$ *and* $(0\{1,2\})^+012$ *with at least two 2s.*
- V. $(01)^*0212$ *and* $(10)^+212$
- VI. $1(20)^+1(20)^*2$ *and* $0(21)^+0(21)^*2$
- VII. $1(02)^+1(02)^+$
- VIII. $1(02)^+12$
- IX. *77 strings of length at most 11, shown in Table 2.*

All other fully ternary strings ending in a 2 are good *strings. Strings of type I-VIII (I-strings ... VIII-strings for short) are called generically bad, or* g-bad *for short.*

This definition makes 0212 a bad string as well. From Lemma 5 we know that 0212 is the only ternary string ending in a 2 with sorting distance $n-1$.

Theorem 5. *String 0212 has sorting distance 3. Any other fully ternary string s of length n with $s_n = 2$ has prefix reversal sorting distance $n-2$ if it is bad and $n-3$ if it is good. A fully ternary string s ending in a 0 or 1 has the same sorting distance as $s2$.*

Table 2. Type IX strings

$Y_1 = 210212$	$Y_{17} = 10210212$	$Y_{33} = 12120102$	$Y_{48} = 010210212$	$Y_{63} = 1021201012$
$Y_2 = 021012$	$Y_{18} = 21021212$	$Y_{34} = 12010212$	$Y_{49} = 010210202$	$Y_{64} = 1020210212$
$Y_3 = 212012$	$Y_{19} = 02102012$	$Y_{35} = 12010202$	$Y_{50} = 010212012$	$Y_{65} = 1010210202$
$Y_4 = 120102$	$Y_{20} = 02101212$	$Y_{36} = 20120102$	$Y_{51} = 202010212$	$Y_{66} = 0202010212$
$Y_5 = 201202$	$Y_{21} = 10212012$	$Y_{37} = 12012012$	$Y_{52} = 121202012$	$Y_{67} = 2120202012$
$Y_6 = 0210202$	$Y_{22} = 02121012$	$Y_{38} = 021021202$	$Y_{53} = 121201202$	$Y_{68} = 2120102012$
$Y_7 = 1021202$	$Y_{23} = 02120102$	$Y_{39} = 102120102$	$Y_{54} = 201021202$	$Y_{69} = 2021021212$
$Y_8 = 0212012$	$Y_{24} = 10102102$	$Y_{40} = 102010212$	$Y_{55} = 120212012$	$Y_{70} = 2010212012$
$Y_9 = 2120102$	$Y_{25} = 02010212$	$Y_{41} = 021202012$	$Y_{56} = 012021212$	$Y_{71} = 1201021202$
$Y_{10} = 0102102$	$Y_{26} = 21202012$	$Y_{42} = 021201012$	$Y_{57} = 120102012$	$Y_{72} = 1201202012$
$Y_{11} = 1212012$	$Y_{27} = 21201012$	$Y_{43} = 020210212$	$Y_{58} = 201202012$	$Y_{73} = 10202010212$
$Y_{12} = 2010212$	$Y_{28} = 21201202$	$Y_{44} = 101020212$	$Y_{59} = 120120212$	$Y_{74} = 02120102012$
$Y_{13} = 0120212$	$Y_{29} = 20210212$	$Y_{45} = 020212012$	$Y_{60} = 201201012$	$Y_{75} = 02021021212$
$Y_{14} = 1201012$	$Y_{30} = 01021202$	$Y_{46} = 212010202$	$Y_{61} = 0210212012$	$Y_{76} = 21201202012$
$Y_{15} = 1201212$	$Y_{31} = 01020212$	$Y_{47} = 212012012$	$Y_{62} = 1021202012$	$Y_{77} = 12120202012$
$Y_{16} = 2012012$	$Y_{32} = 20212012$			

Proof. Directly from Lemmas 6 and 7 below. Note that every sorting sequence for s sorts $s2$ as well while every sorting sequence for $s2$ can be modified to avoid flipping the whole string and thus works for s as well.   □

Lemma 6. $d_s(s) = n - 2$ *for every bad ternary string* $s \neq 0212$ *of length* n.

Proof. Since $d_s(s) \geq n - 3$ and any 1-flip decreases the length of the string by 1, Lemma 5 says it suffices to show that for each type in Definition 2 a 0-flip is necessary.

- For I-strings only 0-flips are possible.
- A 1-flip on a II- or III-string leads to a string of the same type, so that eventually no 1-flip is possible.
- A 1-flip on a IV-string leads either again to a IV-string or (when destroying the 12 suffix) to a III-string.
- A 1-flip on a V-string leads either again to a V-string or (when destroying the suffix with a $...0212$ flip) to a IV-string. Flips $...0212$ and $...0212$ are not possible for lack of more 2's.
- For strings of VI-, VII- and VIII-strings only one 1-flip is possible, leading to II-, III- and IV-strings respectively.
- For IX-strings it is easy (although time consuming) to check that every 1-flip either leads to a shorter IX-string or to a string of type I-VIII [15].   □

Lemma 7. $d_s(s) = n - 3$ *for every good ternary string* s *of length* n.

Proof. The proof is by induction on n and is similar to the proof of Lemma 4. We defer the details to a full version of the paper [15].   □

Theorem 6. *There exists a polynomial time algorithm for optimally sorting ternary strings.*

Proof. Follows rather easily from Theorem 5.   □

Finally, in light of the fact that the complexity of the sorting problem on quaternary (and higher) strings remains open, the following serves as an intermediate result:

Theorem 7. *For every fixed k there is a PTAS for sorting k-ary strings.*

Proof. The proof is very similar to that used in Theorem 3. The details are again deferred to a full version of the paper [15].   □

5   Prefix Reversal Diameter

Let $S(n, k)$ be the set of fully k-ary strings of length n. We define $\delta(n, k)$ as the largest value of $d(s, t)$ ranging over all compatible $s, t \in S(n, k)$. The value $\delta(n, k)$ is called the *prefix reversal diameter* of fully k-ary, length-n strings.

Theorem 8. *For all $n \geq 2$, $\delta(n, 2) = n - 1$.*

Proof. To prove $\delta(n, 2) \geq n - 1$, consider compatible $s, t \in S(n, 2)$ with $s = (10)^{n/2}$ in case n even and $s = 0(10)^{(n-1)/2}$ in case n odd and in both cases $t = I(s)$ i.e. t is the sorted version of s. By Theorem 4, $d(s, t) \geq n - 1$.

The proof that $\delta(n, 2) \leq n - 1$, for all $n \geq 2$ is by induction on n. The lemma is trivially true for $n = 2$. Consider two compatible binary strings of length n: $s = s_1 s_2 \ldots s_n$ and $t = t_1 t_2 \ldots t_n$. If $s_n = t_n$ then by induction $d(s, t) \leq n - 2$. Thus, suppose (wlog) $s_n = 0$ and $t_n = 1$. If $t_1 = 0$ then $f^{(n)}t$ and s both end with a 0, and using induction and symmetry $d(s, t) \leq 1 + d(f^{(n)}t, s) \leq n - 2 + 1 = n - 1$. An analogous argument holds if $s_1 = 1$.

Remains the case $s_1 = s_n = 0$ and $t_1 = t_n = 1$. First, suppose $t_{n-1} = 0$. Since s and t are compatible, there must exist index i such that $s_i = 0$ and $s_{i+1} = 1$. Hence, $f^{(n)}(f^{(i+1)}(s))$ ends with 01 like t and by induction $d(s, t) \leq 2 + d(f^{(n)}(f^{(i+1)}(s)), t) = 2 + n - 3$. Analogously, we resolve the case $s_{n-1} = 1$.

Finally, suppose $s = 0 \ldots 00$ and $t = 1 \ldots 11$. If s contains 11 as a substring, then flipping that 11 (in the same manner as above) to the back of s using 2 flips, gives two strings that both end in 11. Alternatively, if s does not contain 11 as a substring then s has at least two more 0's than 1's, which implies that t must contain 00 as a substring. In that case two prefix reversals on t suffice to create two strings that both end with 00. In both cases, the induction hypothesis gives the required bound.   □

Note that, trivially, $d(s, t) \leq 2n$ for all compatible $s, t \in S(n, k)$, for all k, because two prefix reversals always suffice to increase the maximal common suffix between s and t by at least 1. The following tighter bound gives the best bound known on the diameter of ternary strings.

Lemma 8. *For any two compatible $s, t \in S(n, k)$, for any k, let a be the most frequent symbol in s and α its multiplicity. Then $d(s, t) \leq 2(n - \alpha)$.*

Proof. We prove the lemma, by induction on n. The lemma is trivially true for $n = 2$. Consider $s, t \in S(n, k)$. If $s_n = t_n = a$ then $s_1 s_2 \ldots s_{n-1}$ and $t_1 t_2 \ldots t_{n-1}$

are compatible length-$(n-1)$ strings where the most frequent symbol occurs at least $\alpha - 1$ times. Thus, by induction $d(s,t) \leq 2((n-1) - (\alpha - 1)) = 2(n-\alpha)$. In case $s_n = t_n \neq a$ induction even gives $d(s,t) \leq 2((n-1) - (\alpha)) = 2(n-\alpha) - 2$. Thus, suppose $s_n \neq t_n$ implying (wlog) that $t_n = b \neq a$. Suppose $s_i = b$; after two flips $s' = f^{(n)}(f^{(i)}(s))$ has b at the end; $s'_n = t_n$. Moreover the length $n-1$ suffixes of s' and t still contain α a's. Hence by induction $d(s,t) \leq 2 + d(s',t) \leq 2 + 2((n-1) - \alpha) = 2(n-\alpha)$. □

Lemma 9. *For all $n > 3$, $n - 1 \leq \delta(n,3) \leq (4/3)n$.*

Proof. Since in any ternary case $\alpha \geq \lceil n/3 \rceil$, Lemma 8 implies $\delta(n,3) \leq (4/3)n$. To prove $\delta(n,3) \geq n-1$ we distinguish between n is odd and n is even. For odd $n = 2h + 1$, let s be $2(01)^h$, and for even $n = 2h$ let $s = 01(21)^{h-1}$. In both cases we let $t = I(s)$. We observe that, in the even and in the odd case, $s2$ is a bad I-string and a bad IV-string, respectively, in the sense of Definition 2. Thus, by Theorem 5 we have that $d(s,t) = d(s2,t2) = (n+1) - 2 = n - 1$. (Here $s2$, respectively $t2$, refers to the concatenation of s, respectively t, with an extra 2 symbol.) □

Brute force enumeration has shown that, for $4 \leq n \leq 13$, $\delta(n,3) = n - 1$. (Note that $\delta(3,3) = 3$ because $d(021,012) = 3$.) Proving or disproving the conjecture that $\delta(n,3) = n - 1$ for $n > 3$ remains an intriguing open problem[3].

6   Prefix Reversal Distance

We show that computing flip distance is NP-hard on binary strings. We also point out, using a result from [17], that computing flip distance on *arbitrary* strings is polynomial-time reducible (in an approximation-preserving sense) to computing it on binary strings.

Theorem 9. *The problem of computing the prefix reversal distance of binary strings is NP-hard.*

Proof. We prove NP-completeness of the corresponding decision problem:

Name: BINARY-PD (2PD shortly)
Input: Two compatible strings $s, t \in S(n,2)$, and a bound $B \in \mathbb{Z}^+$.
Question: Is $d(s,t) \leq B$?

2PD$\in$NP, since a certificate for a positive answer consists of at most B flips[4]. To show completeness we use a reduction from 3-PARTITION [10] (cf. [4] and [17]).

Name: 3-PARTITION (3P shortly)
Input: A set $A = \{a_1, a_2, ..., a_{3k}\}$ and a number $N \in \mathbb{Z}^+$. Element a_i has size $r(a_i) \in \mathbb{Z}^+$ satisfying $N/4 < r(a_i) < N/2$, $i = 1, \ldots, 3k$, and $\sum_{i=1}^{3k} r(a_i) = kN$.

---

[3] Interestingly, initial experiments with brute force enumeration have also shown that, for $4 \leq n \leq 10$, $\delta(n,4) = n$, and for $5 \leq n \leq 9$, $\delta(n,5) = n$.
[4] Recall that for all compatible strings $s, t \in S(n,2)$, trivially $d(s,t) \leq 2n$.

Question: Can A be partitioned into k disjoint triplet sets A_1, A_2, ..., A_k such that $\sum_{a \in A_j} r(a) = N$, $j = 1, \ldots, k$?

Given instance $I = (A, N, r)$ of 3P, we create an instance of 2PD by setting $B = 6k$ and building two compatible binary strings s and t:

$$ s = \left(\prod_{1 \leq i \leq 3k} 0001^{r(a_i)} \right) 000 \qquad t = 0^{3(3k+1)-k} (01^N)^k $$

This construction is clearly polynomial in a unary encoding of the 3P instance; we use the *strong* NP-hardness of 3P [10]. We claim that $I = (A, N, r)$ is a positive instance of 3P if and only if $d(s, t) \leq 6k$. We defer the proof to a full version of the paper [15]. □

For studying problems on arbitrary strings, let X and Y be two compatible, length-n strings, where we assume (wlog) that each of the symbols from X and Y are drawn from the set $\{0, 1, ..., n-1\}$. We define $D(X, Y)$ as the smallest number of flips required to transform X to Y. The arity of the strings X and Y does not need to be fixed, and symbols may be repeated. Hence, sorting of a permutation by flips (MIN-SBPR), and the flip distance problem over fixed arity strings, are both special cases of computing D. Given that computing D is a generalisation of computing distance d of binary strings, immediately implies that it is NP-hard. However, an approximation-preserving reduction in the *other* direction is possible, meaning that inapproximability results for one of the problems will be automatically inherited by the other.

Theorem 10. *Given two compatible strings X and Y of length n with each symbol from X and Y drawn from $\{0, 1, ..., n-1\}$, it is possible to compute in time polynomial in n two binary strings x and y of length polynomial in n such that $D(X, Y) = d(x, y)$.*

Proof. This result follows directly from work by Radcliffe, Scott and Wilmer [17]. The proof is deferred to a full version of the paper [15]. □

7   Open Problems

In this study we have unearthed many rich (and surprisingly difficult) combinatorial questions which deserve further analysis. We discuss some of them here. The main unifying, "umbrella" suggestion is that, to go beyond ad-hoc (and case-based) proof techniques, it will be necessary to develop deeper, more structural insights into the action of flips on strings over fixed size alphabets.

Grouping and sorting on higher arity alphabets. We have shown how to group and sort optimally binary and ternary strings, but characterisations and algorithms for quaternary (and higher) alphabets have so far evaded us. As observed in Section 3.2, it *seems* that for $k = 4, 5$ and for sufficiently long strings, the strings with grouping distance $n - 2$ settle into some kind of pattern, but this has not yet offered enough insights to allow either the development of a characterisation or of an algorithm. Related problems include: for all fixed k, are there polynomial algorithms to optimally sort (optimally group) k-ary strings? Is grouping strictly easier than sorting, in a complexity sense? How does grouping function under other operators e.g. reversals, transpositions? An upper bound on the grouping transposition distance has been presented in [7].

Diameter questions. Proving or disproving that $\delta(n, 3) = n - 1$ for $n > 3$ remains the obvious open diameter question. Beyond that, diameter results for quaternary and higher arity alphabets are needed. How does the diameter $\delta(n, k)$ grow for increasing k? (At this point we conjecture that, for sufficiently long strings, the diameter of 3-ary, 4-ary and 5-ary strings is $n - 1$, n, and n respectively.)

The suspicion also exists that, for all k and for all sufficiently long n, there exists a length-n fully k-ary string s such that $d(s, I(s)) = \delta(n, k)$. In other words, the set of all pairs of strings that are $\delta(n, k)$ flips apart includes some instances of the sorting problem. It should be noted however that, following empirical testing, it is apparent that there are also very many pairs of strings s, t with $s \neq I(t)$ and $t \neq I(s)$ that are $\delta(n, k)$ flips apart.

It also seems important to develop diameter results for subclasses of strings, perhaps (as in [17]) characterised by the frequency of their most frequent symbol. It may be that such refined diameter results for k-ary alphabets provide information that is important in determining $\delta(n, k + 1)$.

Note finally that the diameter of strings over fixed size alphabets, i.e. $\delta(n, k)$, is always bounded from above by the diameter of permutations. This is because the distance problem on two length-n, fixed size alphabet strings s, t can easily be re-written as a sorting problem on a length-n permutation π, such that a sequence of prefix reversals sorting the permutation also suffices to transform s into t. Indeed, because of this relabelling property, the flip distance between two fixed size alphabet strings can be viewed as being equal to the minimum permutation sorting distance, ranging over all such relabellings into a permutation. Can this relationship between the fixed size alphabet and permutation world be further specified and exploited?

Signed strings. The problem of sorting signed permutations by flips (the *burnt* pancake flipping problem) is well known [5] [11] [14], but in this paper we have not yet attempted to analyse the action of flips on signed, fixed size alphabet strings. Obviously, analogues of all the problems described in this paper exist for signed strings.

Complexity/approximation. In the presence of hardness results (e.g. Theorem 9) it is interesting to explore the complexity of restricted instances, and to develop algorithms with guaranteed approximation bounds. For example, [17] gives a PTAS for dense instances. The development of approximation algorithms is also a useful intermediate strategy where the complexity of a problem remains elusive. In particular, this requires the development of improved lower bounds.

References

1. Bader, M., Ohlebusch, E.: Sorting by weighted reversals, transpositions, and inverted transpositions. In: Apostolico, A., Guerra, C., Istrail, S., Pevzner, P., Waterman, M. (eds.) RECOMB 2006. LNCS (LNBI), vol. 3909, pp. 563–577. Springer, Heidelberg (2006)
2. Bergeron, A., Mixtacki, J., Stoye, J.: On sorting by translocations. J. Comp. Biol. 13(2), 567–578 (2006)
3. Chen, X., Zheng, J., Fu, Z., Nan, P., Zhong, Y., Lonardi, S., Jiang, T.: Assignment of orthologous genes via genome rearrangement, IEEE/ACM Trans. Comput. Biol. Bioinform. 2(4) (October- December 2005)
4. Christie, D.A., Irving, R.W.: Sorting strings by reversals and transpositions. SIAM J. on Discrete Math. 14(2), 193–206 (2001)
5. Cohen, D.S., Blum, M.: On the problem of sorting burnt pancakes. Discrete Appl. Math. 61(2), 105–120 (1995)
6. Elias, I., Hartman, T.: A 1.375-approximation algorithm for sorting by transpositions. IEEE/ACM Trans. Comput. Biology Bioinform. 3(4), 369–379 (2006)
7. Eriksson, H., Eriksson, K., Karlander, J., Svensson, L., Wastlund, J.: Sorting a bridge hand. Discrete Math. 241, 289–300 (2001)
8. Fischer, J., Ginzinger, S.W.: A 2-approximation algorithm for sorting by prefix reversals. In: Brodal, G.S., Leonardi, S. (eds.) ESA 2005. LNCS, vol. 3669, pp. 415–425. Springer, Heidelberg (2005)
9. Garey, M.R., Johnson, D.S.: Complexity results for multiprocessor scheduling under resource constraints. SIAM J. Comput. 4(4), 397–411 (1975)
10. Garey, M.R., Johnson, D.S.: Computers And Intractability: A Guide To The Theory Of NP-completeness. W.H. Freeman, San Francisco, CA (1979)
11. Gates, W.H., Papadimitriou, C.H.: Bounds for sorting by prefix reversal. Discrete Math. 27, 47–57 (1979)
12. Goldstein, A., Kolman, P., Zheng, J.: Minimum Common String Partition problem: hardness and approximations, Electron. J. Combin. 12, #R50 (2005)
13. Hannenhalli, S., Pevzner, P.A.: Transforming cabbage into turnip: a polynomial algorithm for sorting permutations by reversals. Jour. ACM 46(1), 1–27 (1999)
14. Heydari, M.H., Sudborough, I.H.: On the diameter of the pancake network. J. Algorithms 25, 67–94 (1997)
15. Hurkens, C.A.J., van Iersel, L.J.J., Keijsper, J.C.M., Kelk, S.M., Stougie, L., Tromp, J.T.: Prefix reversals on binary and ternary strings, technical report (2006), http://www.win.tue.nl/bs/spor/2006-10.pdf
16. Morales, L., Sudborough, I.H.: Comparing Star and Pancake Networks. In: The Essence Of Computation: Complexity, Analysis, Transformation, LNCS (2002), Springer, Heidelberg, pp. 18–36 (2566)
17. Radcliffe, A.J., Scott, A.D., Wilmer, E.L.: Reversals and transpositions over finite alphabets. SIAM J. on Discrete Math. 19(1), 224–244 (2005)
18. Scott, A.D.: Personal communication

Toric Ideals of Phylogenetic Invariants for the General Group-Based Model on Claw Trees $K_{1,n}$

Julia Chifman and Sonja Petrović

Department of Mathematics, University of Kentucky, Lexington, KY 40506, USA
{jchifman,petrovic}@ms.uky.edu

Abstract. We address the problem of studying the toric ideals of phylogenetic invariants for a general group-based model on an arbitrary claw tree. We focus on the group $\mathbb{Z}_2$ and choose a natural recursive approach that extends to other groups. The study of the lattice associated with each phylogenetic ideal produces a list of circuits that generate the corresponding lattice basis ideal. In addition, we describe *explicitly* a quadratic lexicographic Gröbner basis of the toric ideal of invariants for the claw tree on an arbitrary number of leaves. Combined with a result of Sturmfels and Sullivant, this implies that the phylogenetic ideal of *every* tree for the group $\mathbb{Z}_2$ has a quadratic Gröbner basis. Hence, the coordinate ring of the toric variety is a Koszul algebra.

1 Introduction

Phylogenetics is concerned with determining genetic relationship between species based on their DNA sequences. First, the various DNA sequences are aligned, that is, a correspondence is established that accounts for their differences. Assuming that all DNA sites evolve identically and independently, the focus is on one site at a time. The data then consists of observed pattern frequencies in aligned sequences. This observed data are used to estimate the true joint probabilities of the observations and, most importantly, to reconstruct the ancestral relationship among the species. The relationship can be represented by a phylogenetic tree.

A *phylogenetic tree* T is a simple, connected, acyclic graph equipped with some statistical information. Namely, each node of T is a random variable with k possible states chosen from the state space S. Edges of T are labeled by transition probability matrices that reflect probabilities of changes of the states from a node to its child. These probabilities of mutation are the parameters for the statistical model of evolution, which is described in terms of a discrete-state continuous-time Markov process on the tree. Since the goal is to reconstruct the tree, the interior nodes are hidden. The relationship between the random variables is encoded by the structure of the tree. At each of the n leaves, we can observe any of the k states; thus there are k^n possible observations. Let p_σ be the joint probability of making a particular observation $\sigma \subset S^n$ at the leaves. Then p_σ is a polynomial in the model parameters.

H. Anai, K. Horimoto, and T. Kutsia (Eds.): AB 2007, LNCS 4545, pp. 307–321, 2007.

A *phylogenetic invariant* of the model is a polynomial in the leaf probabilities which vanishes for every choice of model parameters. The set of these polynomials forms a prime ideal in the polynomial ring over the unknowns p_σ. The objective is to compute this ideal explicitly. Thus we consider a polynomial map $\phi : \mathbb{C}^N \to \mathbb{C}^{k^n}$, where N is the total number of model parameters. The map depends only on the tree T and the number of states k; its coordinate functions are the k^n polynomials p_σ. The map ϕ induces a parametrization of an algebraic variety. The study of these algebraic varieties for various statistical models is a central theme in the field of algebraic statistics ([11]). Phylogenetic invariants are a powerful tool for tree reconstruction ([2], [3], [7]).

There is a specific class of models for which the ideal of invariants is particularly nice. Let M_e be the $k \times k$ transition probability matrix for edge e of T. In the general Markov model, each matrix entry is an independent model parameter. A group-based model is one in which the matrices M_e are pairwise distinct, but it is required that certain entries coincide. For these models, transition matrices are diagonalizable by the Fourier transform of an abelian group. The key idea behind this linear change of coordinates is to label the states (for example, A,C,G, and T) by a finite abelian group (for example, $\mathbb{Z}_2 \times \mathbb{Z}_2$) in such a way that transition from one state to another depends only on the difference of the group elements. Examples of group-based models include the Jukes-Cantor and Kimura's one-parameter models used in computational biology.

Sturmfels and Sullivant in [11] reduce the computation of ideals of phylogenetic invariants of group-based models on an arbitrary tree to the case of claw trees $T_n := K_{1,n}$, the complete bipartite graph from one node (the root) to n nodes (the leaves). The main result of [11] gives a way of constructing the ideal of phylogenetic invariants for any tree *if* the ideal for the claw tree is known. However, in general, it is an open problem to compute the phylogenetic invariants for a claw tree. We consider the ideal for a general group-based model for the group $\mathbb{Z}_2$. Let q_σ be the image of p_σ under the Fourier transform. Assuming the identity labeling function and adopting the notation of [11], the ideal of phylogenetic invariants for the tree T_n is the kernel of the following homomorphism between polynomial rings:

$$\varphi_n : \mathbb{C}[q_{g_1,\ldots,g_n} : g_1,\ldots,g_n \in G] \to \mathbb{C}[a_g^{(i)} : g \in G, i = 1,\ldots,n+1]$$

$$q_{g_1,\ldots,g_n} \mapsto a_{g_1}^{(1)} a_{g_2}^{(2)} \ldots a_{g_n}^{(n)} a_{g_1+g_2+\cdots+g_n}^{(n+1)}, \qquad (*)$$

where G is a finite group with k elements, each corresponding to a state. The coordinate $q_{g_1,\ldots,g_n}$ corresponds to observing the element g_1 at the first leaf of T, g_2 at the second, and so on. The phylogenetic invariants form a *toric ideal* in the Fourier coordinates q_σ, which can be computed from the corresponding lattice basis ideal by saturation. The main result of this paper is a complete description of the lattice basis ideal and a quadratic Gröbner basis of the ideal of invariants for the group $\mathbb{Z}_2$ on T_n for any number of leaves n.

Our paper is organized as follows. In section 2 we lay the foundation for our recursive approach. The ideal of the two-leaf claw tree is trivial, so we begin with the case when the number of leaves is three. Sections 3 and 4 address the problem of describing the lattices corresponding to the toric ideals. We provide a nice lattice basis consisting of circuits. The corresponding lattice basis ideal is generated by circuits of degree two and thus in particular satisfies the Sturmfels-Sullivant conjecture.

The ideal of phylogenetic invariants is the saturation of the lattice basis ideal. However, we do not use any of the standard algorithms to compute saturation (e.g. [8], [10]). Instead, our recursive construction of the lattice basis ideals can be extended to give the full ideal of invariants, which we describe in the final section. The recursive description of these ideals depends only on the number of leaves of the claw tree and it does not require saturation. Finally, and possibly somewhat surprisingly, we show that the ideal of invariants for every claw tree admits a quadratic Gröbner basis with respect to a lexicographic term order. We describe it *explicitly*.

Combined with the main result of Sturmfels and Sullivant in [11], this implies that the phylogenetic ideal of *every* tree for the group $\mathbb{Z}_2$ has a quadratic Gröbner basis. Hence, the coordinate ring of the toric variety is a Koszul algebra. In addition, the ideals for every tree can be computed explicitly. These ideals are particularly nice as they satisfy the conjecture in [11] which proposes that the order of the group gives an upper bound for the degrees of minimal generators of the ideal of invariants. The case of $\mathbb{Z}_2$ has been solved in [11] using a technique that does not generalize. We hope to extend our recursive approach and obtain the result for an arbitrary abelian group.

For a detailed background on phylogenetic trees, invariants, group-based models, Fourier coordinates, labeling functions and more, the reader should refer to [1], [6], [9], [11].

2   Matrix Representation

Fix a claw tree T_n on n leaves and a finite abelian group G of order k. Soon we will specialize to the case $k = 2$. We want to compute the ideal of phylogenetic invariants for the general group-based model on T_n. After the Fourier transform, the ideal of invariants (in Fourier coordinates) is given by $I_n = \ker \varphi_n$, where φ_n is a map between polynomial rings in k^n and $k(n+1)$ variables, respectively, defined by (*). In order to compute the toric ideal I_n, we first compute the lattice basis ideal $I_{L_n} \subset I_n$ corresponding to φ_n as follows. Fixing an order on the monomials of the two polynomial rings, the linear map φ can be represented by a matrix $B_{n,k}$ that describes the action of φ on the variables. Then the lattice $L_n = \ker(B_{n,k}) \subset \mathbb{Z}^{k^n}$ determines the ideal I_{L_n}. It is generated by elements of the form $(\prod q_{g_1,\ldots,g_n})^{v^+} - (\prod q_{g_1,\ldots,g_n})^{v^-}$ where $v = v^+ - v^- \in L_n$. We will give an explicit description of this basis and, equivalently, the ideals I_{L_n}.

Hereafter assume that $G = \mathbb{Z}_2$. For simplicity, let us say that $B_n := B_{n,2}$.

To create the matrix B_n, first order the two bases as follows. Order the $a_g^{(i)}$ by varying the upper index (i) first and then the group element g: $a_0^{(1)}, a_0^{(2)}, \ldots,$

$a_0^{(n+1)}$, $a_1^{(1)}$, ..., $a_1^{(n+1)}$. Then, order the $q_{g_1,...,g_n}$ by ordering the indices with respect to binary counting:

$$q_{0...00} > q_{0...01} > \cdots > q_{1...10} > q_{1...1}.$$

That is, $q_{g_1...g_n} > q_{h_1...h_n}$ if and only if $(g_1 \ldots g_n)_2 < (h_1 \ldots h_n)_2$, where

$$(g_1 \ldots g_n)_2 := g_1 2^{n-1} + g_2 2^{n-2} + \cdots + g_n 2^0$$

represents the binary number $g_1 \ldots g_n$.

Next, index the rows of B_n by $a_g^{(i)}$ and its columns by $q_{g_1,...,g_n}$. Finally, put 1 in the entry of B_n in the row indexed by $a_g^{(i)}$ and column indexed by $q_{g_1,...,g_n}$ if $a_g^{(i)}$ divides the image of $q_{g_1,...,g_n}$, and 0 otherwise.

Example 1. Let $n = 2$. Then we order the q_{ij} variables according to binary counting: $q_{00}, q_{01}, q_{10}, q_{11}$, so that

$$\varphi : \mathbb{C}[q_{00}, q_{01}, q_{10}, q_{11}] \to \mathbb{C}[a_0^{(1)}, a_0^{(2)}, a_0^{(3)}, a_1^{(1)}, a_1^{(2)}, a_1^{(3)}]$$

$$q_{00} \mapsto a_0^{(1)} a_0^{(2)} a_{0+0}^{(3)}$$

$$q_{01} \mapsto a_0^{(1)} a_1^{(2)} a_{0+1}^{(3)}$$

$$q_{10} \mapsto a_1^{(1)} a_0^{(2)} a_{1+0}^{(3)}$$

$$q_{11} \mapsto a_1^{(1)} a_1^{(2)} a_{1+1}^{(3)}.$$

Now we put the $a_i^{(j)}$ variables in order: $a_0^{(1)}$, $a_0^{(2)}$, $a_0^{(3)}$, $a_1^{(1)}$, $a_1^{(2)}$, $a_1^{(3)}$. Thus

$$B_2 = \begin{bmatrix} 1 & 1 & 0 & 0 \\ 1 & 0 & 1 & 0 \\ 1 & 0 & 0 & 1 \\ 0 & 0 & 1 & 1 \\ 0 & 1 & 0 & 1 \\ 0 & 1 & 1 & 0 \end{bmatrix}.$$

The tree T_{n-1} can be considered as a subtree of T_n by ignoring, for example, the leftmost leaf of T. As a consequence, a natural question arises: how does B_n relate to B_{n-1}?

Remark 1. The matrix B_{n-1} for the subtree of T_n with the leaf (1) removed can be obtained as a submatrix of B_n for the tree T_n by deleting rows 1 and $(n+1)+1$ and taking only the first 2^{n-1} columnns. Divide the n-leaf matrix B_n into a 2×2 block matrix with blocks of size $(n+1) \times 2^{n-1}$:

$$B_n = \begin{bmatrix} B_{11} & B_{12} \\ B_{21} & B_{22} \end{bmatrix}.$$

Then, grouping together B_{11}, B_{21} without the first row of each B_{i1}, we obtain the matrix B_{n-1}. This is true because rows 1 and $(n+1)+1$ represent the variables

$a_g^{(1)}$ for $g \in G$ associated with the leaf (1) of T_n. Note that the entries in row $a_g^{(n+1)}$ remain undisturbed as the omitted rows are indexed by the identity of the group.

Example 2. The matrix B_2 is equal to the submatrix of B_3 formed by rows 2,3,4,6,7,8, and first 4 columns.

Remark 2. Fix any observation $\sigma = g_1, \ldots, g_n$ on the leaves. Clearly, at any given leaf $j \in \{1, \ldots, n\}$, we observe exactly one group element, g_j. Since the matrix entry $b_{a_{g_j}^{(j)}, q_\sigma}$ in the row indexed by $a_{g_j}^{(j)}$ and column indexed by q_σ is 1 exactly when $a_{g_j}^{(j)}$ divides the image of q_σ, one has that

$$\sum_{g_j \in G} b_{a_{g_j}^{(j)}, q_\sigma} = 1$$

for a fixed leaf (j) and fixed observation σ. Note that the formula also holds if $j = n+1$ by definition of $a_{g_{n+1}}^{(n+1)} = a_{g_1 + \cdots + g_n}^{(n+1)}$. In particular, the rows indexed by $a_{g_j}^{(j)}$ for a fixed j sum up to the row of ones.

3   Number of Lattice Basis Elements

We compute the dimension of the kernel of B_n by induction on n. We proceed in two steps.

Lemma 1 (Lower bound)

$$\text{rank}(B_n) \geq \text{rank}(B_{n-1}) + 1.$$

Proof. First note that $\text{rank}(B_n) \geq \text{rank}(B_{n-1})$ since B_{n-1} is a submatrix of the first 2^{n-1} columns of B_n. In the block $\begin{bmatrix} B_{11}, B_{12} \end{bmatrix}^T$, the row indexed by $a_1^{(1)}$ is zero, while in the block $\begin{bmatrix} B_{21}, B_{22} \end{bmatrix}^T$, the row indexed by $a_1^{(1)}$ is 1. Choosing one column from $\begin{bmatrix} B_{21}, B_{22} \end{bmatrix}^T$ provides a vector independent of the first 2^{n-1} columns. The rank must therefore increase by at least 1.   □

Lemma 2 (Upper bound)

$$\text{rank}(B_n) \leq n + 2.$$

Proof. B_n has $2(n+1)$ rows. Remark 2 provides n independent relations among the rows of our matrix: varying j from 1 to $n+1$, we obtain that the sum of the rows j and $n+1+j$ is 1 for each $j = 1, \ldots, n+1$. Thus the upper bound is immediate.   □

We are ready for the main result of the section.

Proposition 1 (Cardinality of lattice basis)
Let $n \geq 2$. Then there are $2^n - 2(n+1) + n$ elements in the basis of the lattice L_n corresponding to T_n. That is,

$$\dim \ker(B_n) = 2^n - 2(n+1) + n.$$

Proof. We show $\mathrm{rank}(B_n) = 2(n+1) - n$. It can be checked directly that B_2 has full rank. Assume that the claim is true for $n-1$. Then by Lemmae (1) and (2),

$$2(n+1) - n \geq \mathrm{rank}(B_n) \geq \mathrm{rank}(B_{n-1}) + 1 = 2n - (n-1) + 1,$$

where the last equality is provided by the induction hypothesis. The claim follows since the left- and the right-hand sides agree.   □

4 Lattice Basis

In this section we describe a basis of the kernel of $B_n := B_{n,2}$, in which the binomials corresponding to the basis elements satisfy the conjecture on the degrees of the generators of the phylogenetic ideal. In particular, since the ideal is generated by squarefree binomials and contains no linear forms, these elements are actually circuits. By Proposition 1, we need to find $2^n - (n+2)$ linearly independent vectors in the lattice. The matrix of the tree with $n = 2$ leaves has a trivial kernel, so we begin with the tree on $n = 3$ leaves. The dimension of the kernel is 3 and the lattice basis is given by the rows of the following matrix:

$$\begin{bmatrix} 0\,0\,1 & -1 & -1\,1\,0\,0 \\ 0\,1\,0 & -1 & -1\,0\,1\,0 \\ 1\,0\,0 & -1 & -1\,0\,0\,1 \end{bmatrix}.$$

In order to study the kernels of B_n for any n, it is useful to have an algorithmic way of constructing the matrices.

Algorithm 1. [The construction of B_n]
Input: the number of leaves n of the claw tree T_n.
Output: $B_n \in \mathbb{Z}^{2(n+1) \times 2^n}$.
Initialize B_n to the zero matrix.
Construct the first n rows:
  for k from 1 to n do:
    for c from 0 to $2^k - 1$ with $c \equiv 0 \mod 2$ do:
      for j from $c2^{n-k} + 1$ to $(c+1)2^{n-k}$ do: $b_{k,j} := 1$.
Construct row $n+1$:
  if $n \equiv (\sum_{r=1}^{n} b_{r,j}) \mod 2$, then $b_{n+1,j} := 1$.
Construct rows $n+2$ to $2(n+1)$:
  for i from 1 to $n+1$ do:
    for j from 1 to 2^n do: $b_{n+1+i,j} := 1 - b_{i,j}$.

One checks that this algorithm gives indeed the matrices B_n as defined in Section 3.

The $(n+1+i)^{th}$ row r_{n+1+i} of B_n is by definition the binary complement of the ith row r_i of B_n. Suppose that $r_i \cdot k = 0$ for some vector k. Since all entries of B_n are nonnegative, a subvector of k restricted to the entries where r_i is nonzero must be homogeneous in the sense that the sum of the positive entries equals the sum of the negative entries. But since the ideal I_{L_n} itself is homogeneous ([10]), the same must be true for the subvector of k restricted to the entries where r_i is zero. Hence $r_{n+1+i} \cdot k = 0$. Therefore, it is enough to analyze the top half of the matrix B_n when determining the kernel elements.

Remark 3. There are n copies of B_{n-1} inside B_n. By deleting one leaf at a time, we get n copies of T_{n-1} as a subtree of T_n. Suppose we delete leaf (i) from T_n to get the tree $T_n^{(i)}$ on leaves $1, 2, \ldots, i-1, i+1, \ldots, n$. Ignoring the two rows of B_n that represent the leaf (i) and taking into account the columns of B_n containing nonzero entries of the row indexed by $a_0^{(i)}$ (that is, observing 0 at leaf (i)) gives precisely the matrix B_{n-1} corresponding to $T_n^{(i)}$. Note that the entry indexed by $a_g^{(n+1)}$, for any $g \in G$, will be correct since we are ignoring the identity of the group, as in Remark 1.

This leads to a way of constructing a basis of $\ker(B_n)$ from the one of $\ker(B_{n-1})$. Namely, removing leaf (1) from T_n produces $\dim(\ker(B_{n-1})) = 2^{n-1} - n - 1$ independent vectors in $\ker(B_n)$. Let us name this collection of vectors V_1. Removing leaf (2) produces a collection V_2 consisting of $\dim(\ker B_{n-1}) - \dim(\ker B_{n-2}) = 2^{n-2} - 1$ vectors in $\ker(B_n)$. V_2 is independent of V_1 since the second half of each vector in V_2 has nonzero entries in the columns of B_n where all vectors in V_1 are zero, a direct consequence of the location of the submatrix corresponding to $T_n^{(2)}$. Finally, removing any other leaf (i) of T_n produces a collection V_i of as many new kernel elements as there are new columns involved (in terms of the submatrix structure); namely, 2^{n-i} new vectors. Note that every vector in V_2 has a nonzero entry in at least one new column so that the full collection is independent of V_1.

Using the above procedure, we have obtained

$$(2^{n-1} - n - 1) + (2^{n-2} - 1) + (2^{n-3}) + \cdots + 2^{n-n}$$

independent vectors in the kernel of B_n. This is exactly one less than the desired number, $2^n - n - 2$. Hence to the list of the kernel generators we add one additional vector v that is independent of all the V_i, $i = 1, \ldots, n$ as it has a nonnegative entry in the last column. (Note that no $v \in V_i$ has this property by the observation on the column location of the submatrix associated with each $T_n^{(i)}$.) In particular, $v = [0, \ldots, 0, 1, 0, 0, -1, -1, 0, 0, 1] \in \ker(B_n)$. To see this, we simply notice that the rows of the last 8-column block of B_n are precisely the rows of the first 8-column block of B_n up to permutation of rows, which does not affect the kernel.

The lattice basis we just constructed is directly computed by the following algorithm.

Algorithm 2. [Construction of the lattice basis for T_n]
Input: the number of leaves n of the claw tree T_n.
Output: a basis of ker B_n in form of a $(2^n - n - 2) \times 2^n$ matrix L_n.

$$\text{Let } L_3 := \begin{bmatrix} 0\,0\,1 -1 -1\,1\,0\,0 \\ 0\,1\,0 -1 -1\,0\,1\,0 \\ 1\,0\,0 -1 -1\,0\,0\,1 \end{bmatrix}.$$

Set $k := 4$.
The following subroutine lifts L_{k-1} to L_k:
WHILE $k \leq n$ do:{
Initialize L_k to the zero matrix.
For i from 1 to k do:
  $\text{cols}(i) := \{1..2^{k-i}, (2)2^{k-i} + 1..(3)2^{k-i}, \ldots, (2^i - 2)2^{k-i} + 1..(2^i - 1)2^{k-i}\}.$
Denote by $L_{k,j}[\text{cols}(i)]$ the jth row vector of L_k restricted to columns $\text{cols}(i)$.
Set $i := 1$:
  for j from 1 to $2^{k-1} - k - 1$ do: $L_{k,j}[\text{cols}(i)] := L_{k-1,j}.$
Set $i := 2$:
  for j from 1 to $2^{k-2} - 1$ do :
  $L_{k,(2^{k-1}-k-1)+j}[\text{cols}(i)] := L_{k-1,(2^{k-1}-k-1)-(2^{k-2}-1)+j}.$
For i from 3 to k do:
  for j from 1 to 2^{k-i} do:
  $L_{k,(2^k-2^{k+1-i}-k-2)+j}[\text{cols}(i)] := L_{k-1,(2^{k-1}-k-1)-(2^{k-i})+j}.$
Finally, $L_{k,2^k-k-2}[2^k - 7..2^k] := [1,0,0,-1,-1,0,0,1].$
RETURN L_k. }

Example 3. Consider the tree on $n = 4$ leaves. Then

$$B_4 = \begin{bmatrix} 1\,1\,1\,1\,1\,1\,1\,1\,0\,0\,0\,0\,0\,0\,0\,0 \\ 1\,1\,1\,1\,0\,0\,0\,0\,1\,1\,1\,1\,0\,0\,0\,0 \\ 1\,1\,0\,0\,1\,1\,0\,0\,1\,1\,0\,0\,1\,1\,0\,0 \\ 1\,0\,1\,0\,1\,0\,1\,0\,1\,0\,1\,0\,1\,0\,1\,0 \\ 1\,0\,0\,1\,0\,1\,1\,0\,0\,1\,1\,0\,1\,0\,0\,1 \\ 0\,0\,0\,0\,0\,0\,0\,0\,1\,1\,1\,1\,1\,1\,1\,1 \\ 0\,0\,0\,0\,1\,1\,1\,1\,0\,0\,0\,0\,1\,1\,1\,1 \\ 0\,0\,1\,1\,0\,0\,1\,1\,0\,0\,1\,1\,0\,0\,1\,1 \\ 0\,1\,0\,1\,0\,1\,0\,1\,0\,1\,0\,1\,0\,1\,0\,1 \\ 0\,1\,1\,0\,1\,0\,0\,1\,1\,0\,0\,1\,0\,1\,1\,0 \end{bmatrix}.$$

The lattice basis is given by the rows of the following matrix:

$$L_4 = \begin{bmatrix}
0 & 0 & 1 & -1 & -1 & 1 & 0 & 0 & 0 & 0 & 0 & 0 & 0 & 0 & 0 & 0 \\
0 & 1 & 0 & -1 & -1 & 0 & 1 & 0 & 0 & 0 & 0 & 0 & 0 & 0 & 0 & 0 \\
1 & 0 & 0 & -1 & -1 & 0 & 0 & 1 & 0 & 0 & 0 & 0 & 0 & 0 & 0 & 0 \\
0 & 0 & 1 & -1 & 0 & 0 & 0 & 0 & -1 & 1 & 0 & 0 & 0 & 0 & 0 & 0 \\
0 & 1 & 0 & -1 & 0 & 0 & 0 & 0 & -1 & 0 & 1 & 0 & 0 & 0 & 0 & 0 \\
1 & 0 & 0 & -1 & 0 & 0 & 0 & 0 & -1 & 0 & 0 & 1 & 0 & 0 & 0 & 0 \\
0 & 1 & 0 & 0 & 0 & -1 & 0 & 0 & -1 & 0 & 0 & 0 & 1 & 0 & 0 & 0 \\
1 & 0 & 0 & 0 & 0 & -1 & 0 & 0 & -1 & 0 & 0 & 0 & 0 & 1 & 0 & 0 \\
1 & 0 & 0 & 0 & 0 & 0 & -1 & 0 & -1 & 0 & 0 & 0 & 0 & 0 & 1 & 0 \\
0 & 0 & 0 & 0 & 0 & 0 & 0 & 1 & 0 & 0 & -1 & -1 & 0 & 0 & 1
\end{bmatrix}.$$

The lattice vectors correspond to the relations on the leaf observations in the natural way; namely, the first column corresponds to $q_{0,\dots,0}$, the second to $q_{0,\dots,0,1}$, and so on. Therefore, the lattice basis ideal for T_4 in Fourier coordinates is

$$I_{L_4} = (q_{0010}q_{0101} - q_{0011}q_{0100}, q_{0001}q_{0110} - q_{0011}q_{0100}, q_{0000}q_{0111} - q_{0011}q_{0100},$$
$$q_{0010}q_{1001} - q_{0011}q_{1000}, q_{0001}q_{1010} - q_{0011}q_{1000}, q_{0000}q_{1011} - q_{0011}q_{1000},$$
$$q_{0001}q_{1100} - q_{0101}q_{1000}, q_{0000}q_{1101} - q_{0101}q_{1000},$$
$$q_{0000}q_{1110} - q_{0110}q_{1000}, q_{1000}q_{1111} - q_{1011}q_{1100}).$$

This ideal is contained in the ideal of phylogenetic invariants I_4 for T_4. In the next section, we compute explicitly the generators of the ideal of invariants for any claw three T_n and the group $\mathbb{Z}_2$.

5   Ideal of Invariants

We show that the lattice basis ideals provide basic building blocks for the full ideals of invariants, as expected. However, instead of computing the ideal of invariants as a saturation of the lattice basis ideal in a standard way (e.g. [8],[10]), we use the recursive constructions from the previous section on the saturated ideals directly. We begin with the ideal of invariants for the smallest tree, and build all other trees recursively. The underlying ideas for how to lift the generating sets come from Algorithm 2.

We will denote the ideal of the claw tree on n leaves by $I_n = \ker \varphi_n$. As we have seen, the first nontrivial ideal is I_3.

5.1   The Tree on $n = 3$ Leaves

Claim. The ideal of the claw tree on $n = 3$ leaves is

$$I_3 = (q_{000}q_{111} - q_{100}q_{011}, q_{001}q_{110} - q_{100}q_{011}, q_{010}q_{101} - q_{100}q_{011}).$$

This can be verified by computation. In particular, this ideal is equal to the lattice basis ideal for the tree on three leaves; I_{L_3} is already prime in this case.

Let $< := <_{lex}$ be the lexicographic order on the variables induced by

$$q_{000} > q_{001} > q_{010} > q_{011} > q_{100} > q_{101} > q_{110} > q_{111}.$$

(That is, $q_{ijk} > q_{i'j'k'}$ if and only if $(ijk)_2 < (i'j'k')_2$, where $(ijk)_2$ denotes the binary number ijk.)

Remark 4. The three generators of I_3 above are a Gröbner basis for I_3 with respect to $<$, since the initial terms, written with coefficient $+1$ in the above description, are relatively prime so all the S-paris reduce to zero.

Remark 5. Write the quadratic binomial $q = q^+ - q^-$ as

$$q_{g_1^{(1)} g_1^{(2)} g_1^{(3)}} q_{g_2^{(1)} g_2^{(2)} g_2^{(3)}} - q_{h_1^{(1)} h_1^{(2)} h_1^{(3)}} q_{h_2^{(1)} h_2^{(2)} h_2^{(3)}}.$$

Then $q \in I_3$ if and only if the following two conditions hold:

1. Exchanging the roles of $q_{h_1^{(1)} h_1^{(2)} h_1^{(3)}}$ and $q_{h_2^{(1)} h_2^{(2)} h_2^{(3)}}$ if necessary,

$$g_1^{(1)} + g_1^{(2)} + g_1^{(3)} = h_1^{(1)} + h_1^{(2)} + h_1^{(3)}$$

and

$$g_2^{(1)} + g_2^{(2)} + g_2^{(3)} = h_2^{(1)} + h_2^{(2)} + h_2^{(3)},$$

2. $g_1^{(i)} + g_2^{(i)} = 1 = h_1^{(i)} + h_2^{(i)}$ for $1 \le i \le 3 = n$.

Note that the second condition holds since otherwise the projection of q obtained by eliminating the leaf (i) at which the observations $g_1^{(i)}$ and $g_2^{(i)}$ are both equal to 0 or to 1 produces an element q' in the kernel of the map φ_2 of the 2-leaf tree, which is trivial.

5.2   The Tree on an Arbitrary Number of Leaves

Let us now define a set of maps and a distinguished set of binomials in I_n.

Definition 1. *Let $\pi_i(q)$ be the projection of q that eliminates the ith index of each variable in q.*

For example,

$$\pi_4(q_{0000} q_{1110} - q_{1000} q_{0110}) = q_{000} q_{111} - q_{100} q_{011}.$$

Definition 2. *Assume that $n \ge 4$.*

Let $\mathcal{G}_n$ be the set of quadratic binomials $q \in I_n$ that can be written as

$$q = q^+ - q^- = q_{g_1^{(1)} \ldots g_1^{(n)}} q_{g_2^{(1)} \ldots g_2^{(n)}} - q_{h_1^{(1)} \ldots h_1^{(n)}} q_{h_2^{(1)} \ldots h_2^{(n)}}$$

such that one of the two following properties is satisfied:

Property (i): For some $1 \le i \le n$, $j \in \mathbb{Z}_2$,

$$g_1^{(i)} = g_2^{(i)} = j = h_1^{(i)} = h_2^{(i)} \tag{1}$$

and

$$\pi_i(q) \in I_{n-1}. \tag{2}$$

Property (ii): For each $1 \le k \le n$,

$$g_1^{(k)} + g_2^{(k)} = 1 = h_1^{(k)} + h_2^{(k)} \tag{3}$$

and

$$\pi_k(q) \in I_{n-1}. \tag{4}$$

Example 4. Let $n = 4$. The set of elements $q \in \mathcal{G}_n$ with Property (i) consists of those for which $j = 0$:

$q_{0000}q_{0111} - q_{0100}q_{0011}, \; q_{0001}q_{0110} - q_{0100}q_{0011}, \; q_{0010}q_{0101} - q_{0100}q_{0011},$
$q_{0000}q_{1011} - q_{1000}q_{0011}, \; q_{0001}q_{1010} - q_{1000}q_{0011}, \; q_{0010}q_{1001} - q_{1000}q_{0011},$
$q_{0000}q_{1101} - q_{1000}q_{0101}, \; q_{0001}q_{1100} - q_{1000}q_{0101}, \; q_{0100}q_{1001} - q_{1000}q_{0101},$
$q_{0000}q_{1110} - q_{1000}q_{0110}, \; q_{0010}q_{1100} - q_{1000}q_{0110}, \; q_{0100}q_{1010} - q_{1000}q_{0110};$

and those for which $j = 1$:

$q_{1000}q_{1111} - q_{1100}q_{1011}, \; q_{1001}q_{1110} - q_{1100}q_{1011}, \; q_{1010}q_{1101} - q_{1100}q_{1011},$
$q_{0100}q_{1111} - q_{1100}q_{0111}, \; q_{0101}q_{1110} - q_{1100}q_{0111}, \; q_{0110}q_{1101} - q_{1100}q_{0111},$
$q_{0010}q_{1111} - q_{1010}q_{0111}, \; q_{0011}q_{1110} - q_{1010}q_{0111}, \; q_{0110}q_{1011} - q_{1010}q_{0111},$
$q_{0001}q_{1111} - q_{1001}q_{0111}, \; q_{0011}q_{1101} - q_{1001}q_{0111}, \; q_{0101}q_{1011} - q_{1001}q_{0111}.$

The set of elements $q \in \mathcal{G}_n$ with Property (ii) are:

$q_{0000}q_{1111} - q_{1001}q_{0110}, \; q_{0001}q_{1110} - q_{1000}q_{0111}, \; q_{0011}q_{1100} - q_{1001}q_{0110},$
$q_{0010}q_{1101} - q_{1000}q_{0111}, \; q_{0101}q_{1010} - q_{1001}q_{0110}, \; q_{0100}q_{1011} - q_{1000}q_{0111}.$

Proposition 2. *For $n \ge 4$, the set of binomials in $\mathcal{G}_n$ generates the ideal I_n. That is,*

$$I_n = (q : q^+ - q^- \in \mathcal{G}_n).$$

In addition, this set of generators can be obtained inductively by lifting the generators corresponding to the various phylogenetic ideals on $n - 1$ leaves.

Proof. Condition (3) is simply the negation of (1). Condition (1) can be restated as follows: for some $1 \le i \le n$ and a fixed j,

$$(a_j^{(i)})^2 | \varphi_n(q^+) \text{ and } (a_j^{(i)})^2 | \varphi_n(q^-).$$

Therefore, Property (i) translates to having an observation j fixed at leaf (i) for each of the variables in q. On the other hand, condition (3) means that for any k, not all the kth indices are 0 and not all are 1. Thus Property (ii) means that no leaf has a fixed observation, and can be restated as follows: for every $1 \le i \le n$,

$$a_0^{(i)} a_1^{(i)} | \varphi_n(q^+) \text{ and } a_0^{(i)} a_1^{(i)} | \varphi_n(q^-). \tag{5}$$

By definition, the ideal I_n is toric, so it is generated by binomials. In fact, it is generated by homogeneous binomials, because each row of the matrix B_n used for defining it has row sum $n + 1$ ([10], chapter 4). In addition, Sturmfels and Sullivant in [11] have shown that the ideal I_n is generated in degree 2. Hence it suffices to consider homogeneous quadratic binomials. Let $q = q^+ - q^-$ be a binomial in I_n of degree 2. Then clearly either (1) or (3) holds; that is, either the index corresponding to one leaf is fixed for all the monomials in q, or none of them are.

In the former case, for the index i from equation (1),

$$q \in I_n \iff \varphi_n(q^+) = \varphi_n(q^-)$$
$$\iff \varphi_{n-1}(\pi_i(q^+)) = \varphi_{n-1}(\pi_i(q^-)) \iff \pi_i(q) \in I_{n-1},$$

where the first statement holds by definition of φ_n and the second by definition of the projection π_i.

In the latter case, for each i with $1 \leq i \leq n$,

$$q \in I_n \iff \varphi_n(q^+) = \varphi_n(q^-)$$
$$\iff \varphi_{n-1}(\pi_i(q^+)) = \varphi_{n-1}(\pi_i(q^-)) \iff \pi_i(q) \in I_{n-1},$$

where the second statement holds by definition of π_i and (5). It follows that $I_n = (q : q \in \mathcal{G}_n)$.

In particular, the set of generators for I_n with Property (i) can be obtained from those of I_{n-1} by inserting first 0 at the ith index position for each monomial of $q \in \mathcal{G}_{n-1}$ and then repeating the same process by inserting 1. This operation corresponds to lifting to all the possible preimages of $\pi_i(q)$ that satisfy Property (i) for each $1 \leq i \leq n$ and every $q \in \mathcal{G}_{n-1}$. The set of generators for I_n with Property (ii) can be obtained from those of I_{n-1} by a similar lifting to all preimages of $\pi_i(q)$ for each $q \in \mathcal{G}_{n-1}$ in such a way that Property (ii) is satisfied. Namely, for every $q = q^+ - q^- \in \mathcal{G}_{n-1}$ with Property (ii), one inserts 0 at the ith index position for one monomial of q^+ and for one monomial of q^-, and inserts 1 at the ith index position for the remaining monomials of q^+ and q^-. In addition, by definition of Property (ii), it suffices to lift to the preimages of $\pi_n(q)$ only. $\square$

Remark 6. A different recursion has been proposed by Sturmfels and Sullivant in [12].

Recall ([10]) that a binomial $q = q^+ - q^- \in I$ is said to be *primitive* if there exists no binomial $f = f^+ - f^- \in I$ with the property that $f^+|q^+$ and $f^-|q^-$. A *circuit* is a primitive binomial of minimal support.

Remark 7. The binomials in $\mathcal{G}_n$ are circuits of I_n, since the ideal is generated by squarefree binomials and contains no linear forms.

In general, we can describe the generators of I_n as follows: given n, begin by lifting $\mathcal{G}_3$ recursively to produce $\mathcal{G}_{n-1}$; that is, until the number of indices of each generator reaches $n - 1$. Next, lift $\mathcal{G}_{n-1}$ n times so that Property (i) is satisfied for one of the n index positions. For example,

$$q := q_{0000}q_{1111} - q_{1001}q_{0110} \in \mathcal{G}_4$$

can be lifted to a generator of I_5 in ten different ways: by lifting to preimages of $\pi_1, \ldots, \pi_5$ so that Property (i) is satisfied with either a 0 or a 1:

$$\pi_1^{-1}(q) = \{q_{000000}q_{01111} - q_{01001}q_{00110}, q_{10000}q_{11111} - q_{11001}q_{10110}\},$$
$$\pi_2^{-1}(q) = \{q_{000000}q_{10111} - q_{10001}q_{00110}, q_{01000}q_{11111} - q_{11001}q_{01110}\},$$

and so on. This will be the set of binomials in $\mathcal{G}_n$ with Property (i). Clearly, some generators will repeat during the recursive lifting: lifting by inserting 0 at position (i) allows the 0 to occur at the previous $i - 1$ positions. Also, fixing 1 at any leaf allows 0 to appear on any of the other leaves.

To construct $q^+ - q^-$ with Property (ii), we need not proceed inductively, as all projections of binomials that satisfy this property must satisfy it, too. Instead, we consider two cases corresponding to the parity of n. Namely, recalling the definition of Property (ii), first we fix q^- in such a way to ensure that $in_{<_{lex}}(q) = q^+$.

Suppose n is odd. Fix q^- by taking

$$q^- = q_{01\ldots1}q_{10\ldots0}$$

with n indices in each of the two variables. Then $n - 1$ being even provides that $a_0^{(n+1)}a_1^{(n+1)}|\varphi_n(q^-)$. Thus every choice of q^+ must satisfy the same. To find q^+, we need to choose pairs of n-digit binary numbers with digits complementary to each other, and thus there are $2^{n-1} - 1$ choices for q^+. Specifically, listing the smallest $2^{n-1} - 1$ n-digit binary numbers and pairing them with the largest $2^{n-1} - 1$ n-digit binary numbers in reverse order produces all choices for q^+, and we have a complete list of generators. For example, the first such generator in the list would be $q_{0\ldots0}q_{1\ldots1} - q_{01\ldots1}q_{10\ldots0}$.

If n is even, then we can create q^- such that $(a_0^{(n+1)})^2$ or $(a_1^{(n+1)})^2$ divides $\varphi_n(q^-)$ and $\varphi_n(q^+)$. Namely, the two choices for q^- are

$$q^- = q_{01\ldots1}q_{10\ldots0} \text{ and } q^- = q_{01\ldots10}q_{10\ldots01}.$$

The list of all possible q^+ is obtained in the manner similar to the case when n is odd, except that the odd pairs in the list receive the first choice of q^-, while the even pairs receive the second. The number of such generators $q^+ - q^-$ is $2^{n-1} - 2$, since there are 2^n n-digit binary numbers and thus half as many pairs, and 2 choices are taken by the q^-.

In summary, the number of generators of I_n that satisfy Property (ii) is $(2^{n-1} - 2) + (n \mod 2)$.

Next we strengthen Proposition (2).

Proposition 3. *The set $\mathcal{G}_n$ is a lexicographic Gröbner basis of I_n, for any $n \geq 4$.*

Proof. For the case $n = 3$ this is already shown. Let $n > 3$. Then we can partition the set of $q \in \mathcal{G}_n$ into those satisfying Property (i) or (ii). Note that I_n is prime by definition, and thus radical. Also, Proposition (2) shows it is generated by squarefree quadratic binomials. These facts are used in what follows.

Let $q_i, q_j \in I_n$. If $(q_i^+, q_j^+) = 1$, the S-pair $S(q_i, q_j)$ reduces to zero. Also, if q_i^- and q_j^- are not relatively prime, the cancellation criterion provides that the corresponding S-pair also reduces to zero. Therefore we consider $f := S(q_i, q_j) \in I_n$ with $(q_i^+, q_j^+) \neq 1$ and $(q_i^-, q_j^-) = 1$. In particular, $\deg(f) = 3$. Let us write $q_i = q_{g_1} q_{g_2} - q_{h_1} q_{h_2}$ and $q_j = q_{g_1} q_{g_3} - q_{h_3} q_{h_4}$. Then

$$f = q_{g_3} q_{h_1} q_{h_2} - q_{g_2} q_{h_3} q_{h_4} \in I_n.$$

Case I. Suppose q_i satisfies Property (i) and q_j satisfies Property (ii). Then there exists a k such that $\pi_k(q_i) \in I_{n-1}$. Furthermore, Property (ii) implies that $\pi_k(q_j) \in I_{n-1}$. A very technical argument shows that

$$\pi_k(f) \in I_{n-1}$$

and furthermore, this projection preserves the initial terms. In summary, to check that $\pi_k(f) \in I_{n-1}$, it suffices to ensure that $a_s^{(n)} | \varphi_{n-1}(\pi_k(q_{g_3} q_{h_1} q_{h_2}))$ if and only if $a_s^{(n)} | \varphi_{n-1}(\pi_k(q_{g_2} q_{h_3} q_{h_4}))$, where s is the sum of the observations on the leaves of the $(n-1)$-leaf tree obtained from T by deleting leaf (k). There are two cases corresponding to the parity of n. If n is odd, there are additional subcases determined by the correspondence of the images of the variables in the two monomials of f under φ_{n-1}. The facts that q_i and q_j satisfy Properties (i) and (ii), respectively, play a crucial role in the argument. Checking all the cases then shows that $\pi_k(f) \in I_{n-1}$ and that initial terms are preserved under this projection.

Applying the induction hypothesis then finishes the proof.

Case II. Suppose both q_i and q_j satisfy Property (i). Then there is a $q_k \in \mathcal{G}_n$ satisfying Property (ii) where both $S(q_i, q_k)$ and $S(q_j, q_k)$ reduce to zero. The three-pair criterion ([8]) provides the desired result.

Case III. If both q_i and q_j satisfy Property (ii), then it can be seen from the construction preceding this Proposition that the initial terms are relatively prime, so their S-polynomial need not be considered. □

Proposition 3 has important theoretical consequences. Let S be a polynomial ring over the field K. Recall ([4]) that S/I is *Koszul* if the field K has a linear resolution as a graded S/I-module:

$$\cdots \to (S/I)^{\beta_2}(-2) \to (S/I)^{\beta_1}(-1) \to S/I \to K \to 0.$$

An ideal $I \subset S$ is said to be quadratic if it is generated by quadrics. S/I is *quadratic* if its defining ideal I is quadratic, and it is *G-quadratic* if I has a quadratic Gröbner basis. It is known (e.g. [4]) that if S/I is G-quadratic, then it is Koszul, which in turn implies it is quadratic. The reverse implications do not hold in general. We have just found an infinite family of toric varieties whose coordinate rings S/I are G-quadratic.

Corollary 1. *The coordinate ring of the toric variety whose defining ideal is I_n is Koszul for every n.*

The approach developed here produces the list of generators for the kernel of B_n all of which are of degree two. In addition, by constructing the toric ideals of invariants inductively, we are able to explicitly calculate the quadratic Gröbner bases. In light of the conjecture posed in [11] that the ideal of phylogenetic invariants for the group of order k is generated in degree at most k, we are working on generalizing the above approach to any abelian group of order k. In particular, we want to give a description of the lattice basis ideal I_{L_n} and the ideal of invariants I for $G = \mathbb{Z}_2 \times \mathbb{Z}_2$ with generators of degree at most 4. These phylogenetic ideals are of interest to computational biologists.

Acknowledgment. The authors would like to thank Uwe Nagel for introducing us to the field of phylogenetic algebraic geometry and for his continuous support, motivation and guidance.

References

1. Allman, E., Rhodes, J.: Phylogenetic ideals and varieties for the general Markov model. Advances in Applied Mathematics, Preprint, arXiv.org:math/0410604 (to appear)
2. Allman, E., Rhodes, J.: Identifying evolutionary trees and substitution parameters for the general Markov model with invariable sites. Preprint, arXiv.org:q-bio/0702050 (2007)
3. Casanellas, M., Fernandez-Sanchez, J.: Performance of a new invariants method on homogeneous and non-homogeneous quartet trees. Preprint, arXiv.org:q-bio/0610030 (2006)
4. Conca, A., Rossi, M.E., Valla, G.: Gröbner flags and Gorenstein algebras. Compositio Math. 129 27(1), 95–121 (2001)
5. Eisenbud, D.: Commutative algebra with a view toward algebraic geometry. In: Eisenbud, D. (ed.) Graduate Texts in Mathematics, vol. 150, Springer, Heidelberg (1995)
6. Eriksson, N., Ranestad, K., Sturmfels, B., Sullivant, S.: Phylogenetic Algebraic Geometry. In: Projective varieties with unexpected properties, Ciliberto, C., Geramita, A., Harbourne, B., Roig, R.-M., Ranestad, K. (eds.) De Gruyter, Berlin, pp. 237–255 (2005)
7. Eriksson, N., Yao, Y.: Metric learning for phylogenetic invariants. Preprint, arXiv.org:q-bio/0703034 (2007)
8. Hemmecke, R., Malkin, P.: Computing generating sets of lattice ideals. Preprint, arXiv.org:math/0508359 (2005)
9. Pachter, L., Sturmfels, B.: Algebraic statistics for computational biology. Cambridge University Press, New York, NY, USA (2005)
10. Sturmfels, B.: Gröbner bases and convex polytopes. American Mathematical Society, University Lecture Series 8 (1996)
11. Sturmfels, B., Sullivant, S.: Toric ideals of phylogenetic invariants. J. Comp. Biol. 12, 204–228 (2005)
12. Sturmfels, B., Sullivant, S.: Toric geometry of cuts and splits. Preprint, arXiv.org:math.AC/0606683 (2006)

Inference of Protein-Protein Interactions by Using Co-evolutionary Information

Tetsuya Sato[1], Yoshihiro Yamanishi[2], Katsuhisa Horimoto[3],
Minoru Kanehisa[2], and Hiroyuki Toh[1]

[1] Division of Bioinformatics, Medical Institute of Bioregulation, Kyushu University,
3-1-1, Maidashi, Higashi-ku, Fukuoka 812-8582, Japan
sato@bioreg.kyushu-u.ac.jp, toh@bioreg.kyushu-u.ac.jp
[2] Bioinformatics Center, Institute for Chemical Research, Kyoto University,
Gokasho, Uji, Kyoto 611-0011, Japan
yoshi@kuicr.kyoto-u.ac.jp, kanehisa@kuicr.kyoto-u.ac.jp
[3] Computational Biology Research Center, National Institute of Advanced Industrial
Science and Technology, 2-42 Aomi, Koto-ku, Tokyo, 135-0064, Japan
k.horimoto@aist.go.jp

Abstract. The mirror tree is a method to predict protein-protein interaction by evaluating the similarity between distance matrices of proteins. It is known, however, that predictions by the mirror tree method include many false positives. We suspected that the information about the evolutionary relationship of source organisms may be the cause of the false positives, because the information is shared by the distance matrices. Therefore, we excluded the information from the distance matrices and evaluated the similarity of the residuals as the intensity of co-evolution. We developed two different methods with a projection operation and partial correlation coefficient. The number of false positives were drastically reduced by our methods.

Keywords: protein-protein, co-evolution, projection operation, partial correlation coefficient.

1 Introduction

Information about protein–protein interactions in living cells provides deep insight into the biological functions of proteins at the cellular level. The development of large-scale experimental analyses, such as the yeast 2-hybrid system [7,21] and pull-down method [3,6], has facilitated understanding the protein–protein interaction network in cells. However, such experimental approaches have problems in coverage and accuracy [20,22]. Following the trend, the prediction of protein–protein interactions has become one of the major issues in bioinformatics. The predicted protein–protein interactions can provide complementary or supporting evidence to the large-scale experimental studies on protein–protein interactions although computational analyses also have the same drawbacks as experimental studies, that is, low coverage and low accuracy.

H. Anai, K. Horimoto, and T. Kutsia (Eds.): AB 2007, LNCS 4545, pp. 322–333, 2007.
© Springer-Verlag Berlin Heidelberg 2007

Various computational methods to predict protein–protein interactions have been developed until today. Co-evolutionary behavior between interacting proteins provides useful information for the prediction of protein-protein interaction. The mirror tree method [15] and the *in silico* 2-hybrid system method [14] are two representative methods to predict protein-protein interaction with co-evolutionary information. In this paper, we explain our studies [18,19] aiming at improvement of the mirror tree method. The mirror tree method was developed by Pazos and Valencia [15], although there are several preceding works, such as Goh *et al.* [5]. The mirror tree method predicts protein–protein interactions under the assumption that the interacting proteins show similarity in molecular phylogenetic tree because of the co-evolution through the interaction. To avoid the difficulty to evaluate the similarity between a pair of phylogenetic trees, however, the mirror tree method compares a pair of distance matrices. Consider two proteins, proteins A and B. The orthologous amino acid sequences of protein A are collected from n species. The n sequences of protein A are aligned and the distance matrix, D_A, is calculated. The size of D_A is $n \times n$, and each row or column of the matrix corresponds to a species under consideration. An element of the matrix, $D_A(i,j)$, represents the genetic distance between species i and j, which is calculated by comparing the amino acid sequences of protein A between the two species. A distance matrix is symmetric, and only the upper or lower half of the matrix includes sufficient information for tree construction. Likewise, the orthologous amino acid sequences of protein B are collected from the same n species, and the distance matrix, D_B, is calculated. The intensity of co-evolution between proteins A and B is evaluated as Pearson's correlation coefficient, ρ_{AB}^{MIRROR}, between the distance matrices D_A and D_B, which is calculated as follows:

$$\rho_{AB}^{MIRROR} = \frac{\sum_{i=1}^{n-1} \sum_{j=i+1}^{n} (D_A(i,j) - \text{Ave}(D_A))(D_B(i,j) - \text{Ave}(D_B))}{\sqrt{\text{Var}(D_A)\text{Var}(D_B)}} , \quad (1)$$

where Ave and Var represent the average and the variance of the upper (or lower) half elements of a distance matrix. High correlation between the distance matrices indicates the resemblance of the corresponding phylogenetic trees. Therefore, a pair of proteins are predicted to interact with each other, when the distance matrices of the proteins show high correlation. Because of the simplicity, modification and improvement have been introduced into the mirror tree method by several groups [4,9,16]. On the other hand, it has been recognized that the mirror tree predictions include many false positives. That is, even protein pairs that are known not to interact often show high correlation coefficients. Then, such pairs are predicted to interact in error. The abundance of false positives in the mirror tree prediction reduces the reliability of the method in actual applications. We suspected that the cause of the false positives is the information about the evolutionary relationship among the source organisms of the collected orthologous sequences. The distance matrices of orthologous proteins from the same set of n source organisms are compared in the mirror tree method. Therefore, all of the distance matrices of the proteins are considered to include the same information about the evolutionary relationships among the same n sources. The information shared by the distance matrices would generate high correlation even between the matrices of non-interacting proteins. If our hypothesis is correct, the number of false positives in the predictions could be reduced by excluding such information from the distance matrices. We developed two different methods to exclude the information from the distance matrices for the prediction of protein-protein interaction. One of them uses a projection operator, whereas the other is based on multiple regression. The two methods were applied to physically contacting proteins, to evaluate their performances. Then, it was found that our methods drastically reduced the number of false positives in the predicted protein–protein interactions as expected.

2   Material and Method

2.1   Data Preparation

13 pairs of *Escherichia coli* proteins that are physically in contact were selected from the Database of Interacting Proteins (DIP) [17]. The pairs are listed in the legend for Table 1. Each pair was selected so that neither of the interacting proteins participated in the remaining 12 pairs of interacting proteins. Then, (putative) orthologues corresponding to the 26 proteins were collected from 40 different bacterial species, according to the KEGG KO database [10]. We assumed that a pair of proteins, which are orthologous to the interacting proteins of *E. coli*, are also physically in contact.

2.2   Multiple Sequence Alignment and Distance Matrix

A multiple alignment of each set of orthologous amino acid sequences was made with the alignment software MAFFT [11]. A distance matrix for the orthologous sequences was calculated from the multiple alignment. A genetic distance between every pair of aligned sequences was calculated as a maximum likelihood estimate using the PROTDIST in the PHYLIP package [2]. JTT model [8] was used as a model for the amino acid substitution for the estimation.

2.3   Transformation from Distance Matrix to Phylogenetic Vector [19]

The distance matrix was transformed into a vector. The upper or lower half of the non-diagonal elements of the distance matrix was arranged as a one-dimensional array of the numerical values in a certain order. All of the matrices were transformed into vectors with the same arrangement of the elements. When the matrix has a size of $n \times n$ the dimension of the vector is $n(n-1)/2$. The vector is hereafter referred to as a 'phylogenetic vector'. The dimension of the phylogenetic vector is 820, because n is 41. Consider a pair of phylogenetic vectors, which are transformed from distance matrices D_i and D_j. The subscripts

i and j indicate different sets of orthologues, that is, different proteins. Then, the elements of each vector are normalized with the average and the standard deviation of the elements as follows:

$$\left|\nu_i^{\#}\right\rangle = \frac{|\nu_i\rangle - |\mu\rangle}{\sqrt{\text{Var}(\nu_i)}}, \tag{2}$$

where $|\mu\rangle$ is a vector with the same dimension as $|\nu_i\rangle$. All the elements of $|\mu\rangle$ are constant, and are equal to the arithmetic average over the elements of $|\nu_i\rangle$. $\text{Var}(\nu_i)$ is the variance over all the elements of $|\nu_i\rangle$. The superscript $\#$ in $\left|\nu_i^{\#}\right\rangle$ indicates that the vector is normalized. Then, the inner product between a pair of normalized vectors is the Pearson's correlation coefficient used for the mirror tree method, which is defined by formula (1). Hereafter, the correlation coefficient by the mirror tree method is denoted as $\rho_{ij}^{\text{MIRROR}}$.

$$\rho_{ij}^{\text{MIRROR}} = \langle\nu_i^{\#}|\nu_j^{\#}\rangle. \tag{3}$$

2.4   First Method with Projection Operator [19]

Consider an $n(n-1)/2$-dimensional unit vector $|u\rangle$, which represents the evolutionary relationship of the source species under consideration. Given such a vector, following projection operator P can be defined:

$$P = I - |u\rangle\langle u|. \tag{4}$$

The projection operator is a matrix with the size of $n(n-1)/2 \times n(n-1)/2$. The method to obtain $|u\rangle$ is explained below. I represents an identity matrix with the size of $n(n-1)/2 \times n(n-1)/2$. By applying the projection operator (4) to a phylogenetic vector, say, $|\nu_i\rangle$, the component within $|\nu_i\rangle$, which is orthogonal to $|u\rangle$, is generated:

$$|\varepsilon_i\rangle = P|\nu_i\rangle = |\nu_i\rangle - |u\rangle\langle u|\nu_i\rangle. \tag{5}$$

$|\varepsilon_i\rangle$ is a residual vector obtained by excluding the information about the evolutionary relationship from the phylogenetic vector . The same projection operator was applied to all of the phylogenetic vectors under consideration. Each of the residual vectors was then normalized with the average and the standard deviation of the elements. The inner product between the two residual vectors $\left|\varepsilon_i^{\#}\right\rangle$ and $\left|\varepsilon_j^{\#}\right\rangle$ represents the Pearson's correlation coefficient between the residual vectors:

$$\rho_{ij}^{\text{PROJECTION}} = \langle\varepsilon_i^{\#}|\varepsilon_j^{\#}\rangle \tag{6}$$

was used as a new measure to evaluate the intensity of co-evolution between proteins i and j.

In order to obtain the unit vector representing the phylogenetic relationship of the source organisms, three different methods were considered. In the first method, 16S rRNA was used for the calculation. Basically, at least one copy of the 16S rRNA gene is encoded by each genome. Therefore, the distance matrix or the phylogenetic vector of the 16S rRNAs is considered to represent the evolutionary relationship among the source organisms. The nucleotide sequences of rRNA were collected from the same sources as the proteins under consideration according to the KEGG GENES database [10] and the Ribosomal Database Project-II Release 9 [1]. The nucleotide sequences of the 16S rRNA were aligned, and the distance between every pair of the aligned nucleotide sequences was calculated by using the F84 model [12] with the DNADIST in the PHYLIP package [2]. The distance matrix was then transformed into a phylogenetic vector $|\nu_{16S}\rangle$. Then, a unit vector $|u_{16S}\rangle$ was obtained as $|\nu_{16S}\rangle/\|\nu_{16S}\|$.

In the second method, all of the phylogenetic vectors of proteins under consideration were normalized so that the size of the elements in each protein was '1' at first. Then, they were averaged as

$$|\nu_{\text{AVE}}\rangle = \frac{1}{m} \sum_{i=1}^{m} \frac{|\nu_i\rangle}{\|\nu_i\|}, \tag{7}$$

where m is the number of proteins. So, m was 26 here. The second unit vector $|u_{\text{AVE}}\rangle$, was obtained as $|\nu_{\text{AVE}}\rangle/\|\nu_{\text{AVE}}\|$.

In the third method, the phylogenetic vectors were used again. Let X be a matrix of $n(n-1)/2 \times m$ in which the i-th column corresponds to a normalized phylogenetic vector of protein i. Then, a correlation coefficient matrix Y of $m \times m$ was calculated as $X^T X$. The superscript T indicates the transpose of a matrix. The principal component analysis for the data corresponding to X is equivalent to solving the eigenvalue problem of Y. Then, $|\nu_{\text{PC1}}\rangle$ was obtained as $|\nu_{\text{PC1}}\rangle = X|z_1\rangle$, where $|z_1\rangle$ is a vector corresponding to the first principal component axis. Then, $|\nu_{\text{PC1}}\rangle/\|\nu_{\text{PC1}}\|$ generated the third unit vector, $|u_{\text{PC1}}\rangle$.

The Pearson's correlation coefficients between the residual vectors for a pair of proteins i and j, which were generated by the projection operations constructed with $|u_{16S}\rangle$, $|u_{\text{AVE}}\rangle$ and $|u_{\text{PC1}}\rangle$, were represented by ρ_{ij}^{16S}, ρ_{ij}^{AVE} and ρ_{ij}^{PC1}. The type of correlation coefficient is collectively represented by ρ^* without the subscripts, i and j where the superscript indicates the type of correlation coefficient.

2.5   Second Method with Multiple Regression [18]

Suppose that m proteins are given and we want to predict interacting pairs from them. Consider multiple regressions of $|\nu_i\rangle$ and $|\nu_j\rangle$ with $(m-2)$ phylogenetic vectors:

$$|\nu_i\rangle = \alpha_0 + \sum_{k \neq i,j}^{m} \alpha_k |\nu_k\rangle + |\delta_i\rangle, \tag{8}$$

$$|\nu_j\rangle = \beta_0 + \sum_{l \neq i,j}^{m} \beta_l |\nu_l\rangle + |\delta_j\rangle, \tag{9}$$

where α_i and β_j are parameters. The residual vectors, $|\delta_i\rangle$ and $|\delta_j\rangle$, are expected to lack the evolutionary information of the source organisms. Note that $|\nu_j\rangle$ is excluded from the summation on the right side of the equation (8). Likewise, $|\nu_i\rangle$ is excluded from the summation on the right side of the equation (9). The similarity between the two residual vectors is considered to indicate the intensity of co-evolution between proteins i and j. To evaluate the similarity between the residual vectors, the Pearson's correlation coefficient between $|\delta_i\rangle$ and $|\delta_j\rangle$ was calculated. As described above, the inner product between the normalized residual vectors is equivalent to the Pearson's correlation coefficient between them:

$$\rho_{ij}^{\text{PARTIAL}} = \langle \delta_i^{\#} | \delta_j^{\#} \rangle. \tag{10}$$

The correlation coefficient is called the partial correlation coefficient between $|\nu_i\rangle$ and $|\nu_j\rangle$. In actual practice, the following formula was used to obtain the partial correlation coefficient, instead of performing multiple regression.

$$\rho_{ij}^{\text{PARTIAL}} = \frac{-(R^{-1})_{ij}}{\sqrt{(R^{-1})_{ii}}\sqrt{(R^{-1})_{jj}}}, \tag{11}$$

where R is the correlation coefficient matrix whose (i,j)-th element is $\rho_{ij}^{\text{MIRROR}}$, and the superscript -1 indicates inverse. ρ^{PARTIAL} without subscripts, i and j, collectively represents that the type is partial correlation coefficient.

3   Results and Discussions

We calculated five types of correlation coefficients, ρ^{MIRROR}, ρ^{16S}, ρ^{AVE}, ρ^{PC1} and ρ^{PARTIAL}, for all of the possible pairs of 26 proteins, that is, 325 pairs of proteins. The performance of each correlation coefficient was evaluated with specificity and sensitivity. Out of the 325 pairs, the interactions of 13 pairs have been experimentally identified. Only top 20 of the five types of correlation coefficients are shown in Table 1, where the actually interacting pairs are highlighted with circles. As shown in the table, the top ranks of ρ^{16S}, ρ^{AVE}, ρ^{PC1} and ρ^{PARTIAL} were occupied by pairs of actually interacting proteins. In contrast, non-interacting proteins were present within the top ranks of ρ^{MIRROR}. The decreasing patterns of the five correlation coefficients are seen in this table. The decrease of ρ^{MIRROR} was quite slow, whereas ρ^{AVE}, ρ^{PC1} and ρ^{PARTIAL} decreased rapidly. The rate of the ρ^{16S} decrease was rather moderate. The decreasing patterns shown in Table 1 clearly demonstrates the problem of the original mirror tree method. Even if a high value, e.g. 0.9, is used as a threshold for the correlation coefficient to predict a protein–protein interaction, ρ^{MIRROR} produces many pairs with high correlation, including non-interacting pairs, which likely lead to the generation of many false positives. However, the occupation of the top ranks by interacting proteins and the rapid decreases of ρ^{16S}, ρ^{AVE}, ρ^{PC1} and ρ^{PARTIAL} guarantee the specificity of prediction, if the threshold is set at a sufficiently high value.

Table 1. Comparison of top 20 protein pairs sorted in decreasing order of the correlation coefficients

rank	ρ^{MIRROR}		ρ^{16S}		ρ^{AVE}		ρ^{PC1}		ρ^{PARTIAL}	
1	dnaN-rpoB	0.977	sucD-sucC 0.924	O	sucD-sucC 0.910	O	sucD-sucC 0.915	O	sucD-sucC 0.885	O
2	dnaK-secY	0.963	atpA-atpD 0.803	O	trpA-trpB 0.792	O	trpA-trpB 0.754	O	trpA-trpB 0.744	O
3	dnaK-rpoB	0.963	carA-carB 0.803	O	rpoA-rpoB 0.654	O	carA-carB 0.649	O	tufB-tsf 0.612	O
4	sucD-sucC 0.962	O	dnaK-secY 0.801		carA-carB 0.642	O	atpA-atpD 0.640	O	carA-carB 0.597	O
5	dnaN-dnaK	0.960	trpA-trpB 0.797	O	dnaN-rpoB	0.634	dnaN-rpoB	0.587	rpoA-carA	0.519
6	atpA-atpD 0.959	O	dnaF-secA 0.783		atpA-atpD 0.615	O	dnaK-atpD	0.560	dnaN-iscS	0.505
7	rpoA-rpoB 0.958	O	dnaK-atpD 0.774		iscS-iscU 0.607	O	dnaK-secY	0.560	grpE-rpoB	0.462
8	rpoB-secY	0.955	dnaN-rpoB 0.772		grpE-clpP	0.553	iscS-iscU 0.555	O	dnaK-secY	0.457
9	secY-secA 0.954	O	rpoA-rpoB 0.768	O	dnaK-carB	0.541	grpE-clpP	0.545	rpoA-rpoB 0.450	O
10	dnaK-atpD	0.953	dnaN-carA 0.761		grpE-tsf	0.541	dnaK-carB	0.542	atpA-atpD 0.443	O
11	dnaN-secY	0.953	dnaN-dnaK 0.760		dnaK-secY	0.516	secY-carB	0.531	ruvA-ruvB 0.439	O
12	dnaK-atpA	0.952	dnaK-carB 0.758		dnaK-atpD	0.514	dnaK-rpoB	0.500	dnaA-dnaB 0.358	O
13	dnaE-secA	0.945	dnaN-secY 0.758		ruvA-ruvB 0.511	O	rpoA-rpoB 0.498	O	grpE-dnaN	0.355
14	dnaN-rpoA	0.945	dnaF-secY 0.755		secY-carB	0.501	grpE-tsf	0.497	clpP-atpD	0.345
15	dnaK-secA	0.944	rpoB-secY 0.753		rpoB-secY	0.498	dnaK-atpA	0.496	dnaK-tufB	0.339
16	dnaE-secY	0.944	dnaK-atpA 0.747		tsf-trpB	0.476	ruvA-ruvB 0.489	O	dnaB-sucD	0.336
17	dnaN-clpX	0.943	dnaE-dnaK 0.743		dnaA-ruvB	0.469	dnaE-secA	0.480	secA-carB	0.333
18	dnaN-secA	0.940	secY-carB 0.727		secA-trpB	0.453	dnaE-secY	0.477	dnaK-sucC	0.329
19	clpX-rpoB 0.937	O	iscS-iscU 0.717	O	tufB-tsf 0.445	O	dnaA-ruvB	0.455	atpA-sucC	0.306
20	dnaN-carA	0.937	dnaK-carA 0.716		dnaK-atpA	0.438	secY-secA 0.431	O	clpX-dnaE	0.302

The abbreviated names of the interacting proteins are as follows: sucC-sucD, succinyl-CoA synthetases alpha - beta; atpA-atpD, ATP synthases alpha - beta; rpoA-rpoB, DNA - directed RNA polymerases alpha - beta; secA-secY, preprotein translocase secA - secY; carA-carB, carbamoyl-phosphate synthases small - large; ruvA-ruvB, Holliday junction DNA helicases ruvA - ruvB; iscS-iscU, putative aminotransferase - NifU-like protein; dnaE-dnaN, DNA polymerases III alpha - beta; trpA-trpB, tryptophan synthases alpha - beta; tufB-tsf, elongation factors EF-Tu - EF-Ts; dnaA-dnaB, DNA helicase - dnaA; grpE-dnaK, heat shock protein grpE - dnaK protein; and clpX-clpP, ATP-dependent clp proteases ATP-binding subunit - protease proteolytic subunit.

The unit vector $|u\rangle$ seems to be a crucial factor for the prediction of a protein–protein interaction when a projection operator is used. Therefore, we examined the relationships among $|u_{16S}\rangle$, $|u_{AVE}\rangle$ and $|u_{PC1}\rangle$ by calculating absolute value of Pearson's correlation coefficients $|r|$ among them. $|r|$ between $|u_{16S}\rangle$ and $|u_{AVE}\rangle$ was 0.947, whereas $|r|$ between $|u_{16S}\rangle$ and $|u_{PC1}\rangle$ was 0.946. The highest correlation, $|r| = 0.998$, was observed between $|u_{AVE}\rangle$ and $|u_{PC1}\rangle$. The high correlation between $|u_{16S}\rangle$ and the other unit vectors suggests that the information except for the evolutionary relationship of source organisms can be approximately canceled out by the average operation or principal component analysis.

The ρ^{16S}, ρ^{AVE}, ρ^{PC1} and $\rho^{PARTIAL}$ seem to outperform the ρ^{MIRROR}. That is, the exclusion of the information about the evolutionary relationships among the source organisms from the distance matrices is effective to reduce the number of the false positives from the mirror tree predictions. The specificities and the sensitivities of the five types of correlation coefficients under four different threshold values, 0.9, 0.8, 0.7 and 0.6, are shown in Table 2. When a pair of proteins had a correlation coefficient greater than the threshold the proteins were predicted to interact with each other. Three types of correlation coefficients, ρ^{AVE}, ρ^{PC1} and $\rho^{PARTIAL}$, showed high specificity under any threshold value, whereas ρ^{16S} showed high specificity only when threshold was 0.9 or 0.8. The high specificities of ρ^{16S}, ρ^{AVE}, ρ^{PC1} and $\rho^{PARTIAL}$ mean the drastic reduction of false positives, compared with ρ^{MIRROR} [18,19]. Recently, Pazos et al. [13] have independently developed a method to exclude the information of evolutionary relationship among the source organisms by using 16S rRNA. They adjust the scale of the distance matrix of rRNA to that of the distance matrix of a protein, and simply subtract the former from the latter. Then, correlation coefficient is calculated between the sets of residual elements. Improvement in specificity is also observed by their operation, although the mathematical framework of their method is different from those of ours.

Table 2. Specificity and Sensitivity of the prediction

Method	Specificity				Sensitivity			
	0.9	0.8	0.7	0.6	0.9	0.8	0.7	0.6
ρ^{MIRROR}	13.79	6.21	4.96	4.17	61.54	84.62	100.00	100.00
ρ^{16S}	100.00	75.00	28.57	24.32	7.14	21.43	42.86	64.29
ρ^{AVE}	100.00	100.00	100.00	85.71	7.14	7.14	14.29	42.86
ρ^{PC1}	100.00	100.00	100.00	100.00	7.14	7.14	14.29	28.57
$\rho^{PARTIAL}$	–	100.00	100.00	100.00	0.00	7.14	14.29	21.43

$$\text{Specificity} = \frac{\text{true positive}}{(\text{true positive} + \text{false positive})} \times 100\%,$$

$$\text{Sensitivity} = \frac{\text{true positive}}{(\text{true positive} + \text{false negative})} \times 100\%.$$

When threshold was set to 0.9, no interacting pair was predicted with $\rho^{PARTIAL}$, and specificity was not calculated in the case.

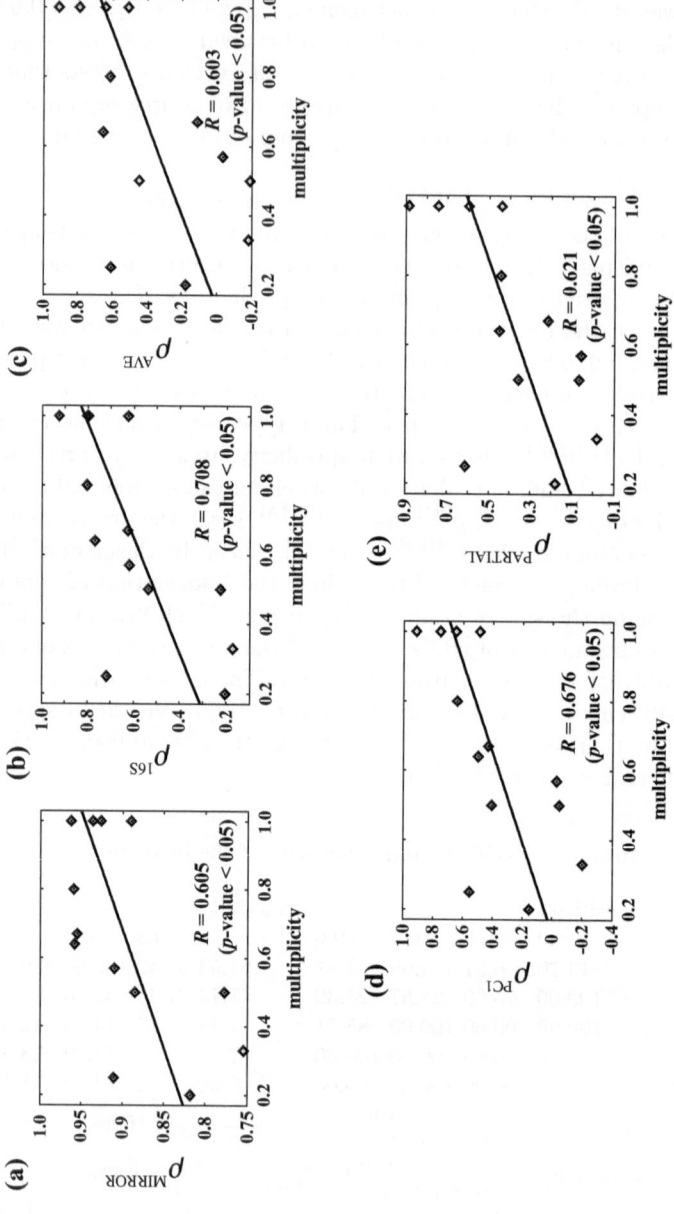

Fig. 1. The relationship between multiplicity and five types of correlation coefficients, ρ^{MIRROR}, ρ^{16S}, ρ^{AVE}, ρ^{PC1} and $\rho^{PARTIAL}$

Despite the improvement described above, the sensitivities of ρ^{16S}, ρ^{AVE}, ρ^{PC1} and $\rho^{PARTIAL}$ were lower than that of ρ^{MIRROR}. This means that a pair of proteins i and j did not always show high ρ_{ij}^{16S}, ρ_{ij}^{AVE}, ρ_{ij}^{PC1} and $\rho_{ij}^{PARTIAL}$ even when proteins, i and j, interact with each other. In other words, the number of false negatives increased when our methods were used, compared with the original mirror tree method. Here, we calculated the intensity of co-evolution between a pair of proteins as the correlation coefficient after excluding the information about the evolutionary relationship among the source organisms from the phylogenetic vectors. However, the pairs may also interact with other proteins. If such proteins exist, it would be difficult to detect the interaction with the pair, because the co-evolution with the other partners may function as noise for the prediction of interaction of a pair. To examine this hypothesis, we investigated the relationship between the multiplicity of the interaction [19] and the correlation coefficient (Fig 1). The multiplicity, or a modified Jaccard coefficient, is a measure defined between a pair of interacting proteins. Consider an interacting pair of proteins A and B. Let $\mathbf{M}$ and $\mathbf{N}$ be the sets of interaction partners of proteins A and B. The information about the interaction partners were obtained from the DIP database [17]. Protein B belongs to $\mathbf{M}$, whereas $\mathbf{N}$ includes protein A. The multiplicity between proteins, A and B, is defined as follows:

$$\text{Multiplicity (modified Jaccard coefficient)} = \frac{|\mathbf{M} \cap \mathbf{N}| + 1}{|\mathbf{M} \cup \mathbf{N}| - 1}. \qquad (12)$$

When proteins A and B interact each other without other interaction partners, multiplicity takes a value 1. When proteins A and B have other interaction partners, the multiplicity decreases. However, when proteins A and B share the other interaction partners, the multiplicity takes a value close to 1. In contrast, when proteins A and B have their own interaction partners respectively, the multiplicity is close to 0. As shown in Fig. 1, the intensities of co-evolution calculated by any method show positive correlation with the multiplicity. That is, the intensities of co-evolution were high when proteins A and B formed a complex without other interaction partners or share the other interaction partners. When proteins A and B had their own interaction partners, that is, the multiplicity was low, the intensities of co-evolution were low. The observation suggests that the false negatives are generated by the presence of unshared interaction partners. Further accumulation of experimental knowledge is required to ascertain this hypothesis.

4 Conclusion

The mirror tree method is a simple approach for the prediction of protein–protein interactions. Here, we reviewed our methods to improve the performance of the original mirror tree method. In the experiment, we confirmed that our methods could drastically reduce the number of false positives in the prediction. Our method, however, generated more false negatives than the original mirror tree method. Our analysis suggested that the presence of unshared interaction partners may be the cause of the false negatives. However, if we select protein pairs with a high correlation coefficient, e.g. > 0.8, by any one of our methods, we can predict interacting protein pairs with high reliability.

Acknowledgments. This work was supported by Grants-in-Aid for Scientific Research on Priority Areas 'Systems Genomics' (K.H.), 'Comprehensive Genomics' (M.K.) and 'Membrane Interface' (H.T.) from the Ministry of Education, Culture, Sports, Science and Technology of Japan. The computational resource was provided by the Bioinformatics Center, Institute for Chemical Research, Kyoto University.

References

1. Cole, J.R., Chai, B., Farris, R.J., Wang, Q., Kulam-Syed-Mohideen, A.S., McGarrell, D.M., Bandela, A.M., Cardenas, E., Garrity, G.M., Tiedje, J.M.: The ribosomal database project (RDP-II): introducing myRDP space and quality controlled public data. Nucleic Acids Res. 35, D169–D172 (2007)
2. Felsenstein, J.: PHYLIP (Phylogeny Inference Package) version 3.6. Distributed by the author. Department of Genome Sciences, University of Washington, Seattle (2004)
3. Gavin, A.C., Bosche, M., Krause, R., Grandi, P., Marzioch, M., Bauer, A., Schultz, J., Rick, J.M., Michon, A.M., Cruciat, C.M., Remor, M., Hofert, C., Schelder, M., Brajenovic, M., Ruffner, H., Merino, A., Klein, K., Hudak, M., Dickson, D., Rudi, T., Gnau, V., Bauch, A., Bastuck, S., Huhse, B., Leutwein, C., Heurtier, M.A., Copley, R.R., Edelmann, A., Querfurth, E., Rybin, V., Drewes, G., Raida, M., Bouwmeester, T., Bork, P., Seraphin, B., Kuster, B., Neubauer, G., Superti-Furga, G.: Functional organization of the yeast proteome by systematic analysis of protein complexes. Nature 415, 141–147 (2002)
4. Gertz, J., Elfond, G., Shustrova, A., Weisinger, M., Pellegrini, M., Cokus, S., Rothschild, B.: Inferring protein interactions from phylogenetic distance matrices. Bioinformatics 19, 2039–2045 (2003)
5. Goh, C.S., Bogan, A.A., Joachimiak, M., Walther, D., Cohen, F.E.: Co-evolution of proteins with their interaction partners. J. Mol. Biol. 299, 283–293 (2000)
6. Ho, Y., Gruhler, A., Heilbut, A., Bader, G.D., Moore, L., Adams, S.L., Millar, A., Taylor, P., Bennett, K., Boutilier, K., Yang, L., Wolting, C., Donaldson, I., Schandorff, S., Shewnarane, J., Vo, M., Taggart, J., Goudreault, M., Muskat, B., Alfarano, C., Dewar, D., Lin, Z., Michalickova, K., Willems, A.R., Sassi, H., Nielsen, P.A., Rasmussen, K.J., Andersen, J.R., Johansen, L.E., Hansen, L.H., Jespersen, H., Podtelejnikov, A., Nielsen, E., Crawford, J., Poulsen, V., Sorensen, B.D., Matthiesen, J., Hendrickson, R.C., Gleeson, F., Pawson, T., Moran, M.F., Durocher, D., Mann, M., Hogue, C.W., Figeys, D., Tyers, M.: Systematic identification of protein complexes in *Saccharomyces cerevisiae* by mass spectrometry. Nature 415, 180–183 (2002)
7. Ito, T., Chiba, T., Ozawa, R., Yoshida, M., Hattori, M., Sakaki, Y.: A comprehensive two-hybrid analysis to explore the yeast protein interactome. In: Proc. Natl. Acad. Sci. USA 98, pp. 4569–4574 (2001)
8. Jones, D.T., Taylor, W.R., Thornton, J.M.: The rapid generation of mutation data matrices from protein sequences. Comput. Appl. Biosci. 8, 275–282 (1992)

9. Jothi, R., Cherukuri, P.F., Tasneem, A., Przytycka, T.M.: Co-evolutionary analysis of domains in interacting proteins reveals insights into domain-domain interactions mediating protein-protein interactions. J. Mol. Biol. 362, 861–875 (2006)

10. Kanehisa, M., Goto, S., Hattori, M., Aoki-Kinoshita, K.F., Itoh, M., Kawashima, S., Katayama, T., Araki, M., Hirakawa, M.: From genomics to chemical genomics: new developments in KEGG. Nucleic Acids Res. 34, D354–D357 (2006)

11. Katoh, K., Kuma, K., Toh, H., Miyata, T.: MAFFT version 5: improvement in accuracy of multiple sequence alignment. Nucleic Acids Res. 33, 511–518 (2005)

12. Kishino, H., Hasegawa, M.: Evaluation of the maximum likelihood estimate of the evolutionary tree topologies from DNA sequence data, and the branching order in hominoidea. J. Mol. Evol. 29, 170–179 (1989)

13. Pazos, F., Ranea, J.A., Juan, D., Sternberg, M.J.: Assessing protein co-evolution in the context of the tree of life assists in the prediction of the interactome. J. Mol. Biol. 352, 1002–1015 (2005)

14. Pazos, F., Valencia, A.: In silico two-hybrid system for the selection of physically interacting protein pairs. Proteins 47, 219–227 (2002)

15. Pazos, F., Valencia, A.: Similarity of phylogenetic trees as indicator of protein–protein interaction. Protein Eng. 14, 609–614 (2001)

16. Ramani, A., Marcotte, E.M.: Exploiting the co-evolution of interacting proteins to discover interaction specificity. J. Mol. Biol. 327, 273–284 (2003)

17. Salwinski, L., Miller, C.S., Smith, A.J., Pettit, F.K., Bowie, J.U., Eisenberg, D.: The Database of Interacting Proteins: 2004 update. Nucleic Acids Res. 32, D449–D451 (2004)

18. Sato, T., Yamanishi, Y., Horimoto, K., Kanehisa, M., Toh, H.: Partial correlation coefficient between distance matrices as a new indicator of protein-protein interactions. Bioinformatics 22, 2488–2492 (2006)

19. Sato, T., Yamanishi, Y., Kanehisa, M., Toh, H.: The inference of protein-protein interactions by co-evolutionary analysis is improved by excluding the information about the phylogenetic relationships. Bioinformatics 21, 3482–3489 (2005)

20. Sprinzak, E., Sattath, S., Margalit, H.: How reliable are experimental protein–protein interaction data? J. Mol. Biol. 327, 919–923 (2003)

21. Uetz, P., Giot, L., Cagney, G., Mansfield, T.A., Judson, R.S., Knight, J.R., Lockshon, D., Narayan, V., Srinivasan, M., Pochart, P., Qureshi-Emili, A., Li, Y., Godwin, B., Conover, D., Kalbfleisch, T., Vijayadamodar, G., Yang, M., Johnston, M., Fields, S., Rothberg, J.M.: A comprehensive analysis of protein–protein interactions in Saccharomyces cerevisiae. Nature 403, 623–627 (2000)

22. von Mering, C., Krause, R., Snel, B., Cornell, M., Oliver, S.G., Fields, S., Bork, P.: Comparative assessment of large-scale data sets of protein-protein interactions. Nature 417, 399–403 (2002)

A Short Survey of Automated Reasoning

John Harrison

Intel Corporation, JF1-13
2111 NE 25th Avenue, Hillsboro OR 97124, USA
johnh@ichips.intel.com

Abstract. This paper surveys the field of automated reasoning, giving some historical background and outlining a few of the main current research themes. We particularly emphasize the points of contact and the contrasts with computer algebra. We finish with a discussion of the main applications so far.

1 Historical Introduction

The idea of reducing reasoning to mechanical calculation is an old dream [75]. Hobbes [55] made explicit the analogy in the slogan 'Reason [...] is nothing but Reckoning'. This parallel was developed by Leibniz, who envisaged a 'characteristica universalis' (universal language) and a 'calculus ratiocinator' (calculus of reasoning). His idea was that disputes of all kinds, not merely mathematical ones, could be settled if the parties translated their dispute into the *characteristica* and then simply calculated. Leibniz even made some steps towards realizing this lofty goal, but his work was largely forgotten.

1.1 The Characteristica Universalis

The dream of a truly universal language in Leibniz's sense remains unrealized and probably unrealizable. But over the last few centuries a language that is at least adequate for (most) mathematics has been developed.

Boole [11] developed the first really successful symbolism for logical and set-theoretic reasoning. What's more, he was one of the first to emphasize the possibility of applying formal calculi to several different situations, and doing calculations according to formal rules without regard to the underlying interpretation. In this way he anticipated important parts of the modern axiomatic method. However Boole's logic was limited to propositional reasoning (plugging primitive assertions together using such logical notions as 'and' and 'or'), and it was not until the much later development of *quantifiers* that formal logic was ready to be applied to general mathematics.

The introduction of formal symbols for quantifiers, in particular the *universal quantifier* 'for all' and the *existential quantifier* 'there exists', is usually credited independently to Frege, Peano and Peirce. Logic was further refined by Whitehead and Russell, who wrote out a detailed formal development of the foundations of mathematics from logical first principles in their *Principia Mathematica* [109]. In a short space of time, stimulated by Hilbert's foundational programme (of which more below), the usual logical language as used today had been developed.

H. Anai, K. Horimoto, and T. Kutsia (Eds.): AB 2007, LNCS 4545, pp. 334–349, 2007.

English	Symbolic	Other symbols
false	$\bot$	$0, F$
true	$\top$	$1, T$
not p	$\neg p$	$\bar{p}, -p, \sim p$
p and q	$p \wedge q$	$pq, p\&q, p \cdot q$
p or q	$p \vee q$	$p + q, p \mid q, p$ or q
p implies q	$p \Rightarrow q$	$p \rightarrow q, p \supset q$
p iff q	$p \Leftrightarrow q$	$p = q, p \equiv q, p \sim q$
for all x, p	$\forall x.\, p$	$(x)p$
there exists x such that p	$\exists x.\, p$	$(Ex)p$

At its simplest, one can regard this just as a convenient shorthand, augmenting the usual mathematical symbols with new ones for logical concepts. After all, it would seem odd nowadays to write 'the sum of a and b' instead of '$a+b$', so why not write '$p \wedge q$' instead of 'p and q'? However, the consequences of logical symbolism run much deeper: arriving at a precise formal syntax means that we can bring deeper logical arguments within the purview of mechanical computation.

1.2 Hilbert's Programme

At various points in history, mathematicians have become concerned over apparent problems in the accepted foundations of their subject. For example, the Pythagoreans tried to base mathematics just on the rational numbers, and so were discombobulated by the discovery that $\sqrt{2}$ must be irrational. Subsequently, the apparently self-contradictory treatment of infinitesimals in Newton and Leibniz's calculus disturbed many, as later did the use of complex numbers and the discovery of non-Euclidean geometries. Later still, when the theory of infinite sets began to be pursued for its own sake and generalized, mainly by Cantor, renewed foundational worries appeared.

Hilbert [53] suggested an ingenious programme to give mathematics a reliable foundation. In the past, new and apparently problematic ideas such as complex numbers and non-Euclidean geometry had been given a foundation based on some well-understood concepts, e.g. complex numbers as points on the plane. However it hardly seems feasible to justify infinite sets in this way based on finite sets. Hilbert's ingenious idea was to focus not on the mathematical structures themselves but on the *proofs*. Given a suitable formal language, mathematical proofs could themselves become an object of mathematical study — Hilbert called it *metamathematics*. The hope was that one might be able to show in this way that concrete conclusions reached using some controversial abstract concepts could nevertheless be show still to be valid or even provable without them.

1.3 The Calculus Ratiocinator

Gödel's famous incompleteness theorems [40,98,37] show that formal systems for deducing mathematics have essential weaknesses. For example, his first theorem is that any given formal system of deduction satisfying a few natural conditions is incomplete in the sense that some formally expressible and true statement is not formally provable.

It is generally agreed that Gödel's results rule out the possibility of realizing Hilbert's programme as originally envisaged, though this is a subtle question [65,96]. What is certainly true is that Gödel's theorem was the first of a variety of 'impossibility' results that only really become possible when the notion of mathematical proof is formalized.

Inspired by techniques used in Gödel's incompleteness results, Church [23] and Turing [106] proposed definitions of 'mechanical computability' and showed that one famous logical decision question, Hilbert's *Entscheidungsproblem* (decision problem for first-order logic) was unsolvable according to their definitions. Although this showed the limits of mechanical calculation, Turing machines in particular were an important inspiration for the development of real computers. And before long people began to investigate actually using computers to formalize mathematical proofs.

In the light of the various incompleteness and undecidability results, there are essential limits to what can be accomplished by automated reasoning. However, Gödel's results apply to human reasoning too from any specific set of axioms, and in principle most present-day mathematics can be expressed in terms of sets and proven from the axioms of Zermelo-Fraenkel set theory (ZF). Given any conventional set of mathematical axioms, e.g. a finite set, or one described by a finite set of schemas, such as ZF, there is at least a *semi-decision* procedure that can in principle verify any logical consequence of those axioms. Moreover many suitably restricted logical problems *are* decidable. For example, perhaps the very first computer theorem prover [29] could prove formulas involving quantifiers over natural numbers, but with a linearity restriction ensuring decidability [85].

2   Theorem Provers and Computer Algebra Systems

Before we proceed to survey the state of automated reasoning, it's instructive to consider the similarities and contrasts with computer algebra, which is already an established tool in biology as in many other fields of science. In some sense theorem provers (TPs) and computer algebra systems (CASs) are similar: both are computer programs to help people with formal symbolic manipulations. Yet there is at present surprisingly little common ground between them, either as regards the internal workings of the systems themselves or their respective communities of implementors and users. A theorem prover might be distinguished by a few features, which we consider in the following sections.

2.1   Logical Expressiveness

The typical computer algebra system supports a rather limited style of interaction [27]. The user types in an expression E; the CAS cogitates, usually not for very long, before returning another expression E'. The implication is that we should accept the theorem $E = E'$. Occasionally some slightly more sophisticated data may be returned, such as a set of possible expressions $E'_1, \ldots, E'_n$ with corresponding conditions on validity, e.g.

$$\sqrt{x^2} = \begin{cases} x & \text{if } x \geq 0 \\ -x & \text{if } x \leq 0 \end{cases}$$

However, the simple equational style of interaction is by far the most usual. By contrast, theorem provers have the logical language available to express far more sophisticated mathematical concepts such as the $\epsilon - \delta$ definition of continuity:

$$\forall x \in \mathbb{R}. \, \forall \epsilon > 0. \, \exists \delta > 0. \, \forall x'. \, |x - x'| < \delta \Rightarrow |f(x) - f(x')| < \epsilon$$

In particular, the use of a full logical language with quantifiers often unifies and generalizes existing known concepts from various branches of mathematics. For instance, the various algorithms for quantifier elimination in real-closed fields starting with Tarski's work [103] can be considered a natural and far-reaching generalization of Sturm's algorithm for counting the number of real roots of a polynomial. At the same time, quantifier elimination is another potentially fruitful way of viewing the notion of projection in Euclidean space. Chevalley's constructibility theorem in algebraic geometry 'the projection of a constructible set is constructible', and even some of its generalizations [45], are really just quantifier elimination in another guise.

2.2 Clear Semantics

The underlying semantics of expressions in a computer algebra system is often unclear, though some are more explicit than others. For example, the polynomial expression $x^2 + 2x + 1$ can be read in several ways: as a member of the polynomial ring $\mathbb{R}[x]$ (not to mention $\mathbb{Z}[x]$ or $\mathbb{C}[x]$ ...), as the associated function $\mathbb{R} \to \mathbb{R}$, or as the value of that expression for some particular $x \in \mathbb{R}$. Similarly, there may be ambiguity over which branch of various complex functions such as square root, logarithm and power is considered, and it may not really be clear in what sense 'integral' is meant to be understood. (Riemann? Lebesgue? Just antiderivative?) Such ambiguities are particularly insidious since in many situations it doesn't matter which interpretation is chosen (we have $x^2 + 2x + 1 = (x + 1)^2$ for any of the interpretations mentioned above), but there are situations where the distinction matters.

By contrast, theorem provers usually start from a strict and precisely defined logical foundation and build up other mathematical concepts by a sequence of definitions. For example, the HOL system [42] starts with a few very basic axioms for higher-order logic and a couple of set-theoretic axioms, and these are given a rather precise semantics in the documentation. From that foundation, other concepts such as natural numbers, lists and real and complex numbers are systematically built up without any new axioms.

2.3 Logical Rigour

Even when a CAS can be relied upon to give a result that admits a precise mathematical interpretation, that doesn't mean that its answers are always right. With a bit of effort, it's not very hard to get incorrect answers out of any mainstream computer algebra system. Particularly troublesome are simplifications involving functions with complex branch cuts. It's almost irresistible to apply simplifications such as $\log(xy) - \log(x) + \log(y)$ and $\sqrt{x^2} = x$, and many CASs will do this kind of thing freely. Although systematic approaches to keeping track of branch cuts are possible, most mainstream systems don't use them. For example, using the concept of 'unwinding number' $u(z)$ [28], we can express rigorously simplification rules such as:

$$w \neq 0 \wedge z \neq 0 \Rightarrow \log(wz) = \log(w) + \log(z) - 2\pi i u(\log(w) + \log(z))$$

Most users probably find such pedantic details as branch cut identification a positively unwelcome distraction. They often know (or at least think they know) that the obvious simplifications are valid. In any case, if a CAS lacks the expressiveness to produce a result that distinguishes possible cases, it is confronted with the unpalatable choice of doing something that isn't strictly correct or doing nothing. Many users would prefer the former.

By contrast, most theorem provers take considerable care that all alleged 'theorems' are deduced in a rigorous way, and all conditions made explicit. Indeed, many such as HOL actually construct a complete proof using a very simple kernel of primitive inference rules. Although nothing is ever completely certain, a theorem in such a system is very likely to be correct.

2.4   What's Wrong with Theorem Provers?

So far, we have noted several flaws of the typical computer algebra systems and the ways in which theorem provers are better. However, on the other side of the coin, CASs are normally easier to use and much more efficient. Moreover, CASs implement many algorithms useful for solving real concrete problems in applied (and even pure) mathematics, e.g. factoring polynomials and finding integrals. By contrast, theorem provers emphasize proof search in logical systems, and it's often non-trivial to express high-level mathematics in them. Thus, it is not surprising that CASs are more or less mainstream tools in various fields, whereas interest in theorem provers is mainly confined to logicians and computer scientist interested in formal correctness proofs for hardware, software and protocols and the formalization of mathematics.

Since the strengths and weaknesses of theorem provers and CASs are almost perfectly complementary, a natural idea is to somehow get the best of both worlds. One promising idea [50] is to use the CAS as an 'oracle' to compute results that can then be rigorously *checked* in the theorem prover. This only works for problems where checking a result is considerably easier than deriving it, but this does take in many important applications such as factoring (check by multiplying) and indefinite integration in the sense of antiderivatives (check by differentiating).

3   Research in Automated Reasoning

We can consider various ways of classifying research in automated reasoning, and perhaps some contrasts will throw particular themes into sharp relief.

3.1   AI Versus Logic-Oriented

Some researchers have attacked the problem of automated theorem proving by attempting to emulate the way humans reason. Crudely we can categorize this as the 'Artificial

Intelligence' (AI) approach. For example in the 1950s Newell and Simon [81] designed a program that could prove many of the simple logic theorems in *Principia Mathematica* [109], while Gelerntner [38] designed a prover that could prove facts in Euclidean geometry using human-style diagrams to direct or restrict the proofs. A quite different approach was taken by other pioneers such as Gilmore [39], Davis and Putnam [31], and Prawitz [84]. They attempted to implement proof search algorithms inspired by results from logic (e.g. the completeness of Gentzen's cut-free sequent calculus), often quite remote from the way humans prove theorem.

Early indications were that machine-oriented methods performed much better. As Wang [107] remarked when presenting his simple systematic program for the AE fragment of first order logic that was dramatically more effective than Newell and Simon's:

> The writer [...] cannot help feeling, all the same, that the comparison reveals a fundamental inadequacy in their approach. There is no need to kill a chicken with a butcher's knife. Yet the net impression is that Newell-Shore-Simon failed even to kill the chicken with their butcher's knife.

Indeed, in the next few decades, far more attention was paid to systematic machine-oriented algorithms. Wos, one of the most successful practitioners of automated reasoning, attributes the success of his research group in no small measure to the fact that they play to a computer's strengths instead of attempting to emulate human thought [111].

Today, there is still a preponderance of research on the machine-oriented side, but there have been notable results based on human-oriented approaches. For example Bledsoe attempted to formalize methods often used by humans for proving theorems about limits in analysis [10]. Bledsoe's student Boyer together with Moore developed the remarkable NQTHM prover [13] which can often perform automatic generalization of arithmetic theorems and prove the generalizations by induction. The success of NQTHM, and the contrasting difficulty of fitting its methods into a simple conceptual framework, has led Bundy [20] to reconstruct its methods in a general science of reasoning based on *proof planning*. Depending on one's point of view, one can regard the considerable interest in proof planning as representing a success of the AI approach, or the attempt to present aspects of human intelligence in a more machine-oriented style.

3.2   Automated vs. Interactive

Thanks to the development of effective algorithms, some of which we consider later, automated theorem provers have become quite powerful and have achieved notable successes. Perhaps the most famous case is McCune's solution [76], using the automated theorem prover EQP, of the longstanding 'Robbins conjecture' concerning the axiomatization of Boolean algebra, which had resisted human mathematicians for some time. This success is just one particularly well-known case where the Argonne team has used Otter and other automated reasoning programs to answer open questions. Some more can be found in the monograph [77].

However, it seems at present that neither a systematic algorithmic approach nor a heuristic human-oriented approach is capable of proving a wide range of difficult mathematical theorems automatically. Besides, one might object that even if it were possible, it is hardly desirable to automate proofs that humans are incapable of developing themselves [35]:

> [...] I consider mathematical proofs as a reflection of my understanding and 'understanding' is something we cannot delegate, either to another person or to a machine.

A more easily attained goal, and if one agrees with the sentiments expressed in that quote perhaps a more desirable one, is to create a system that can verify a proof found by a human, or assist in a more limited capacity under human guidance. At the very least the computer should act as a humble clerical assistant checking the correctness of the proof, guarding against typical human errors such as implicit assumptions and forgotten special cases. At best the computer might help the process substantially by automating certain parts of the proof. After all, proofs often contain parts that are just routine verifications or are amenable to automation, such as algebraic identities. This idea of a machine and human working together to prove theorems from sketches was already envisaged by Wang [107]:

> [...] the writer believes that perhaps machines may more quickly become of practical use in mathematical research, not by proving new theorems, but by formalizing and checking outlines of proofs, say, from textbooks to detailed formalizations more rigorous than *Principia* [Mathematica], from technical papers to textbooks, or from abstracts to technical papers.

The idea of a proof assistant began to attract particular attention in the late 1960s, perhaps because the abilities of fully automated systems were apparently starting to plateau. Many proof assistants were based on a batch model, the machine checking in one operation the correctness of a proof sketch supplied by a human. But a group at the Applied Logic Corporation who developed a sequence of theorem provers in the SAM (Semi-Automated Mathematics) family made their provers interactive, so that the mathematician could work on formalizing a proof with machine assistance. As they put it [46]:

> Semi-automated mathematics is an approach to theorem-proving which seeks to combine automatic logic routines with ordinary proof procedures in such a manner that the resulting procedure is both efficient and subject to human intervention in the form of control and guidance. Because it makes the mathematician an essential factor in the quest to establish theorems, this approach is a departure from the usual theorem-proving attempts in which the computer *unaided* seeks to establish proofs.

In 1966, the fifth in the series of systems, SAM V, was used to construct a proof of a hitherto unproven conjecture in lattice theory [19]. This was certainly a success for the semi-automated approach because the computer automatically proved a result now called "SAM's Lemma" and the mathematician recognized that it easily yielded a proof of the open conjecture. Not long after the SAM project, the AUTOMATH [32,33], Mizar [104,105] and LCF [43] proof checkers appeared, and each of them in its way has been profoundly influential. Many of the most successful interactive theorem provers around today are directly descended from one of these.

Nowadays there is active and vital research activity in both 'automated' and 'interactive' provers. Automated provers for first-order logic compete against each other in annual competitions on collections of test problems such as TPTP [102], and the Vampire system has usually come out on top for the last few years. There is also intense interest in special provers for other branches of logic, e.g. 'SAT' (satisfiability of purely propositional formulas), which has an amazing range of practical applications. More recently a generalization known as 'SMT' (satisfiability modulo theories), which uses techniques for combining deduction in certain theories [80,93], has attracted considerable interest. Meanwhile, interactive provers develop better user interfaces and proof languages [48], incorporate ideas from automated provers and even link to them [58], and develop ever more extensive libraries of formalized mathematics. For a nice survey of some of the major interactive systems, showing a proof of the irrationality of $\sqrt{2}$ in each as an example, see [110].

3.3   Proof Search vs. Special Algorithms

Right from the beginning of theorem proving, some provers were customized for a particular theory or fragment of logic (such as Davis's prover for linear arithmetic [29]), while others performed general proof search in first-order logic from a set of axioms. The explicit introduction of unification as part of Robinson's resolution method [88] made it possible for the machine to instantiate variables in an entirely algorithmic way which nevertheless has an almost "intelligent" ability to focus on relevant terms. This gave a considerable impetus to general first-order proof search, and for a long time special algorithms were subordinated to resolution or similar principles rather than being developed in themselves. There are numerous different algorithms for general proof search, such as tableaux [7,54], model elimination [70] as well as resolution [88] and its numerous refinements [64,71,72,34,87,97]. Despite the general emphasis on pure first-order logic, there has also been research in automating higher-order logic [1], which allows quantification over sets and functions as part of the logic rather than via additional axioms.

However, there have been some successes for more specialized algorithms. In particular, there has always been strong interest in effective algorithms for purely equational reasoning. Knuth-Bendix completion [63] led to a great deal of fruitful research [3,4,56]. Automated proof of geometry problems using purely algebraic methods has also attracted much interest. The first striking success was by Wu [108] using his special triangulation algorithm, and others have further refined and applied this approach [22] as well as trying other methods such as resultants and Gröbner bases [61,89]. Incidentally Gröbner bases [16,17] are more usually considered a part of computer algebra, but as a tool for testing ideal membership they give a powerful algorithm for solving various logical decision problems [95,60].

4   Applications of Automated Reasoning

At present there are two main applications of automated reasoning.

4.1 Formal Verification

One promising application of formalization, and a particularly easy one to defend on utilitarian grounds, is to verify the correct behaviour of computer systems, e.g. hardware, software, protocols and their combinations. We might wish to prove that a sorting algorithm really does always sort its input list, that a numerical algorithm does return a result accurate to within a specified error bound, that a server will under certain assumptions always respond to a request, etc.

In typical programming practice, programs are usually designed with clear logical ideas behind them, but the final properties are often claimed on the basis of intuitive understanding together with testing on a variety of inputs. As programmers know through bitter personal experience, it can be very difficult to write a program that always performs its intended function. Most large programs contain 'bugs', i.e. in certain situations they do not behave as intended. And the inadequacy of even highly intelligent forms of testing for showing that programs are bug-free is widely recognized. There are after all usually far too many combinations of possibilities to exercise more than a tiny fraction. The idea of rigorously *proving* correctness is attractive, but given the difficulty of getting the formal proof right, one might wish to check the proof by machine rather than by hand.

Formal verification first attracted interest in the 1970s as a response to the perceived "software crisis", the fundamental difficulty of writing correct programs and delivering them on time, as well as interest in computer security; see [74] for a good discussion. But over the last couple of decades there has been increased interest in formal verification in the *hardware* domain. This is partly because hardware is usually a more amenable target for highly automated techniques. Such techniques include SAT (propositional satisfiability testing), using new algorithms or high-quality implementations of old ones [14,100,92,79,41], sophisticated forms of symbolic simulation [15,91], and temporal logic model checking [24,86,25]. Also, hardware verification is particularly attractive because fixing errors is often invasive and potentially expensive. For example, in response to an error in the FDIV (floating-point division) instruction of some early Intel® Pentium® processors in 1994 [83], Intel set aside approximately $475M to cover costs.

Since the 1980s there has been extensive research in formal verification of microprocessor designs using traditional theorem proving techniques [57,26,44,59,99]. Generally there has been more emphasis on the highly automated techniques like model checking that lie somewhat apart from the automated reasoning mainstream. However, recently there has been something of a convergence, as interest in SMT (satisfiability modulo theories) leads to the incorporation of various theorem-proving methods into highly automated tools. There has also been renewed interest in applications to software, particularly partial verification or sophisticated static checking rather than complete functional verification [5]. And for certain applications, especially implementations of mathematically sophisticated algorithms, more general and interactive theorem proving is needed. A particularly popular and successful target is the verification of floating-point algorithms [78,90,82,49].

4.2   The Formalization of Mathematics

The *formalizability in principle* of mathematical proof is widely accepted among professional mathematicians as the final arbiter of correctness. Bourbaki [12] clearly says that 'the correctness of a mathematical text is verified by comparing it, more or less explicitly, with the rules of a formalized language', while Mac Lane [73] is also quite explicit (p377):

> As to precision, we have now stated an absolute standard of rigor: A Mathematical proof is rigorous when it is (or could be) written out in the first-order predicate language $L(\in)$ as a sequence of inferences from the axioms ZFC, each inference made according to one of the stated rules. [...] When a proof is in doubt, its repair is usually just a partial approximation to the fully formal version.

However, before the advent of computers, the idea of actually formalizing proofs had seemed quite out of the question. (Even the painstaking volumes of proofs in *Principia Mathematica* are for extremely elementary results compared with even classical real analysis, let alone mathematics at the research level.) But computerization can offer the possibility of *actually* formalizing mathematics and its proofs. Apart from the sheer intellectual interest of doing so, it may lead to a real increase in reliability. Mathematical proofs are subjected to peer review before publication, but there are plenty of well-documented cases where published results turned out to be faulty. A notable example is the purported proof of the 4-colour theorem by Kempe [62], the flaw only being noticed a decade later [51], and the theorem only being conclusively proved much later [2]. The errors need not be deep mathematical ones, as shown by the following [69]:

> Professor Offord and I recently committed ourselves to an odd mistake (Annals of Mathematics (2) 49, 923, 1.5). In formulating a proof a plus sign got omitted, becoming in effect a multiplication sign. The resulting false formula got accepted as a basis for the ensuing fallacious argument. (In defence, the final result was known to be true.)

A book written 70 years ago by Lecat [68] gave 130 pages of errors made by major mathematicians up to 1900. With the abundance of theorems being published today, often emanating from writers who are not trained mathematicians, one fears that a project like Lecat's would be practically impossible, or at least would demand a journal to itself! Moreover, many proofs, including the modern proof of the four-colour theorem [2] and the recent proof of the Kepler conjecture [47], rely on extensive computer checking and it's not clear how to bring them within the traditional process of peer review [66].

At present we are some way from the stage where most research mathematicians can pick up one of the main automated theorem provers and start to formalize their own research work. However, substantial libraries of formalized mathematics have been built up in theorem provers, notably the mathematical library in Mizar, and a few quite substantial results such as the Jordan Curve Theorem, the Prime Number Theorem and the Four-Colour Theorem have been completely formalized. As mathematical libraries are further built up and interactive systems become more powerful and user-friendly, we can expect to see more mathematicians starting to use them.

5   Conclusion

Automated reasoning is already finding applications in formal verification and the formalization of mathematical proofs. At present, applications to mainstream applied mathematics are limited, and so it may be premature to seek applications in computational biology. However, theorem proving has sometimes been applied in unexpected ways. For instance, many combinatorial problems are solved better by translating to SAT than by customized algorithms! Perhaps this short survey will lead some readers to find applications of automated reasoning in the biological sciences. In any case, we hope it has given some flavour of this vital and fascinating research field.

References

1. Andrews, P.B., Bishop, M., Issar, S., Nesmith, D., Pfenning, F., Xi, H.: TPS: A theorem proving system for classical type theory. Journal of Automated Reasoning 16, 321–353 (1996)
2. Appel, K., Haken, W.: Every planar map is four colorable. Bulletin of the American Mathematical Society 82, 711–712 (1976)
3. Baader, F., Nipkow, T.: Term Rewriting and All That. Cambridge University Press, Cambridge (1998)
4. Bachmair, L., Dershowitz, N., Plaisted, D.A.: Completion without failure. In: Aït-Kaci, H., Nivat, M. (eds.) Resolution of Equations in Algebraic Structures Rewriting Techniques, vol. 2, pp. 1–30. Academic Press, San Diego (1989)
5. Ball, T., Bounimova, E., Cook, B., Levin, V., Lichtenberg, J., McGarvey, C., Ondrusek, B., Rajamani, S., Ustuner, A.: Thorough static analysis of device drivers. In: Proceedings of EuroSys'06, the European Systems Conference (2006)
6. Benacerraf, P., Putnam, H.: Philosophy of mathematics: selected readings, 2nd edn. Cambridge University Press, Cambridge (1983)
7. Beth, E.W.: Semantic entailment and formal derivability. Mededelingen der Koninklijke Nederlandse Akademie van Wetenschappen, new series 18, 309–342 (1955)
8. Biggs, N.L., Lloyd, E.K., Wilson, R.J.: Graph Theory, pp. 1736–1936. Clarendon Press, Oxford (1976)
9. Birtwistle, G., Subrahmanyam, P.A. (eds.): VLSI Specification, Verification and Synthesis. International Series in Engineering and Computer Science, vol. 35. Kluwer, Dordrecht (1988)
10. Bledsoe, W.W.: Some automatic proofs in analysis. In: Bledsoe, W.W., Loveland, D.W. (eds.) Automated Theorem Proving: After 25 Years, Contemporary Mathematics, vol. 29, pp. 89–118. American Mathematical Society (1984)
11. Boole, G.: The calculus of logic. The Cambridge and Dublin Mathematical Journal 3, 183–198 (1848)
12. Bourbaki, N.: Theory of sets. In: Elements of mathematics Translated from French Théorie des ensembles in the series Eléments de mathématique, originally published by Hermann, Addison-Wesley, London (1968)
13. Boyer, R.S., Moore, J.S.: A Computational Logic. ACM Monograph Series. Academic Press, San Diego (1979)
14. Bryant, R.E.: Graph-based algorithms for Boolean function manipulation. IEEE Transactions on Computers C-35, 677–691 (1986)
15. Bryant, R.E.: A method for hardware verification based on logic simulation. Journal of the ACM 38, 299–328 (1991)

16. Buchberger, B.: Ein Algorithmus zum Auffinden der Basiselemente des Restklassenringes nach einem nulldimensionalen Polynomideal. PhD thesis, Mathematisches Institut der Universität Innsbruck, 1965. English translation to appear in Journal of Symbolic Computation (2006)

17. Buchberger, B.: Ein algorithmisches Kriterium fur die Lösbarkeit eines algebraischen Gleichungssystems. Aequationes Mathematicae, vol. 4, 374–383, 1970, English translation An Algorithmical Criterion for the Solvability of Algebraic Systems of Equations in [18] pp. 535–545 (1970)

18. Buchberger, B., Winkler, F. (eds.): Gröbner Bases and Applications. London Mathematical Society Lecture Note Series, vol. 251. Cambridge University Press, Cambridge (1998)

19. Bumcrot, R.: On lattice complements. In: Proceedings of the Glasgow Mathematical Association 7, 22–23 (1965)

20. Bundy, A.: A science of reasoning. In: Lassez, J.-L., Plotkin, G. (eds.) Computational Logic: Essays in Honor of Alan Robinson, pp. 178–198. MIT Press, Cambridge (1991)

21. Caviness, B.F., Johnson, J.R. (eds.): Quantifier Elimination and Cylindrical Algebraic Decomposition, Texts and monographs in symbolic computation. Springer, Heidelberg (1998)

22. Chou, S.-C.: An introduction to Wu's method for mechanical theorem proving in geometry. Journal of Automated Reasoning 4, 237–267 (1988)

23. Church, A.: An unsolvable problem of elementary number-theory. American Journal of Mathematics 58, 345–363 (1936)

24. Clarke, E.M., Emerson, E.A.: Design and synthesis of synchronization skeletons using branching-time temporal logic. In: Kozen, D. (ed.) Logics of Programs. LNCS, vol. 131, pp. 52–71. Springer, Heidelberg (1981)

25. Clarke, E.M., Grumberg, O., Peled, D.: Model Checking. MIT Press, Cambridge (1999)

26. Cohn, A.: A proof of correctness of the VIPER microprocessor: The first level. In: Birtwistle and Subrahmanyam [9], pp. 27–71

27. Corless, R.M., Jeffrey, D.J.: Well... it isn't quite that simple. SIGSAM Bulletin 26(3), 2–6 (1992)

28. Corless, R.M., Jeffrey, D.J.: The unwinding number. SIGSAM Bulletin 30(2), 28–35 (1996)

29. Davis, M.: A computer program for Presburger's algorithm. In: Summaries of talks presented at the Summer Institute for Symbolic Logic, Cornell University, pp. 215–233. Institute for Defense Analyses, Princeton, NJ, Reprinted in [94], pp. 41–48 (1957)

30. Davis, M. (ed.): The Undecidable: Basic Papers on Undecidable Propositions, Unsolvable Problems and Computable Functions. Raven Press, NY (1965)

31. Davis, M., Putnam, H.: A computing procedure for quantification theory. Journal of the ACM 7, 201–215 (1960)

32. de Bruijn, N.G.: The mathematical language AUTOMATH, its usage and some of its extensions. In: Laudet, et al. [67], pp. 29–61

33. de Bruijn, N.G.: A survey of the project AUTOMATH. In: Seldin, J.P., Hindley, J.R. (eds.) To H. B. Curry: Essays in Combinatory Logic, Lambda Calculus, and Formalism, pp. 589–606. Academic Press, San Diego (1980)

34. de Nivelle, H.: Ordering Refinements of Resolution. PhD thesis, Technische Universiteit Delft (1995)

35. Dijkstra, E.W.: Formal techniques and sizeable programs (EWD563). In: Dijkstra, E.W. (ed.) Selected Writings on Computing: A Personal Perspective Paper prepared for Symposium on the Mathematical Foundations of Computing Science, Gdansk, pp. 205–214. Springer, Heidelberg (1976)

36. Feigenbaum, E.A., Feldman, J. (eds.): Computers & Thought. AAAI Press / MIT Press (1995)

37. Franzén, T.: Gödel's Theorem. An Incomplete Guide to its Use and Abuse. A. K. Peters (2005)

38. Gelerntner, H.: Realization of a geometry-theorem proving machine. In: Proceedings of the International Conference on Information Processing, UNESCO House, pp. 273–282, 1959 Also appears in [94], pp. 99–117 and in [36], pp. 134–152 (1959)

39. Gilmore, P.C.: A proof method for quantification theory: Its justification and realization. IBM Journal of research and development 4, 28–35 (1960)

40. Gödel, K.: Über formal unentscheidbare Sätze der Principia Mathematica und verwandter Systeme, I. Monatshefte für Mathematik und Physik, vol. 38, 173–198, English translation, On Formally Undecidable Propositions of Principia Mathematica and Related Systems, I In: [52], pp. 592–618 or [30], pp. 4–38 (1931)

41. Goldberg, E., Novikov, Y.: BerkMin: a fast and robust Sat-solver. In: Kloos, C.D., Franca, J.D. (eds.) Design, Automation and Test in Europe Conference and Exhibition (DATE 2002), Paris, France, pp. 142–149. IEEE Computer Society Press, Los Alamitos (2002)

42. Gordon, M.J.C., Melham, T.F.: Introduction to HOL: a theorem proving environment for higher order logic. Cambridge University Press, Cambridge (1993)

43. Gordon, M.J.C., Milner, R., Wadsworth, C.P.: Edinburgh LCF. LNCS, vol. 78. Springer, Heidelberg (1979)

44. Graham, B.T.: The SECD Microprocessor: A verification case study. Kluwer international series in engineering and computer science, vol. 178. Kluwer Academic Publishers, Boston (1992)

45. Grothendieck, A.: Éléments de Géométrie Algébraique IV: Étude locale de schémas et des morphismes de schémas, vol. 20 of Publications Mathématiques. IHES (1964)

46. Guard, J.R., Oglesby, F.C., Bennett, J.H., Settle, L.G.: Semi-automated mathematics. Journal of the ACM 16, 49–62 (1969)

47. Hales, T.C.: The Kepler conjecture (1998), Available at http://front.math.ucdavis.edu/math.MG/9811078

48. Harrison, J.: Proof style. In: Giménez, E., Paulin-Mohring, C. (eds.) TYPES 1996. LNCS, vol. 1512, pp. 154–172. Springer, Heidelberg (1998)

49. Harrison, J.: Floating-point verification using theorem proving. In: Bernardo, M., Cimatti, A. (eds.) SFM 2006. LNCS, vol. 3965, pp. 211–242. Springer, Heidelberg (2006)

50. Harrison, J., Théry, L.: A sceptic's approach to combining HOL and Maple. Journal of Automated Reasoning 21, 279–294 (1998)

51. Heawood, P.J.: Map-colour theorem. Quarterly Journal of Pure and Applied Mathematics Reprinted in [8] 24, 332–338 (1890)

52. Heijenoort, J.v. (ed.): From Frege to Gödel: A Source Book in Mathematical Logic 1879–1931. Harvard University Press, Cambridge (1967)

53. Hilbert, D.: Die logischen Grundlagen der Mathematik. Mathematische Annalen 88, 151–165 (1922)

54. Hintikka, J.: Form and content in quantification theory. Acta Philosophica Fennica — Two papers on Symbolic Logic 8, 8–55 (1955)

55. Hobbes, T.: Leviathan. Andrew Crooke (1651)

56. Huet, G.: A complete proof of correctness of the Knuth-Bendix completion procedure. Journal of Computer and System Sciences 23, 11–21 (1981)

57. Hunt, W.A.: A Verified Micrprocessor. PhD thesis, University of Texas, 1985. In: Hunt Jr., W.A. (ed.) FM8501: A Verified Microprocessor. LNCS, vol. 795, Springer, Heidelberg (1994)

58. Hurd, J.: Integrating Gandalf and HOL. In: Bertot, Y., Dowek, G., Hirschowitz, A., Paulin, C., Théry, L. (eds.) TPHOLs 1999. LNCS, vol. 1690, pp. 311–321. Springer, Heidelberg (1999)

59. Joyce, J.J.: Formal verification and implementation of a microprocessor. In: Birtwistle and Subrahmanyam [9], pp. 129–158

60. Kandri-Rody, A., Kapur, D., Narendran, P.: An ideal-theoretic approach to word problems and unification problems over finitely presented commutative algebras. In: Jouannaud, J.-P. (ed.) Rewriting Techniques and Applications. LNCS, vol. 202, pp. 345–364. Springer, Heidelberg (1985)

61. Kapur, D.: Automated geometric reasoning: Dixon resultants, Gröbner bases, and characteristic sets. In: Wang, D. (ed.) Automated Deduction in Geometry. LNCS, vol. 1360, Springer, Heidelberg (1998)

62. Kempe, A.B.: On the geographical problem of the four colours. American Journal of Mathematics Reprinted in [8] 2, 193–200 (1879)

63. Knuth, D., Bendix, P.: Simple word problems in universal algebras. In: Leech, J. (ed.) Computational Problems in Abstract Algebra, Pergamon Press, New York (1970)

64. Kowalski, R.A., Kuehner, D.: Linear resolution with selection function. Artificial Intelligence 2, 227–260 (1971)

65. Kreisel, G.: Hilbert's programme. Dialectica Revised version in [6] 12, 346–372 (1958)

66. Lam, C.W.H.: How reliable is a computer-based proof? The Mathematical Intelligencer 12, 8–12 (1990)

67. Laudet, M., Lacombe, D., Nolin, L., Schützenberger, M. (eds.): Symposium on Automatic Demonstration. Lecture Notes in Mathematics, vol. 125. Springer, Heidelberg (1970)

68. Lecat, M.: Erreurs de Mathématiciens des origines à nos jours. Ancne Libraire Castaigne et Libraire Ém Desbarax, Brussels (1935)

69. Littlewood, J.E.: Littlewood's Miscellany Edited by Bela Bollobas. Cambridge University Press, Cambridge (1986)

70. Loveland, D.W.: Mechanical theorem-proving by model elimination. Journal of the ACM 15, 236–251 (1968)

71. Loveland, D.W.: A linear format for resolution. In: Laudet, et al. [67], pp. 147–162

72. Luckham, D.: Refinements in resolution theory. In: Laudet, et al.[67], pp. 163–190

73. Mac Lane, S.: Mathematics: Form and Function. Springer, Heidelberg (1986)

74. MacKenzie, D.: Mechanizing Proof: Computing, Risk and Trust. MIT Press, Cambridge (2001)

75. Marciszewski, W., Murawski, R.: Mechanization of Reasoning in a Historical Perspective, vol. 43 of Poznań Studies in the Philosophy of the Sciences and the Humanities. Rodopi, Amsterdam (1995)

76. McCune, W.: Solution of the Robbins problem. Journal of Automated Reasoning 19, 263–276 (1997)

77. McCune, W., Padmanabhan, R.: Automated Deduction in Equational Logic and Cubic Curves. LNCS, vol. 1095. Springer, Heidelberg (1996)

78. Moore, J.S., Lynch, T., Kaufmann, M.: A mechanically checked proof of the correctness of the kernel of the $AMD5_K86$ floating-point division program. IEEE Transactions on Computers 47, 913–926 (1998)

79. Moskewicz, W., Madigan, C.F., Zhao, Y., Zhang, L., Malik, S.: Chaff: Engineering an efficient SAT solver. In: Proceedings of the 38th Design Automation Conference (DAC 2001), pp. 530–535. ACM Press, New York (2001)

80. Nelson, G., Oppen, D.C.: Simplification by cooperating decision procedures. ACM Transactions on Programming Languages and Systems 1, 245–257 (1979)

81. Newell, A., Simon, H.A.: The logic theory machine. IRE Transactions on Information Theory 2, 61–79 (1956)

82. O'Leary, J., Zhao, X., Gerth, R., Seger, C.-J.H.: Formally verifying IEEE compliance of floating-point hardware. Intel Technology Journal, 1999-Q1, 1–14 (1999), Available on the Web as http://developer.intel.com/technology/itj/q11999/articles/art_5.htm

83. Pratt, V.R.: Anatomy of the Pentium bug. In: Mosses, P.D., Schwartzbach, M.I., Nielsen, M. (eds.) CAAP 1995, FASE 1995, and TAPSOFT 1995. LNCS, vol. 915, pp. 97–107. Springer, Heidelberg (1995)

84. Prawitz, D., Prawitz, H., Voghera, N.: A mechanical proof procedure and its realization in an electronic computer. Journal of the ACM 7, 102–128 (1960)

85. Presburger, M.: Über die Vollständigkeit eines gewissen Systems der Arithmetik ganzer Zahlen, in welchem die Addition als einzige Operation hervortritt. In: Sprawozdanie z I Kongresu metematyków slowiańskich, Warszawa 1929, pp. 92–101, 395. Warsaw, 1930. Annotated English version by [101] (1929)

86. Queille, J.P., Sifakis, J.: Specification and verification of concurrent programs in CESAR. In: Dezani-Ciancaglini, M., Montanari, U. (eds.) International Symposium on Programming. LNCS, vol. 137, pp. 195–220. Springer, Heidelberg (1982)

87. Robinson, J.A.: Automatic deduction with hyper-resolution. International Journal of Computer Mathematics 1, 227–234 (1965)

88. Robinson, J.A.: A machine-oriented logic based on the resolution principle. Journal of the ACM 12, 23–41 (1965)

89. Robu, J.: Geometry Theorem Proving in the Frame of Theorema Project. PhD thesis, RISC-Linz (2002)

90. Rusinoff, D.: A mechanically checked proof of IEEE compliance of a register-transfer-level specification of the AMD-K7 floating-point multiplication, division, and square root instructions. LMS Journal of Computation and Mathematics 1, 148–200 (1998), Available on the Web at http://www.onr.com/user/russ/david/k7-div-sqrt.html

91. Seger, C.-J.H., Bryant, R.E.: Formal verification by symbolic evaluation of partially-ordered trajectories. Formal Methods in System Design 6, 147–189 (1995)

92. Sheeran, M., Stålmarck, G.: A tutorial on Stålmarck's proof procedure for propositional logic. In: Gopalakrishnan, G.C., Windley, P. (eds.) FMCAD 1998. LNCS, vol. 1522, pp. 82–99. Springer, Heidelberg (1998)

93. Shostak, R.: Deciding combinations of theories. Journal of the ACM 31, 1–12 (1984)

94. Siekmann, J., Wrightson, G. (eds.): Automation of Reasoning — Classical Papers on Computational Logic, vol. I, pp. 1957–1966. Springer, Heidelberg (1983)

95. Simmons, H.: The solution of a decision problem for several classes of rings. Pacific Journal of Mathematics 34, 547–557 (1970)

96. Simpson, S.: Partial realizations of Hilbert's program. Journal of Symbolic Logic 53, 349–363 (1988)

97. Slagle, J.R.: Automatic theorem proving with renamable and semantic resolution. Journal of the ACM 14, 687–697 (1967)

98. Smullyan, R.M.: Gödel's Incompleteness Theorems. Oxford Logic Guides, vol. 19. Oxford University Press, Oxford (1992)

99. Srivas, M.K., Miller, S.P.: Applying formal verification to the AAMP5 microprocessor: A case study in the industrial use of formal methods. Formal Methods in System Design 8, 31–36 (1993)

100. Stålmarck, G., Säflund, M.: Modeling and verifying systems and software in propositional logic. In: Daniels, B.K. (ed.) Safety of Computer Control Systems, 1990 (SAFECOMP '90), Gatwick, UK, pp. 31–36. Pergamon Press, New York (1990)

101. Stansifer, R.: Presburger's article on integer arithmetic: Remarks and translation. Technical Report CORNELLCS:TR84-639, Cornell University Computer Science Department (1984)

102. Suttner, C.B., Sutcliffe, G.: The TPTP problem library. Technical Report AR-95-03, Institut für Infomatik, TU München, Germany, Also available as TR 95/6 from Dept. Computer Science, James Cook University, Australia, and on the Web (1995)

103. Tarski, A.: A Decision Method for Elementary Algebra and Geometry. University of California Press, 1951. Previous version published as a technical report by the RAND Corporation, 1948; prepared for publication by J. C. C. McKinsey. Reprinted in [21], pp. 24–84 (1951)

104. Trybulec, A.: The Mizar-QC/6000 logic information language. ALLC Bulletin (Association for Literary and Linguistic Computing) 6, 136–140 (1978)

105. Trybulec, A., Blair, H.A.: Computer aided reasoning. In: Parikh, R. (ed.) Logics of Programs. LNCS, vol. 193, pp. 406–412. Springer, Heidelberg (1985)

106. Turing, A.M.: On computable numbers, with an application to the Entscheidungsproblem. Proceedings of the London Mathematical Society (2) 42, 230–265 (1936)

107. Wang, H.: Toward mechanical mathematics. IBM Journal of research and development 4, 2–22 (1960)

108. Wen-tsün, W.: On the decision problem and the mechanization of theorem proving in elementary geometry. Scientia Sinica 21, 157–179 (1978)

109. Whitehead, A.N., Russell, B.: Principia Mathematica (3 vols). Cambridge University Press, Cambridge (1910)

110. Wiedijk, F. (ed.): The Seventeen Provers of the World. LNCS, vol. 3600. Springer, Heidelberg (2006)

111. Wos, L., Pieper, G.W.: A Fascinating Country in the World of Computing: Your Guide to Automated Reasoning. World Scientific, Singapore (1999)

Inference of Complex Regulatory Network for the Cell Cycle System in *Saccharomyces Cerevisiae*

Sachiyo Aburatani

Computational Biology Research Center (CBRC), AIST, AIST Tokyo Waterfront
Bio-IT Reseach Builiding, 2-42, Aomi, Koto-ku, Tokyo, 135-0064, Japan

Abstract. Recently we developed a graphical chain modeling procedure
to infer a network from expression profiles between different stages, according to the natural order. In this study, this procedure was applied
to 796 gene expression profiles of cell-cycle related genes, which were
experimentally identified to be transcribed among five cell-cycle phases.
These phases were well characterized by the inferred network. In the G1
and S phases, almost all of the genes were related among them, even
though limited numbers of genes were related to each other in the other
phases. Furthermore, the inferred network indicated that some genes in
the former phases have long-distance regulation throughout the cell cycle; however, the main regulatory system in the cell cycle occurred step
by step in a manner. This approach provides us a comprehensive analysis
of serial gene regulation in the cell cycle system.

Keywords: graphical chain modeling, cell cycle, gene expression profile.

1   Introduction

The cell cycle is a fundamental system for the proliferation of eukaryotes [1], and
it is known to be regulated at multiple levels. Among the regulatory systems,
gene transcriptional control is one of the most important regulatory processes
for cell cycle progression [2], and periodic transcription seems to be a universal feature of cell cycle regulation [2,8]. This periodic transcription of genes is
required for the onset of S phase [3], which is the important phase for DNA
synthesis. The onset of S phase is considered to be regulated by transcription
factors in G1 phase, and these genes are activated earlier in G1 phase, so S phase
can begin. Hundreds of genes with periodic transcription have been identified by
recent genome-wide studies of cells undergoing growth and division [4,5]. Furthermore, some transcription factors are known to regulate a small set of cell
cycle-dependent genes [2,6]. To reveal the transcriptional control in the cell cycle, extrapolation of the regulatory relationships between transcription factors
and other cell cycle-dependent genes is a critical subject [1].

In budding the yeast *Saccharomyces cerevisiae*, the expression profiles of the
cell cycle-dependent genes demonstrated that approximately 800 genes are transcribed periodically during the five phases in the cell cycle [4,5]. This work has led

H. Anai, K. Horimoto, and T. Kutsia (Eds.): AB 2007, LNCS 4545, pp. 350–364, 2007.
© Springer-Verlag Berlin Heidelberg 2007

to the proposal of that serial regulation of cell cycle-dependent genes, whereby the expression of the genes in one cell cycle phase regulates the gene expression in the next phase [7,8]. Thus, the expression profiles analyzed by computational methods have made it possible to define the global manner of transcriptional control in the cell cycle.

To reveal the regulatory networks from genome-wide expression profile data, we have developed an approach in combination with graphical Gaussian modeling (GGM) and hierarchical clustering [9,10]. Among the graphical models, GGM is the simplest structure in a mathematical sense; only the inverse of the correlation coefficient between the variables is needed. GGM infers only the undirected graph, instead of the directed graph showing the causality in the Boolean and Bayesian models, and therefore, GGM can be easily applied to a wide variety of data. Our method provides a framework of gene regulatory relationships by inferring the relationships between the clusters [9,10,11], and provides clues toward estimating the global relationships between genes on a genomic scale.

In the present study, we applied the graphical chain modeling (GCM) to the expression profiles related to cell cycle regulation. In a previous study, GCM was useful for analyzing a limited set of profiles of cell-cycle related genes [11]. Here, we have analyzed a full set of cell-cycle-related gene profiles, to reveal the entire regulatory network of the cell cycle. In the present study, the data are composed of 796 gene profiles in five cell phases, while 619 profiles in four phases were analyzed in the previous study. The newly discovered linkages between the phases with the full set of profiles will provide novel insights of into the global system of cell cycle regulation.

2   Material and Methods

2.1   Graphical Gaussian Model

The concept of conditional independence is fundamental to graphical Gaussian modeling. The conditional independence structure of the data is characterized by a conditional independence graph. In this graph, each variable is represented by a vertex, and two vertices are connected by an edge if there is a direct association between them. In contrast, any pair of vertices that are not connected in the graph is conditionally independent.

In the procedure for applying the GGM to the profile data [12], a graph, $G = (V, E)$, is used to represent the relationship among the M clusters, where V is a finite set of nodes, each corresponding to the M clusters, and E is a finite set of edges between the nodes. E consists of the edges between cluster pairs with averaged expression levels that are conditionally dependent, given the rest. The conditional independence is estimated by the partial correlation coefficient, expressed by

$$r_{ij|rest} = -\frac{r^{ij}}{\sqrt{r^{ii}}\sqrt{r^{jj}}}$$

where $r_{ij|rest}$ is the partial correlation coefficient between variables i and j, given the rest, and r^{ij} is the (i, j) element in the reverse of the correlation coefficient matrix.

In order to evaluate which pair of clusters is conditionally independent, we applied the covariance selection [13], which was attained by the stepwise and iterative algorithm developed by Wermuth and Scheidt [14]. When the partial correlation coefficient for a cluster pair is equal to 0, the cluster pair is conditionally independent, and the relationship is expressed as no edge between the nodes corresponding to the clusters in the independence graph. In other words, the graph represents the gene systems network of the M clusters.

2.2 Graphical Chain Model

The graphical chain model is one of the probability models for multivariate random observations, in which the independence of the structure can be represented by a graph. The graph $\Gamma = (V, E)$ consists of a set of vertices V, representing the variables, and a set of edges E, representing the associations between pairs of variables. The chain graph is based on the partitioning of V into disjointed subsets: $V = V_1 \cup V_2 \cup, \cdots, \cup V_T$. The subsets are called blocks or chain components. Edges within blocks are undirected, reflecting the systematic associations, and the edges between blocks are arrows pointing from blocks with lower index numbers to those with higher indices A graphical chain model displays the independence between variables conditioned on all of the other variables in the current and previous blocks. In a graphical chain model, any direct association between two variables in the same block is assumed to be non-causal, and is represented by an undirected edge (line) in a graph. Any direct association between two variables from different blocks is assumed to be potentially causal, and is represented by a directed edge (arrow). The absence of a line or arrow between two variables in the graph indicates that there is no direct association between the variables, i.e, the variables are independent, after controlling for all of the other variables in the same and previous blocks.

The graphical chain model is fitted in a number of stages. When fitting a graphical chain model, the first step is to partition the variables into a number of ordered blocks. Then, the significant direct associations between the variables in the first block are determined. For each pair of variables, the null hypothesis when tested shows that the variables are independent, given all of the other variables in the first block, and the deviance statistics in graphical Gaussian modeling is used [15,16,17], as described below. The maximized log likelihood under the full model and that under a reduced model for the samples with a multivariate normal distribution are

$$\hat{l}_f = -N_q \ln(2\pi)/2 - n \ln|\hat{S}|/n - Nq/2$$

and

$$\hat{l}_r = -N_q \ln(2\pi)/2 - n \ln|\hat{\Sigma}|/n - Nq/2$$

where N, q, $\hat{S}$, and $\hat{\Sigma}$ are the number of samples, the sum of the diagonal elements of a square matrix $(tr(\Sigma^{-1}\hat{S}) = tr(\Sigma^{-1}\hat{\Sigma}))$, the covariance matrix of the full model, and that of the reduced model. Then, the deviance of the model is thus given by

$$G^2 = 2(\hat{l}_f - \hat{l}_r) = N \ln |\hat{\Sigma}/\hat{S}|$$

and the deviance difference for $M_1 \subseteq M_0$ by

$$d = N \ln |\hat{\Sigma}_0/\hat{\Sigma}_1|$$

where $\hat{\Sigma}_0$ and $\hat{\Sigma}_1$ are the estimates of Σ under M_0 and M_1, respectively. Under M_1, d has an asymptotic χ^2 distribution of freedom, given as the difference in the number of free parameters (number of edges) between M_0 and M_1.

Next, the significant direct associations between the variables in the second block and between the first and second blocks are determined. For each pair of variables, the null hypothesis when tested shows that the variables are independent, given all of the other variables in the first and second blocks, and again the deviance statistics is used. The fitting continues, block by block, by determining all of the significant direct associations between the variables in the current block and between all of the variables in the current and previous blocks. The null hypothesis is now independence, given the other variables in the current and previous blocks, and again the deviance statistics is used. All of these tests were carried out at the 5% level, using the χ^2 distribution in deviance statistics.

2.3   Application of the Graphical Chain Model to Gene Expression Profiles

The block in the graphical chain model simply corresponds to the cell cycle phase that is defined by biological information, which is used in the present study. By the intact correspondence to graphical chain modeling, the variable is the gene that has an expression profile with numerical values. However, since the expression profiles often show similar patterns, the genes are highly related to one another. Thus, hierarchical clustering is performed for the genes within each block as a preprocessing for the graphical chain modeling, and then, each gene cluster corresponds with the variable in the present procedure. The details of the procedure are as follows.

(a) The genes within each block are grouped into some clusters. Since the metrics and the techniques in the clustering depend on the data and interests [18], the pair of the metric and the distance in hierarchical clustering for these expression profiles was determined in our previous study. The pair of Euclidean distance between correlation coefficients and Ward's method seems to be suitable for the hierarchical clustering of the present data. Ward's method is summarized as follows. If two groups, G_i and G_j, amalgamate to form a new group, then the dissimilarity between this group and any other group can be expressed by

$$d_{k(i,j)} = \alpha_i d_{ki} + \alpha_j d_{kj} + \beta d_{ij} + \gamma |d_{ki} - d_{kj}|$$

where $\alpha_{i(j)}$, β, and γ are parameters specifying the particular clustering strategy employed: $\alpha_{i(j)} = ((n_i(j)) + n_k)/(n_i + n_j + n_k)$ and $\beta = -n_k/(n_i + n_j + n_k)$.
(b) The cluster numbers are estimated in the dendrograms in four blocks, which are constructed in (a), respectively. In the estimation, the variance inflation factor (VIF) is adopted as a stopping rule for the hierarchical clustering of expression profiles [19]. The VIF is utilized to diagnose the variables that are involved in the multicollinearity in the multiple regression analysis, and is defined by

$$VIF_i = r_{ii}^{-1}$$

where r_{ii}^{-1} is the ith diagonal element of the inverse of the correlation coefficient matrix (CCM) between explanatory variables [20]. In a CCM for m explanatory variables, therefore, m VIF's are calculated. The VIF is applied to estimate the cluster boundaries in the expression profile data. When the explanatory variables in the above equation correspond to the gene profiles, the VIF expresses the degree of linear relationship between the profiles. In the diagnosis of multicollinearity, the popular cutoff value of 10.0 [20] was adopted as a threshold in the present analysis: when VIF_i is larger than 10.0, the linear relationship of the ith variable exists. The m VIF's are assessed with the following condition:

$$\max VIF_i \le 10.0 \; for \; i = 1, 2, \; \cdots, m$$

If the condition is satisfied, then no linear relationship exists in the m sets of profiles. In contrast, if the condition is not satisfied, then the linear relationship still exists in the profiles. Thus, the maximum number of clusters with no linear relationship is searched along the dendrogram. As a result, in this step, we obtain clusters in each block, and the clusters are regarded as the variables in the blocks for further analyses.
(c) The average expression profiles are calculated over the members of each cluster in the respective blocks, according to the cluster estimation in (b). Then, the average correlation coefficient matrix between the clusters in all blocks is calculated from the average profiles.
(d) The average correlation coefficient matrix between the clusters is subjected to the graphical chain modeling. In the chain modeling, the association of the variables (clusters) within and between blocks is inferred by the covariance selection in the graphical Gaussian modeling.

All of the calculations for the clustering, the estimation of cluster number, and the graphical Gaussian modeling were performed by our ASIAN site (web http://eureka.cbrc.jp/asian) [10].

2.4   Rearrangement of the Inferred Graph

In the examples of the application of GGM to actual profiles, the intact networks by GGM showed complicated forms with many edges [9,12]. Actually, many edges also remained among the clusters in the present study. Since drawing all of the associations produces a messy pattern, rearrangement of the inferred graph with significant strong associations is useful to clarify the a graph. In the graphical chain modeling, the magnitude of the partial correlation coefficient indicates the strength of association between clusters. Thus, the intact network can be rearranged according to the partial correlation coefficient value, to interpret the association between clusters. The strength of the association can be assigned by the t test for the partial correlation coefficient [21]. In the present study, the significance levels in the t test were set to 5 %.

2.5   Expression Profile Data

We used the expression profiles of 796 genes measured under 77 conditions [4], which were identified as an objective minimum criterion for cell cycle regulation. In the present analysis, the 796 genes are divided into 298 in G1 phase, 71 in S phase, 120 in SG2 phase, 195 in G2M phase, and 112 in MG1 phase. These genes have a periodic expression pattern, and the peaks of expression were specific to each phase.

3   Results and Discussion

3.1   Clustering of Genes in Five Phases

The cell cycle related genes were classified into 66 clusters by hierarchical clustering, as a preprocessing step for the network inference by GCM. The number of clusters in each phase is: 20 clusters in G1 phase, 13 clusters in S, 17 clusters in SG2 phase, 6 clusters in G2M phase, and 10 clusters in MG1 phase. Table 1 shows the number of genes included within each cluster in each phase. In this Table, the number of clusters in each phase is independent from the number of genes within the phase. Indeed, the smallest number of clusters is for the genes within G2M phase, although the number of genes in G2M phase is not the smallest. On the other hand, the number of genes in S phase is the smallest in the five phases, but the number of clusters in S phase is not the smallest. This feature reveals that the decision of the cluster numbers for each phase is not dependent on the number of genes.

The decision of the number of clusters is related to gene expression, and in general, gene expression is controlled by transcription factors. Therefore, the decision of the number of clusters in each phase is considered to be related to transcription factors. The numbers of transcription factors are similar among the five phases, even though the numbers of genes differ among the five phases. The number of transcription factors within each cluster is indicated in parentheses in Table 1.

To estimate the relationship between the number of transcription factors and the number of clusters in each phase, the fraction of transcription factors and the average number of genes included in the clusters were calculated in each phase.

Fig.1 shows the relationship between the fraction of transcription factors and the average number of genes included in the clusters in each phase. This figure shows that the fraction of transcription factors is related to the average number

Table 1. The number of genes in each cluster

Cluster	G1	S	Phase SG2	G2M	MG1
1	19	3	8(1)	6	12(1)
2	20(3)	2	5	27(4)	16
3	11	2	3	43(1)	5(2)
4	19	6	3	36(2)	5
5	10(1)	7	7(1)	48(4)	7
6	10(1)	4(1)	4	35(4)	13
7	5(2)	96			15(3)
8	10	4(2)	7(1)		15(1)
9	12	6(1)	1		11(1)
10	8	3	13(2)		13(1)
11	30(5)	13(1)	7		
12	25	3	14(3)		
13	23(1)	9(9)	10		
14	25		11(2)		
15	12		3		
16	17		12(1)		
17	10		6(1)		
18	14(1)				
19	9				
20	9				
Sum	298(14)	71(15)	120(12)	195(15)	112(9)

of genes included in the clusters. There is a tendency for a high fraction of transcription factors to occur with a low average number of genes in the clusters. Indeed, the fraction of transcription factors is the highest in S phase(0.21), where the average number of genes in clusters is the lowest(5.46).

The low average number of genes in the clusters means that there is a large number of clusters in the phase, since the number of clusters is derived from the number of genes in each phase and the average number of genes in the clusters. Thus, the relationship between the fraction of transcription factors and the average of the numbers of genes in clusters also indicates the relationships between the fraction of transcription factors and the number of clusters in each phase.

3.2 Functional Category Estimation with a Significant Probability

To evaluate the results of the hierarchical clustering of the genes, the correspondence between the members of each cluster and the biological information of the gene function was obtained. To determine the cluster function from the component genes, each member of the cluster was mapped to 161 functional categories at the second level in the Comprehensive Yeast Genome Database(CYGD). The chance probability for observing the frequencies of genes in particular functional categories within a cluster was estimated with the use of the hypergeometric distribution [26].

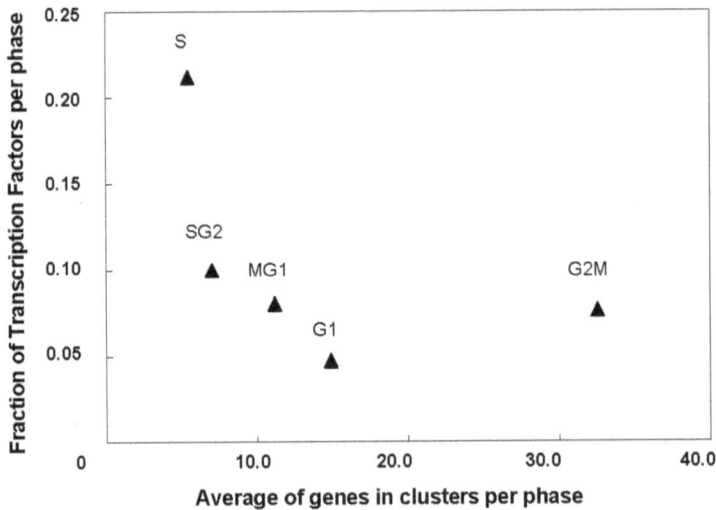

Fig. 1. Relationship between the fraction of transcription factors per phase and the decision of the number of clusters in each phase

(The fraction of transcription factors per phase is plotted on the vertical axis. The average numbers of cluster members per phase are plotted on the horizontal axis. The number of clusters in each phase can be calculated from the number of genes and the average number of cluster members in each phase. The fraction of transcription factors indicates a low average number of cluster members, which means a large number of clusters.)

The details of the clusters that were significantly enriched for genes with similar functions are shown in Table 2. The features, which were extracted from Table 2, are consistent with the general cell cycle information. First, the gene groups that are related to the cell cycle or DNA processing were found in every phase. Second, the characteristic functions were found at specific parts of the cell cycle. The functions related to metabolism or transcription were found in the earlier phases, such as G1 and S and SG2. This feature is consistent with known biological information. In general, DNA synthesis is the main event of the earlier phases in the cell cycle, and thus these phases are considered to be related to transcription and metabolism, to produce energy for DNA synthesis. The other functions, which are related to cellular transport or cell differentiation, were found in the latter phases, such as G2M and MG1. The latter phases in the cell cycle involve mitotic division, and thus these phases are considered to be related to intracellular transport and differentiation. The results of the hierarchical clustering are quite consistent with the known biological information.

Table 2. The clusters with statistically significant functional category enrichment in each phase

phase	Cluster number	Member in a cluster	MIPS category	Members in a category	P
G1	1	19	DNA processing	12	4.3E-04
	2	20	DNA processing	11	3.9E-03
	9	12	cell wall	3	6.9E-03
	10	8	C-compound and carbohydrate metabolism	4	3.9E-03
	10		lipid, fatty acid and isoprenoid metabolism	3	5.0E-03
	10		metabolism of energy reserves (e.g. glycogen, trehalose)	2	9.0E-03
	10		cell wall	3	1.9E-03
	11	30	RNA synthesis	5	7.7E-03
	17	10	complex cofactor/cosubstrate binding	2	1.0E-03
	17		nitrogen and sulfur metabolism	2	1.0E-03
S	2	2	RNA processing	2	4.0E-04
	5	7	cytoskeleton	4	9.1E-03
	6	4	amino acid metabolism	3	7.7E-03
	7	9	complex cofactor/cosubstrate binding	3	5.5E-03
	7		amino acid metabolism	7	3.0E-06
	7		nitrogen and sulfur metabolism	5	1.8E-04
	9	6	cell cycle	4	4.3E-03
	13	9	DNA processing	9	2.7E-08
	13		RNA synthesis	9	2.7E-08
	13		nucleic acid binding	9	2.7E-08
SG2	2	5	cell cycle	4	9.0E-03
	3	3	amino acid metabolism	3	7.8E-04
G2M	2	27	cellular sensing and response	7	8.0E-04
	3	43	respiration	4	2.1E-03
	3		cell cycle	1	5.8E-04
	3		transported compounds (substrates)	17	3.4E-03
	6	35	fungal/microorganismic cell type differentiation	9	1.1E-03
	6		cell cycle	17	4.5E-06
	6		transported compounds (substrates)	1	6.1E-04
MG1	3	5	cellular sensing and response	5	4.6E-05
	3		fungal/microorganismic development	4	1.2E-05
	3		protein binding	3	2.3E-03
	5	7	cell cycle	7	3.4E-04
	8	15	ionic homeostasis	2	8.8E-03
	9	11	transported compounds (substrates)	6	4.6E-04
	9		transport routes	6	1.8E-05
	10	13	nucleus	3	8.9E-03
	10		DNA processing	6	1.8E-03

3.3   Overview of the Inferred Network

To apply GCM for inferring serial regulation during the cell cycle, the linear order of the five phases was determined. In general, the cell cycle works as a circuit, and the natural order of the five phases during this circuit mechanism is known. To infer the network by GCM, the circuit mechanism should be opened, to create a linear mechanism. Basically, the start of DNA replication is decided during G1 phase, and this point is called START in the cell cycle. Thus, G1 phase is considered as the first phase. On the assumption that G1 phase is the first of the 5 phases, MG1 is considered to be the last phase during the cell cycle.

The inferred associations between the 66 clusters are schematically shown in Fig.2. In this figure, a closed cell indicates an inferred association between clusters. In GCM, the inferred associations between the variables within the same phase have no direction, even though the inferred associations between the different phases have direction. In Fig.2, the inferred associations between the clusters within the phase have no direction, while on the other hand, the inferred associations between the different phases have direction, from the former phase to the latter phase.

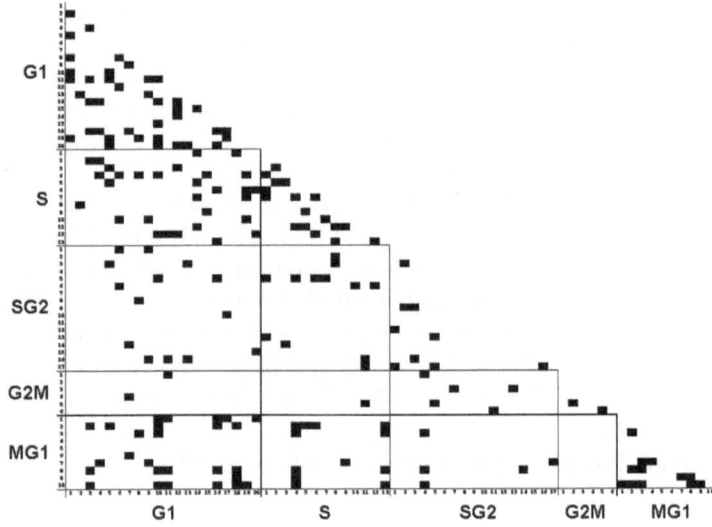

Fig. 2. Overview of the inferred network by GCM

(Overview of the inferred associations between the clusters. The connections between clusters are shown in the 66 clusters: 20 in G1 phase, 13 in S phase, 17 in SG2 phase, 6 in G2M phase and 10 in MG1 phase. Open and closed cells indicate the significant associations and non-significant associations between the clusters, respectively. Each phase is divided with lines.)

In Fig.2, 310 associations remained with significant probability from the 2145 available connections among the 66 clusters. The fraction of inferred associations is 0.1445(310/2145). The inferred associations between the clusters within the phases showed that the fraction of associations remaining within the same phase is 0.2475(114/464). On the other hand, the fraction of inferred associations between the different phases is 0.1166(196/1681). The fraction of inferred associations within the same phase is at least two-fold higher than that between the different phases. Therefore, the interactions between the gene expression mainly occurred within the phases during the cell cycle.

3.4   Inferred Network Within the Phase

The inferred associations among the clusters within each phase int Fig.2 revealed that all of the clusters had some associations with the other clusters in G1 and S. On the other hand, a limited number of clusters had associations within the same phase in SG2, G2M and MG1. In this study, the number of inferred associations indicates the complexity of the cooperation or the regulation of the gene expression within the phase. Thus, the cooperation or regulation of the gene expression in G1 and S phase is more complicated than those of the other phases.

The consideration of the complexity of gene expression within the phase is related to the general events that occur during the cell cycle. The inferred complex cooperation or regulation of gene expression occurred during the event of DNA replication. This complexity of gene expression is quite consistent with the biological information, since many genes are expressed and multiple systems function during DNA replication.

Furthermore, the sparseness of cooperation or regulation of the gene expression at SG2, G2M and MG1 is considered to be related to the events of mitotic division, such as the transport of cellular components or the monitoring of replicated DNA. Actually, the clusters related to transport or differentiation, which were estimated in the former section, have some associations with the other clusters in our study. The specific genes are considered to interact with each other to operate the intracellular components during mitotic division.

3.5   Inferred Network Between the Different Phases

To reveal the main regulatory system during the cell cycle, the fractions of the inferred associations to the number of all combinations between the clusters between the phases were obtained. The fractions of the inferred associations between the phases are shown in Table 3. For example, the number of inferred associations between G1 and S is 30, and the number of all combinations between G1 and S is 260. Thus, the fraction of inferred associations between G1 and S was 0.115(30/260). The obtained fractions of inferred associations between the phases indicate the majority of regulation between the phases for cell cycle control.

Table 3. The fractions of inferred associations between the five phases

	G1	S	SG2	G2M	MG1
G1	0.205				
S	0.115	0.218			
SG2	0.041	0.054	0.066		
G2M	0.017	0.013	0.049	0.133	
MG1	0.135	0.092	0.029	0.000	0.267

The fractions of inferred associations between the different phases displayed in Table 3 show that almost all of the fractions between the neighbor phases are higher than those between the non-neighbor phases. For example, the fractions of inferred associations between G1 and the other phases are: 0.115 with S phase, 0.041 with SG2 phase and 0.017 with G2M phase. A high fraction of the inferred association indicates the high density of regulation between phases. Thus, this feature provide us with the insight that the cell cycle is mainly controlled by serial regulation, depending the distance between the two phases.

The inferred association between the different phases by GCM indicates the regulatory relationship from the former phase to the latter phase. A look at the regulation from the first three phases to the latter phases revealed that some clusters had remarkable features. Some clusters regulated only neighbor phases (neighbor regulation clusters), even though some other clusters regulated only non-neighbor phases (non-neighbor regulation clusters). The details of these specifically regulated clusters are shown in Table 4. The numbers of neighbor regulation clusters are: 4 in G1 phase, 6 in S phase, and 4 in SG2 phase. On the other hand, the numbers of non-neighbor regulation clusters are: 3 in G1 phase, 3 in S phase, and 2 in SG2 phase. A comparison between the numbers of these specific regulation clusters shows that the numbers of neighbor regulation clusters are larger than those of non-neighbor regulation clusters in all phases. Furthermore, the numbers of genes included in the neighbor regulation clusters were also larger than those in the non-neighbor regulation clusters.

Table 4. The fractions of inferred associations between the five phases

		Num. of clusters	Num. of genes (a)	Num. of TFs (b)	Fraction of TFs (b/a)
G1	Neighbor	4	82	3	0.03
	Non-neighbor	3	38	4	0.11
S	Neighbor	6	24	2	0.08
	Non-neighbor	3	22	10	0.45
SG2	Neighbor	4	30	1	0.03
	Non-neighbor	2	17	3	0.17

An analysis of the fraction of transcription factors included in the specific regulation clusters generates the opposite feature. Table 4 shows the numbers of transcription factors and the fraction of transcription factors in those specific regulation clusters. Interestingly, the number of transcription factors included in the non-neighbor regulation clusters are larger than those in the neighbor regulation clusters in each phase. Furthermore, the fractions of transcription factors in the non-neighbor regulation clusters are higher than those in the neighbor regulation clusters. This table shows that the long-distance regulation working in the cell cycle system is controlled by transcription factors.

To reveal the effects of this specific regulation on the other phases, the cell cycle is distinguished as three known cycles. The cell cycle includes three minor cycles, such as the Chromosome cycle, the Cytoplasmic cycle, and the Centrosome cycle, and some phenomena were published about these three cycles [22,23,24]. These phenomena are observed at specific phases during the cell cycle, and thus the control mechanisms of these phenomena are considered to be related to the specific regulation between the phases. To reveal the relationships between the specific regulation clusters and the biological phenomena, the fractions of the three cycle-related genes, which are included in the specific regulation clusters, were obtained.

Fig. 3 demonstrates the fractions of the genes related with the three minor cycles in each phase. The fractions of the three cycle-related genes at neighbor regulation clusters are indicated in Fig. 3A, and those at non-neighbor regulation clusters are indicated in Fig. 3B. Fig. 3A shows that the fractions of

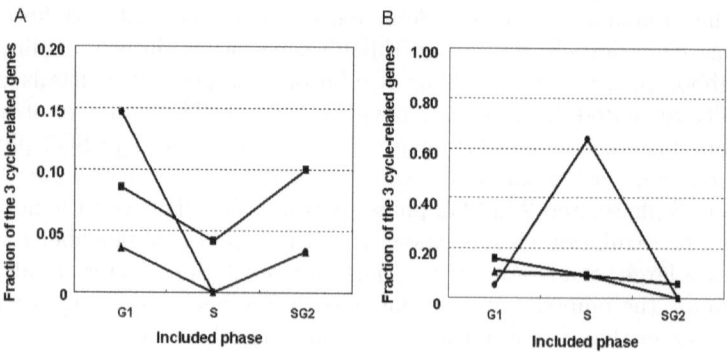

Fig. 3. Comparison between the different phase regulation and the fraction of three minor cycle-related genes

(The fraction of three cycle-related genes in neighbor regulation clusters (A). The fraction of three cycle-related genes in non-neighbor regulation clusters (B). The fraction of chromosome-cycle related genes, cytoplasmic cycle-related genes and centrosome cycle-related genes are indicated by circles, rectangles and triangles, respectively. Among the neighbor regulation clusters, the fractions of all 3 cycle-related genes become lower in S phase.)

three cycle-related genes become lower especially at S phase. This means that a small number of genes related with the three cycles is included in the neighbor regulation cluster in S phase. This feature is consistent with some biological phenomena. In each cycle, the biological phenomena were observed at early S phase [22,23,25], but no phenomena were observed at late S and the next G2 phase. These observations are considered to be related to the small fraction of three cycle-related genes in the neighbor regulation clusters in S phase.

There is a remarkable feature in Fig.3b. The fraction of chromosome cycle-related genes is especially higher in S phase. Actually, many histone-related genes are included in the non-neighbor regulation clusters in S phase. In the chromosome cycle, the replication of the spindle pole body(SPB) and the SPB separation are known to occur during S phase [25], and the next observed phenomenon is the separation of the nucleus during M phase. Indeed, the chromosome cycle phenomena are only observed at the non-neighbor phase in S phase, and this observation is considered to reflect the high fraction of chromosome cycle-related genes, which are included in the non-neighbor regulation clusters in S phase.

To application of the GCM to whole cell cycle-related genes provides us with the main scheme of cell cycle regulation. First of all, the inferred network in our study is consistent with the general biological information about the cell cycle. Since the analyzed expression profiles do not include some known cell cycle-related genes, some known regulations between the genes were not identified in our inferred network. The cell cycle system has been extensively and many genes are known to be related to this system. The genes analyzed in our study are considered to represent a portion of all cell cycle-related genes. Thus, our inferred network is considered to reconstruct part of the cell cycle regulatory mechanism. To reveal the complete mechanism of the cell cycle system, comprehensive analysesof different empirical data will be required.

Acknowledgement

Thanks are due to Dr. Horimoto (CBRC, AIST) and Mr. Saito (INFOCOM, CORPORATION) for valuable discussions. This study was carried out as part of "The Project for Development of Analysis Technology for Gene Functions with Cell Arrays" which was entrusted by the New Energy and Industrial Technology Development Organization (NEDO). The author was partiallly supported by a Grant-in-Aid for Scientific Research (grant 18681031) from the Minisitry of Education, Culture, Sports, Science and Technology of Japan.

References

1. Nurse, P.: A long twentieth century of the cell cycle and beyond. Cell 100, 71–78 (2000)
2. Breeden, L.L.: Periodic transcription: a cycle within a cycle. Curr. Biol 13, R31–R38 (2003)
3. Muller, R.: Transcriptional regulation during the mammalian cell cycle. Trends Genet. 11, 173–178 (1995)
4. Spellman, P.T., Sherlock, G., Zhang, M.Q., Iyer, V.R., Anders, K., Eisen, M.B., Brown, P.O., Botstein, D., Futcher, B.: Comprehensive identification of cell cycle-regulated genes of the yeast Saccharomyces cerevisiae by microarray hybridization. Mol. Biol. Cell 9, 3273–3297 (1998)
5. Cho, R.J., Campbell, M.J., Winzeler, E.A., Steinmetz, L., Conway, A., Wodicka, L., Wolfsberg, T.G., Gabrielian, A.E., Landsman, D., Lockhart, D.J., Davis, R.W.: A genome-wide transcriptional analysis of the mitotic cell cycle. Mol. Cell 2, 65–73 (1998)
6. Koranda, M., Scheleiffer, A., Endler, L., Ammerer, G.: Forkhead-like transcription factors recruit Ndd1 to the chromatin of G2/M-specific promoters. Nature 406, 94–98 (2000)
7. Simon, I., Barnet, J., et al.: Serial Regulation of Transcriptional Regulators in the Yeast Cell Cycle. Cell 106, 697–708 (2001)
8. Rustici, G., Mata, J., et al.: Periodic gene expression program of the fission yeast cell cycle. Nat. Gen 36, 809–817 (2004)
9. Aburatani, S., Kuhara, S., Toh, H., Horimoto, K.: Deduction of a gene regulatory relationship framework from gene expression data by the application of graphical Gaussian modeling. Signal Processing 83, 777–788 (2003)
10. Aburatani, S., Goto, K., Saito, S., Toh, H., Horimoto, K.: ASIAN: a web server for inferring a regulatory network framework from gene expression profiles. Nucleic Acids Res. 33, 659–664 (2005)
11. Aburatani, S., Saito, S., Horimoto, K.: A graphical chain model for inferring regulatory system networks from gene expression profiles. Stat. Method 3, 17–29 (2006)
12. Toh, H., Horimoto, K.: Inference of a genetic network by a combined approach of cluster analysis and graphical Gaussian modeling. Bioinformatics 18, 287–297 (2002)

13. Dempster, A.P.: Covariance selection. Biometrics 28, 157–175 (1972)
14. Wermuth, N., Scheidt, E.: Fitting a covariance selection to a matrix. Algorithm AS 105, Appl. Statist. 26, 88–92 (1977)
15. Edwards, D.: Introduction to Graphical Modelling. Springer, New York (1995)
16. Frydenberg, M.: The Chain Graph Markov property. Scan. J. Stats. 17, 143–153 (1990)
17. Whittaker, J.: Graphical Models in Applied Multivariate Statistics. Wiley, Chichester (1990)
18. Gordon, D.: Classification. Chapman and Hall, London (1981)
19. Horimoto, K., Toh, H.: Statistical estimation of cluster boundaries in gene expression profile data. Bioinformatics 17, 1143–1151 (2001)
20. Freund, R.J,, Wilson, W.J.: Regression Analysis. Academic Press, San Diego (1998)
21. Anderson, T.W.: An Introduction to Multivariate Statistical Analysis, 2nd edn. John Wiley & Sons, New York (1984)
22. Stillman, B.: Origin recognition and the chromosome cycle. FEBS Letters 579, 877–884 (2005)
23. Doxsey, S., Zimmerman, W., Mikule, K.: Centrosome control of the cell cycle. Trends in Cell Biology 15, 303–311 (2005)
24. Blow, J.J., Tanaka, T.U.: The chromosome cycle: coordinating replication and segregation. EMBO reports 6, 1028–1034 (2005)
25. Pereira, G., Schirbel, E.: The role of the yeast spindle pole body and the mammalian centrosome in regulating late mitotic events. Current Opinion in Cell Biol 13, 762–769 (2001)
26. Tavazoie, S., Hughes, J.D., Campbell, M.J., Cho, R.J., Church, G.M.: Systematic determination of genetic network architecture. Nat. Genet. 22, 281–285 (1999)

Manifestation and Exploitation of Invariants in Bioinformatics

Limsoon Wong

School of Computing, National University of Singapore
3 Science Drive 2, Singapore 117543
`wongls@comp.nus.edu.sg`

Abstract. Whenever a programmer writes a loop, or a mathematician does a proof by induction, an invariant is involved. The discovery and understanding of invariants often underlies problem solving in many domains. I discuss in this tutorial powerful invariants in some problems relevant to biology and medicine. In the process, we learn several major paradigms (invariants, emerging patterns, guilt by association), some important applications (active sites, key mutations, origin of species, protein functions, disease diagnosis), some interesting technologies (sequence comparison, multiple alignment, machine learning, signal processing, microarrays), and the economics of bioinformatics.

1 Introduction

The frontier of biological and medical sciences is exciting and full of opportunities today, due to the accumulation of huge amount of biomedical data and the imminent need to turn such data into useful knowledge [31]. There are numerous techniques for dealing with each of the broad spectrum of bioinformatics problems that have emerged, and more are being proposed everyday. There have been a number of useful reviews and tutorials written on various bioinformatics problems. In general, these reviews and tutorials are focused on a specific bioinformatics problem [5], or on a specific technology [19], or both [16].

In this tutorial, I do not focus on a single problem or a single technology. Instead, I present a large varieties of problems and techniques, and try to highlight a fundamental property that is common to all of them. Specifically, I observe that these problems are characterized by *invariants* that emerge naturally from the causes and/or effects of these problems, and show that the techniques for their solutions are essentially exploitation of these invariants.

Before I provide more detail, let me first use an example to illustrate the concept of invariants. We are given a bag of x red beans and y green beans. We are to repeatedly remove two beans from the bag. If both beans are red, we discard both of them. If both beans are green, we discard one and return the other one to the bag. If one is green and one is red, we discard the green bean and return the red bean to the bag. Suppose there is a single bean left in the bag at the end of this process. Can we predict the color of this last remaining bean? The solution is simple: This last remaining bean is red if and only if x is

H. Anai, K. Horimoto, and T. Kutsia (Eds.): AB 2007, LNCS 4545, pp. 365–377, 2007.

odd. The simplicity of this solution arises from a property of the process: The parity of the red beans is preserved—i.e., invariant—at each step of the process. We thus see that invariants are fundamental properties of a problem and can be exploited to provide surprisingly simple solutions to the problem.

As mentioned earlier, the problems presented in this tutorial are all manifestations of invariants. Specifically,

– Section 2 and Section 3 look at the problems of recognizing the active sites of an enzyme, finding the mutations that reduces the efficiency of a protein function, and determining the origin of Polynesians. These problems are manifestations of invariants in the process of Evolution—in particular, sequence features that are conserved during evolution.
– Section 4 looks at the problem of protein function prediction. The process of Evolution has also preserved and/or imposed a number of invariant characteristics on proteins with different functions. The invariant characteristics of a protein is naturally useful for prediction of its function.
– Section 5 looks at the problem of disease subtype diagnosis. Each disease and its various subtypes have their underlying causes. The causes are often difficult to decipher due to the complexity of molecular circuitries and gene-environment interactions. Nevertheless, different causes have different invariant down-stream effects that are useful as diagnostic indicators.

I also show that the techniques for their solutions are essentially exploitation of these invariants.

2   Invariants in Evolution

Let me begin with the problem of finding active sites of an enzyme. An "active site" is a region of an enzyme that a substrate binds to, so that a biochemical reaction can occur. Such sites must be conserved through the evolution process, because the function of the enzyme would be disabled, severely reduced, or completely changed if the physico-chemical properties of the amino acid sequence at these sites were changed. That is, the physico-chemical properties of the amino acid sequence required at these sites are the invariants of the enzyme that must be preserved during the evolution process in order for the protein to retain its specific enzyme function.

Figure 1 illustrates the evolution of a hypothetical enzyme. The function f of ancestor enzyme #1 is characterized by active site "A". Enzyme #2 is evolved from enzyme #1 by having a different physico-chemical property "a_1" at the site "A"; thus it no longer has function f. Enzyme #3 is also evolved from enzyme #1, but by having a different physico-chemical property at site "B"; thus it may have a new function g in additional to f. Enzymes #4, #5, #6, and #7 are similarly evolved. It is clear that "A" is the only property common between all enzymes that have function f. Similarly, "A" and "B" are the only properties common between all proteins that have both functions f and g.

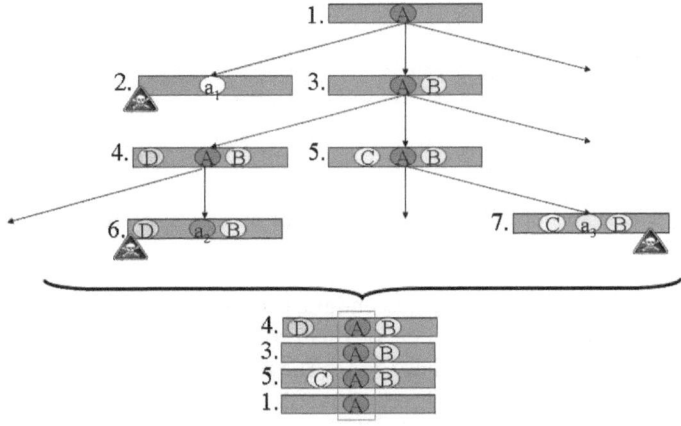

Fig. 1. The evolution of a hypothetical enzyme

The effect of this type of requirement is that the amino acid sequence at these sites is also under pressure to be invariant. This invariant is an indirect and imperfect one, because a limited amount of changes at the amino acid level is permissible as different amino acid sequences can result in very compatible physicochemical properties. In spite of its indirectness and imperfect conservation, it gives rise to the simplest computational solution—multiple alignment [29]—to the problem of finding the active sites of an enzyme.

A multiple alignment can be thought of as a way of writing two or more sequences across the page. Some gaps may be inserted into the sequences in such a way that the number of columns having characters that are identical or that are representing similar physico-chemical properties is maximized. The positions corresponding to these columns are called "conserved positions". The most conserved positions in a multiple alignment are good candidates of active sites of the enzyme, provided the sequences used in the multiple alignment are from suitably diverged species. That is, the sequences should be sufficiently diverged so that enough mutations have accumulated in positions that do not correspond to active sites. At the same time, the sequences should not be so wildly diverged that they no longer have the required enzyme function. Figure 2 shows a multiple alignment of several protein tyrosine phosphatase sequences. The candidate active sites are the conserved consecutive positions indicated by "*" and ".".

3   A Couple of Interesting Twists

An interesting twist in the tale of active sites is the problem of finding key mutations that cause a protein to reduce the efficiency of its function. Here, one of the ancestor proteins with a function f has a mutation in one of its active sites for function f. This mutation reduces the efficiency of the protein. The

```
gi|126467|    FHFTSWPDFGVPFTPIGMLKFLKKVKACNP--QYAGAIVVHCSAGVGRTGTFVVIDAMLD
gi|2499753    FHFTGWPDHGVPYHATGLLSFIRRVKLSNP--PSAGPIVVHCSAGAGRTGCYIVIDIMLD
gi|462550|    YHYTQWPDMGVPEYALPVLTFVRRSSAARM--PETGPVLVHCSAGVGRTGTYIVIDSMLQ
gi|2499751    FHFTSWPDHGVPDTTDLLINFRYLVRDYMKQSPPESPILVHCSAGVGRTGTFIAIDRLIY
gi|1709906    FQFTAWPDHGVPEHPTPFLAFLRRVKTCNP--PDAGPMVVHCSAGVGRTGCFIVIDAMLE
gi|126471|    LHFTSWPDFGVPFTPIGMLKFLKKVKTLNP--VHAGPIVVHCSAGVGRTGTFIVIDAMMA
gi|548626|    FHFTGWPDHGVPYHATGLLSFIRRVKLSNP--PSAGPIVVHCSAGAGRTGCYIVIDIMLD
gi|131570|    FHFTGWPDHGVPYHATGLLGFVRQVKSKSP--PNAGPLVVHCSAGAGRTGCFIVIDIMLD
gi|2144715    FHFTSWPDHGVPDTTDLLINFRYLVRDYMKQSPPESPILVHCSAGVGRTGTFIAIDRLIY
              ..* *** ***       . *           ..****** ****... ** ..
```

Fig. 2. A snapshot of a multiple alignment of several protein sequences

mutation is passed to a group of descendant proteins with function f at a lower efficiency, and becomes an invariant of this group.

Thus, to find key mutations that reduce the efficiency of a protein for function f, we proceed as illustrated in Figure 3. We first identify a group D_1 of proteins having function f at the normal level of efficiency. Then we identify a group D_2 of proteins having function f at the reduced level of efficiency. Then we identify a common active site in two groups of proteins so that two different invariants—one for each of the groups—are observed at the site. That is, the change in efficiency is traced to mutations in specific active sites in the first group which are inherited and conserved in the second group. This takes us from the concept of invariants to the concept of emerging patterns—patterns which are invariant in one group and are changed in a contrast group [12,8]. A beautiful illustration of this logical solution can be found in the study of protein tyrosine phosphatases [15].

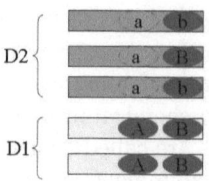

Fig. 3. The site "B" is consistently conserved in the D_1 group of sequences, but is not consistently mutated in the D_2 group. It is thus not a likely cause of D_2's reduced efficiency; otherwise, the second sequence in the D_2 group which has an unmutated site "B" should have normal efficiency. The site "A" is consistently conserved in the D_1 group, and is consistently mutated in the D_2 group. Thus it is a possible cause of D_2's reduced efficiency.

An important invariant of mutations underlies the twist in the tale of active sites above: Mutations are cumulative. That is, a mutation is passed on to future generations unless there is another mutation at the same site that replaces it. This invariant can be exploited in problems concerning the origin of species. The human mitochondrial control region accumulates about 1 mutation every 10,000 years [27]. Given the short length of human history, the length of the mitochondrial control region, and each position in it has an equal chance to mutate, it is reasonable to assume that any position has a negligible likelihood of being mutated twice. In other words, a mutation in the mitochondrial control region that is observed in all instances of the ancestor species must also be observed in descendant species. Thus a link from an ancestor species to its descendant species can be traced.

A beautiful illustration of this idea can be found in the story of the origin of Polynesians [27], depicted in Figure 4. All indigenous Taiwanese have two mutations referred to as #189 and #217 in their mitochondrial control region. Indigenous Solomon Islanders have mutations #189, #217, and #261. Thus, we conclude that an indigenous Taiwanese or his descendant with the #261 mutation somehow travelled to the Solomon Islands, and all indigenous Solomon Islanders are his descendants. All Rarotongans have mutations #189, #217, #261, and #247. Similarly, we infer that a Solomon Islander or his descendant with the #247 mutation somehow reached Rarotonga, and present-day Rarotongans are his descendents.

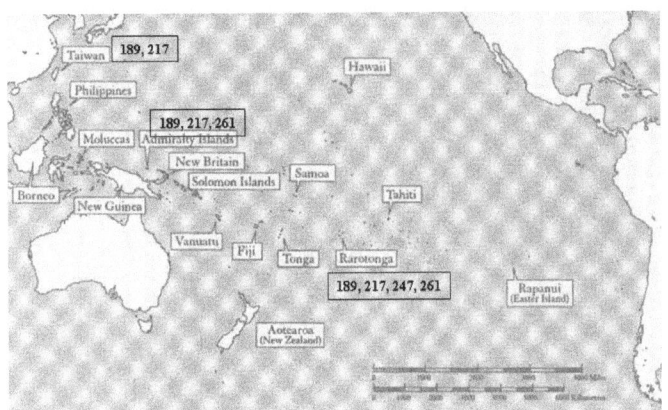

Fig. 4. Origin of Polynesian. Image credit: Sykes [27].

4   Invariants in Protein Function

There are two main invariants that determine the function of a protein: The three-dimensional conformation of the protein and the environment the protein is in. These invariants impose important constraints on the amino acid sequence of protein. For example, mutations in the sequence may completely change the three-dimensional conformation of the sequence. Thus the sequence of the protein is also under pressure to be invariant. However, this invariant is indirect and does not have to be perfect. For example, a limited amount of changes at the amino acid level is permissible without severely affecting the three-dimensional conformation of the protein. Nevertheless, one can perform an abductive inference to predict that two proteins that exhibit a high level of sequence similarity are likely to have the same or similar function. This is the so-called "guilt by association" of similarity of sequences, exemplified by the classic paper of Doolittle and others [9].

The procedure of "guilt by association" is depicted in Figure 5. We compare the sequence of the unknown protein T with a database of protein sequences with known functions. Those proteins in the database that have high sequence identities or sequence alignment scores when compared to T are predicted to be homologs of T; and T is predicted to have functions identical or similar to those of these homologous proteins. A pairwise alignment algorithm [20,25] should be used for sensitive search of homologs. Due to the rapid increase in sequence database sizes, it is also common to sacrifice some amount of sensitivity in favour of significantly increased speed by first using short perfect matches to select likely candidate sequences before performing pairwise alignments [1].

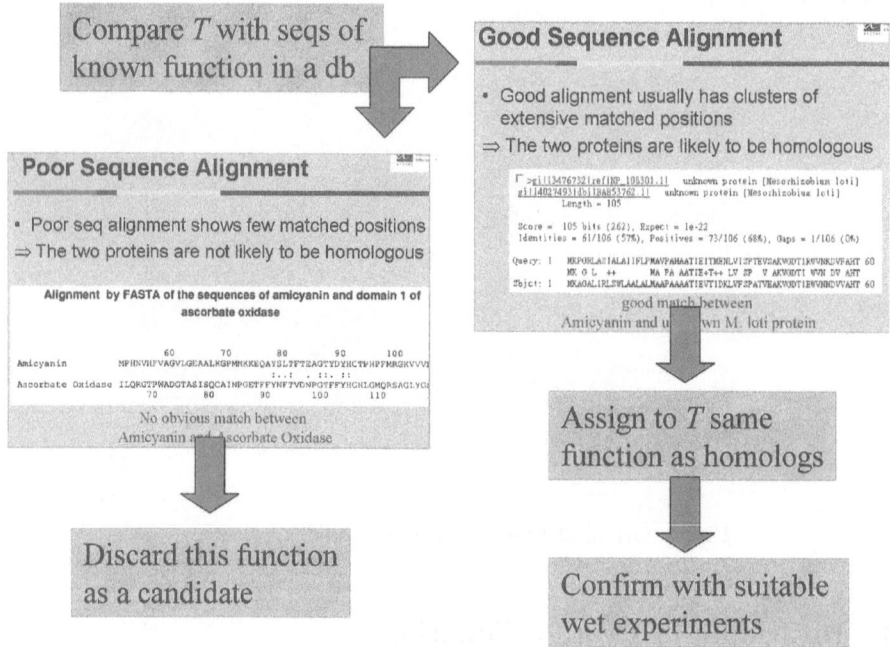

Fig. 5. Protein function prediction using "guilt by association" of sequence similarity

However, there are many protein sequences that have very low sequence similarity to all proteins of known functions. In such a situation, we have to appeal to additional consequences of the two invariants of three-dimensional conformation and operating environment required for a protein function. I describe one such consequence below.

The invariant on a protein sequence, though indirect and imperfect, has an interesting and subtle consequence. Proteins exhibiting a function f, proteins exhibiting a different function g, and proteins exhibiting a function h have different three-dimensional conformations and possibly operate in different environments. So the sequences of these three groups of proteins have distinct invariant compositional characteristics. However, the differences of the invariant compositional characteristics of any two proteins of functions f and g are very likely to be very similar to the differences of the invariant compositional characteristics of any other two proteins of functions f and g! On the other hand, these differences are very likely to be very different from the differences of the invariant compositional characteristics of two proteins of functions f and h, or of functions g and h.

In short, the differences of the invariant characteristics of one group of proteins compared to another group are also invariant, and are emerging patterns when contrasted with the differences compared to a third group. This logic is best illustrated by the comparison of apples to oranges and bananas in Figure 6, where the fruit X is deduced as an apple because its differences with $orange_1$, $banana_1$, and other fruits are identical to that of $apple_1$.

	$orange_1$	$banana_1$	$\cdots$
$apple_1$	color=red vs orange	color=red vs yellow	$\cdots$
	skin =smooth vs rough	skin=smooth vs smooth	$\cdots$
	shape=round vs round	shape=round vs oblong	$\cdots$
$orange_2$	color=orange vs orange	color=orange vs yellow	$\cdots$
	skin =rough vs rough	skin=rough vs smooth	$\cdots$
	shape=round vs round	shape=round vs oblong	$\cdots$
fruit X	color=red vs orange	color=red vs yellow	$\cdots$
	skin =smooth vs rough	skin=smooth vs smooth	$\cdots$
	shape=round vs round	shape=round vs oblong	$\cdots$
$\cdots$	$\cdots$	$\cdots$	$\cdots$

Fig. 6. Comparing apples vs oranges vs bananas. The fruit X is likely to be an apple because its differences with $orange_1$, $banana_1$, etc. are identical to that of $apple_1$.

To wit, we can associate two proteins as having the same or similar function by the similarity of the differences of their sequences compaired to all other sequences. This is precisely the strategy followed by SVM Pairwise [14]. Here, a feature vector is generated for each protein by recording its pairwise alignment score with each sequence in the database. To create a classifier for distinguishing proteins of function f from the rest, the feature vectors are divided into f vs non-f, and a support vector machine classifier is then trained. Given a new unknown protein, a feature vector is first generated by recording its pairwise alignment score with each sequence in the database. The feature vector is then given to the classifier for prediction. SVM Pairwise has much greater sensitivity and precision than the more direct guilt by association of sequence similarity described earlier. SVM Pairwise succeeds for two main reasons. Guilt by association of sequence similarity cannot be applied if a sequence has low similarity with the database and it does not make use of contrast groups. In contrast, SVM Pairwise does not care about the level of sequence similarity, so long as the sequence alignment scores have consistent differences between f vs non-f.

5   Invariants in Diseases

One of the popular problems in bioinformatics is the analysis of gene expression profiles for disease subtype diagnosis. Each disease and its various subtypes have their underlying causes. The causes are often difficult to decipher due to the complexity of molecular circuitries and gene-environment interactions. Nevertheless, different causes have different invariant down-stream effects that are useful as diagnostic indicators. These invariant down-stream effects are often— but not always—manifested as consistent gene expression profile differences in a large number of target genes over the different disease subtypes.

This type of invariant down-stream effects can be discovered in a variety of ways [18]. For example, in an unsupervised setting, one discards those genes with low variants, performs a bi-clustering of the remaining genes vs patient samples, and identifies the invariant gene expression profiles for each disease subtype. As another example, in a supervised setting, one groups the patient samples based on disease subtypes, computes a test statistics such as χ^2 for each gene to determine how well it separates one disease subtype from the rest, and identifies those genes that best distinguishes a subtype. Figure 7 is a beautiful illustration based on the gene expression profiles of childhood acute lymphoblastic leukemia samples [33].

Childhood acute lymphoblastic leukaemia (ALL) is the most common form of childhood cancer. It has as many as 6 different subtypes with differing treatment outcome. To avoid under-treatment, which causes relapse and eventual death, or over-treatment, which causes severe long-term side effects, accurate diagnostic subgroup must be assigned upfront so that the correct intensity of therapy can be delivered to ensure that the child is accorded the highest chance for cure [22]. Contemporary approaches to the diagnosis of childhood ALL require an extensive range of procedures including morphology, immunophenotyping, cytogenetics, and molecular diagnostics [22]. Such a multi-specialist expertise requirement is generally unsatisfiable in developing countries. Thus, even though childhood ALL is a great success story of modern cancer therapy with survival rates of 75–80% in major advanced hospitals [23], it is still a fatal disease in developing countries with survival rates of 5–20%.

Our microarray gene expression profiling followed by computational analyses described above accurately identifies each of the known clinically important subgroups of childhood ALL [33]. We achieve an exceedingly accurate overall diagnostic accuracy of 96% in a blinded test set illustrating the robustness of the invariants identified.

It is worth noting that about 2000 new cases of childhood ALL are diagnosed in ASEAN countries each year. About 50% of these cases need low-intensity therapy, 40% need intermediate intensity, and 10% need high intensity. This is a

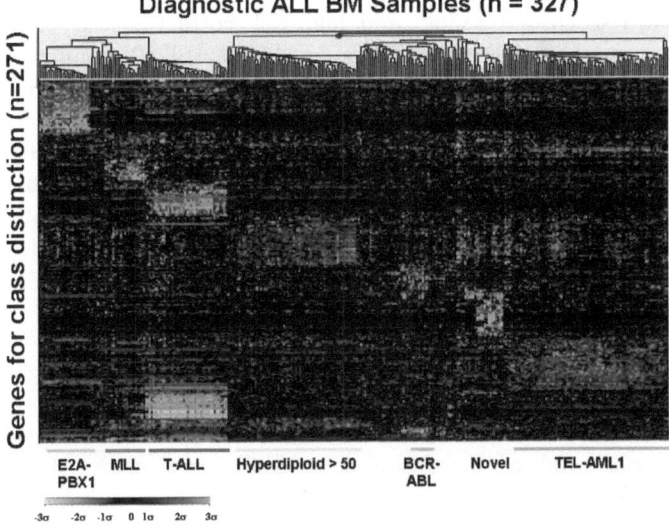

Fig. 7. Gene expression profiles of childhood ALL. Each row is gene. Each column is a patient. Image credit: Yeoh and others [27].

disease with a cure rate of >75% in Singapore. But in ASEAN countries, except Malaysia and Singapore, childhood ALL patients have a dismal 5–20% cure rates. This is mainly due to these countries' inability to deliver the correct intensity of

Treatment	Cost–new cases	Cost–relapses	Total cost
Low-intensity treatment for everyone	$36K * 2000	$150K * 1000	$222M
Intermediate-intensity treatment for everyone	$60K * 2000	$150K * 200	$150M and 50% of patients have side effects
High-intensity treatment for everyone	$72K * 2000	$0	$144M and 90% of patients have side effects
Risk-stratified treatment; viz., low intensity to 50%, intermediate intensity to 40%, high intensity to 10%	$36K * 1000 + $60K * 800 + $72K * 200	$0	$98M

Fig. 8. Costs of treatment options for childhood ALL in ASEAN countries

therapy. Treatment for childhood ALL over 2 years for intermediate-risk costs US$60k, good-risk costs US$36k, and high-risk costs US$72k. Treatment for relapse cases costs US$150k. As the less developed ASEAN countries generally lack the ability to diagnose the subtypes of their childhood ALL patients, the treatment for intermediate risk case is conventionally applied for everyone, as it maximizes the expected benefit in such a situation; see Figure 8. If our single-test platform becomes broadly available, they can then adopt a more accurate risk-stratified treatment strategy. As shown in Figure 8, this can result in savings of US$52M a year yet with better cure rates and much reduced side effects, as the correct intensity of therapy is applied upfront.

6   Remarks

Let me now summarize the key learnings of this tutorial.

I have considered several common bioinformatics applications such as recognizing active sites and key mutations, determining the origin of Polynesians, predicting the function of proteins, and diagnosing a disease and optimizing its treatment. We have seen that there are invariants underlying these problems, and the exploitation of such invariants and/or their consequences yield logical solutions to these problems.

I have used three paradigms in the exploitation of invariants here. The first paradigm is a direct search of an invariant in a group. An example of such a direct search is the application of finding active sites with the use of a multiple alignment algorithm. The second paradigm is a search of "emerging patterns", where we look for patterns that are invariant in one group but are changed in a contrast or control group. The use of a contrast group helps isolate invariants that are fundamental to the target group, as opposed to invariants that are observed in a general population. An example of a search for emerging patterns is the application of finding key mutations that cause a group of proteins to reduce the efficiency in their function. The third paradigm is the concept of "guilt by association", where we deduce that two objects belong to the same type if they exhibit specific common invariants associated with that type. An example of this is the inference of protein function. These three paradigms are also used in combination. An example is the identification of gene expression profiles for diagnosing childhood ALL subtypes. Here, we look for gene expression profiles as emerging patterns that distinguish one ALL subtype from the other subtypes, and use such gene expression profiles to classify patients into the associated ALL subtypes.

I have also discussed the softer but still very important aspect of economics of bioinformatics. This is illustrated in the treatment optimization of childhood ALL. In particular, we have explained why the intermediate-intensity treatment is conventionally applied if the ALL subtype cannot be applied, and why a risk-stratified treatment based on bioinformatics analysis is a superior strategy.

I have briefly mentioned four kinds of computational techniques here. The first kind is that of multiple sequence alignment, where we determine how to best match up several sequences, as illustrated in the application of finding active sites. The second kind is that of sequence comparison, where we determine if two sequences are sufficiently similar, as illustrated in the application of protein function inference. The third and fourth kinds are those of statistical testing and machine learning, as illustrated in the analysis of childhood ALL gene expression data.

Paper length constraints do not allow a more detailed exposition of the above. The reader is encouraged to consult the following articles and references therein for more information. In particular, for sequence comparison, Waterman [30] provides an excellent theoretical background, Gusfield [10] provides an excellent algorithmic background, and Li et. al. [13] present the exciting recent development of using spaced seeds for extremely sensitive and efficient sequence comparison. For multiple sequence alignment, Thompson et. al. [29,28] describe one of the most popular multiple alignment tool packages, and Chin et. al. [6] present a recent improvement in efficient multiple sequence alignment with performance guarantee. For protein function prediction, Altschul et. al. [1] describe the extremely popular BLAST approach to guilt by association of sequence similarity, Bateman et. al. [2] describe guilt by association of domain similarity as embodied in PFAM domains, Liao and Noble [14] describe guilt by association of similarity of dissimilarities as embodied in SVM Pairwise, Wu et. al. [32] describe guilt by association of similarity of phylogenetic profiles, Ma et. al. [17] describe guilt by association of secondary structures, Kung et. al. [11] describe guilt by association of similarity in gene expression profiles, and Chua et. al. [7] present the exciting recent development of guilt by association of similarity of interaction partners. For gene expression analysis, Slonim et. al. [24] is the classic paper that started the field, Miller et. al. [18] is an excellent overview of the issues and techniques, Broberg [4] is a good discussion on several popular test statistics, Breitling and Herzyk [3] describe new rank-based test statistics, Niijima and Kuhara [21] describe new kernel subspace methods for multiclass classification, and Subramanian et. al. [26] present the exciting recent development of the gene set enrichment analysis approach.

Acknowledgements

I would like to thank Prof. Bruno Buchberger and the organizers of AB2007 for inviting me to present this tutorial paper.

References

1. Altschul, S.F., Madden, T.L., et al.: Gapped BLAST and PSI-BLAST: A new generation of protein database search programs. Nucleic Acids Research 25(17), 3389–3402 (1997)
2. Bateman, A., Birney, E., et al.: Pfam 3.1: 1313 multiple alignments and profile HMMs match the majority of proteins. Nucleic Acids Research 27(1), 260–262 (1999)
3. Breitling, R., Herzyk, P.: Rank-based methods as a non-parametric alternative of the T-statistic for the analysis of biological microarray data. Journal of Bioinformatics and Computational Biology 3(5), 1171–1190 (2005)

4. Broberg, P.: Statistical methods for ranking differentially expressed genes. Genome Biology, 4 R41.1–R41.9 (2003)
5. Brown, D.G., Li, M., Ma, B.: A tutorial of recent developments in the seedings of local alignment. Journal of Bioinformatics and Computational Biology 2(4), 819–842 (2004)
6. Chin, F.Y.L., Ho, N.L., et al.: Efficient constrained multiple sequence alignment with performance guarantee. Journal of Bioinformatics and Computational Biology 3(1), 1–18 (2005)
7. Chua, H.N., Sung, W.-K., Wong, L.: Exploiting indirect neighbours and topological weight to predict protein function from protein-protein interactions. Bioinformatics 22, 1623–1630 (2006)
8. Dong, G., Li, J.: Efficient mining of emerging patterns: Discovering trends and differences. In: Proc. 5th ACM SIGKDD Intl Conf on Knowledge Discovery & Data Mining, pp. 15–18, San Diego (1999)
9. Doolittle, R.F., Hunkapiller, M.W.: Simian sarcoma virus onc gene, v-sis, is derived from the gene (or genes) encoding a platelet-derived growth factor. Science 221, 275–277 (1983)
10. Gusfield, D.: Algorithms on Strings, Trees, and Sequences. Cambridge University Press, Cambridge (1997)
11. Kung, S.-Y., Mak, M.-W., Tagkopoulos, I.: Symmetric and asymmetric multi-modality biclustering analysis for microarray data matrix. Journal of Bioinformatics and Computational Biology 4(2), 275–298 (2006)
12. Li, J., Wong, L.: Identifying good diagnostic genes or genes groups from gene expression data by using the concept of emerging patterns. Bioinformatics 18, 725–734 (2002)
13. Li, M., Ma, B., et al.: PatternHunter II: Highly sensitive and fast homology search. Journal of Bioinformatics and Computational Biology 2(3), 417–440 (2004)
14. Liao, L., Noble, W.S.: Combining pairwise sequence similarity and support vector machines for remote protein homology detection. In: Proc. 6th Annual Intl Conf on Research in Computational Molecular Biology, pp. 225–232 (2002)
15. Lim, K.L., Kolatkar, P.R., et al.: Interconversion of kinetic identities of the tandem catalytic domains of receptor-like protein-tyrosine phosphatase PTP-α by two point mutations is synergistic and substrate-dependent. Journal of Biological Chemistry 273(44), 28986–28993 (1998)
16. Liu, H., Wong, L.: Data mining tools for biological sequences. Journal of Bioinformatics and Computational Biology 1(1), 139–168 (2003)
17. Ma, B., Wu, L., Zhang, K.: Improving the sensitivity and specificity of protein homology search by incorporating predicted secondary structures. Journal of Bioinformatic and Computational Biology 4(3), 709–720 (2006)
18. Miller, L.D., Long, P.M., et al.: Optimal gene expression analysis by microarrays. Cancer Cell 2, 353–361 (2002)
19. Mukherjee, S., Mitra, S.: Hidden Markov models, grammars, and biology: A tutorial. Journal of Bioinformatics and Computational Biology 3(2), 491–526 (2005)
20. Needleman, S.B., Wunsch, C.D.: A general method applicable to the search for similarities in the amino acid sequence of two proteins. Journal of Molecular Biology 48, 444–453 (1970)
21. Niijima, S., Kuhara, S.: Multiclass molecular cancer classification by kernel subspace methods with effective kernel parameter selection. Journal of Bioinformatics and Computational Biology 3(5), 1071–1088 (2005)
22. Pui, C.H., Evans, W.E.: Acute lymphoblastic leukemia. New England Journal of Medicine 339, 605–615 (1998)

23. Schrappe, M., Reiter, A., et al.: Improved outcome in childhood acute lymphoblastic leukemia despite reduced use of anthracyclines and cranial radiotherapy: Results of trial ALL-BFM 90. Blood 95, 3310–3322 (2000)

24. Slonim, D.K., Tamayo, P., et al.: Class prediction and discovery using gene expression data. In: Proc. 4th Intl Conf on Computational Molecular Biology, pp. 262–271 (2000)

25. Smith, T.F., Waterman, M.S.: Identification of common molecular subsequences. Journal of Molecular Biology 147, 195–197 (1981)

26. Subramanian, A., Tamayo, P., et al.: Gene set enrichment analysis: A knowledge-based approach for interpreting genome-wide expression profiles. Proc. Nat. Acad. Sci. USA 102(43), 15545–15550 (2005)

27. Sykes, B.: The Seven Daughters of Eve. Gorgi Books (2002)

28. Thompson, J.D., Gibson, T.J., et al.: The CLUSTAL-X windows interface: Flexible strategies for multiple sequence alignment aided by quality analysis tools. Nucleic Acids Research 25(24), 4876–4882 (1997)

29. Thompson, J.D., Higgins, D.G., Gibson, T.J.: CLUSTAL W: Improving the sensitivity of progressive multiple sequence alignment through sequence weighting, position-specific gap penalties, and weight matrix choice. Nucleic Acids Research 22, 4673–4680 (1994)

30. Waterman, M.S.: Introduction to Computational Biology: Maps, Sequences, and Genomes. CRC Press, Boca Raton (2000)

31. Wooley, J.C., Lin, H.S. (eds.): Catalyzing Inquiry at the Interface of Computing and Biology. National Academy Press, Washington (2005)

32. Wu, J., Kasif, S., DeLisi, C.: Identification of functional links between genes using phylogenetic profiles. Bioinformatics 19(12), 1524–1530 (2003)

33. Yeoh, E.-J., Ross, M.E., et al.: Classification, subtype discovery, and prediction of outcome in pediatric acute lymphoblastic leukemia by gene expression profiling. Cancer Cell 1, 133–143 (2002)

Author Index

Aburatani, Sachiyo 350
Anai, Hirokazu 110
Asarin, Eugene 81

Baldan, Paolo 262
Bockmayr, Alexander 36
Boulier, François 66
Bracciali, Andrea 262
Brodo, Linda 262
Bruni, Roberto 262

Cachat, Thierry 81
Casagrande, Alberto 51
Casey, Kevin 51
Chebiryak, Yury 140
Chifman, Julia 307
Chorukova, Elena 202

Diop, Sette 202

Emiris, Ioannis Z. 217

Falchi, Rachele 51

Graça, Ana 125
Guerriero, Maria Luisa 247

Harrison, John 334
Horimoto, Katsuhisa 110, 322
Hurkens, Cor 292

Jarrah, Abdul S. 15

Kanehisa, Minoru 322
Keijsper, Judith 292
Kelk, Steven 292
Knapp, Merrill 155
Kovács, Levente 95
Kroening, Daniel 140
Kuttler, Céline 232

Laderoute, Keith 155
Laubenbacher, Reinhard 15
Lefranc, Marc 66
Lemaire, François 66
Lhoussaine, Cédric 232
Lincoln, Patrick 155
Lynce, Inês 125
Lyubetsky, Vassily 81

Marques-Silva, João 125
Mishra, Bud 1, 51, 170
Morant, Pierre-Emmanuel 66
Mysore, Venkatesh 170

Nakagawa, Koji 110
Niehren, Joachim 232

Oliveira, Arlindo L. 125

Paláncz, Béla 95
Pantos, Sotirios I. 217
Păun, Gheorghe 23
Petrović, Sonja 307
Piazza, Carla 51
Priami, Corrado 247

Romanel, Alessandro 247
Ruperti, Benedetto 51

Sato, Tetsuya 322
Sedoglavic, Alexandre 277
Seliverstov, Alexander 81
Senachak, Jittisak 185
Siebert, Heike 36
Simeonov, Ivan 202
Stougie, Leen 292

Talcott, Carolyn 155
Tiwari, Ashish 155
Toh, Hiroyuki 322
Touili, Tayssir 81
Tromp, John 292

Ürgüplü, Aslı 66

van Iersel, Leo 292
Vestergaard, Mun'delanji 185
Vestergaard, René 185
Vizzotto, Giannina 51

Wong, Limsoon 365

Yamanishi, Yoshihiro 322
Yoshida, Hiroshi 110

Zinovik, Igor 140

Lecture Notes in Computer Science

For information about Vols. 1–4486

please contact your bookseller or Springer

Vol. 4600: H. Comon-Lundh, C. Kirchner, H. Kirchner (Eds.), Rewriting, Computation and Proof. XVI, 273 pages. 2007.

Vol. 4595: D. Bošna\vcki, S. Edelkamp (Eds.), Model Checking Software. X, 285 pages. 2007.

Vol. 4592: Z. Kedad, N. Lammari, E. Métais, F. Meziane, Y. Rezgui (Eds.), Natural Language Processing and Information Systems. XIV, 442 pages. 2007.

Vol. 4591: J. Davies, J. Gibbons (Eds.), Integrated Formal Methods. IX, 660 pages. 2007.

Vol. 4590: W. Damm, H. Hermanns (Eds.), Computer Aided Verification. XV, 562 pages. 2007.

Vol. 4588: T. Harju, J. Karhumäki, A. Lepistö (Eds.), Developments in Language Theory. XI, 423 pages. 2007.

Vol. 4587: R. Cooper, J. Kennedy (Eds.), Data Management. XIII, 259 pages. 2007.

Vol. 4584: N. Karssemeijer, B. Lelieveldt (Eds.), Information Processing in Medical Imaging. XIII, 775 pages. 2007.

Vol. 4583: S.R. Della Rocca (Ed.), Typed Lambda Calculi and Applications. XI, 395 pages. 2007.

Vol. 4582: J. Lopez, P. Samarati, J.L. Ferrer (Eds.), Public Key Infrastructure. XI, 375 pages. 2007.

Vol. 4581: A. Petrenko, M. Veanes, J. Tretmans, W. Grieskamp (Eds.), Testing of Software and Communicating Systems. XII, 379 pages. 2007.

Vol. 4578: F. Masulli, S. Mitra, G. Pasi (Eds.), Fuzzy Logic and Applications. XVIII, 693 pages. 2007. (Sublibrary LNAI).

Vol. 4577: N. Sebe, Y. Liu, Y. Zhuang (Eds.), Multimedia Content Analysis and Mining. XIII, 513 pages. 2007.

Vol. 4574: J. Derrick, J. Vain (Eds.), Formal Techniques for Networked and Distributed Systems – FORTE 2007. XI, 375 pages. 2007.

Vol. 4573: M. Kauers, M. Kerber, R. Miner, W. Windsteiger (Eds.), Towards Mechanized Mathematical Assistants. XIII, 407 pages. 2007. (Sublibrary LNAI).

Vol. 4572: F. Stajano, C. Meadows, S. Capkun, T. Moore (Eds.), Security and Privacy in Ad-hoc and Sensor Networks. X, 247 pages. 2007.

Vol. 4570: H.G. Okuno, M. Ali (Eds.), New Trends in Applied Artificial Intelligence. XXI, 1194 pages. 2007. (Sublibrary LNAI).

Vol. 4569: A. Butz, D. Fisher, A. Krüger, P. Olivier, S. Owada (Eds.), Smart Graphics. IX, 237 pages. 2007.

Vol. 4565: D.D. Schmorrow, L.M. Reeves (Eds.), Foundations of Augmented Cognition. XIX, 450 pages. 2007. (Sublibrary LNAI).

Vol. 4564: D. Schuler (Ed.), Online Communities and Social Computing. XVII, 520 pages. 2007.

Vol. 4561: V.G. Duffy (Ed.), Digital Human Modeling. XXIII, 1068 pages. 2007.

Vol. 4560: N. Aykin (Ed.), Usability and Internationalization, Part II. XVIII, 576 pages. 2007.

Vol. 4559: N. Aykin (Ed.), Usability and Internationalization, Part I. XVIII, 661 pages. 2007.

Vol. 4549: J. Aspnes, C. Scheideler, A. Arora, S. Madden (Eds.), Distributed Computing in Sensor Systems. XIII, 417 pages. 2007.

Vol. 4548: N. Olivetti (Ed.), Automated Reasoning with Analytic Tableaux and Related Methods. X, 245 pages. 2007. (Sublibrary LNAI).

Vol. 4547: C. Carlet, B. Sunar (Eds.), Arithmetic of Finite Fields. XI, 355 pages. 2007.

Vol. 4546: J. Kleijn, A. Yakovlev (Eds.), Petri Nets and Other Models of Concurrency – ICATPN 2007. XI, 515 pages. 2007.

Vol. 4545: H. Anai, K. Horimoto, T. Kutsia (Eds.), Algebraic Biology. XIII, 379 pages. 2007.

Vol. 4544: S. Cohen-Boulakia, V. Tannen (Eds.), Data Integration in the Life Sciences. XI, 282 pages. 2007. (Sublibrary LNBI).

Vol. 4543: A.K. Bandara, M. Burgess (Eds.), Inter-Domain Management. XII, 237 pages. 2007.

Vol. 4542: P. Sawyer, B. Paech, P. Heymans (Eds.), Requirements Engineering: Foundation for Software Quality. IX, 384 pages. 2007.

Vol. 4541: T. Okadome, T. Yamazaki, M. Makhtari (Eds.), Pervasive Computing for Quality of Life Enhancement. IX, 248 pages. 2007.

Vol. 4539: N.H. Bshouty, C. Gentile (Eds.), Learning Theory. XII, 634 pages. 2007. (Sublibrary LNAI).

Vol. 4538: F. Escolano, M. Vento (Eds.), Graph-Based Representations in Pattern Recognition. XII, 416 pages. 2007.

Vol. 4537: K.C.-C. Chang, W. Wang, L. Chen, C.A. Ellis, C.-H. Hsu, A.C. Tsoi, H. Wang (Eds.), Advances in Web and Network Technologies, and Information Management. XXIII, 707 pages. 2007.

Vol. 4536: G. Concas, E. Damiani, M. Scotto, G. Succi (Eds.), Agile Processes in Software Engineering and Extreme Programming. XV, 276 pages. 2007.

Vol. 4534: I. Tomkos, F. Neri, J. Solé Pareta, X. Masip Bruin, S. Sánchez Lopez (Eds.), Optical Network Design and Modeling. XI, 460 pages. 2007.

Vol. 4531: J. Indulska, K. Raymond (Eds.), Distributed Applications and Interoperable Systems. XI, 337 pages. 2007.

Vol. 4530: D.H. Akehurst, R. Vogel, R.F. Paige (Eds.), Model Driven Architecture- Foundations and Applications. X, 219 pages. 2007.

Vol. 4529: P. Melin, O. Castillo, L.T. Aguilar, J. Kacprzyk, W. Pedrycz (Eds.), Foundations of Fuzzy Logic and Soft Computing. XIX, 830 pages. 2007. (Sublibrary LNAI).

Vol. 4528: J. Mira, J.R. Álvarez (Eds.), Nature Inspired Problem-Solving Methods in Knowledge Engineering, Part II. XXII, 650 pages. 2007.

Vol. 4527: J. Mira, J.R. Álvarez (Eds.), Bio-inspired Modeling of Cognitive Tasks, Part I. XXII, 630 pages. 2007.

Vol. 4526: M. Malek, M. Reitenspieß, A. van Moorsel (Eds.), Service Availability. X, 155 pages. 2007.

Vol. 4525: C. Demetrescu (Ed.), Experimental Algorithms. XIII, 448 pages. 2007.

Vol. 4524: M. Marchiori, J.Z. Pan, C.d.S. Marie (Eds.), Web Reasoning and Rule Systems. XI, 382 pages. 2007.

Vol. 4523: Y.-H. Lee, H.-N. Kim, J. Kim, Y. Park, L.T. Yang, S.W. Kim (Eds.), Embedded Software and Systems. XIX, 829 pages. 2007.

Vol. 4522: B.K. Ersbøll, K.S. Pedersen (Eds.), Image Analysis. XVIII, 989 pages. 2007.

Vol. 4521: J. Katz, M. Yung (Eds.), Applied Cryptography and Network Security. XIII, 498 pages. 2007.

Vol. 4519: E. Franconi, M. Kifer, W. May (Eds.), The Semantic Web: Research and Applications. XVIII, 830 pages. 2007.

Vol. 4517: F. Boavida, E. Monteiro, S. Mascolo, Y. Koucheryavy (Eds.), Wired/Wireless Internet Communications. XIV, 382 pages. 2007.

Vol. 4516: L. Mason, T. Drwiega, J. Yan (Eds.), Managing Traffic Performance in Converged Networks. XXIII, 1191 pages. 2007.

Vol. 4515: M. Naor (Ed.), Advances in Cryptology - EUROCRYPT 2007. XIII, 591 pages. 2007.

Vol. 4514: S.N. Artemov, A. Nerode (Eds.), Logical Foundations of Computer Science. XI, 513 pages. 2007.

Vol. 4513: M. Fischetti, D.P. Williamson (Eds.), Integer Programming and Combinatorial Optimization. IX, 500 pages. 2007.

Vol. 4511: C. Conati, K. McCoy, G. Paliouras (Eds.), User Modeling 2007. XVI, 487 pages. 2007. (Sublibrary LNAI).

Vol. 4510: P. Van Hentenryck, L. Wolsey (Eds.), Integration of AI and OR Techniques in Constraint Programming for Combinatorial Optimization Problems. X, 391 pages. 2007.

Vol. 4509: Z. Kobti, D. Wu (Eds.), Advances in Artificial Intelligence. XII, 552 pages. 2007. (Sublibrary LNAI).

Vol. 4508: M.-Y. Kao, X.-Y. Li (Eds.), Algorithmic Aspects in Information and Management. VIII, 428 pages. 2007.

Vol. 4507: F. Sandoval, A. Prieto, J. Cabestany, M. Graña (Eds.), Computational and Ambient Intelligence. XXVI, 1167 pages. 2007.

Vol. 4506: D. Zeng, I. Gotham, K. Komatsu, C. Lynch, M. Thurmond, D. Madigan, B. Lober, J. Kvach, H. Chen (Eds.), Intelligence and Security Informatics: Biosurveillance. XI, 234 pages. 2007.

Vol. 4505: G. Dong, X. Lin, W. Wang, Y. Yang, J.X. Yu (Eds.), Advances in Data and Web Management. XXII, 896 pages. 2007.

Vol. 4504: J. Huang, R. Kowalczyk, Z. Maamar, D. Martin, I. Müller, S. Stoutenburg, K.P. Sycara (Eds.), Service-Oriented Computing: Agents, Semantics, and Engineering. X, 175 pages. 2007.

Vol. 4501: J. Marques-Silva, K.A. Sakallah (Eds.), Theory and Applications of Satisfiability Testing – SAT 2007. XI, 384 pages. 2007.

Vol. 4500: N. Streitz, A. Kameas, I. Mavrommati (Eds.), The Disappearing Computer. XVIII, 304 pages. 2007.

Vol. 4499: Y.Q. Shi (Ed.), Transactions on Data Hiding and Multimedia Security II. IX, 117 pages. 2007.

Vol. 4498: N. Abdennahder, F. Kordon (Eds.), Reliable Software Technologies – Ada Europe 2007. XII, 247 pages. 2007.

Vol. 4497: S.B. Cooper, B. Löwe, A. Sorbi (Eds.), Computation and Logic in the Real World. XVIII, 826 pages. 2007.

Vol. 4496: N.T. Nguyen, A. Grzech, R.J. Howlett, L.C. Jain (Eds.), Agent and Multi-Agent Systems: Technologies and Applications. XXI, 1046 pages. 2007. (Sublibrary LNAI).

Vol. 4495: J. Krogstie, A. Opdahl, G. Sindre (Eds.), Advanced Information Systems Engineering. XVI, 606 pages. 2007.

Vol. 4494: H. Jin, O.F. Rana, Y. Pan, V.K. Prasanna (Eds.), Algorithms and Architectures for Parallel Processing. XIV, 508 pages. 2007.

Vol. 4493: D. Liu, S. Fei, Z. Hou, H. Zhang, C. Sun (Eds.), Advances in Neural Networks – ISNN 2007, Part III. XXVI, 1215 pages. 2007.

Vol. 4492: D. Liu, S. Fei, Z. Hou, H. Zhang, C. Sun (Eds.), Advances in Neural Networks – ISNN 2007, Part II. XXVII, 1321 pages. 2007.

Vol. 4491: D. Liu, S. Fei, Z.-G. Hou, H. Zhang, C. Sun (Eds.), Advances in Neural Networks – ISNN 2007, Part I. LIV, 1365 pages. 2007.

Vol. 4490: Y. Shi, G.D. van Albada, J. Dongarra, P.M.A. Sloot (Eds.), Computational Science – ICCS 2007, Part IV. XXXVII, 1211 pages. 2007.

Vol. 4489: Y. Shi, G.D. van Albada, J. Dongarra, P.M.A. Sloot (Eds.), Computational Science – ICCS 2007, Part III. XXXVII, 1257 pages. 2007.

Vol. 4488: Y. Shi, G.D. van Albada, J. Dongarra, P.M.A. Sloot (Eds.), Computational Science – ICCS 2007, Part II. XXXV, 1251 pages. 2007.

Vol. 4487: Y. Shi, G.D. van Albada, J. Dongarra, P.M.A. Sloot (Eds.), Computational Science – ICCS 2007, Part I. LXXXI, 1275 pages. 2007.